FUNDAMENTALS OF CELL IMMOBILISATION BIOTECHNOLOGY

FOCUS ON BIOTECHNOLOGY

Volume 8A

Series Editors

MARCEL HOFMAN

Centre for Veterinary and Agrochemical Research, Tervuren, Belgium

JOZEF ANNÉ

Rega Institute, University of Leuven, Belgium

Volume Editors

VIKTOR NEDOVIĆ

University of Belgrade,
Belgrade, Serbia and Montenegro

RONNIE WILLAERT

Free University of Brussels,
Brussels, Belgium

COLOPHON

Focus on Biotechnology is an open-ended series of reference volumes produced by Kluwer Academic Publishers BV in co-operation with the Branche Belge de la Société de Chimie Industrielle a.s.b.l.

The initiative has been taken in conjunction with the Ninth European Congress on Biotechnology. ECB9 has been supported by the Commission of the European Communities, the General Directorate for Technology, Research and Energy of the Wallonia Region, Belgium and J. Chabert, Minister for Economy of the Brussels Capital Region.

Fundamentals of Cell Immobilisation Biotechnology

Edited by

VIKTOR NEDOVIĆ
University of Belgrade,
Belgrade, Serbia and Montenegro

and

RONNIE WILLAERT
Free University of Brussels,
Brussels, Belgium

KLUWER ACADEMIC PUBLISHERS
DORDRECHT / BOSTON / LONDON

A C.I.P. Catalogue record for this book is available from the Library of Congress.

ISBN 1-4020-1887-8 (HB)
ISBN 1-4020-2298-0 (e-book)

Published by Kluwer Academic Publishers,
P.O. Box 17, 3300 AA Dordrecht, The Netherlands.

Sold and distributed in North, Central and South America
by Kluwer Academic Publishers,
101 Philip Drive, Norwell, MA 02061, U.S.A.

In all other countries, sold and distributed
by Kluwer Academic Publishers,
P.O. Box 322, 3300 AH Dordrecht, The Netherlands.

Printed on acid-free paper

Printed in the Netherlands.

EDITOR'S PREFACE

A considerable progress in the field of cell immobilisation/encapsulation biotechnology has been made during recent years as a result of extensive research that has been carried out. This is clearly reflected in the voluminous publications of original research, patents, and symposia, and the development of successful commercial applications. As cell immobilisation biotechnology is a multidisciplinary discipline which showed to have an important impact on many scientific subdisciplines – including biomedicine, pharmacology, cosmetics, food and agricultural sciences, beverage production, industrial and municipal waste treatment, analytical applications, biologics production – it is indisputable that many biotechnological processes could benefit from this technology.

The "Cell Immobilisation Biotechnology" book is the outcome of the editors' intention to bundle the extensive widespread information on fundamental aspects and applications of immobilisation/encapsulation biotechnology into a comprehensive reference book and to give the reader an overview of the most recent results and developments that have been realised in this domain. The editors hope that the multidisciplinary concept of this book will encourage the reader to explore and benefit from other subdisciplines than the readers' own specialist field. This book is also addressed to newcomers to the field, since the chosen concept gives a rather extensive overview.

"Cell immobilisation biotechnology" is divided into two volumes. The first volume is dedicated to fundamental aspects of cell immobilisation while the second will deal with its diverse applications. This first volume consists of 26 chapters that are arranged into 4 parts: (1) "Materials for cell immobilisation/encapsulation", (2) "Methods and technologies for cell immobilisation/encapsulation", (3) "Carrier characterisation and bioreactor design", and (4) "Physiology of immobilised cells: experimental characterisation and mathematical modelling". In each part various topics are presented in detail.

Specifically, the different chapters explore a vast range of materials for cell immobilisation and micro- and macro- encapsulation (with especial emphasis on alginate, polyvinyl alcohol, starches, proteins, polyelectrolyte complexes, pre-formed carriers and microcarriers), carrier design and selection, biocompatibility, different encapsulation and immobilisation methods and technologies and their applicability to be scaled-up to industrial level, characterisation of carriers, mass transport phenomena in immobilised cell systems, biorector design and selection, considerations of immobilised cells physiology and mathematical modelling of immobilised cell processes. Over 50 carefully selected experts in the field of immobilisation/encapsulation biotechnology from 34 different research laboratories all over the world contributed to this volume and collectively provided a unique, rich expertise and knowledge. Today, this book presents the most comprehensive, complete, up-to-date source of information on all aspects of cell immobilisation/encapsulation fundamentals.

This book is intended to cover the needs and to be the essential resource for both the academic and industrial communities involved in the study of cell immobilisation biotechnology, as well as specialists in biochemistry, microbiology, biology, medicine, chemical, biochemical and tissue engineering who seek a broad view on cell immobilisation/encapsulation fundamentals and applications. A combination of "biological" and "engineering/technology" aspects is included to reach an even wider audience. Due to the in depth treatment of the fundamentals of cell immobilisation, this volume – in combination with the second volume – can be used as a "handbook" of "Cell Immobilisation Biotechnology".

We express our gratitude and appreciation to our many colleagues, who as experts in their fields, have contributed to this volume. We would also like to thank the series editor, Marcel Hofman, and the publisher, Kluwer Academic Publishers, for giving us the opportunity to edit this book and for their excellent support in assuring the high quality of this publication.

Viktor A. Nedović and Ronnie Willaert

Belgrade/Brussels, July 2003

TABLE OF CONTENTS

PART 1

MATERIALS FOR CELL IMMOBILISATION/ENCAPSULATION

BIOMATERIALS FOR CELL IMMOBILIZATION

A look at carrier design

KATHRYN W. RIDDLE AND DAVID J. MOONEY
University of Michigan, Chemical Engineering, 2300 Hayward, 3074 H.H. Dow Ann Arbor, MI 48105 – Fax: 734-763-0459
Email: mooneyd@umich.edu

1. Introduction

There are several major biomedical applications where the transplantation of immobilized cells is being employed to restore, maintain or improve tissue function. These strategies can be split into two main categories: the replacement of biochemical function only or the replacement of structurally functional tissue. As only chemical communication (*e.g.*, diffusion of molecules) is required in the former, it is possible to deliver cells encapsulated in a nanoporous, immunoisolatory polymer membrane. The membranes is constructed such that there are pores large enough to allow for nutrients, waste and the bioactive factor to diffuse but not large enough as to allow immune cells to attack the cells within [1]. This strategy has mainly been employed to temporarily or permanently replace biochemical functions of the liver [2,3], pancreas [4,5], and provide local protein delivery in neurological disorders [6]. The second major strategy involves entrapping cells on a micro or macroporous polymer scaffold and promoting the formation of a new tissue that is structurally and functionally integrated with the surrounding tissue. The scaffold is constructed with a biocompatible material that degrades over time to leave only the integrated tissue in its place. Researchers have attempted to use this strategy with a variety of tissues, including skin [7,8,9], liver [10,11,12], pancreas [13], cornea [14], blood vessels [15,16], cartilage [17,18], heart [19], and bone [20,21].

The biomaterial component of these therapies must provide the appropriate mass transport properties, membrane or scaffold stability and desirable cellular interactions depending on the location and desired function of the implant. In the case of nanoporous immunoisolated strategies, the main goal of the biomaterial is to provide a barrier to the host defences while allowing the essential molecules to diffuse through (Figure 1). In cases where cells are seeded onto a micro or macroporous biomaterial and then implanted, the main function of the biomaterial is to provide cells with a synthetic

V. Nedović and R. Willaert (eds.), Fundamentals of Cell Immobilisation Biotechnology, 15-32.

extracellular matrix (ECM) that directs multiple cell functions, including cell proliferation, migration and gene expression (Figure 2).

This chapter will provide an overview of carrier design of biomaterials for cell immobilization. For the purposes of this chapter, only applications involving mammalian cells will be discussed. A brief overview of cell sourcing is followed by a section on applications of nano-, micro- and macroporous materials, and the common materials used for these applications are described in the last section.

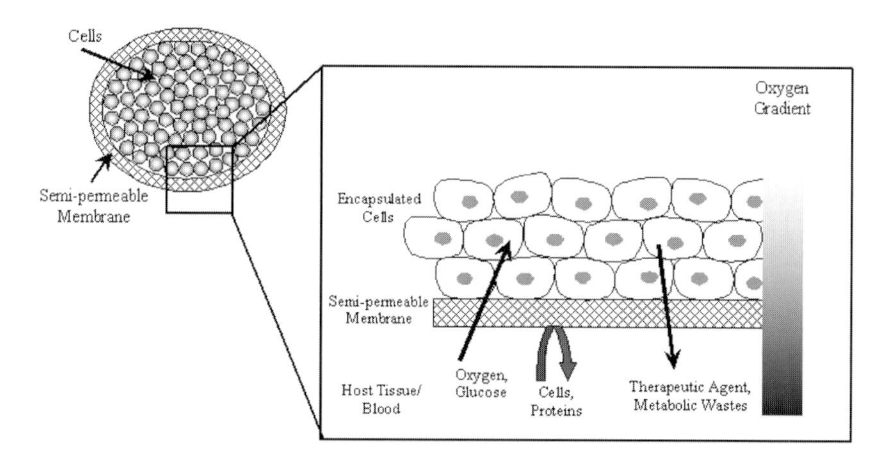

Figure 1. Cartoon of a nanoporous immunoisolated device.

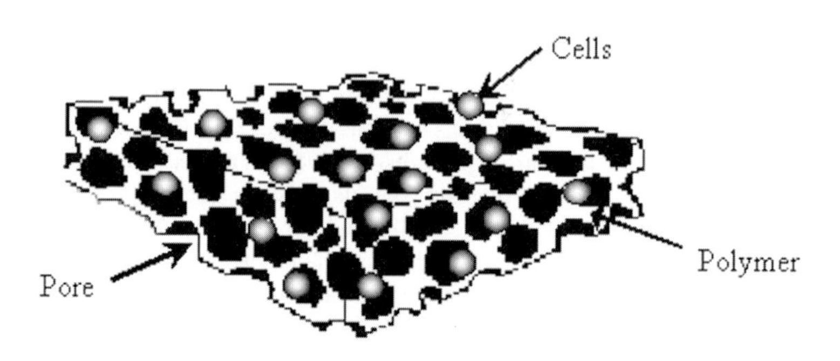

Figure 2. Cartoon of a macroporous scaffold seeded with cells.

2. Cell sourcing

There are three main sources of cells for cell immobilization biotechnology: autologous (obtained from the same individual for whom the cells are intended), allogeneic (obtained from an individual of the same species for whom the cells are intended), and xenogeneic (obtained from a species different from which the cells are intended). The appropriate source of cells for a specific application depends on many factors, including functional requirements, availability, ease of collection, processing and storage, and economic factors.

Autologous cells potentially offer the most immunocompatible source of cells, and for this reason are frequently used for reconstruction of structural tissues [22]. However, the time required and costs incurred to generate sufficient numbers of these cells from a tissue biopsy is often a deterrent to their application. Specifically, the cells must typically be significantly expanded within a short (*e.g.*, days to weeks) time frame before returning to the patient [23]. In addition, the scarcity of healthy tissue from which to harvest the cells or an inability to expand specific cell types in culture, as with hepatocytes or β-islets, can make other cell sourcing options necessary.

An attractive alternative to autologous cells in many situations is to instead use allogeneic cells. These cells can potentially be expanded in large batches for treating many patients and stored before they are needed [24]. Each batch can be screened for safety and function, decreasing the cost associated with patient-by-patient screening of autologous cells. In addition, genetically modified immortalized cells or cell lines can be extensively studied and characterized for use in cell immobilization strategies [25]. Standardized allogeneic cell sources may also allow the construction of complex tissues that could otherwise be prohibitively expensive in time, money, and effort if they were to be produced by custom manufacture from autologous sources [24]. However, the use of allogeneic cell therapy may require some form of immunoprotection to prevent rejection by the host. For this reason, allogeneic cells have mainly been used in applications where they will eventually be replaced by native cells (*e.g.*, skin replacement), or in immunoisolatory devices that are intended to provide a solely biochemical function. In addition, the limited availability of certain cell types and/or an inability to expand in culture, limits this cell source in certain applications (*e.g.*, treating diabetes with β-islets).

Xenogeneic cells are highly attractive for replacement of biochemical functions in certain situations, as they are available in unlimited supply. Two of the most widely researched applications for xenogeneic cell transplantation are the replacement of liver and pancreatic function [3,4,26,27]. Autologous and allogeneic cells from these organs are scarce and difficult to expand *in vitro*, leading to great difficulty in identifying good human sources for these strategies [28]. A main challenge in the application of xenogeneic cells is rejection by the host immune system. To overcome these difficulties, these cells must be encapsulated within nanoporous membranes that protect them from rejection. Xenogeneic cells tend to elicit a greater immune response from the host than allogeneic cells in these applications, possibly due to the release of antigenic products from deteriorated and dying cells within the device in both cases [29]. Immunoisolated xenogeneic cells may be useful for both implantation in the body, and as a component of extracorporeal tissue support systems [30].

3. Material applications

Applications for nano, micro and macroporous materials are widely varied from temporary replacement of biochemical function to formation of structurally functional tissues that are completely integrated with the surrounding host tissue. Nanoporous materials are utilized when it is desired to replace a biochemical function of a tissue by entrapping viable cells within a semi-permeable membrane that prevents cells from being recognized and therefore destroyed by the host immune system [31]. This characteristic is especially useful when allogeneic or xenogeneic cells are used in this strategy. Microporous and macroporous scaffolds provide a synthetic extracellular matrix (ECM) for attachment and proliferation of anchorage dependant cells. These types of scaffolds can be formed using natural or synthetic biomaterials, and have been investigated for the replacement and repair of many tissue types. In all cell encapsulation strategies, it is generally accepted that an appropriate biocompatible material must be isolated or synthesized, and manufactured into the desired shape and dimensions [32].

3.1. NANOPOROUS

Potential applications of nanoencapsulation technology include replacement of the biochemical functions of major organs and transplantation of engineered cells for gene therapy [29]. In many of these situations, the scarcity of healthy autologous tissue from which to harvest cells requires the use of xenogeneic and allogeneic cells requiring immunoisolation. Pancreatic function may potentially be replaced with transplanted porcine islets of Langerhans [5,26] and have been widely investigated as a treatment for diabetes. Similarly, porcine hepatocytes [2,27] have been investigated for temporary replacement of liver function. Other disorders that have been addressed by this method include chronic pain [33,34], hypocalcaemia [35], dwarfism [36], anaemia [37,38], and haemophilia [39,40].

A major limitation to this form of therapy is the limited number of cells that can be delivered, and thus a limited dose of the therapeutic factor secreted by the cells. In some cases, the sheer number of cells needed to replace the function of the damaged organ makes this form of therapy impractical. For example, it has been estimated that approximately 7×10^9 hepatocytes are required to adequately replace the function of the liver [41]. Considering that the average hepatocyte is approximately 25 microns in diameter [42] and that the average hollow fibre nanoporous device has a diameter of approximately three millimetres [43], it would take an implant 8 meters in length to accommodate the number of cells necessary to adequately replace liver function. In many of these cases implantation of the device is impractical and it is preferable to pursue extracorporeal strategies reviewed elsewhere [41].

3.1.1. Design considerations

Key considerations that must be addressed for this approach to be successful include the required mass transport properties, membrane stability, and cellular distribution with the encapsulation material.

3.1.1.1. Mass transport. Survival of the entrapped cells as well as efficacy of the therapeutic agent will be greatly affected by the mass transport properties of the encapsulation membrane. Ideally, the pore size in the membrane should be carefully controlled to allow bi-directional diffusion of the bioactive factor and molecules essential for cell survival (*e.g.*, oxygen and glucose), while staying impervious to larger molecules (host complement and antibodies) and immunogeneic cells. One of the major hurdles of this technology is the fact that it is difficult to make a polymer membrane with uniform pore sizes [44]. A distribution of pore sizes leads to a gradient of diffusional resistances to molecules of varying molecular weight, instead of an absolute cut-off. Another limitation to transport can be the fibrous capsule that often forms around the implant due to the foreign body response. For example, a fibrous capsule thickness of 160 microns is sufficient to cause hypoxia for transplanted islets of Langerhans [45]. Due to this issue, materials that don't elicit a significant foreign body response are required for these applications. Other factors that influence the mass transport properties of a nanoporous membrane are the size and shape of the device, as the surface area:volume ratio is a key parameter for diffusional transport [39]. Since the metabolic requirements of each cell type are different, optimal membrane permeability will depend on choice of cells [46]. A variety of structures including, intravascular devices, spherical microcapsules, cylindrical hollow-fibre macrocapsules, flat sheets [47] and planar macrocapsules [48] have been utilized. The spherical shape provides the most surface area to volume ratio, which provides for efficient transfer of nutrients, bioactive factors and waste. However, microspheres can be difficult to retrieve at a later time due to their small size. If retrieval of the device is desired or large numbers of cells are required, cylindrical and plate designs have been employed, in spite of their inferior mass transport properties [31].

3.1.1.2. Membrane stability. The need to periodically replenish encapsulated cells, due to limitations on cell longevity, leads to specific requirements for membrane biostability. Following cell death, the implant may be retrieved, or the implant may be left in place and allowed to degrade over time. When short term drug delivery or temporary replacement of biochemical function is desired, it may be desirable for the membrane to slowly erode and allow the body's defences to destroy the implant. For example, a microcapsule system has been designed to degrade and allow clearance of the encapsulated cells following their death or loss of function [49]. These hydrogel-based microcapsules, made from alginate and poly (L-lactide), can be made to degrade over several weeks or months and can eliminate the need for surgery to remove old capsules. If retrieval of the implant is necessary or the implant is to be replenished *in vivo*, a non-biodegradable membrane would be preferable. Having a non-degradable membrane is also particularly useful in applications where genetically engineered cells and cell lines are encapsulated, as the breakdown of the membrane could lead to escape of the cells and potential tumour development [50].

3.1.1.3. Cellular distribution. Nanoporous materials must be designed to optimally organize the encapsulated cells, as a common observation is the appearance of dead or necrotic cells within biomaterial membranes (*e.g.* hollow fibre devices) [46]. Cells loaded as dilute suspensions in aqueous growth media into these materials presumably settle to create high-density aggregates that hinder diffusional transport of essential nutrients. To overcome this problem, a variety of cell immobilization matrices, including collagen, chitosan and alginate have been examined to more uniformly distribute the cells within the membrane [51].

3.2. MICROPOROUS AND MACROPOROUS

Potential applications for micro and macroporous cell immobilization include the replacement of tissues with structural and mechanical functions, including heart, blood vessels, bone and cartilage [52,53,54,55]. This contrasts with nanoporous cell encapsulation technologies, which are typically utilized to replace only biochemical functions of tissues. Micro and macroporous matrices, or scaffolds, can be fabricated from polymers or other materials, and act as a temporary ECM for the forming tissue. Example polymers used in these applications include both natural materials (*e.g.*, hyaluronate), and synthetic polymers such as the polyanhydrides and polylactides. These scaffolds mechanically support the formation of tissue, and serve to either transplant cells and/or allow for cells from the surrounding tissue to migrate into this space.

3.2.1. Design considerations

Ideally, a micro or macroporous cell immobilization scaffold should have the following characteristics:

- Adequate porosity and pore sizes to control mass transport properties
- Appropriate mechanical properties for the desired application
- Controllable degradation and resorption to match tissue replacement.
- Suitable cell-interactions to allow for appropriate tissue development.

3.2.1.1. Mass transport properties. The pore size in micro or macroporous scaffolds modulates the ability of cells and biological molecules to pass into and out of the scaffold. In general, very small pores (*e.g.*, d < 1μm), while allowing free diffusion of molecules, will not allow cellular migration, while pores in the 10 – 100's of microns readily allow transplanted or host cells to migrate through the scaffold volume. In addition, the rate of tissue ingrowth is dependent on the pore size and overall porosity of the scaffold [56]. Other characteristics that can control transport properties with an implanted scaffold include surface area and volume. A large surface area allows significant cell attachment and growth on the material, a high porosity allows significant host tissue infiltration. In general, it is believed that a porosity of greater than 90% will provide an appropriate surface area for cell-polymer interactions, sufficient space for ECM regeneration, and minimal diffusion constraints [57].

3.2.1.2. Mechanical design issues. Whereas nanoporous cell encapsulation aims at only replacing a biochemical function, micro and macroporous scaffolds often provide temporary structural support during tissue repair, and therefore require significant mechanical stability. In soft tissue and non-load bearing applications, the mechanical properties (*e.g.* elastic modulus) of the scaffold are often not required to be extremely high, and a variety of natural polymers (*e.g.*, collagen) and synthetic materials (*e.g.*, poly(lactide-co-glycolide)) have been employed. In contrast, in load bearing applications high strength synthetic materials or reinforced natural polymers (*e.g.* chitosan/calcium phosphate composites) are likely required.

Mechanical stimuli conveyed to cells *via* the scaffold can also regulate the development of many tissues. For example, under cyclical mechanical strain of the scaffold, smooth muscle cells (SMCs) show increased proliferation, and elastin and collagen synthesis [58]. This phenotypic control over SMCs has also been demonstrated using collagen gels [59,60] and polyglycolide scaffolds [61]. Similarly, the development of engineered cartilage has also been found to depend on the mechanical environment [62]. These results indicate scaffolds must be designed to appropriately convey mechanical signals to interacting cells.

3.2.1.3. Scaffold degradation. For each application, the desired rate and method of biodegradation will play a large role in the choice of material. Generally, there are two methods by which biodegradable polymers erode in the body: non-specific hydrolysis and enzymatic degradation. These methods of erosion lead to varying degrees of local vs. pre-programmed control over resorption of the biopolymer. For example, hydrolysis of the ester linkage in poly(lactide-co-glycolide) polymers leads to bulk degradation following the cleavage of ester bonds at random sites along the polymer chains until eventually, lactic acid and/or glycolic acid is produced [63]. This bulk degradation can bring about a rapid loss in mechanical properties and mass loss after prolonged periods of reactive hydrolysis. In contrast, polymers that degrade by surface erosion *via* enzymatic activity provide a more gradual resorption rate. Enzymatic degradation relies on certain classes of enzymes (*e.g.* collagenase), secreted by cells in the body, which have the ability to cleave the biopolymer [64]. Since polymer degradation in this case is proportional to the amount of enzyme present, the polymer may erode at a rate controlled by the local cell activity. One interesting approach to controlling polymer degradation is crosslinking synthetic polymers with enzymatically degradable molecules [65]. This concept combines the benefits of the highly controllable properties of synthetic materials with local cellular control over degradation provided by the enzymatically labile cross-links. Regardless of the material, scaffolds should ideally biodegrade at a rate that is comparable to the rate of tissue regeneration so that the scaffold is completely replaced and integrated into the host tissue.

3.2.1.4. Cellular interaction. The surface chemistry of scaffolds can be modified to affect specific cellular responses. These responses may include the rate of host cell invasion, the ingrowth of specific cell types, and cell differentiation. In the past, cell adhesion to biomaterials was frequently based on the cells binding to non-specifically adsorbed proteins (*e.g.*, vitronectin) present in body fluids [65]. A more controlled approach is to covalently, or physiochemically incorporate adhesion-promoting oligopeptides and oligosaccharides on the biomaterial surface. One of the most extensively studied adhesion-promoting peptide sequences is the arginine-glycine-aspartic acid (RGD) sequence, which has been shown to promote the adhesion and spreading of a variety of anchorage dependant cell types [66,67,68]. Specific cell types may also be targeted by utilizing peptides, which only those cells can recognize and bind [69]. In addition to cell adhesion peptides, growth factors [70] and plasmid DNA [71] may also be immobilized on the surface of a biomaterial to modify cellular behaviour within and surrounding the implant.

4. Material chemistry

In order to provide the varying requirements for each specific application, a variety of naturally derived and synthetic biomaterials are used for cell encapsulation. These materials can be processed into many different physical forms and geometries. A representative group of these materials, many of the most commonly used, are discussed in this section.

4.1. NATURAL POLYMERS

Natural polymers are attractive for cell immobilization due to their abundance and apparent biocompatibility. These polymers can provide a wide range of physical properties that offer unique characteristics for cell encapsulation technologies.

4.1.1. Collagen

Collagen is the major component of mammalian connective tissue and has been used in cell immobilization due to its biocompatibility, biodegradability, abundance in nature, and natural ability to bind cells. It is found in high concentrations in tendon, skin, bone, cartilage and, ligament, and these tissues are convenient and abundant sources for isolation of this natural polymer. Collagen can be readily processed into porous sponges, films and injectable cell immobilization carriers. Collagen may be gelled utilizing changes in pH, allowing cell encapsulation in a minimally traumatic manner [72,73]. It may also be processed into fibres and macroporous scaffolds [74,75]. Its natural ability to bind cells makes it a promising material for controlling cellular distribution within immunoisolated devices, and its enzymatic degradation can provide appropriate degradation kinetics for tissue regeneration in micro and macroporous scaffolds. Challenges to using collagen as a material for cell immobilization includes its high cost to purify, the natural variability of isolated collagen, and the variation in enzymatic degradation depending on the location and state of the implant site. [76]. Collagen has been used to engineer a variety of tissues, including skin [77,78], bone [79,80], heart valves [81], and ligaments [82].

4.1.2. Alginate

Alginate is a polysaccharide extracted from seaweed, and has widely been used for cell immobilization due to its biocompatibility and simple gelation with divalent cations (*e.g.* calcium ions). The polymer is made of the two sugars: D-mannuronate (M) and L-guluronate (G) (Figure 3). The M to G ratio in alginate and their distribution will dictate the gelling and mechanical properties of the resulting gel [83,84,85]. Alginate can also be covalently crosslinked using bi-functional molecules such as adipic dihydrazide, methyl ester L-lysine, and polyethylene glycol (PEG) [86]. A potential limitation to the use of alginate is its often uncontrollable and unpredictable dissolution, which occurs by a process involving the loss of divalent ions into surrounding fluids [87]. In addition, alginate gels do not promote high levels of protein adsorption [88], and have therefore been modified with lectin or RGD-containing cell adhesion ligands to control cell adhesion [89]. When used as a biomaterial for nanoporous immunoisolated devices, alginate beads are typically coated with a polyamino acid such as poly-L-lysine (PLL) or poly-L-ornithine (PLO) to establish selective permeability and maintain the capsule durability and biocompatibility *in vivo* [46]. Alginate has also been examined as a carrier for a variety of cell types, including chondrocytes [85,90], for engineering structurally integrated tissues. Its ability to be injected in a minimally invasive manner into tissues is a significant attraction in this latter application.

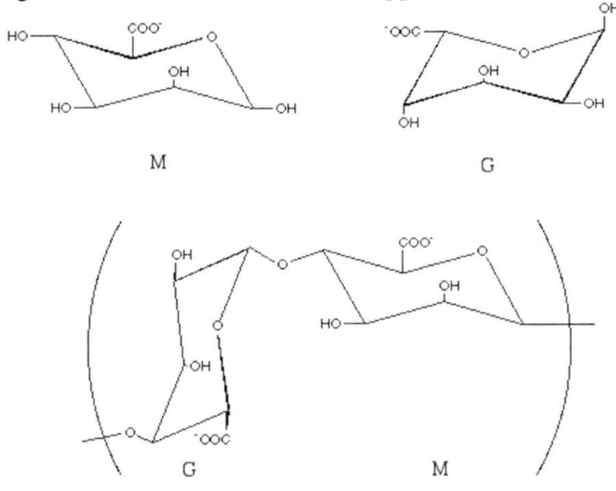

Figure 3. Structure of alginate.

4.1.3. Hyaluronic acid

Hyaluronic acid, which is the largest glycosaminoglycan (GAG) found in nature, has many attractive features for cell encapsulation (Figure 4). These include ready isolation from abundant natural sources and the minimal inflammatory or foreign body reaction it elicits following implantation [91]. Hyaluronic acid gels can be formed by covalently cross-linking with various hydrazide derivatives [92], and these gels are enzymatically degraded by hyaluronidase [93]. Potential limitations to the use of hyaluronic acid are its long residence time in the body, and the limited range of mechanical properties available from its gels. Moreover, hyaluronic acid requires thorough purification prior to use to remove impurities and endotoxins that may potentially transmit disease or elicit an immune response [87]. Hyaluronic acid has been used in micro and macroporous cell encapsulation as a delivery vehicle for bone-marrow-derived mesenchymal progenitors [94], or as an injectable microporous cell carrier for soft tissue augmentation [91]. It has also been combined with collagen to create an osteoconductive scaffold for bone regeneration [95].

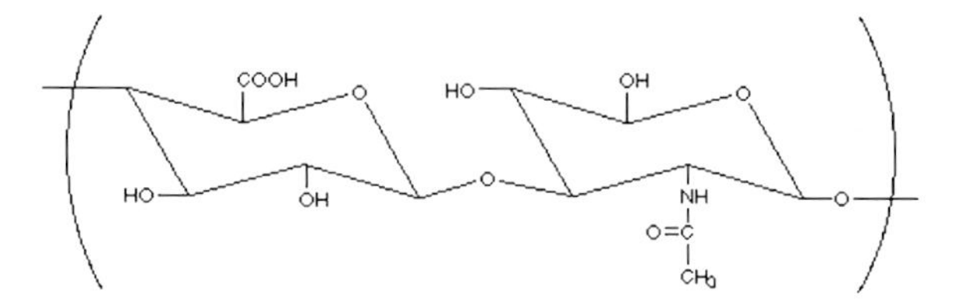

Figure 4. Structure of hyaluronic acid.

4.1.4. Chitosan

Chitosan is a deacetylated derivative of chitin, which is widely found in crustacean shells, fungi, insects, and molluscs (Figure 5). Chitosan forms hydrogels by ionic [96] or chemical cross-linking with glutaraldehyde [97], and degrades *via* enzymatic hydrolysis [98]. Chitosan and some of its complexes have been employed in a number of biological applications including wound-dressings [8], drug delivery systems [99] and space-filling implants [100]. Due to its weak mechanical properties and lack of bioactivity, chitosan is often combined with other materials to achieve more desirable mechanical properties. Specifically, chitosan has been combined with calcium phosphate to increase its mechanical strength for micro and macroporous scaffold applications [100], and has been combined with collagen to provide a more biomimetic microenvironment in nanoporous cell encapsulation applications [101].

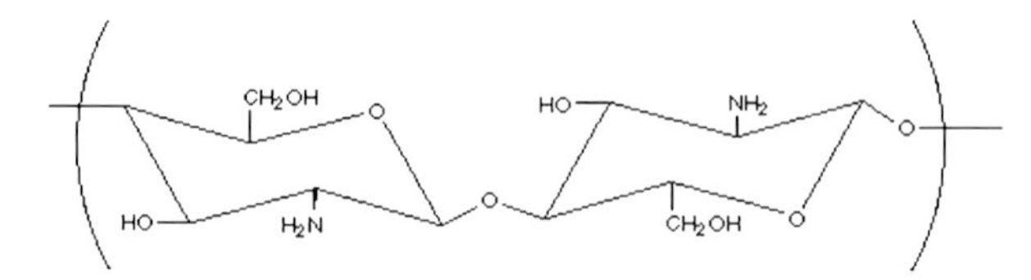

Figure 5. Structure of chitosan.

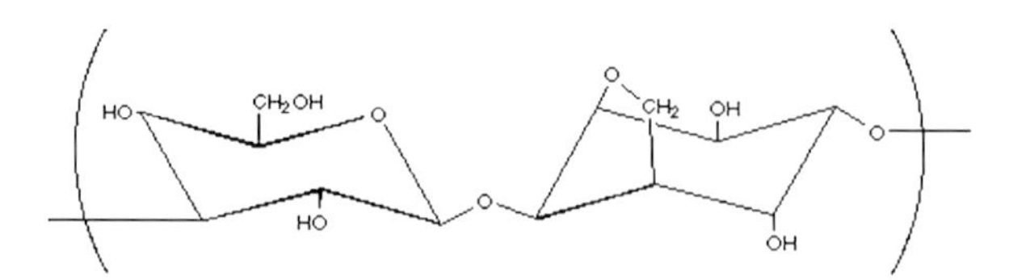

Figure 6. Structure of agarose.

4.1.5. Agarose

Agarose, similar to alginate, is a seaweed derived polysaccharide, but one that has the ability to form thermally reversible gels [46] (Figure 6). Mainly used for nanoencapsulation of cells, agarose/cell suspensions can be transformed into microbeads by utilizing a reduction in temperature [102]. A possible drawback to its use in this application is cellular protrusion through the membrane after gelation [46]. Other uses of agarose in cell immobilization include the fabrication of microporous gels seeded with chondrocytes for the repair of cartilage defects [103].

4.2. SYNTHETIC POLYMERS

Synthetic polymers are attractive for cell immobilization due to the high degree of control over the structure, and resultant properties that is possible with this class of materials. Unlike natural polymers, whose properties can vary from batch to batch, synthetic polymers can also provide highly consistent starting materials for scaffold fabrication.

4.2.1. Poly(glycolide) and poly(lactide)

Poly (α-hydroxy esters), specifically poly(glycolide) (PGA), poly(lactide) (PLA) and their copolymers, poly(lactide-co-glycolide) (PLGA), are the most widely used synthetic degradable polymers in medicine (Figure 7). Formed by a ring opening polymerization [104], these polymers degrade *via* non-specific hydrolysis of the ester linkage. The degradation products are lactic and glycolic acid, which are naturally metabolized and easily cleared by the body [57]. The methyl pendant group that differentiates PLA from PGA accounts for the differing properties (*e.g.* degradation rate) of these two polymers. Scaffolds made from these polymers have been used to repair a variety of tissues, including bone [53,105], liver [106], and nerve [107]. A possible limitation to the use of these polyesters is the bulk degradation mechanism, which typically results in a rapid loss of mechanical properties.

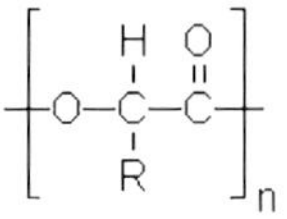

$$\left[-O-\overset{H}{\underset{R}{C}}-\overset{O}{\overset{\|}{C}}- \right]_n$$

PLA [R = CH_3]

PGA [R = H]

Figure 7. Structure of poly(a-hydroxy esters).

4.2.2. Polyanhydrides

Polyanhydrides have found their place in cell immobilization technologies due to their high strength and surface eroding degradation mechanism. Polyanhydrides are the polymeric products resulting from the dehydration of di-acid molecules, and are generally synthesized by melt polycondensation [108] (Figure 8). Polyanhydride degradation is controlled by surface erosion, which makes them attractive for certain micro and macroporous cell immobilization technologies. Surface erosion may be beneficial in load bearing applications where maintaining structural integrity is important for the lifetime of the implant [109,110]. Modifications to polyanhydrides, using methylacrylate anhydride monomers, allows for *in-situ* polymerizing materials [111]. Advantages to *in-situ* polymerization include spatial and temporal control of polymerization, and the ability for minimally invasive surgical techniques [112]. Limitations for the use of polyanhydrides include the local build-up of acidic by-products when the polymer degrades. As low-pH environments are deleterious to some proteins and nucleic acids and since some polymers degrade by hydrolysis that is acid-catalyzed, the environment caused by polymer degradation may disrupt cellular function in the area around the implant and affect scaffold stability [113].

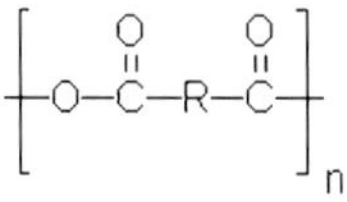

$$\left[-O-\overset{\overset{O}{\|}}{C}-R-\overset{\overset{O}{\|}}{C}-\right]_n$$

Figure 8. Structure of polyanhydrides.

4.2.3. Poly(ethylene oxide) and poly(ethylene glycol)

Due to their biocompatibility and low toxicity, poly(ethylene oxide) (PEO) and poly(ethylene glycol) (PEG) have been approved by the FDA for several medical applications. PEG and PEO are used synonymously in this section and differ only in that PEG has two hydroxyl end groups, whereas PEO has one (Figure 9). These polymers are synthesized by anionic or cationic polymerization of ethylene oxide. These polymers have been investigated as materials for cell immobilization due to their protein-repellent surface characteristics, which may decrease the amount of non-specific protein adsorption to the surface and possibly prevent biofouling of the implant [114]. This attribute could also be useful when trying to pattern the biomaterial surface with specific signals that promote cellular attachment. PEG has also been used as a surface coating on alginate-PLL nanoporous membranes to reduce fibrous encapsulation [115]. PEG can be co-polymerized with poly(lactic acid) (PLA) to form a degradable, hydrophilic polymer that can be used as a hydrogel for cell immobilization [116]. These hydrogels are attractive materials for tissue regeneration because of their high water content, which resembles to some degree the environment of native tissues [117]. PEG has also been used for nanoporous encapsulation of pancreatic islets [118], and as microporous scaffolds for cell immobilization in orthopaedic applications [17]. Although these polymers have found many uses in cell encapsulation technologies, they have limitations for biomedical uses mainly due to their lack of biodegradation. To address this issue, PEG has been copolymerized with poly(lactic acid) to provide degradable linkages [116].

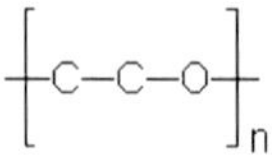

$$\left[-C-C-O-\right]_n$$

Figure 9. Structure of polyethylene glycol.

5. Summary and future directions

Cell immobilization biotechnology has involved the use of natural as well as synthetic biomaterials for applications that vary from complete immunoisolation within a nanoporous membrane, to cell-seeded micro and macroporous scaffolds that allow for rapid vascularization and tissue ingrowth. Immunoisolated devices show promise for temporary or permanent replacement of biochemical organ function while porous scaffolds can promote the formation of new structurally functional tissue. Together, these nano, micro and macroporous cell immobilization strategies offer promising therapeutic strategies to restore, maintain, and improve tissue function.

Although significant progress has been made in cell encapsulation biotechnologies, a variety of challenges remain. These include obtaining a reliable cell source that resists host rejection, controlling cell behaviour with the biomaterial, and the need for more highly controlled pore structures. The first major challenge involves the creation of a reliable cell source. Recent developments in stem cell research may provide a solution to cell sourcing problems [119,120,121]. However, it will be critical to control the differentiation of stem cell populations into the desired cell types if they are to be useful [32]. It may be possible to resolve the cell rejection challenges by genetically altering the transplanted cells to mask the histocompatibility proteins on their surface that normally cause them to be recognized as foreign [122]. Another major challenge for cell immobilization technologies is to design the biomaterial to promote specific cellular responses. For example, in immunoisolatory devices it may be beneficial to incorporate specific signals for the encapsulated cells to promote desired behaviours (*e.g.* increased production of the therapeutic agent). Another area of intense study is the creation of materials with pre-defined pore structures and appropriate degradation and mechanical properties to promote sufficient vascularization [12,123]. Through new manufacturing techniques (*e.g.* three-dimensional printing), it may be possible to create scaffolds with pre-defined and tightly controlled pore structures [124]. Advances in each of these areas will undoubtedly have a significant impact on the development of new therapeutics based on cell encapsulation.

References

[1] Langer, R. (2000) Tissue Engineering: Status and Challenges. E-biomed. 1: 5-6.

[2] Elcin, Y.M.; Dixit, V.; Lwein, K. and Gitnick, G. (1999) Xenotransplantation of fetal porcine hepatocytes in rats using a tissue engineering approach. Artif. Organs 23(2): 146-152.

[3] Dixit, V. (1995) Transplantation of isolated hepatocytes and their role in extrahepatic life support systems. Scand. J. Gastroenterol. Suppl. 208: 101-110.

[4] Lanza, R.P.; Sullivan, S.J. and Chick, W.L. (1992) Islet transplantation with immunoisolation. Diabetes. 41: 1503-1510.

[5] O'Shea, G.M.; Goosen, M.F. and Sun, A.M. (1984) Prolonged survival of transplanted islets of Langerhans encapsulated in a biocompatible membrane. Biochim. Biophys. Acta 804(1): 133-136.

[6] Sagen, J.; Bruhn, S.L.; Rein, D.H.; Li, R.H. and Carpenter, M.K. (1999) Transplantation of encapsulated cells into the central nervous system. In: Kuhtreiber, W.M.; Lanza, R.P. and Chick, W.L. (Eds.) Cell Encapsulation Technology and Therapeutics. Birkhauser, Boston; pp. 351-378.

[7] Hansbrough, J.F.; Cooper, M.L.; Cohen, R.; Spielvogel, R.; Greenleaf, G.; Bartel, R.L. and Naughton, G. (1992) Evaluation of a biodegradable matrix containing cultured human fibroblasts as a dermal replacement beneath meshed skin grafts on athymic mice. Surgery. 111(4): 438-446.

[8] Ma, J.; Wang, H.; He, B. and Chen, J. (2001) A preliminary *in vitro* study on the fabrication and tissue engineering applications of a novel chitosan bilayer material as a scaffold of human neofetal dermal fibroblasts. Biomaterials 22(4): 331-336.
[9] Galassi, G.; Brun, P.; Radice, M.; Cortivo, R.; Zanon, G.F.; Genovese, P. and Abatangelo, G. (2000) *In vitro* reconstructed dermis implanted in human wounds: degradation studies of the HA-based supporting scaffold. Biomaterials 21(21) 2183-2191.
[10] Watanabe, F.D.; Mullon, C.J.; Hewitt, W.R.; Arkadopoulos, N.; Kahaku, E.; Eguchi, S.; Khalili, T.; Arnaout, W.; Shackleton, C.R.; Rozga, J.; Solomon, B. and Demetriou, AA. (1997) Clinical experience with a bioartificial liver in the treatment of severe liver failure. A phase I clinical trial. Ann Surg 225(5): 484-494.
[11] Cima, L.G.; Vacanti, J.P.; Vacanti, C.; Ingber, D.; Mooney, D. and Langer, R. (1991) Tissue engineering by cell transplantation using degradable polymer substrates. J. Biomech. Eng. 113(2): 143-151.
[12] Kim, S.S.; Utsunomiya, H.; Koski, J.A.; Wu, B.M.; Cima, M.J.; Sohn, J.; Mukai, K.; Griffith, L.G. and Vacanti, J.P. (1998) Survival and function of hepatocytes on a novel three-dimensional synthetic biodegradable polymer scaffold with an intrinsic network of channels. Ann. Surg. 228(1): 8-13.
[13] Sullivan, S.J.; Maki, T.; Borland, K.M.; Mahoney, M.D.; Solomon, B.A.; Muller, T.E.; Monaco, A.P. and Chick, W.L. (1991) Biohybrid artificial pancreas: long-term implantation studies in diabetic, pancreatectomized dogs. Science 252(5006): 718-721.
[14] Kobayashi, H.; Ikada, Y.; Moritera, T.; Ogura, Y. and Honda, Y. (1991) Collagen-immobilized hydrogel as a material for lamellar keratoplasty. J. Appl. Biomater. 2(4): 261-267.
[15] Niklason, L.E.; Gao, J.; Abbott, W.M.; Hirschi, K.K.; Houser, S.; Marini, R.; Langer, R. (1999) Functional arteries grown *in vitro*. Science 284(5413): 489-493.
[16] Miwa, H.; Matsuda, T. and Iida, F. (1993) Development of a hierarchically structured hybrid vascular graft biomimicking natural arteries. ASAIO J. 39(3): M273-277.
[17] Metters, A.T.; Anseth, K.S. and Bowman, C.N. (1999) Fundamental studies of biodegradable hydrogels as cartilage replacement materials. Biomed. Sci. Instrum. 35: 33-38.
[18] Vacanti, C.A.; Langer, R.; Schloo, B. and Vacanti, J.P. (1991) Synthetic polymers seeded with chondrocytes provide a template for new cartilage formation. Plast. Reconstru. Surg. 88(5): 753-759.
[19] Sodian, R.; Sperling, J.S.; Martin, D.P.; Stock, U.; Mayer, J.E., Jr. and Vacanti, J.P. (1999) Tissue engineering of a trileaflet heart valve-early *in vitro* experiences with a combined polymer. Tissue Eng. 5(5): 489-494.
[20] Crane, G.M.; Ishaug, S.L.; Mikos, A.G. (1995) Bone tissue engineering. Nat. Med. 1(12): 1322-1324.
[21] Solchaga, L.A.; Dennis, J.E.; Goldberg, V.M. and Caplan, A.I. (1999) Hyaluronic acid-based polymers as cell carriers for tissue-engineered repair of bone and cartilage. J. Orthop. Res. 17(2): 205-213.
[22] Brittberg, M.; Lindahl, A.; Nilsson, A.; Ohlsson, C.; Isaksson, O. and Peterson, L. (1994) Treatment of deep cartilage defects in the knee with autologous chondrocyte transplantation. N. Engl. J. Med. 331(14): 889-895.
[23] Auger, F.A.; Pouliot, R.; Tremblay, N.; Guignard, R.; Noel, P.; Juhasz, J.; Germain, L. and Goulet, F. (2000) Multistep production of bioengineered skin substitutes: sequential modulation of culture conditions. *In Vitro* Cell Dev. Biol. Anim. 36(2): 96-103.
[24] Hardin-Young, J.; Teumer, J.; Ross, R.N. and Parenteau, N.L. (2000) Approaches to transplanting engineered cells and tissues. In: Lanza, R.P.; Langer, R. and Vacanti, J. (Eds.) Principles of Tissue Engineering. Academic Press, San Diego; pp. 281-292.
[25] MacKay, S.M.; Funke, A.J.; Buffington, D.A. and Humes, H.D. (1998) Tissue engineering of a bioartificial renal tubule. ASAIO J. 44(3): 179-183.
[26] Wandrey, C. and Vidal, D.S. (2001) Purificaiton of polymeric biomaterials. Ann. N.Y. Acad. Sci. 944: 187-198.
[27] Rozga, J. and Demetriou, A.A. (1995) Artificial liver. Evolution and future perspectives. ASAIO J. 41(4): 831-837.
[28] Khalil, M.; Shariat-Panahi, A.; Tootle, R.; Ryder, T.; McCloskey, P.; Roberts, E.; Hodgson, H. and Selden, C. (2001) Human hepatocyte cell lines proliferating as cohesive spheroid colonies in alginate markedly upregulate both synthetic and detoxificatory liver function. J. Hepatol. 34(1): 68-77.
[29] Rihova, B. (2000) Immunocompatibility and biocompatibility of cell delivery systems. Adv. Drug Deliv. Rev. 42(1-2) 65-80.
[30] Hui, T.; Rozga, J. and Demetriou, A.A. (2001) Bioartificial liver support. J. Hepatobiliary Pancreat. Surg. 8(1): 1-15.
[31] Zielinski, B.A. and Lysaght, M.J. (2000) Immunoisolation. In: Lanza, R.P.; Langer, R. and Vacanti, J. (Eds.) Principles of Tissue Engineering. Academic Press, San Diego; pp. 321-330.

[32] Langer, R. (2000) Tissue Engineering. Mol. Ther. 1(1): 12-15.
[33] Sagen, J.; Hama, A.T.; Winn, S.R. *et al.*, (1993) Pain reduction by spinal implantation of xenogeneic chromaffin cells immunologically-isolated in polymer capsules. Neurosci. Abstracts. 19: 234.
[34] Joseph, J.M.; Goddard, M.B.; Mills, J.; Padrun, V.; Zurn, A.; Zielinski, B.; Favre, J.; Gardaz, J.P.; Mosimann, F.; Sagen, J.; Christenson, L. and Aebischer, P. (1994) Transplantation of encapsulated bovine chromaffin cells in the sheep subarachnoid space: a preclinical study for the treatment of cancer pain. Cell Transplant. 3(5): 355-364.
[35] Aebischer, P.; Russell, P.C.; Christenson, L.; Panol, G.; Monchik, J.M. and Galletti, P.M. (1986) A bioartificial parathyroid. ASAIO Trans. 32(1): 134-137.
[36] Chang, P.L.; Shen, N. and Westcott, A.J. (1993) Delivery of recombinant gene products with microencapsulated cells *in vivo*. Hum. Gene Ther. 4(4): 433-440.
[37] Koo, J. and Chang, T.M. (1993) Secretion of erythropoietin from microencapsulated rat kidney cells: preliminary results. Int. J. Artif. Organs 16(7): 557-560.
[38] Rinsch, C.; Regulier, E.; Deglon, N.; Dalle, B.; Beuzard, Y. and Aebischer, P. (1997) A gene therapy approach to regulated delivery of erythropoietin as a function of oxygen tension. Hum. Gene Ther. 8(16): 1881-1889.
[39] Colton, C.K. (1995) Implantable biohybrid artificial organs. Cell Transplant. 4(4): 415-436.
[40] Brauker, J.; Frost, G.H.; Dwarki, V.; Nijjar, T.; Chin, R.; Carr-Brendel, V.; Jasunas, C.; Hodgett, D.; Stone, W.; Cohen, L.K. and Johnson, R.C. (1998) Sustained expression of high levels of human factor IX from human cells implanted within an immunoisolation device into athymic rodents. Hum. Gene Ther. 9(6): 879-888.
[41] Mullon, C. and Solomon, B.A. (2000) HepatAssist liver support system. In: Lanza, R.P., Langer, R. and Vacanti, J. Principles of Tissue Engineering, Academic Press, San Deigo; pp 553-558.
[42] Alberts, B.; Bray, D.; Lewis, J.; Raff, M.; Roberts, K. and Watson, J.D. (1989) Molecular Biology of the Cell. Garland Publishing Inc., New York, New York.
[43] Lee, M.K. and Bae, Y.H. (2000) Cell transplantation for endocrine disorders. Adv. Drug Deliv. Rev. 42(1-2): 103-120.
[44] Wang, T.G. (1999) Polymer membranes for cell encapsulation. In: Kuhtreiber, W.M.; Lanza, R.P. and Chick, W.L. (Eds.) Cell Encapsulation Technology and Therapeutics. Birkhauser, Boston; pp. 29-39.
[45] Iwata, H.; Park, G. and Ikada, Y. (1998) A model for oxygen transport in microencapsulated islets. In Thomson, R.C.; Mooney, D.J.; Healy, K.E.; Ikada, Y. and Mikos, A.G. (Eds.) Biomaterials Regulating Cell Function and Tissue Development. (530) MRS Symposium Preceedings, MRS, Warrendale; pp. 19-24.
[46] Uludag, H.; De Vos, P. and Tresco, P.A. (2000) Technology of mammalian cell encapsulation. Adv. Drug Deliv. Rev. 42(1-2) 29-64.
[47] Geller, R.L.; Loudovaris, T.; Neuenfeldt, S.; Johnson, R.C. and Brauker, J.H. (1997) Use of an immunoisolation device for cell transplantation and tumor immunotherapy. Ann. N.Y. Acad. Science 831: 438-451.
[48] Ezzell, C. (1995) Tissue Engineering and the Human Body Shop: Encapsulated-Cell Transplants Enter the Clinic. J. NIH Res. 7: 47-51.
[49] Lanza, R.P.; Jackson, R.; Sullivan, A.; Ringeling, J.; McGrath, C.; Kuhtreiber, W. and Chick, W.L. (1999) Xenotransplantation of cells using biodegradable microcapsules. Transplantation 67(8) 1105-1111.
[50] Aebischer, P.; Buchser, E.; Joseph, J.M.; Favre, J.; de Tribolet, N.; Lysaght, M.; Rudnick, S. and Goddard, M. (1994) Transplantation in humans of encapsulated xenogeneic cells without immunosuppression. A preliminary report. Transplantation 58(11): 1275-1277.
[51] Zielinski, B.A. and Aebischer, P. (1994) Chtiosan as a matrix for mammalian cell encapsulation. Biomaterials 15(13): 1049-1056.
[52] Suh, J.K. and Matthew, H.W. (2000) Application of chitosan-based polysaccharide biomaterials in cartilage tissue engineering: a review. Biomaterials 21(24): 2589-2598.
[53] Shea, L.D.; Wang, D.; Franceschi, R.T. and Mooney, D.J. (2000) Engineered bone development from a pre-osteoblast cell line on three-dimensional scaffolds. Tissue Eng. 6(6): 605-617.
[54] Mann, B.K. and West J.L. (2001) Tissue engineering in the cardiovascular system: progress toward a tissue engineered heart. Anat Rec. 263(4). 367-371.
[55] Wildevuur, C.R.; van der Lei, B. and Schakenraad, J.M. (1987) Basic aspects of the regeneration of small-calibre neoarteries in biodegradable vascular grafts in rats. Biomaterials 8(6): 418-422.
[56] Mooney, D.J. and Langer, R.S. (2000) Engineering Biomaterials for Tissue Engineering: The 10-100 micron size scale. In: Bronzino, J.D. (Ed.) The Biomedical Engineeing Handbook. CRC Press LLC; p. 112.
[57] Agrawal, C.M. and Ray, R. B. (2001) Biodegradable polymeric scaffolds for musculoskeletal tissue engineering. J. Biomed. Mater. Res. 55(2): 141-150.

[58] Kim, B.S.; Nikolovski, J.; Bonadio, J. and Mooney, D.J. (1999) Cyclic mechanical strain regulates the development of engineered smooth muscle tissue. Nat. Biotechnol. 17(10): 979-983.
[59] Ziegler, T.; Alexander, R.W. and Nerem, R.M. (1995) An endothelial cell-smooth muscle cell co-culture model for use in the investigation of flow effects on vascular biology. Ann. Biomed. Eng. 23(3): 216-225.
[60] Hunter, C.J.; Imler, S.M.; Malaviya, P.; Nerem, R.M. and Levenston, M.E. (2000) Mechanical compression alters gene expression and extracellular matrix synthesis by chondrocytes cultured in collagen I gels. Biomaterials 23(4): 1249-1259.
[61] Niklason, L.E. and Langer, R. S. (1997) Advances in tissue engineering of blood vessels and other tissues. Transpl. Immunol. 5(4): 303-306.
[62] Carver, S.E. and Heath, C.A. (1999) Influence of intermittent pressure, fluid flow, and mixing on the regenerative properties of articular chondrocytes. Biotechnol. Bioeng. 65(3): 274-281.
[63] Andriano, K.P.; Tabata, Y.; Ikada, Y. and Heller, J. (1999) *In vitro* and *in vivo* comparison of bulk and surface hydrolysis in absorbable polymer scaffolds for tissue engineering. J. Biomed. Mater. Res. 48(5): 602-612.
[64] Iordanskii, A.L.; Rudakova, T.E. and Zaikov, G.E. (1994) Interaction of Polymers with Bioactive and Corrosive Media. In series: New Concepts in Polymer Science, VSP, Utrecht.
[65] Hubbell, J.A. (1999) Bioactive Biomaterials. Curr. Opin. Biotechnol. 10(2): 123-129.
[66] Hern, D.L. and Hubbell, J.A. (1998) Incorporation of adhesion peptides into nonadhesive hydrogels useful for tissue resurfacing. J. Biomed. Mater. Res. 39(2): 266-276.
[67] Alsberg, E.; Anderson, K.W.; Albeiruti, A.; Franceschi, R.T. and Mooney, D.J. (2000) Cell-interactive alginate hydrogels for bone tissue engineering. J. Dent. Res. 80(11): 2025-2029.
[68] Cook, A.D.; Hrkach, J.S.; Gao, N.N.; Johnson, I.M.; Pajvani, U.B.; Cannizzaro, S.M. and Langer, R. (1997) Characterization and development of RGD-peptide-modified poly(lactic acid-co-lysine) as an interactive, resorbable biomaterial. J. Biomed. Mater. Res. 35(4): 513-523.
[69] Yu, X.; Dillon, G.P. and Bellamkonda, R.B. (1999) A laminin and nerve growth factor-laden three-dimensional scaffold for enhanced neurite extension. Tissue Eng. 5(4): 291-304.
[70] Ito, Y.; Chen, G. and Imanishi, Y. (1998) Artificial juxtacrine stimulation for tissue engineering. J. Biomater. Sci. Polym. Ed. 9(8): 879-890.
[71] Han, S.; Mahato, R.I.; Sung, Y.K. and Kim, S.W. (2000) Development of biomaterials for gene thereapy. Mol. Ther. 2(4): 302-317.
[72] Rosenblatt, J.; Devereux, B. and Wallace, D.G. (1994) Injectable collagen as a pH-sensitive hydrogel. Biomaterials 15(12): 985-995.
[73] Senuma, Y.; Franceschin, S.; Hilborn, J.G.; Tissieres, P.; Bisson, I. and Frey, P. (2000) Bioresorbable microspheres by spinning disk atomization as injectable cell carrier: from preparation to *in vitro* evaluation. Biomaterials 21(11): 1135-1144.
[74] Chevallay, B. and Herbage, D. (2000) Collagen-based biomaterials as 3D scaffold for cell cultures: applications for tissue engineering and gene therapy. Med. Biol. Eng. Comput. 38(2): 211-218.
[75] Roche, S.; Ronziere, M.C.; Herbage, D. and Freyria, A.M. (2001) Native and DPPA cross-linked collagen sponges seeded with fetal bovine epiphyseal chondrocytes used for cartilage tissue engineering. Biomaterials 22(1): 9-18.
[76] Lee, C.H.; Singla, A. and Lee, Y. (2001) Biomedical applications of collagen. Int. J. Pharm. 221(1-2): 1-22.
[77] Yannas, I.V.; Lee, E.; Orgill, D.P.; Skrabut, E.M. and Murphy, G.F. (1989) Synthesis and characterization of a model extracellular matrix that induces partial regeneration of adult mammalian skin. Proc. Natl. Acad. Sci. USA 86(3): 933-937.
[78] Yamada, N.; Uchinuma, E. and Kuroyanagi, Y. (1999) Clinical evaluation of an allogeneic cultured dermal substitute composed of fibroblasts within a spongy collagen matrix. Scand. J. Plast. Reconstr. Surg. Han. Surg. 33(2): 147-154.
[79] Nimni, M.E.; Bernick, S.; Cheung, D.T.; Ertl, D.C.; Nishimoto, S.K.; Paule, W.J.; Salka, C. and Strates, B.S. (1988) Biochemical differences between dystrophic calcification of cross-linked collagen implants and mineralization during bone induction. Calcif. Tiddue Int. 42(5): 313-320.
[80] Murata, M.; Huang, B. Z.; Shibata, T.; Imai, S.; Nagai, N. and Arisue, M. (1999) Bone augmentation by recombinant human BMP-2 and collagen on adult rat parietal bone. Int. J. Oral Maxillofac. Surg. 28(3): 232-237.
[81] Ratcliffe, A. (2000) Tissue engineering of vascular grafts. Matrix Biol. 19(4): 353-357.
[82] Butler, D.L. and Awad, H.A. (1999) Perspectives on cell and collagen composites for tendon repair. Clin. Orthop. 367 Suppl: S324-332.

[83] Shapiro, L. and Cohen, S. (1997) Novel alginate sponges for cell culture and transplantation. Biomaterials 18(8): 583-590.
[84] Matsumoto, T.; Kawai, M. and Masuda, T. (1992) influence of concentration and mannuronate/guluronate [correction of gluronate] ratio on steady flow properties of alginate aqueous systems. Biorheology 29(4): 411-417.
[85] Eiselt, P.; Lee, K.Y. and Mooney, D.J. (1999) Rigidity of two-component hydrogels prepared from alginate and poly(ethylene glycol)-diamines. Macromolecules 32(17): 5561-5566.
[86] Lee, K.Y.; Rowley, J.A; Eiselt, P.; Moy, E.M.; Bouhadir, K.H. and Mooney, D.J. (2000) Controlling mechanical and swelling properites of alginate hydrogels independently by cross-linker type and cross-linking density. Macromolecules 33(11): 4291-4294.
[87] Lee, K.Y. and Mooney, D.J. (2001) Hydrogels for tissue engineering. Chem. Rev. 101(7): 1869-1879.
[88] Smetana, K., Jr. (1993) Cell biology of hydrogels. Biomaterials 14(14): 1046-1050.
[89] Rowley, J.A. and Mooney, D.J. (2002) Alginate type and RGD density control myoblast phenotype. J. Biomed. Mater. Res. 60(2): 217-223.
[90] Paige, K.T.; Cima, L.G.; Yaremchuk, M.J.; Vacanti, J.P. and Vacanti, C.A. (1995) Injectable cartilage. Plast. Reconstr. Surg. 96(6): 1390-1400.
[91] Gutowska, A.; Jeong, B. and Jasionowski, M. (2001) Injectable gels for tissue engineering. Anat. Rec. 263(4): 342-349.
[92] Pouyani, T. and Prestwich, G.D. (1994) Functionalized derivatives of hyaluronic acid oligosaccharides: drug carriers and novel biomaterials. Bioconjug. Chem. 5(4): 339-347.
[93] Afify, A.M.; Stern, M.; Guntenhoner, M. and Stern, R. (1993) Purification and characterization of human serum hyaluronidase. Arch. Biochem. Biophys. 305(2): 434-441.
[94] Radice, M.; Brun, P.; Cortivo, R.; Scapinelli, R.; Battaliard, C. and Abatangelo, G. (2000) Hyaluronan-based biopolymers as delivery vehicles for bone-marrow-derived mesenchymal progenitors. J. Biomed. Mater. Res. 50(2): 101-109.
[95] Liu, L.S.; Thompson, A.Y.; Heidaran, M.A.; Poser, J.W. and Spiro, R.C. (1999) An osteoconductive collagen/hyaluronate matrix for bone regeneration. Biomaterials 20(12) 1097-1108.
[96] Chenite, A.; Chaput, C.; Wang, D.; Combes, C.; Buschmann, M.D.; Hoemann, C.D.; Leroux, J.C.; Atkinson, B.L.; Binette, F. and Selmani, A. (2000) Novel injectable neutral solutions of chitosan form biodegradable gels *in situ*. Biomaterials 21(21): 2155-2161.
[97] Monteiro, O.A., Jr. and Airoldi, C. (1999) Some studies of crosslinking chitosan-glutaraldehyde interaction in a homogeneous system. Int. J. Biol. Macromol. 26(2-3): 119-128.
[98] Varum, K.M.; Myhr, M.M.; Hjerde, R.J. and Smidsrod, O. (1997) *In vitro* degradation rates of partially N-acetylated chitosans in human serum. Carbohydr. Res. 299(1-2): 99-101.
[99] Aiedeh, K.; Gianasi, E.; Orienti, I. and Zecchi, V. (1997) Chitosan microcapsules as controlled release systems for insulin. J. Microencapsul. 14(5): 567-576.
[100] Muzzarelli, R.; Baldassarre, V.; Conti, F.; Ferrara, P.; Biagini, G.; Gazzanelli, G. and Vasi, V. (1988) Biological activity of chitosan: ultrastructural study. Biomaterials 9(3): 247-252.
[101] Tan, W.; Krishnaraj, R. and Desai, T.A. (2001) Evaluation of nanostructured composite collagen--chitosan matrices for tissue engineering. Tissue Eng. 7(2): 203-210.
[102] Iwata, H.; Amemiya, H.; Matsuda, T.; Takano, H.; Hayashi, R. and Akutsu, T. (1989) Evaluation of microencapsulated islets in agarose gel as bioartificial pancreas by studies of hormone secretion in culture and by xenotransplantation. Diabetes. 38(Suppl. 1): 224-225.
[103] Rahfoth, B.; Weisser, J.; Sternkopf, F.; Aigner, T.; von der Mark, K. and Brauer, R. (1998) Transplantation of allograft chondrocytes embedded in agarose gel into cartilage defects of rabbits. Osteoarthritis Cartilage 6(1): 50-65.
[104] Seppala, J.V.; Korhonen, H.; Kylma, J. and Tuominen, J. (2002) General Methodology for Chemical Synthesis of Polyesters. In: Yoshiharu, D. and Steinbüchel, A. (Eds.) Biopolymers, Vol. 3b: Polyesters II - Properties and Chemical Synthesis. Wiley-VCH, Weinheim; pp. 327-370.
[105] Hollinger, J.O. and Schmitz, J.P. (1987) Restoration of bone discontinuities in dogs using a biodegradable implant. J. Oral Maxillofac. Surg. 45(7): 594-600.
[106] Mayer, J.; Karamuk, E.; Akaike, T. and Wintermantel, E. (2000) Matrices for tissue engineering-scaffold structure for a bioartificial liver support system. J. Control. Release 64(1-3): 81-90.
[107] Widmer, M.S.; Gupta, P.K.; Lu, L.; Meszlenyi, R.K.; Evans, G.R.; Brandt, K.; Savel, T.; Gurlek, A.; Patrick, C.W., Jr. and Mikos, A.G. (1998) Manufacture of porous biodegradable polymer conduits by an extrusion process for guided tissue regeneration. Biomaterials 19(21): 1945-1955.
[108] Domb, A.J.; Amselem, S.; Langer, R. and Maniar, M.(1994) Polyanhydrides as carriers of drugs. In: Shalaby, W.S. (Ed.) Biomedical Polymers. Hanser Publishers, Munich (Germany); pp. 69-96.

[109] Ibim, S.E.; Uhrich, K.E.; Attawia, M.; Shastri, V.R.; El-Amin, S.F.; Bronson, R.; Langer, R. and Laurencin, C.T. (1998) Preliminary *in vivo* report on the osteocompatibility of poly(anhydride-co-imides) evaluated in a tibial model. J. Biomed. Mater. Res. 43(4): 374-379.

[110] Ibim, S.M.; Uhrich, K.E.; Bronson, R.; El-Amin, S.F.; Langer, R.S. and Laurencin, C.T. (1998) Poly(anhydride-co-imides): *in vivo* biocompatibility in a rat model. Biomaterials 19(10): 941-951.

[111] Anseth, K.S.; Svaldi, D.C.; Laurencin, C.T. and Langer, R. (1997) Potopolymerization of novel degradable networks for orthopaedic applications. In: Scranton, A.B.; Christopher, N.B. and Peiffer, R.W. (Eds.) Photopolymerizaion: fundamentals and applications. American Chemical Society, Washington D.C.; pp. 189-202.

[112] Burkoth, A.K. and Anseth, K.S. (2000) A review of photocrosslinked polyanhydrides: *in situ* forming degradable networks. Biomaterials 21(23): 2395-2404.

[113] Fu, K.; Pack, D.W.; Klibanov, A.M. and Langer, R. (2000) Visual evidence of acidic environment within degrading poly(lactic-co-glycolic acid) (PLGA) microspheres. Pharm. Res. 17(1): 100-106.

[114] Sawhney, A.S. (1999) Poly(ethylene glycol). In: Kuhtreiber, W.M.; Lanza, R.P. and Chick, W.L. (Eds.) Cell encapsulation technology and therapeutics. Birkhauser, Boston; pp. 108-116.

[115] Sawhney, A.S. and Hubbell, J.A. (1992) Poly(ethylene oxide)-graft-poly(L-lysine) copolymers to enhance the biocompatibility of poly(L-lysine)-alginate microcapsule membranes. Biomaterials 13(12) 863-870.

[116] Metters, A.T.; Bowman, C.N. and Anseth, K.S. (2000) A statistical kinetic model for the bulk degradation of PLA-b-PEG-b-PLA hydrogel networks. Journal of Physical Chemistry 104: 7043-7049.

[117] Bryant, S.J. and Anseth, K.S. (2002) Hydrogel properties influence ECM production by chondrocytes photoencapsulated in poly(ethylene glycol) hydrogels. J. Biomed. Mater. Res. 59(1) 63-72.

[118] Dupuy, B.; Gin, H.; Baquey, C. and Ducassou, D. (1988) *In situ* polymerization of a microencapsulating medium round living cells. J. Biomed. Mater. Res. 22(11) 1061-1070.

[119] Zuk, P.A.; Zhu, M.; Mizuno, H.; Huang, J.; Futrell, J.W.; Katz, A.J.; Benhaim, P.; Lorenz, H.P. and Hedrick, M.H. (2001) Multilineage cells from human adipose tissue: implications for cell-based therapies. Tissue Eng. 7(2): 211-228.

[120] Heath, C.A. (2000) Cells for tissue engineering. Trends Biotechnol. 18(1): 17-19.

[121] Oreffo, R.O. and Triffitt, J.T. (1999) Future potentials for using osteogenic stem cells and biomaterials in orthopedics. Bone 25(2 Suppl): 5S-9S.

[122] Fontaine, M.; Hansen, L.K.; Thompson, S.; Uyama, S.; Ingber, D.E.; Langer, R. and Vacanti, J.P. (1993) Transplantation of genetically altered hepatocytes using cell-polymer constructs. Transplant. Proc. 25(1 Pt 2): 1002-1004.

[123] Giordano, R.A.; Wu, B.M.; Borland, S.W.; Cima, L.G.; Sachs, E.M. and Cima, M.J. (1996) Mechanical properties of dense polylactic acid structures fabridated by three dimensional printing. J. Biomater. Sci. Polym. Ed. 8(1): 63-75.

[124] Griffith, L.G.; Wu, B.; Cima, M.J.; Powers, M.J.; Chaignaud, B. and Vacanti, J.P. (1997) *In vitro* organogenesis of liver tissue. Ann. N.Y. Acad. Sci. 831: 382-397.

ALGINATE AS A CARRIER FOR CELL IMMOBILISATION

JAN EGIL MELVIK AND MICHAEL DORNISH
FMC BioPolymer, Gaustadalléen 21, 0349 Oslo, Norway –
Fax: 47 22696470 – Email: jan_egil_melvik@fmc.com

1. Introduction

Alginates, which are natural occurring marine polymers, have been used for several decades in the food and pharmaceutical industries as emulsifying, thickening, film forming and gelling agents [1]. Within the biomedical field alginates are now also well known as immobilisation materials for cells, tissue or macromolecules. Immobilisation (entrapment) in insoluble alginate gel is recognized as a rapid, non-toxic and versatile method for macromolecules and cells. The replacement of cell products lost due to defects by immobilised enzymes and cells was first suggested by Chang [2] in 1964 and this potential has become more and more actualised through increased knowledge about diseases that are caused by the inability of the body to produce critical molecules such as growth factors, hormones or enzymes. Therefore, alginates are now widely used as immobilising materials for cells or tissue in the development of artificial organs and with the potential to be used in treatment of a variety of diseases, including Parkinson's disease, chronic pain, liver failure and hypocalcaemia.

Table 1. Applications involving alginate as a matrix for living cells.

Application	References
Insulin regulation in diabetes	[3-13]
Treatment of haemophilia B	[14,15]
Growth factor for treatment of dwarfism	[16-19]
Treatment of kidney failure	[20,21]
Treatment of liver failure	[22-24]
Treatment of Parkinson's disease	[25,26]
Treatment of parathyroid failure	[27]
Treatment of chronic pain	[28]
Treatment of brain cancer	[29-32]
Bone and cartilage regeneration	[33-38]
Nerve regeneration	[39-41]
Treatment of urethral reflux	[42]
Study of tumour angiogenesis	[43,44]

V. Nedović and R. Willaert (eds.), Fundamentals of Cell Immobilisation Biotechnology, 33-51.

The perhaps most well-known example of this is the immobilisation of Islet of Langerhans utilized as an artificial pancreas in the treatment of Type I diabetes. In Table 1 are listed different biomedical applications utilizing cell immobilisation technology with alginate as a carrier.

However, since alginates are a heterogeneous group of polymers, with a wide range of functional properties, their success will rely on an appropriate choice of materials and methodology for each application. This must be based on knowledge of their chemical composition, the correlation between structure and functional properties, and a sufficient understanding of the behaviour of polysaccharide formulations on a macroscopic as well as a molecular level. Furthermore, it is important to recognize the need for working with sufficiently purified and well-characterised materials in order to obtain reproducible properties. This review summarizes some basic information regarding the use of alginate as an immobilizing agent for cells within the biomedical field.

2. Chemistry and physical properties

Alginic acid is an unbranched binary copolymer of 1-4 glycosidically linked α-L-guluronic acid (G) and its C-5 epimer β-D-mannuronic acid (M). It is, as with cellulose, composed of several (100-3,000) building blocks linked together in a stiff and partly flexible chain. The proportion as well as the distribution of the two monomers determines to a large extent the physiochemical properties. The salts (and esters) of these polysaccharides are generally named alginates. It has been found that the M- and G-residues are joined together in a blockwise fashion (Figure 1). This implies that three types of blocks may be found, homopolymeric M-blocks (M-M-M), homopolymeric G-blocks (G-G-G) and heteropolymeric, sequentially alternating MG-blocks (G-M-G-M). The major source of alginate is brown seaweed, however, certain bacteria also produce alginate. In all alginate producing organisms C-5-epimerisation of M into G is the final step of alginate biosynthesis, an in-chain conversion which is catalyzed by one or more different mannuronan C-5-epimerases [45,46].

The relative amounts of the two uronic acids and the sequential arrangements of them along the polymer chain vary widely, depending on the origin of the alginate. The functional properties of alginate molecules within an immobilisation matrix strongly correlate with the composition and block structure. Alginates do normally not have any regular repeating units. Accordingly, the sequence characteristics are not determined by the monomer composition (monad frequencies) alone, but by the content of diad, triad and higher order frequencies. The four diad (nearest neighbour) frequencies F_{GG}, F_{GM}, F_{MG} and F_{MM}, and the eight possible triad frequencies F_{GGG}, F_{GGM}, F_{MGG}, F_{MGM}, F_{MMM}, F_{MMG}, F_{GMM} and F_{GMG} can be measured by NMR techniques [47]. Knowledge of the diad and triad frequencies permits the calculation of average block length.

Dissolving alginate in water gives a viscous solution of which the viscosity will increase with the length of the alginate molecules, *i.e.* the number of monomers. Upon dissolving alginate in water the molecules start to hydrate and the solution gains viscosity. The hydrated and dissolved molecules are dominated by intramolecular electrostatic repulsion between neighbouring negative charges giving an extended chain confirmation. The chains possess a certain flexibility, but their rotation is somewhat hindered at the glycosidic bond, resulting in a certain stiffening of the chain. The intrinsic inflexibility of the alginate molecules in solution is dependent on block structure and increases in the order MG<MM<GG. Solutions of stiff macromolecules will be highly viscous, as a result of this chain stiffness. Also longer chains will give higher viscosity at similar concentrations.

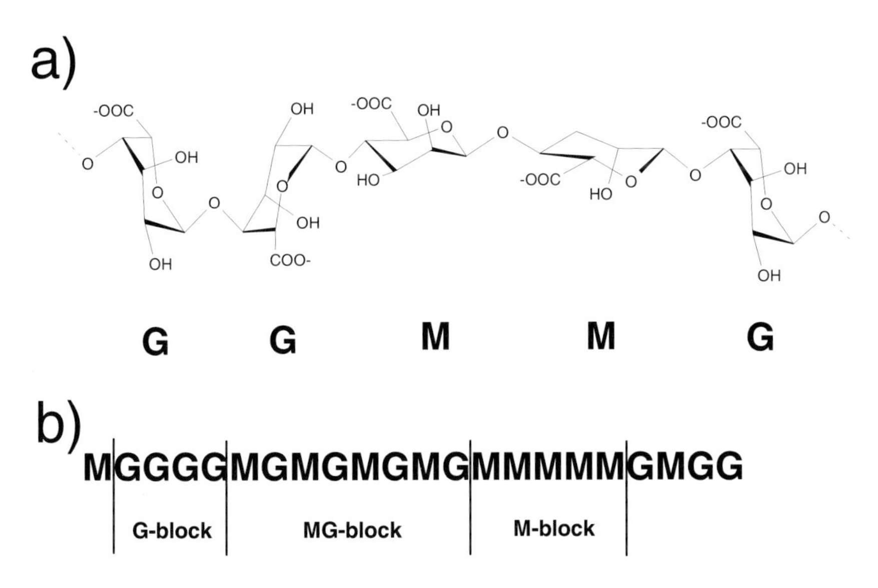

Figure 1. Alginate structure. a) conformation of alginate chain with 1-4-linked ß-D-mannuronic acid (M) and α-L-guluronic acid (G). b) alginate chain sequences.

The molecular chain structure of alginate and therefore also the viscosity characteristics are dependent on the total ionic environment. Because alginate is also a polyelectrolyte, an ionic strength dependent electrostatic repulsion between the charged groups on the polymer chain will increase the chain extension and therefore viscosity [48]. Not surprisingly, seaweed alginate shows an optimal viscosity for salt concentrations equivalent to seawater. At higher ion concentrations alginate molecules may also precipitate, which results in a viscosity reduction ("salt-out" effect). Alginates have a strong affinity for cations which decreases in the following orders [49]:

$$Pb^{2+} > Cu^{2+} = Ba^{2+} > Sr^{2+} > Cd^{2+} > Ca^{2+} > Zn^{2+} > Co^{2+} > Ni^{2+}$$

The presence of divalent ions such as Ca^{2+}, Ba^{2+} or Sr^{2+} may lead to cross-linking of alginate molecules (gelling). Alginates may also precipitate or form gels under acidic conditions. The pKa values for mannuronic and guluronic acid are 3.38 and 3.65, respectively [50], and as a result alginates are polyanionic at neutral pH. In pharmaceutical applications, the change from a soluble alginate to an insoluble alginic acid at low pH is actually the property behind the volume-wise largest product, namely an anti-reflux remedy which, when swallowed, forms an alginic acid raft on the top of the stomach contents.

The uniqueness of alginate as an immobilizing agent in biomedical applications rests in its ability to form heat-stable gels that can develop and set at physiologically relevant temperatures. It is the alginate gel formation with calcium ions that has been mostly used. However, alginate forms gels with most di- and multivalent cations. Monovalent cations and Mg^{2+} ions do not induce gelation [51] while ions like Ba^{2+} and Sr^{2+} will produce stronger alginate gels than Ca^{2+} [52]. The gelling occur when divalent cations take part in the interchain binding between G-blocks giving rise to a three-dimensional network in the form of a gel. The binding zone between the G-blocks is often described by the so-called "egg-box model" [53] (see Figure 2). The process is dependent on both the alginate and gelling ion concentration. When the gelling takes place in presence of excess gelling ions, a modified egg-box model has been suggested [54], involving more than two alginate chains in the gelling zone. This may have an impact on the porosity of the gel. If no shrinking of the gel occurs, there should be more space in between the chains, leading to an increased porosity. Evidence for this has been demonstrated in two different ways [55], by an increased binding of a coating material and an increased release of blue dextran for gel beads made with excess amounts of calcium chloride. The mechanical properties of an alginate gel will to a large extent vary with the alginate composition [56]. The gel strength will depend upon the guluronic content (FG) and also of the average number of G-units in the G-blocks (NG>1). Consequently gels made of alginates with a high G-block content will be stronger and more stable as compared to low G-block content alginate gels.

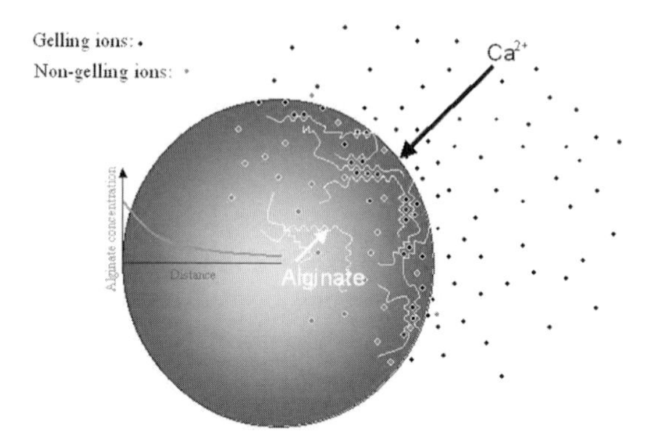

Figure 2. Diffusion gelling of an alginate bead. An alginate-containing mix, is turned into a gel by being dripped into, or sprayed with a calcium salt (or other appropriate salt) solution. When the gelling ions diffuse into the alginate-containing system the gel will form immediately within the G-block units ("egg boxes"). Increasing the concentration of gelling ions can accelerate the diffusion process and the presence of non-gelling ions will slow down the gelling process. This will strongly influence gel homogeneity and porosity as it will influence rearrangement of alginate molecules within the beads.

Earlier it was assumed that the alginate gel concentration was always the same throughout the gel. This is definitely not true for alginate gels formed by diffusion of cross-linking ions into the alginate solution [57,58]. When a droplet of alginate falls into a bath containing divalent cations like Ca^{2+} the gelling process will start immediately at the contact zone between the alginate droplet and the aqueous gelling ions generating a gel bead. The gel formation process, briefly illustrated in Figure 2, and more thoroughly explained elsewhere in this book (Strand *et al.*), may be controlled and gels with different properties may be formed. In particular bead homogeneity is dependent upon the way in which the gelling ions are added. The distribution of the gel within the bead may thus to a large extent be controlled and beads even with a capsular structure can be made [58].In order to produce more inhomogeneous beads while retaining physiologic conditions sodium may be replaced by non-charged molecules like mannitol to prevent osmotic stress to the cells [59]. During the gelling of an alginate bead, the interface between the alginate solution and gel may be observed in a light microscope [55]. The gelling speed was found to decrease somewhat first and then to accelerate somewhat near the centre. The gelling speed is roughly around 100 μm/min in 50 mM $CaCl_2$, which means that a 500 μm bead is gelled in approximately 2.5 minutes. This is near the rate of freely diffusing calcium ions [55].

3. Entrapment of cells in alginate beads

Most methods for immobilisation of biomaterials in alginate beads basically involve two main steps. The first step is the formation of an internal phase where the alginate solution containing biological materials is dispersed into small droplets. In the second step the droplets are solidified by gelling or membrane formation at the droplet surface. The droplet formation method is most commonly used but there are also alternative methods for immobilisation of cells in alginate like the entrapment of cells in a gel sheet [13].

Some critical parameters that need to be accounted for when immobilizing cells in alginate beads are listed in Table 2.

Table 2. Critical parameters to be considered when immobilizing cells in an alginate gel.

- Alginate concentration
- Alginate MW distribution
- Alginate composition and sequential distribution (M/G)
- Presence of impurities
- Concentration of biological material (cells or others)
- Bead size
- Concentration of gelling ions
- Concentration of non-gelling ions and chelating compounds

When using alginate beads as an immobilisation system the appropriate bead size will often be a compromise. The bead itself must be large enough to contain the biological material. Larger beads may also be easier to handle during washing or other treatments. Depending of the bead formation and immobilisation procedure the cells will be more or less randomly distributed within the internal matrix. Also, when generating beads the desired mean size and acceptable size distribution should be accounted for. The bead size is controlled primarily by controlling droplet size. Various techniques are used to form droplets as described in detail by Dulieu *et al.* [60]. For immobilizing small samples of cells in spherical beads down to sizes from about 200 to 1000 μm, the electrostatic bead generator is a convenient alternative [61-65] and is suitable for the immobilisation of islets [66]. The desired bead size is mostly obtained simply by selecting an appropriate nozzle size. When using droplet generators, size and sphericity of the beads will also depend on the viscosity of the alginate solution and the fall distance for the gel droplets above the gelling solution. Also, the final size of the beads will be dependent of the gelling conditions used. The mechanical and swelling property of alginate gel beads are strongly dependent on the monomeric composition, block structure as well as size and size distribution of the alginate molecules. A Ca^{2+} alginate gel will shrink during gelling, and thereby loose water and increase the polymer concentration. Beads made from an alginate with a low G-content will be more susceptible to volume changes than beads with a higher G-content [56,61]. For cells immobilised in alginate beads, shrinkage will increase the relative concentration of cells in immobilised cell matrixes.

Because diffusion of molecules within the alginate gel network may be limited, knowledge about the diffusion characteristics, pore size and pore size distribution is important. Effective systems of immobilised cells or enzymes requires that the transport of substrates and product within the alginate network should be as free as possible. The highest diffusion rate of proteins, indicating the most open pore structures, is found in high G alginates [56,67]. Diffusion coefficients also increase when lowering the alginate concentration. Bead homogeneity, *i.e.* gel forming kinetics, is an important factor for bead porosity and diffusion properties. For homogeneous beads efflux of proteins is faster than for inhomogeneous beads where the alginate is concentrated at the surface [67,68]. Smaller molecules are to a lesser extent influenced by the gel network and diffusion of molecules such as glucose and ethanol has been reported to be as high as about 90 % of the diffusion rate in water. Tanaka *et al.* [69] found no reduction in diffusion coefficients for solutes with MW $< 2 \times 10^4$ in calcium alginate gel beads as compared with free diffusion in water.

The alginate gel as an immobilisation matrix is sensitive to chelating compounds commonly present in biological systems such as phosphate, lactate, citrate and other anti-gelling cations such as Na^+ or Mg^{2+}. To retain gel stability alginate under physiologic conditions alginate beads may be kept in a medium containing a few millimolar free calcium ions and by keeping the $Na^+ : Ca^{2+}$ ratio less than 25:1 for high G alginates and 3:1 for low G alginates [56]. An alternative is also to replace Ca^{2+} with other divalent cations with a higher affinity for alginate. Using stronger gelling ions will reduce gel porosity. However, in applications involving immobilisation of living cells toxicity is a limiting factor in the use of most ions, and only Sr^{2+}, Ba^{2+} and Ca^{2+} are considered as acceptable for these purposes [70]. However, leakage of barium from the gel may possibly induce negative effect on cells, and a recent comparative study suggested best biocompatibility when using strontium as gelling ions [71]. The sensitivity of alginate gels to chelating compounds may easily be used to dissolve the gel to release cells or other entrapped materials. For calcium alginate gels sequestrants like citrate or EDTA are commonly used. However, if the gel is cross-linked with stronger ions like barium resistance towards the sequestrant will be stronger.

The need for improving the properties of alginate beads has promoted development of coating techniques (see also Strand *et al.* in this book). Coated beads are stronger than non-coated beads for *in vivo* applications [72], and coating materials may also be used to reduce the porosity [73]. Alginates may form strong complexes with polycations such as chitosan or polypeptides (*e.g.* poly-L-lysine [74,75] or polyornithine [4,76,77]), or synthetic polymers such as polyethylenimine [9,78] which may be used to stabilize the gel. The most commonly used material for coating of alginate beads are polypeptides like poly-L-lysine (PLL) [9,16,79-86], but other polycations like chitosan are also commonly used [87-94]. Such polycation complexes will stabilize beads in the presence of calcium chelators or non-gelling cations. For coated beads the alginate core may also be dissolved within the capsules by using a calcium chelating agent (citrate or EDTA) or antigelling cations. This will give a polyanion-polycation complex that behaves as a semipermeable membrane with the entrapped cells within the liquid core. This treatment, although frequently used may, however, often damage some of the microcapsules [95] and seems to have little advantage compared to a solid core as mechanical strength of these capsules relies only on the outer membrane and a high internal osmotic pressure [75].

4. Other biostructures with alginates

Although entrapment of biologic materials in alginate gel beads is most commonly used in biomedical application development with this biopolymer, other biostructures made of alginate may also have a promising future. Dependent of the manufacturing process alginates may take various forms such as aqueous solutions, pastes and cross-linked gels, and solid structures such as membranes, fibres, non-woven felts, powders, rods and tubes (Figure 3-4). Porous alginate sponges are being studied as useful materials for cell transplantation [38,96] and nerve regeneration [39,40]. Furthermore, implants of chondrocytes in alginate gelled *in situ* by the addition of calcium chloride solution directly into articular cartilage defects have been promising [36] and alginate pastes containing chondrocytes have been injected in children as successful treatment of urethral reflux problems [42]. Different biostructures can be made either using pure alginate, or as integrated products where another functional ingredient is included.

The functionality will depend both on the innate characteristics of the alginate from which it is manufactured as well as the way of processing. Furthermore, integrated products may be more complex in functionality than pure alginate structures, due to physico-chemical interactions between the alginate and the other ingredients involved. Alginate biostructures in the form of solid bolts and tubes can be made by extrusion from a thick alginate solution or an alginate paste. Furthermore, alginate biostructures in the form of films and membranes can easily be made by evaporation of water from an aqueous alginate solution. Dependent of the alginate used, the surface of such a membrane can be varied with respect to hydrophilic/hydrophobic properties. While a more hydrophilic surface is obtained using alginates with a low or high content of guluronate, a surface with more hydrophobic properties is obtained with a medium-G content alginate.

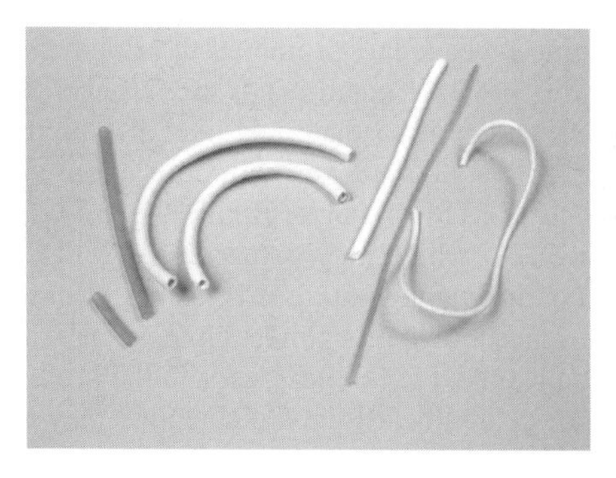

Figure 3-4. Different biostructures made from alginates.

5. Biocompatibility

The major reason for using alginate gel as an implantation device is the protection afforded by the gel network against the immune system of the host [97]. The encapsulation technology using alginates is promising and provides a safe and simple technique for implanting cells into various sites of the human body. It is mostly the gelling properties as well as the inert properties of alginates that make this polysaccharide a very good candidate for immobilisation of living cells. The use of an encapsulation/immobilisation system protects cells or tissue from immune rejection. In many applications microcapsules often have advantages compared to other possible encapsulation techniques such as macrocapsules and intravascular devices. The large surface area of small beads results in enhanced survival of tissue due to better nutrition and oxygen supply, and in addition microcapsules can be implanted with minimal invasive surgery.

For any immobilisation material, a good biocompatibility towards the immobilised cells and for implantation purposes also towards the host is of vital importance. For applications involving implantation, the recognition of beads as foreign bodies by the host may result in the beads eventually becoming coated in a fibrous sheath consisting of giant cells and fibrous tissues [98]. This may decrease the mass transfer of nutrients and metabolites, leading to the eventual death of the immobilised cells. Much research has therefore been focused on the improvement of the biocompatibility of the beads.

Several studies have shown that immobilised cells maintain good morphology and metabolism during long-term *in vitro* culture [3,99,100] and excretion of desired products by encapsulated cell lines entrapped in alginate seem to follow proliferation and cell number [101]. It has been reported that the viability of encapsulated cells may be between 60% and 85% [102,103]. For cells entrapped within an alginate gel there must necessarily be direct contact between the gel network and cells or tissue material. Ideally the gel network should mimic the normal surrounding tissue for the encapsulated cells or in other ways give the cells optimum conditions for the specific application. Cells, and in particular cells from multicellular organisms like animals are highly specialised in responding to adjacent cells and extracellular matrices, a process of which the expression of specific genes are involved. As an example of the cell matrix influence it has been found that the presence of collagen, a major normal extracellular matrix component, inhibits cells from entering into apoptosis and thereby provide a substrate for cell survival and differentiation [104]. The cellular response mechanism of animal cells in contact with the alginate matrix must, however, be regarded as unknown. A recent study using MR-image technique has also indicated that alginate encapsulated islet cells may secrete an extracellular matrix within the gel network [105]. It is observed that using different alginates and gelling procedures may have different effects upon cells and growth and such effects will also be dependent of each specific cell type. Clearly highly differentiated cells like in Islets of Langerhans, with a very limited proliferation, will need to be regarded differently as compared to cell lines of which cells do not differentiate into a nongrowing state. For established cell lines, cell growth and behaviour within a gel network as compared to normal cell monolayer culture conditions, is clearly different. In general, using an alginate with a high M-content will give the cells a softer environment allowing the cells to proliferate and migrate more easily within the gel network [101]. The lower porosity of high M alginates could in some applications also be an advantage in order to avoid immune responses [13]. However, it is a general recommendation to use high G-block alginates as the first choice in application development as they will give stronger gels that will tolerate more physical stress before breakage, be less susceptible to osmotic swelling and be better substrates for coating materials. Comparable to the use of high M alginates using a lower high G alginate concentration and beads with a high degree inhomogeneity could possibly also give immobilised cells improved conditions. The optimal gel network for a particular cell type, however, will be a compromise. It is also our experience that many cell types are able to proliferate readily even within a strong high G alginate network.

For cells immobilised within an alginate gel it seems that the cells often adapt to the new environment. This is in particular seen for proliferating cell lines with a more rapid proliferation of certain clones. In Figure 5 encapsulated animal cells growing in alginate beads are shown. As seen, the cells may also grow out of the beads after some days. *In vivo* applications with such cell systems could be disadvantageous as the cells will then be openly subjected to the host immune system. It, however, also seems that some cell lines do not grow or grow very slowly within the alginate gel, and as mentioned this is dependent on the alginate gel composition and also on other growth conditions. For therapeutic use of alginate to immobilise proliferating cells *in vivo,* the selection of appropriate cells is therefore crucial. Better understanding of the mechanisms involved in regulation of cell growth within the alginate gel matrix is therefore needed.

In applications involving immobilisation of cells, diffusion properties of different molecules within the beads will also depend strongly on the load of cells. As a consequence of diffusion limitations cells surrounded by other cells within the gel network may therefore be strongly influenced by the metabolism of the surrounding cells. As a result surrounded cells may be trapped in a microenvironment lacking essential nutrients like oxygen. This may typically result in cell death in the centre of the beads with an outer rim of viable cells [101]. For lower cell concentrations this problem can be partly solved by keeping the bead size sufficiently low, *i.e.* well below one millimetre.

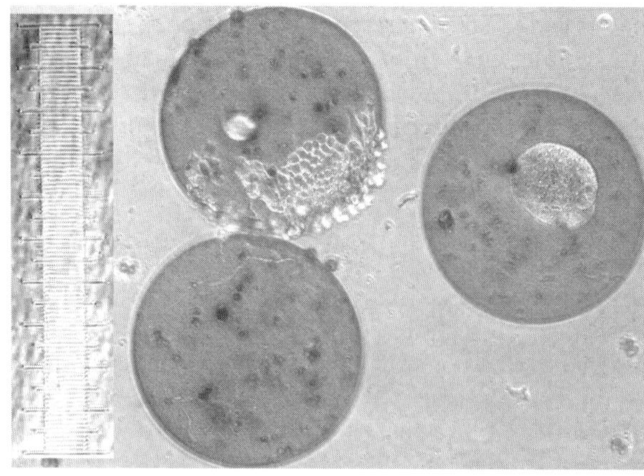

Figure 5. Light microscopy picture of cells of cervix carcinoma cell line (NHIK 3025) 20 days after immobilisation in PRONOVA UP LVM alginate in vitro. *Upper bead: Cells are growing at the bead surface. Bead to the right: Large colony of growing cells in the bead centre. Total scale length on left side is 1 mm.*

The major problem with the design of alginate beads for implantation purposes is reactions from the immune system of the host towards either bead material or immobilised cells. The mammalian immune system is complex involving both humoral and cellular responses toward foreign antigens. However, for cells immobilised in uncoated gel beads a marked prolongation of cells in several animal models has been seen [106]. This is unexpected when only considering the high porosity of the alginate gel, but may perhaps be explained by the negative charges in the alginate gel network counteracting the influx of molecules. It has also been postulated that the negatively charged gel network may permit the entry but also affect the physiologic state of proteins like immunoglobulins and complements through electrostatic interaction, and thus interfere with their action cascade [106]. Thus, the diffusion as well as physiological function of proteins in alginate beads may seem to be interfered by ionic interactions with the gel.

Other reasons for immune rejection of alginate microcapsules may be reactions towards polycation. Also the presence of defective beads either as a result of inappropriate production methods or harmful handling can expose immobilised materials to the immune system. Also, inadequate immobilisation methods or cellular growth may, as a result, give beads with cells present at the bead surface and thereby exposing them to the immune system. The major cause of breakage of encapsulated beads under physiological conditions is probably the osmotic swelling of the alginate core due to the Donnan equilibrium of the non-cooperatively bound counter-ions [107,108]. To avoid this more stable capsules can be made by increasing the strength of the polyanion/polycation membrane or alternatively by keeping a less swelling gel network in the core of the bead by forming inhomogeneous gels [57]. Inhomogeneity may be a preferred structure in alginate/polycation capsules by having the most solid part and highest charge bordering the polyanion/polycation membrane. The stability of capsules can therefore be enhanced without increasing the average polymer concentration and thereby avoiding an increased osmotic pressure [57].

It has been demonstrated that smaller beads show higher biocompatibility than larger beads [79]. This may be due to the fact that smaller beads are less fragile [79,109]. Also larger beads will occupy more space at the site of implantation and this could therefore also possibly give more stress to the surrounding tissue and thereby promoting adverse host reactions. However, if the beads are too small in relation to the size and load of the biologic material they may also be less biocompatible. As a consequence isolated islets will require larger beads than for single cell suspensions. Also if the beads are to small they may more easily collapse when coated with polycation [62].

In order to solve biocompatibility problems production methods and materials must therefore be carefully selected for each application [110]. However, even with sufficiently strong and invariable beads with a selected porosity the choice of bead properties will probably always be a compromise. Secretion of production substance or other permeable molecules might induce immunologic responses in the host [97,98]. This may include proteins or other cellular constituents, but also immunogenic coating materials. It is also very likely that many problems related to immune rejection of beads are related to impurities of the alginate material. Furthermore, it should also be noted that some alginates are immunogenic and their presence may therefore also be responsible for bead rejection. It is known that purified M-alginates (M-content is higher than about 80%), but not high G-alginates, efficiently stimulates monocytes to produce TNF, IL1 and IL6 [111,112] and furthermore that the CD14 receptor is involved in this response [111]. The immune response to alginate was discovered as also empty alginate beads were found to induce fibrosis [113]. It was found that alginate with 96-100% mannuronic acid had the highest immune stimulating effect (cytokine production from monocytes). Leakage of non-gelled alginate molecules are likely responsible for the reported rejection of empty beads *in vivo*. Antibodies against high M but not against high G alginate capsules have also been reported [75]. To ensure that all alginate molecules take part in the gel network, and thereby avoiding low molecular weight material enriched in mannuronic acid to leak out of the beads, a narrow molecular weight distribution is important [114]. This could be important to avoid unwanted immune reactions when alginate gels are used for implantation.

6. Selection of alginates for biomedical use

Alginate is the most abundant marine biopolymer and, after cellulose, the most abundant biopolymer in the world. The major source of alginate is found in the cell walls and the intracellular spaces of brown seaweed. The alginate molecules provide the plant with both flexibility and strength, which are necessary for plant growth in the sea. The first scientific studies on the extraction of alginates from brown seaweed were made by the British chemist E.C.Stanford at the end of the 19th century. He found that the extracted substance, which he named algin, possessed several interesting properties. These included the ability to thicken solutions, to make gels, and to form films. From these, he proposed several industrial applications. However, large-scale industrial production of alginate was not introduced until 50 years later.

Large amounts of seaweed are now harvested along the coastlines in different parts of the world. The seaweed grows naturally mainly in the temperate zone, but large amounts are also cultivated in the Far East, on the coast of China and Japan in particular. An example of large scale harvesting is the utilisation of large "forests" of *Laminaria hyperborea* growing naturally in the arctic water at the west coast of Norway. All current industrial manufacture of alginate is based on extraction of the polymer from brown algae. The M/G composition will vary from one species of brown algae to another. Similar variations are also found in the plant during the growth season, and between different parts of the plant. Generally speaking, the stiffness of the plant reflects the content of G-blocks, as a result of the G-blocks' ability to form strong gels by cross-linking with calcium.

Commercial alginates (Table 3) are produced mainly from *Laminaria hyperborea, Macrocystis pyrifera,* and to a lesser extent, from *Laminaria digitata, Laminaria japonica, Eclonia maxiama, Lessonia negrescens* and *Sargassum* sp. The highest content of guluronic acid residues is usually found in alginates prepared from stipes of old *Laminaria hyperborea* plants. Alginates from *Ascophyllum nodosum* and *Laminaria japonica* are characterised by a low content of G blocks and a low gel strength. The alginate from *Macrocystis pyrifera*, which has frequently been used for immobilisation, yields gels with lower strength and stability than those made from the stem of *Laminaria hyperborea.*

In contrast to other marine algal polysaccharides, alginate is also synthesized as an exocellular material by some bacteria [115]. Fermentation processes for production have been developed and successfully run, but so far only specialty grades of alginate have been found feasible to manufacture by fermentation. Through genetic manipulation and use of enzymatic modifications new types of alginates are promising as materials in special biomedical applications [115].

Table 3. Typical values for M and G content in seaweed used for alginate production

Seaweed	M/G	%M	%G	%MM	%GG
Laminaria hyperborea (stem)	0.45	30	70	18	58
Laminaria hyperborea (leaf)	1.22	55	45	36	26
Laminaria digitata	1.22	55	45	39	29
Macrocystis pyrifera	1.50	60	40	40	20
Lessonia nigrescens	1.50	60	40	43	23
Ascophyllum nodosum	1.86	65	35	56	26
Laminaria japonica	1.86	65	35	48	18
Durvillea antarctica	2.45	71	29	58	16
Durvillea potarum	3.33	77	23	69	13

Alginates in general fulfil the requirements as additives in food and pharmaceutical products and the general opinion is also that they are non-toxic when exposed to animal cells. Alginate based products are well known for their traditional uses in the treatment of topical wounds [116-118] as an antireflux remedy [119,120] and as a tablet excipient [121]. The purity level and the current means of manufacture, however, make it unlikely that commodity alginates will find use in implantable devices and drug formulations for parenteral administration. There are numerous studies using uncharacterised commodity alginates to immobilise cells that have resulted in inconsistent outcomes. This is mainly related to different levels of fibroblastic overgrowth. It is known that alginates may contain pyrogens, polyphenols, proteins and complex carbohydrates. The presence of polyphenols might possibly be harmful to the immobilised cells, and the presence of pyrogens, proteins and complex carbohydrates may induce immunological reactions by the host. It has been demonstrated that perfectly spherical and smooth alginate droplets can only be formed by using a highly purified alginates [122]. The mechanical properties may also be improved when using purified alginates [123]. To avoid such problems it is therefore of crucial importance to use highly purified alginates [106]. To obtain repeatable results in all respects the chosen alginates should also be well-characterised with respect to all critical properties (impurities, M/G content, molecular weight etc.). Highly purified alginates, made in accordance with Good Manufacturing Practice / ISO 9000 guidelines, have been successfully employed for applications inside the human body [42,124], and several products containing purified alginates are in the process of being clinically evaluated. There also exist some data regarding the safety and pharmacological properties of alginates, which confirms their biocompatibility [125]. We have established single dose toxicity studies (IP) in the mouse and rat for purified high M and high G alginates showing no mortality or abnormal clinical signs at relevant dosage. In our lab we have also established for the same alginates, that little or no effect is seen on cell survival or colony forming ability in *in vitro* cell studies. For intended use in biomedical applications a complete safety profile is now a part of a US Drug Master File for cGMP (current Good Manufacturing Practice) alginates.

Characterisation parameters for alginates to be used in biomedical and tissue engineered medical products are now thoroughly described in the ASTM guide F 2064 (American Society for Testing and Materials) of the ASTM Book of Standards [126,127]. The alginate composition and sequential structure together with molecular weight and molecular conformation are highly important key characteristics in determining product properties and functionality. High-resolution NMR is used to determine the composition and sequential structure of alginates and the molecular weight is determined mostly from intrinsic viscosity measurements or size-exclusion chromatography with light-scattering measurements [126]. It is also essential that alginates to be used in the human body are extremely pure and chemically well-defined. This is in particular ensured by measurements of endotoxins, microbial contamination and protein content. For applications using alginates inside the human body the level of endotoxins (pyrogen) needs to be kept very low. According to FDA requirements for device implantations in general the content of endotoxin must below 350 EU per patient (below 15 EU per patient for CNS applications). As the chemical properties of endotoxins are very similar to alginates their removal have been a challenging task. Commercial alginates without a specified endotoxin content must therefore be expected to be less purified. However, purified alginates with a specified endotoxin content below 100 EU/gram are now commercially available. For GMP alginates all selected characterisation parameters must be measured by validated methods and every product release documented with its individual laboratory certificate.

The standardisation of alginates as immobilisation materials also needs to be followed consequently by standardisation and validation of procedures for cell selection and immobilisation [110]. This will be necessary to fulfil the regulatory requirements to be used in applications as bioartificial organ devices. Currently a standard guide for immobilisation of living cells in alginate gels is also under development as a standard at the American Society for Testing and Materials.

References

[1] Onsøyen, E. (1996) Commercial applications of alginates. Carbohydrates in Europe 14: 26-31.

[2] Chang, T.M.S. (1964) Semipermeable microcapsules. Science 146: 524-525.

[3] Lanza, R.P.; Kuhtreiber, W.M.; Ecker, D.; Staruk, J.E. and Chick, W.L. (1995) Xenotransplantation of porcine and bovine islets without immunosuppression using uncoated alginate microspheres. Transplantation 59: 1377-1384.

[4] Bugarski, B.; Sajc, L.; Plavsic, M.; Goosen, M. and Jovanovic, G. (1997) Semipermeable alginate-PLO microcapsules as a bioartificial pancreas. In: Funatsu, K. (Ed.) Animal Cell Technology. Kluwer Academic Publishers; pp. 479-486.

[5] Park, Y.G.; Iwata, H. and Ikada, Y. (1998) Microencapsulation of islets and model beads with a thin alginate-Ba2+ gel layer using centrifugation. Polymers for Adv. Technol. 9: 734-739.

[6] de Vos, P.; De Haan, B.; Wolters, G.H.J. and van Schilfgaarde, R. (1996) Factors influencing the adequacy of microencapsulation of rat pancreatic islets. Transplantation 62: 888-893.

[7] Weber, C.J.; Hagler, M. and Konieczny, B. (1995) Encapsulated islet iso-, allo-, and xenografts in diabetic NOD mice. Transplant. Proc. 27: 3308-3311.

[8] Brissová, M.; Lacík, I.; Powers, A.C.; Anilkumar, A.V. and Wang, T. (1998) Control and measurement of permeability for design of microcapsule cell delivery system. J. Biomed. Mater. Res. 39: 61-70.

[9] Orlowski, T.; Sitarek, E.; Tatarkiewicz, K.; Sabat, M. and Antosiak, M. (1997) Comparison of two methods of pancreas islets immunoisolation. Int. J. Artif. Organs 20: 701-703.

[10] Tze, W.J.; Cheung, S.C.; Tai, J. and Ye, H. (1998) Assessment of the *in vivo* function of pig islets encapsulated in uncoated alginate microspheres. Transplant. Proc. 30: 477-478.

[11] Sandler, S.; Andersson, A.; Eizirik, D.L.; Hellerstrøm, C.; Espevik, T.; Kulseng, B.; Thu, B.; Pipeleers, B. and Skjåk-Bræk, G. (1997) Assessment of insulin secretion *in vitro* from microencapsulated fetal porcine islet-like cell clusters and rat, mouse, and human pancreatic islets. Transplantation 63: 1712-1718.

[12] Wang, T.; Lacík, I.; Brissová, M.; Anilkumar, A.V.; Prokop, A.; Hunkeler, D.; Green, R.; Shahrokhi, K. and Powers,A.C. (1997) An encapsulation system for the immunoisolation of pancreatic islets. Nat. Biotechnol. 15: 358-362.

[13] Storrs, R.; Dorian, R.; King, S.R.; Lakey, J. and Rilo, H. (2001) Preclinical development of the Islet Sheet. Ann. N.Y. Acad. Sci. 944: 252-266.

[14] Hortelano, G.; al-Hendy, A.; Ofosu, F.A. and Chang, P.L. (1996) Delivery of human factor IX in mice by encapsulated recombinant myoblasts: a novel approach towards allogeneic gene therapy of hemophilia B. Blood 87: 5095-5103.

[15] Hortelano, G. and Stockley, T. (1999) Implantable Microcapsules for Gene Therapy for Hemophilia. In: Kühtreiber, W.M.; Lanza, R.P. and Chick W.L. (Eds.) Cell Encapsulation Technology and Therapeutics. Birkhäuser, Boston; pp. 3-17.

[16] Peirone, M.A.; Delaney, K.; Kwiecin, J.; Fletch, A. and Chang, P.L. (1998) Delivery of Recombinant Gene Product to Canines with Nonautologous Microencapsulated Cells. Hum. Gene Ther. 9: 195-206.

[17] al-Hendy, A.; Hortelano, G.; Tannenbaum, G.S. and Chang,P.L. (1996) Growth retardation--an unexpected outcome from growth hormone gene therapy in normal mice with microencapsulated myoblasts. Hum. Gene Ther. 7: 61-70.

[18] Chang, P.L.; Shen, N. and Westcott, A.J. (1993) Delivery of recombinant gene products with microencapsulated cells in *in vivo*. Hum. Gene Ther. 4: 433-440.

[19] Chang, P.L.; Hortelano, G.; Tse, M. and Awrey, D.E. (1994) Growth of recombinant fibroblasts in alginate microcapsules. Biotechnol. Bioeng. 43: 925-933.

[20] Chang, T.M.S. and Prakash, S. (1998) Therapeutic uses of microencapsulated genetically engineered cells. Molecular Medicine Today 5: 221-227.

[21] Chang, T.M. and Malave, N. (2000) The development and first clinical use of semipermeable microcapsules (artificial cells) as a compact artificial kidney. Ther. Apher. 4: 108-116.
[22] Bruni, S. and Chang, T.M.S. (1991) Encapsulated hepatocytes for controlling hyperbilirubinemia in Gunn rats. Int. J. Artificial Organs 14: 239-242.
[23] Miura, Y.; Yoshikawa, N.; Akimoto, T. and Yagi, K. (1990) Therapeutic effect of hepatocytes entrapped within Ca-alginate. Ann. N.Y. Acad. Sci. 613: 475-478.
[24] Cai, Z.H.; Shi, Z.Q.; Sherman, M. and Sun, A.M. (1989) Development and evaluation of a system of microencapsulation of primary rat hepatocytes. Hepatology 10: 855-860.
[25] Winn, S.R.; Tresco, P.A.; Zielinski, B. and Greene, L.A. (1991) Behavioral recovery following intrastriatal implantation of microencapsulated PC12 cells. Exp. Neurol. 113: 327-329.
[26] Emerich, D.F.; Winn, S.R.; Christenson, L.; Palmatier, M.A ; Gentile, F.T. and Sanberg, P.R. (1992) A novel approach to neural transplantation in Parkinson's disease: use of polymer-encapsulated cell therapy. Neurosci. Biobehav. Rev. 16: 437-447.
[27] Fu, X.W. and Sun, A.M. (1989) Microencapsulated parathyroid cells as a bioartificial parathyroid. *In vivo* studies. Transplantation 47: 432-434.
[28] Hagihara, Y.; Saitoh, Y.; Iwata, H.; Taki, T.; Hirano, S. and Hayakawa, T. (1997) Transplantation of xenogeneic cells secreting beta-endorphin for pain treatment: analysis of the ability of components of complement to penetrate through polymer capsules. Cell Transplant. 6: 527-530.
[29] Read, T.-A.; Sørensen, D.R.; Mahesparan, R.; Enger, P.Ø.; Timpl, R.; Olsen, B.R.; Hjelstuen, M.H.B.; Haraldseth, O. and Bjerkvig, R. (2001) Local endostatin treatment of gliomas administered by microencapsulated producer cells. Nat. Biotechnol. 19: 29-34.
[30] Joki, T.; Machluf, M.; Atala, A.; Zhu, J.; Seyfried, N.T.; Dunn, I.F.; Abe, T.; Carroll, R.S. and Black, P.M. (2001) Continuous release of endostatin from microencapsulated engineered cells for tumor therapy. Nat. Biotechnol. 19: 35-39.
[31] Read, T.-A.; Farhadi, M.; Bjerkvig, R.; Olsen, B.R.; Rokstad, A.M.; Huszthy, P.C. and Vajkoczy, P. (2001) Intravital microscopy reveals novel antivascular and antitumor effects of endostatin delivered locally by alginate-encapsulated cells. Cancer Res. 61: 6830-6837.
[32] Visted, T.; Bjerkvig, R. and Enger, P.O. (2001) Cell encapsulation technology as a therapeutic strategy for CNS malignancies. Neuro-oncol. 3: 201-210.
[33] Maruyama, M.; Terayama, K.; Ito, M.; Takei, T. and Kitagawa, E. (1995) Hydroxyapatite clay for gap filling and adequate bone ingrowth. J. Biomed. Mater. Res. 29: 329-336.
[34] Kenley, R.; Marden, L.; Turek, T.; Jin, L.; Ron, E. and Hollinger, J.O. (1994) Osseous regeneration in the rat calvarium using novel delivery systems for recombinant human bone morphogenetic protein-2 (rhBMP-2). J. Biomed. Mater. Res. 28: 1139-1147.
[35] Diduch, D.R.; Jordan, L.C.M.; Mierisch, C.M. and Balian, G. (2000) Marrow stromal cells embedded in alginate for repair of osteochondral defects. J. Arthroscopic and Related Surgery 16: 571-577.
[36] Fragonas, E.; Valente, M.; Pozzi-Mucelli, M.; Toffanin, R.; Rizzo, R.; Silvestri, F. and Vittur, F. (2000) Articular cartilage repair in rabbits by using suspensions of allogenic chondrocytes in alginate. Biomaterials 21: 795-801.
[37] Knight, M.M.; Bravenboer, J.V.D.B.; Lee, D.A.; van Osch, G.J.V.M.; Weinans, H. and Bader, D.L. (2002) Cell and nucleus deformation in compressed chondrocyte-alginate constructs: temporal changes and calculation of cell modulus. Biochim. Biophys. Acta 1570: 1-8.
[38] Miralles, G.; Baudoin, R.; Dumas, D.; Baptiste, D.; Hubert, P.; Stoltz, J.F.; Dellacherie, E.; Mainard, D.; Netter, P. and Payan, E. (2001) Sodium alginate sponges with or without sodium hyaluronate: *In vitro* engineering of cartilage. J. Biomed. Mater. Res. 57: 268-278.
[39] Sufan, W.; Suzuki, Y.; Tanihara, M.; Ohnishi, K.; Suzuki, K.; Endo, K. and Nishimura, Y. (2001) Sciatic nerve regeneration through alginate with tubulation or nontubulation repair in cat. J. Neurotrauma 18: 329-338.
[40] Kataoka, K.; Suzuki, Y.; Kitada, M.; Ohnishi, K ; Suzuki, K.; Tanihara, M.; Ide, C.; Endo, K. and Nishimura, Y. (2001) Alginate, a bioresorbable material derived from brown seaweed, enhances elongation of amputated axons of spinal cord in infant rats. J. Biomed. Mater. Res. 54: 373-384.
[41] Winn, S.R.; Hammang, J.P.; Emerich, D.F.; Lee, A.; Palmiter, R.D. and Baetge, E.E. (1994) Polymer-encapsulated cells genetically modified to secrete human nerve growth factor promote the survival of axotomized septal cholinergic neurons. Proc. Natl. Acad. Sci. USA 91: 2324-2328.
[42] Diamond, D.A. and Caldamone, A.A. (1999) Endoscopic correction of vesicoureteral reflux in children using autologous chondrocytes: preliminary results. J. Urol. 162: 1185-1188.
[43] Plunkett, M.L. and Hailey, J.A. (1990) An *in vivo* quantitative angiogenesis model using tumor cells entrapped in alginate. Lab. Invest. 62: 510-517.

[44] Hoffman, J.; Schirner, M.; Menrad, A. and Schneider, M.R. (1998) A higly sensitive model for quantification of *in vivo* tumor angiogenesis induced by alginate-encapsulated tumor cells. Cancer Res. 57: 3847-3851.

[45] Hartmann, M.; Holm, O.B.; Johansen, G.A.; Skjåk-Bræk, G. and Stokke, B.T. (2002) Mode of action of recombinant Azotobacter vinelandii mannuronan C-5 epimerases AlgE2 and AlgE4. Biopolymers 63: 77-88.

[46] Ertesvåg, H.; Doseth, B.; Larsen, B.; Skjåk-Bræk, G. and Valla, S. (1994) Cloning and expression of an Azotobacter vinelandii mannuronan C- 5-epimerase gene. J. Bacteriol. 176: 2846-2853.

[47] Grasdalen, H. (1983) High-field, H-n.m.r. spectroscopy of alginate: sequential structure and linkage conformations. Carbohydr. Res. 118: 255-260.

[48] Smidsrød, O. and Draget, K.I. (1996) Chemistry and physical properties of alginates. Carbohydrates in Europe 14: 6-13.

[49] Smidsrød, O. (1974) Molecular basis for some physical properties of alginates in the gel state. Faraday Discussions of the Chemical Society 57: 263-274.

[50] Haug, A. (1964) Report No.30, PhD Thesis, Norwegian Univeristy of Science and Technology, Trondheim (Norway).

[51] Sutherland, I.W. (1991) Alginates. In: Byrom, D. (Ed.) Biomaterials; Novel materials from biological sources. Macmillan, New York; pp. 309-331.

[52] Clark, A.H. and Ross-Murphy, S.B. (1987) Structural and mechanical properties of biopolymer gels. Adv. Polymer Sci. 83: 57-192.

[53] Grant, G.T.; Morris, E.R.; Rees, D.A.; Smith, P.J.C. and Thom, D. (1973) Biological interactions between polysaccharides and divalent cations: The egg-box model. FEBS Lett. 32: 195-198.

[54] Stokke, B.T.; Draget, K.I. and Yuguchi, Y. (1997) Small-angle X-ray scattering and rheological characterization of alginate gels. Macromolec. Symp. 120: 97-101.

[55] Gåserød, O. (1998) Microcapsules of alginate-chitosan: A study of capsule formation and functional properties. PhD Thesis, Norwegian Univeristy of Science and Technology, Trondheim (Norway). ISBN 82-471-0273-0.

[56] Martinsen, A.; Skjåk-Bræk, G. and Smidsrød, O. (1989) Alginate as an immobilization material: I. Correlation between chemical and physical properties of alginate gel beads. Biotechnol. Bioeng. 33: 79-89.

[57] Skjåk-Bræk, G.; Grasdalen, H. and Smidsrød, O. (1989) Inhomogeneous polysaccharide ionic gels. Carbohydr. Res. 10: 31-54.

[58] Thu, B.; Gåserød, O.; Paus, D.; Mikkelsen, A.; Skjåk-Bræk, G.; Toffanin, R.; Vittur, F. and Rizzo, R. (2000) Inhomogeneous alginate gel spheres: an assessment of the polymer gradients by synchrotron radiation-induced X-ray emission, magnetic resonance microimaging, and mathematical modeling. Biopolymers 53: 60-71.

[59] Thu, B.; Bruheim, P.; Espevik, T.; Smidsrød, O.; Soon-Shiong, P. and Skjak-Braek, G. (1996) Alginate polycation microcapsules. II. Some functional properties. Biomaterials 17: 1069-1079.

[60] Dulieu, C.; Poncelet, D. and Neufeld, R.J. (1999) Encapsulation and Immobilization Techniques. In: Kühtreiber, W.M.; Lanza, R.P. and Chick, W.L. (Eds.) Cell Encapsulation Technology and Therapeutics. Birkhäuser, Boston; pp. 3-17.

[61] Klokk, T.I. and Melvik, J.E. (2002) Controlling the size of alginate beads by use of a high electrostatic potential. J. Microencapsulation 19: 415-424.

[62] Strand, B.L.; Gåserød, O.; Kulseng, B.; Espevik, T. and Skjåk-Bræk, G. (2002) Alginate-polylysine-alginate microcapsules - Effect of size reduction on capsule properties. J. Microencapsulation 19: 615-630.

[63] Pjanovic, R.; Goosen, M.F.A.; Nedovic, V. and Bugarski, B. (2001) Immobilization/encapsulation of cells using electrostatic droplet generation. Minerva Biotecnologica 12: 241-248.

[64] Poncelet, D.; Bugarski, B.; Amsden, B.G.; Zhu, J.; Neufeld, R. and Goosen, M.F.A. (1994) A parallel plate electrostatic droplet generator: parameters affecting the microbead size. Appl. Microbiol. Biotechnol. 42: 251-255.

[65] Halle, J.P.; Leblond, F.A.; Pariseau, J.F.; Jutras, P.; Brabant, M.J. and Lepage, Y. (1994) Studies on small (<300 μm) microcapsules: II-Parameters governing the production of alginate beads by high voltage electrostatic pulses. Cell Transplant. 3: 365-372.

[66] King, A.; Sandler, S.; Anderson, A.; Hellestrøm, B.; Kulseng, B. and Skjåk-Bræk, G. (1999) Glucose metabolism *in vitro* of cultured and transplanted mouse pancreatic islets microencapsulated by means of a high-voltage electrostatic field. Diabetes Care 22: 121-126.

[67] Martinsen, A.; Storrø, I. and Skjåk-Bræk, G. (1992) Alginate as immobilization material: III. Diffusional properties. Biotechnol. Bioeng. 39: 186-194.

[68] Poncelet, D. (2001) Production of alginate beads by emusification/internal gelation. Ann. N.Y. Acad. Sci. 944: 74-82.

[69] Tanaka, H.; Matsumura, M. and Veliky, I.A. (1984) Diffusion Characteristics of Substrates in Ca-Alginate Gel Beads. Biotechnol. Bioeng. 26: 53-58.

[70] Smidsrød, O. and Skjåk-Bræk, G. (1990) Alginate as Immobilization Matrix for Cells. TIBTECH 8: 71-78.

[71] Wideroe, H. and Danielsen, S. (2001) Evaluation of the use of Sr2+ in alginate immobilization of cells. Die Naturwissenschaften 88: 224-228.

[72] Brodelius, P. and Vandamme, E.J. (1987) Immobilized cell systems. In: Rehm, H.J. and Reed, G. (Eds.) Biotechnology. Verlag Chemie, Weinheim; pp. 405-464.

[73] Schnabl, H. and Zimmermann, U. (1989) Immobilization of plant protoplasts. In: Bajay, Y.P.S. (Ed.) Plant protoplasts and genetic engeneering. Springer Verlag, New York; pp. 63-96.

[74] Bunger, C.M.; Jahnke, A.; Stange, J.; de Vos, P. and Hopt, U.T. (2002) MTS Colorimetric Assay in Combination with a Live-Dead Assay for Testing Encapsulated L929 Fibroblasts in Alginate Poly-l-Lysine Microcapsules *In Vitro*. Artif. Organs 26: 111-116.

[75] Kulseng, B.; Skjåk-Bræk, G.; Ryan, L.; Andersson, A.; King, A.; Faxvaag, A. and Espevik, T. (1999) Transplantation of alginate microcapsules: generation of antibodies against alginates and encapsulated porcine islet-like cell clusters. Transplantation 67: 978-984.

[76] Calafiore, R.; Basta, G.; Boselli, C.; Bufalari, A.; Giustozzi, G.M.; Luca, G.; Tortoioli, C. and Brunetti, P. (1997) Effects of alginate/polyaminoacidic coherent microcapsule transplantation in adult pigs. Transplant. Proc. 29: 2126-2127.

[77] Calafiore, R.; Luca, G.; Calvitti, M.; Neri, L.M.; Basta, G.; Capitani, S.; Becchetti, E. and Brunetti, P. (2001) Cellular support systems for alginate microcapsules containing islets, as composite bioartificial pancreas. Ann. N.Y. Acad. Sci. 944: 240-251.

[78] Tanaka, H.; Kurosawa, H.; Kokufuta, E. and Veliky, I.A. (1984) Preparation of Immobilized Glucoamylase Using Ca-Alginate Gel Coated with Partially Quaternized Poly(ethyleneimine). Biotechnol. Bioeng. 26: 1393-1394.

[79] Robitaille, R.; Pariseau, J.F.; Leblond, F.A.; Lamoureux, M.; Lepage, Y. and Halle, J.P. (1999) Studies on small (<350 μm) alginate-poly-L-lysine microcapsules. III. Biocompatibility of smaller versus standard microcapsules. J. Biomed. Mater. Res. 44: 116-120.

[80] Chen, J.P.; Chu, I.M.; Shiao, M.Y.; Hsu, B.R. and Fu, S.-H. (1998) Microencapsulation of islets in PEG-amine modified alginate-poly(L-lysine)-alginate microcapsules for constructing bioartificial pancreas. J. Ferm. Bioeng. 86: 185-190.

[81] de Vos, P.; De Haan, B. and van Schilfgaarde, R. (1997) Effect of the alginate composition on the biocompatibility of alginate-polylysine microcapsules. Biomaterials 18: 273-278.

[82] Kulseng, B.; Thu, B.; Espevik, T. and Skjåk-Bræk, G. (1997) Alginate polylysine microcapsules as immune barrier: permeability of cytokines and immunoglobulins over the capsule membrane. Cell Transplant. 6: 387-394.

[83] Okada, N.; Miyamoto, H.; Yoshioka, T.; Sakamoto, K.; Katsume, A.; Saito, H.; Nakagawa, S.; Ohsugi, Y. and Mayumi, T. (1997) Immunological studies of SK2 hybridoma cells microencapsulated with alginate-poly(L)lysine-alginate (APA) membrane following allogeneic transplantation. Biochem. Biophys. Res. Commun. 230: 524-527.

[84] Thu, B.; Kulseng, B.; Espevik, T. and Skjåk-Bræk, G. (1997) Diffusion of hormones, immunoglobulins and cytokines into alginate / polylysine capsules. Cell Transplant. 6: 378-394.

[85] Abraham, S.M.; Vieth, R.F. and Burgess, D.J. (1996) Novel technology for the preparation of sterile alginate -poly-l-lysine microcapsules in a bioreactor. Pharm. Dev. Technol. 1: 63-68.

[86] Chang, P.L. (1996) Microencapsulation - an alternative approach to gene therapy. Transfus. Sci. 17: 35-43.

[87] Gåserød, O.; Sannes, A. and Skjåk-Bræk, G. (1999) Microcapsules of alginate-chitosan. II. A study of capsule stability and permeability. Biomaterials 20: 773-783.

[88] Gåserød, O.; Smidsrød, O. and Skjåk-Bræk, G. (1998) Microcapsules of alginate and chitosan I. A quantitative study of the interaction between alginate and chitosan. Biomaterials 19: 1815-1825.

[89] Gåserød, O.; Jolliffe, I.G.; Hampson, F.C.; Dettmar, P.W. and Skjåk-Bræk, G. (1998) The enhancement of the bioadhesive properties of calcium alginate gel beads by coating with chitosan. Int. J. Pharm. 175: 237-246.

[90] Yan, C.; Zhang, H.; Lambert, D.M.; Ussery, M.A. and Nielsen, C.J. (1998) *In Vitro* study of alginate/chitosan microspheres for controlled release of the anti-HIV drug T20. In: 1998 Proceedings Book. The Controlled Release Society, Inc., Deerfield, US; pp. 510-511.

[91] Quong, D.; Groboillot, A.; Darling, G.D., Poncelet, D. and Neufeld, R.J. (1997) Microencapsulation within cross-linked chitosan membranes. In: Muzzarelli, R.A.A. and Peter, M.G. (Eds.) Chitin Handbook. Atec Edizioni, Grottammare; pp. 405-410.
[92] Li, X. (1996) The use of chitosan to increase the stability of calcium alginate beads with entrapped yeast cells. Biotechnol. Appl. Biochem. 23: 269-271.
[93] Okhamafe, A.O.; Amsden, B.; Chu, W. and Goosen, M.F. (1996) Modulation of protein release from chitosan-alginate microcapsules using the pH-sensitive polymer hydroxypropyl methylcellulose acetate succinate. J. Microencapsulation 13: 497-508.
[94] Alexakis, T.; Boadi, D.K.; Quong, D.; Groboillot, A.; O'Neill, I.; Poncelet, D. and Neufeld, R.J. (1995) Microencapsulation of DNA within alginate microspheres and crosslinked chitosan membranes for *in vivo* application. Appl. Biochem. Biotechnol. 50: 93-106.
[95] Zimmermann, U.; Hasse, C.; Rothmund, M. and Kühtreiber, W.M. (1999) Biocompatible encapsulation materials: Fundamentals and application. In: Kühtreiber, W.M.; Lanza R.P. and Chick, W.L. (Eds.) Cell Encapsulation Technology and Therapeutics. Birkhäuser, Boston; pp. 40-52.
[96] Shapiro, L. and Cohen, S. (1997) Novel alginate sponges for cell culture and transplantation. Biomaterials 18: 583-590.
[97] Gray, D.W. (2001) An overview of the immune system with specific reference to membrane encapsulation and islet transplantation. Ann. N.Y. Acad. Sci. 944: 226-239.
[98] Rokstad, A.M.; Kulseng, B.; Strand, B.L.; Skjåk-Bræk, G. and Espevik, T. (2001) Tranplantation of alginate microcapsules with proliferating cells in mice. Capsular overgrowth and survival of encapsulated cells of mice and human origin. Ann. N.Y. Acad. Sci. 944: 216-225.
[99] Falorni, A.; Basta, G.; Santeusanio, F.; Brunetti, P. and Calafiore, R. (1996) Culture maintenance of isolated adult porchine pancreatic islets in three-dimensional gel matrices: morphological and functional results. J. Cell. Physiol. 152: 422-429.
[100] Fraser, R.B.; MacAulay, M.A.; Wright, J.R.J.; Sun, A.M. and Rowden, G. (1995) Migration of macrophage-like cells within encapsulated islets of Langerhans maintained in tissue culture. Cell Transplant. 4: 529-534.
[101] Constantinidis, I.; Rask, I.; Long, R.C., Jr. and Sambanis, A. (1999) Effects of alginate composition on the metabolic, secretory, and growth characteristics of entrapped beta.TC3 mouse insulinoma cells. Biomaterials 20: 2019-2027.
[102] Papas, K.K.; Long, Jr., R.C.; Constantinidis, I. and Sambanis, A. (1997) Role of ATP and Pi in the mechanism of insulin secretion in the mouse insulinoma betaTC3 cell line. Biochem. J. 326: 807-814.
[103] Benson, J.P.; Papas, K.K.; Constantinidis, I. and Sambanis, A. (1997) Towards the development of a bioartificial pancreas: effects of poly-L-lysine on alginate beads with BTC3 cells. Cell Transplant. 6: 395-402.
[104] O'Connor, S.M.; Stenger, D.A.; Shaffer, K.M. and Ma, W. (2001) Survival and neurite outgrowth of rat cortical neurons in three- dimensional agarose and collagen gel matrices. Neurosci. Lett. 304: 189-193.
[105] Constantinidis, I.; Long Jr., R.C.; Weber, C.J.; Safley, S. and Sambanis, A. (2001) Non-Invasive monitoring of a bioartificial pancreas *in vitro* and *in vivo*. Ann. N.Y. Acad. Sci. 944: 83-95.
[106] Yang, H. and Wright, J.R.J. (1999) Calcium Alginate. In: Kühtreiber, W.M.; Lanza, R.P. and Chick, W.L. (Eds.) Cell Encapsulation Technology and Therapeutics. Birkhäuser, Boston; pp. 3-17.
[107] Espevik, T.; Skjåk-Bræk, G.; Smidsrød, O.; Soon-Shiong, P. and Thu, B. (1996) Alginate poly-lysine capsules. II: Some functional properties. Biomaterials 17: 1069-1079.
[108] Thu, B.; Espevik, T.; Smidsrød, O.; Soon-Shiong, P. and Skjåk-Bræk, G. (1996) Alginate poly-cation microcapsules I: Interactions between alginate and polycation. Biomaterials 17: 1031-1040.
[109] Ma, X.; Vacek, I. and Sun, A. (1994) Generation of alginate-poly-l-lysine-alginate (APA) biomicrocapsules: the relationship between the membrane strength and the reaction conditions. Artif. Cells Blood Substit. Immobil. Biotechnol. 22: 43-69.
[110] Hunkeler, D.; Rehor, A.; Ceausoglu, I.; Schuldt, U.; Canaple, L.; Bernhard, P.; Renken, A.; Rindisbacher, L. and Angelova, N. (2001) Objectively assessing bioartificial organs. Ann. N.Y. Acad. Sci. 944: 456-471.
[111] Espevik, T.; Otterlei, M.; Skjåk-Bræk, G.; Ryan, L.; Wright, S.D. and Sundan, A. (1993) The involvement of CD14 in stimulation of cytokine production by uronic acid polymers. Eur. J. Immunol. 23: 255-261.
[112] Kulseng, B.; Skjåk-Bræk, G.; Følling, I. and Espevik, T. (1996) TNF production from peripheral blood mononuclear cells in diabetic patients after stimulation with alginate and lipopolysaccharide. Scand. J. Immunol. 43: 335-340.

[113] Skjåk-Bræk, G. and Espevik, T. (1996) Application of alginate gels in biotechnology and biomedicine. Carbohydrates in Europe 14: 19-25.
[114] Stokke, B.T.; Smidsrød, O.; Zanetti, F.; Skjåk-Bræk, G. and Strand, W. (1993) Distribution of uronate residues in alginate chains in relation to alginate gelling properties - 2: Enrichment of β-D-mannurinoc acid and depletion of α-L-guluronic acid in sol fraction. Carbohydr. Res. 21: 39-46.
[115] Ertesvåg, H.; Høidal, H.K.; Hals, I.K.; Rian, A.; Doseth, B. and Valla, S. (1995) A family of modular type mannuronan C-5-epimerase genes controls alginate structure in *Azotobacter vinelandii*. Molecular Microbiology 16: 719-731.
[116] Thomas, S. (2000) Alginate dressings in surgery and wound managment - part 1. J. Wound. Care 9: 56-60.
[117] Thomas, S. (2000) Alginate dressings in surgery and wound management: part 2. J. Wound. Care 9: 115-119.
[118] Thomas, S. (2000) Alginate dressings in surgery and wound management: part 3. J. Wound. Care 9: 163-166.
[119] Mandel, K.G.; Daggy, B.P.; Brodie, D.A. and Jacoby, H.I. (2000) Review article. Alginate-raft formulations in the treatment of heartburn and acid reflux. Aliment. Pharmacol. Ther. 14: 669-690.
[120] Hagstam, H. (1986) Alginates and heartburn - Evaluation of a medicine with a mechanical mode of action. In: Phillips, G.O. (Ed.) Gums and stabilizers in the food industry. Elsevier, Amsterdam; pp. 363-370.
[121] Onsøyen, E. (1995) Hydration induced swelling of alginate based matrix tablets at GI-tract pH conditions. In: Karse, D.R. and Stephenson, R.A. (Eds.) Excipients and delivery systems for pharmaceutical formulations. The Royal Society of Chemistry, Cambridge; pp. 108-122.
[122] Goosen, M.F.A.; O'Shea, G.M. and Gharapetian, H.M. (1985) Optimization of Microencapsulation Parameters: Semipermeable Microcapsules as a Bioartificial Pancreas. Biotechnol. Bioeng. 27: 146-150.
[123] Wandrey, C. and Vidal, D.S. (2001) Purification of polymeric biomaterials. Ann. N.Y. Acad. Sci. 944: 187-198.
[124] Soon-Shiong, P. (1994) Insulin independence in a type I diabetic patient after encapsulated islet transplantation. Lancet 343: 950-951.
[125] Skaugrud, Ø.; Hagen, A.; Borgersen, B. and Dornish, M. (1999) Biomedical and pharmaceutical applications of alginate and chitosan. Biotechnol. Genet. Eng. Rev. 16: 23-40.
[126] Dornish, J.M.; Kaplan, D. and Skaugrud, Ø. (2001) Standards and guidelines for biopolymers in tissue-engineered medical products. ASTM alginate and chitosan standard guides. Ann. N.Y. Acad. Sci. 944: 388-397.
[127] (2002) F 2064 -Standard guide for characterization and testing of alginates as starting materials intended for use in biomedical and tissue-engineered medical products application. Annual Book of ASTM Standards. ASTM International, West Conshohocken, PA, Vol. 13.01; pp. 1595-1602.

ENTRAPMENT IN LENTIKATS®

Encapsulation of various biocatalysts – bacteria, fungi, yeast or enzymes into polyvinyl alcohol based hydrogel particles

PETER WITTLICH[1], EMINE CAPAN[2], MARC SCHLIEKER[2], KLAUS-DIETER VORLOP[2] AND ULRICH JAHNZ[1]
[1]geniaLab®BioTechnologie - Produkte und Dienstleistungen GmbH, Bundesallee 50, 38116 Braunschweig, Germany – Fax: +49-531-23210-22 – Email: info@geniaLab.com
[2]Federal Agricultural Research Centre (FAL), Institute of Technology and Biosystems Engineering, Braunschweig, Germany – Fax: +49-531-59641-99 – Email: klaus.vorlop@fal.de

Summary

In this paper LentiKats® are presented as an efficient and powerful alternative to traditional and established encapsulation carriers.

The introduction describes LentiKats® regarding appearance, properties and preparation. They are generally compared to other types of immobilised biocatalysts and potential advantages for application in bioconversion processes are depicted. Afterwards a number of examples for successful application of LentiKats® for various valorisation processes and also elimination processes are presented, including encapsulation of enzymes, bacteria, fungi, and yeasts.

1. Introduction

1.1. HETEROGENISATION

For a large variety of biological processes the benefit of applying encapsulated biocatalysts is generally accepted. The reasons for entrapment are manifold: since the catalytic activity is brought from microscopic to macroscopic particles (heterogenisation) the retention of this activity during and after the process is facilitated which means that operation costs can be lowered. Often productivity can be enhanced by allowing high concentrations of biocatalyst in a given reaction volume.

V. Nedović and R. Willaert (eds.), Fundamentals of Cell Immobilisation Biotechnology, 53-63.

Entrapment often leads to enhanced stability of microbial cells or enzymes. Moreover they can be protected effectively against contamination and other exterior influences, which lead to improved process stability. But for all that possible disadvantages of encapsulation like costs for entrapment, lowered specific activity of the biocatalyst, diffusion problems, or bad mechanical stability of particles have to be considered and met by choosing the right entrapment technology.

1.2. NATURAL AND SYNTHETIC MATERIALS

Many immobilisation techniques are based on biopolymers, *e.g.* alginate, pectinate, κ-carrageenan, chitosan, or cellulose. These biopolymers allow for very gentle immobilisation procedures but the materials themselves are easily subject to biodegradation. Instead of being inert they can act as media compounds for potentially contaminating organisms (mostly moulds) and are thus cleaved and metabolised. This leads to more or less rapid deterioration of the complete immobilised biocatalyst when operating under non-sterile conditions. The efforts necessary to keep sterile conditions are very often too costly, especially in the field of bulk products. In contrast, hydrogels of polyvinyl alcohol (PVA) are hardly biodegradable which lowers demands on the process' sterility.

1.3. POLYVINYL ALCOHOL

Polyvinyl alcohol is a hydrophilic polymer whose aqueous solutions, when stored for prolonged time, are capable of gelling on-their-own. Hydrogen bonds between hydroxyl groups of neighbouring polymer chains form a non-covalent network (Figure 1). However, this process takes days and the gels obtained at temperatures above 0°C usually are rather weak and thus not suitable.

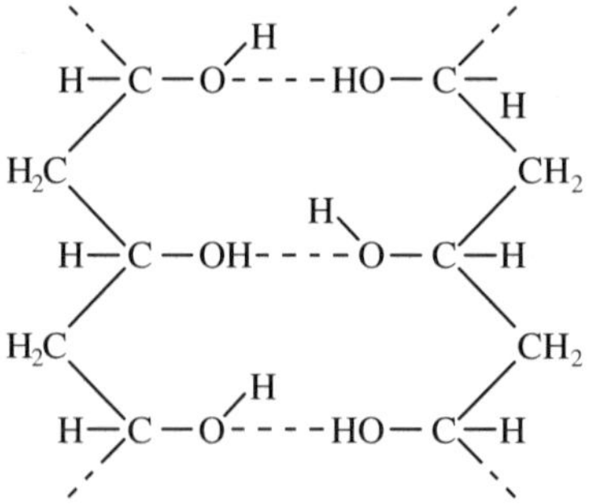

H—C—O(H)- - - -HO—C—H
H_2C CH_2
H—C—OH- - - -(H)O—C—H
H_2C CH_2
H—C—O(H)- - - -HO—C—H

Figure 1. Chemical structure of polyvinyl alcohol, gelated by hydrogen bonds.

A different effect is obtained when PVA solution is subjected to freezing [1,2]. Due to phase separation during the freezing process the formation of hydrogen bonds is enhanced and the resulting hydrogel is significantly stronger. Hydrogels from PVA by this cryogelation are mechanically very stable and show more or less no abrasion when employed in stirred reactors. In gelated form PVA is hardly biodegradable and thus can be used when working under non-sterile conditions. Chemically the hydrogels can be utilised with any physiological compound since they do not dissolve. Merely by heating to above 60°C the hydrogels can be dissolved by melting.

Parameters influencing the gel-strength are the degree of deacetylation of the used PVA, the chain length of the polymer, its concentration in the solution, and the rate of thawing. Usually concentrations of 7 to 15% (w/w) of polymer with a molecular weight of about 80 to 100 kDa are used. The slower the thawing process, the more rigid the resulting gel will be. Especially the temperature range of about -15 to +5°C is crucial for a satisfactory stability and the emerging hydrogel should be thawed with only a few degrees per hour in this scope.

An alternative to the slow-thawing method for the reinforcement of the hydrogel is the multiple freezing-thawing, which is scarcely used since it consumes significant amounts of energy. Another known means of gelating a PVA-solution is the dripping into boric acid, but the resulting hydrogel is less stable and rather brittle.

However, in all cases the formation of the gel causes significant loss of microbial activity when living cells are to be immobilised. To counteract this loss of biological activity the method of room-temperature gelation was developed by Vorlop and Jekel [3] and the resulting particles were named LentiKats®.

1.4. SHAPE OF LENTIKATS®

Many different shapes were proposed for immobilised biocatalysts – many of them determined by the means of preparation that are feasible with the particular polymer system and gelatinisation mechanism. Besides fibres, cubes, irregular shapes, or sheets, primarily beads, *i.e.* a spherical shape, is most frequently availed.

The spheres result from dripping a polymer solution into a hardening solution. Reviewing the publications from the area of bioencapsulation the bead shape is clearly dominating. Last but not least this is due to the fact that theoretical models are considerably simpler for the plain geometric shape of an ideal sphere. A detailed comparison of different methods for producing these spherical immobilised biocatalysts was given by Jahnz *et al.* [4].

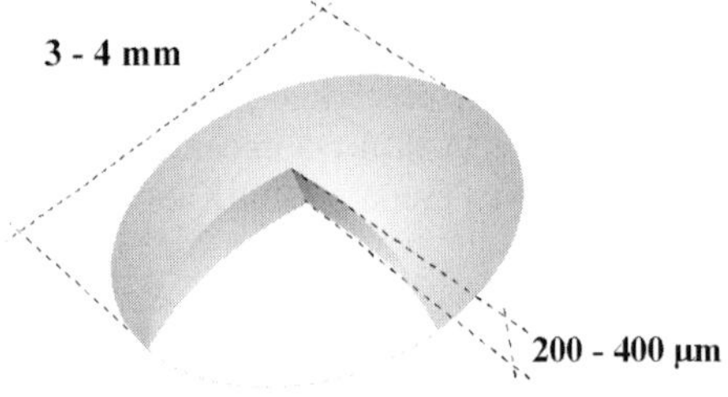

Figure 2. Schematic view of a LentiKat®, a lens-shaped hydrogel particle based on PVA.

No matter how the beads are produced they either have a rather small or rather large diameter and one has to find a more or less satisfactory compromise between good diffusion properties (possible with small diameter) and easy retention of beads (in case of large diameter). Due to their lenticular shape (Figure 2) LentiKats® combine the advantages of small and large beads as a matter of principle.

2. Preparation of lentikats®

2.1. PRINCIPLE OF ENTRAPMENT IN LENTIKATS®

LentiKat®Liquid solution is mixed with the biocatalyst preparation, and small droplets are floored on a suitable surface. For first trials it is sufficient to form the droplets by a simple lab-syringe on a standard petri dish (Figure 3). By exposing these droplets to air, the water starts to evaporate and thus leads to enhanced formation of hydrogen bonds between the PVA molecules. Once about 70% of the polymer-biocatalyst-solution have been removed by evaporation, the hydrogel is stable enough and can be re-swollen in a stabilising solution before the ready LentiKats® are employed.

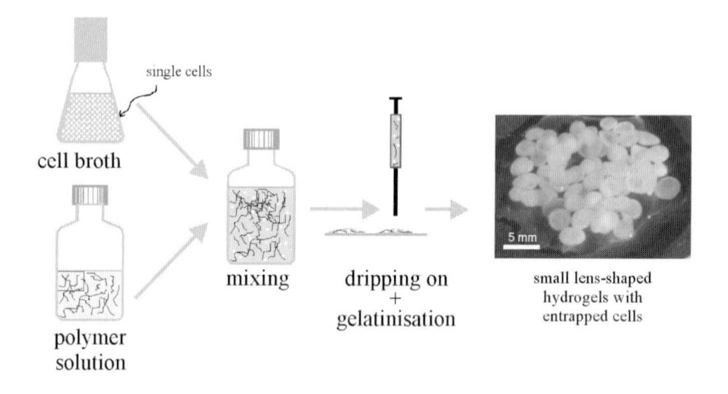

Figure 3. Manual production of LentiKats® using a standard syringe.

The conditions during the entrapment procedure regarding pH value and temperature are quite gentle and no potentially harmful chemicals have to be added at any stage. More detailed information on how to prepare LentiKats® and on their basic properties can be found on the Internet [5].

2.2. WORK ON LAB-SCALE

Once the results show that the method is in principle compatible to the tested biocatalyst, it is advisable to change the method of preparing the LentiKats® to keep identical conditions for different experiments. The mechanical properties of the produced LentiKats® and the survival rates of entrapped microorganisms strongly depend on the time of gelatinisation and the uniformity in size of the droplets. Using a syringe creates droplets differing in size and, even more crucial, this successive approach leads to unequally distributed times of gelatinisation. Moreover the amount of particles, which can be manually produced, is very limited and even for lab-scale applications it is a tedious work.

As an improvement a device was developed for R&D work to produce more than 400 identical droplets simultaneously in one step. Based on the well-known principle of printing technology a special printer head was designed for multiple transfer of equal amounts of PVA solution to a special surface (Figure 4). Handling the printer is a cyclic sequence of the following steps: loading each tip with the same amount of polymer solution by dripping into, replacing the stock of polymer solution with a standard 145-mm petri-dish, lowering the printer head and flooring the droplets on the petri-dish and finally exposing the fresh droplets to gelation conditions.

After appropriate time gelation is abruptly terminated by flooding with a stabilising solution. The described procedure guarantees that all particles are generated in the same way and no delays occur. This is especially important when the data of consecutively run experiments have to be compared and differences caused by the immobilisation process itself have to be precluded. Up to approx. 100 grams of identically immobilised material can be prepared in acceptable time for lab-scale applications under sterile or non-sterile conditions.

Figure 4. LentiKat® Printer and detailed view of the printer head.

2.3. PRODUCTION OF LENTIKATS® ON TECHNICAL SCALE

Polyvinyl alcohol as the core compound of the LentiKat® matrix is a low-cost commodity chemical available in large amounts in a wide range of qualities [6]. The use of this industrial feedstock and the observed long-term stability of LentiKats® make them capable also for large-scale industrial applications. This determines the need for a bulk production technology of immobilised biocatalysts. This is realised by a continuous production procedure employing a conveyor-belt system.

The polyvinyl alcohol solution containing the biocatalysts is delivered to a nozzle-holder carrying a series of nozzles that drop the biocatalyst suspension on a continuously progressing plastic belt. The belt loaded with the droplets is subsequently moving through a channel where the gelatinisation takes place by partial controlled drying (Figure 5). After passing through the chamber the ready LentiKats® are scraped off and can be employed. Production units have an output between 2 and 50 kg/h.

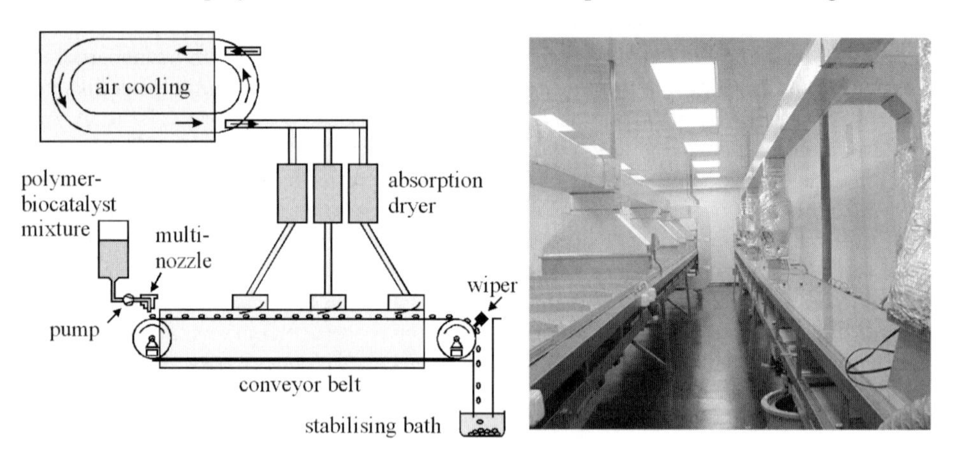

Figure 5. Schematic view and picture of an industrial production line for LentiKats®.

3. Recent developments in biocatalyst encapsulation in LentiKats®

The need to develop a new method for entrapment using PVA arose from the fact, that the other techniques described above were too harsh on the biocatalysts. The first microorganisms tested with the new technology were nitrifiers, *i.e.* a mixed culture of bacteria oxidising ammonia *via* the intermediate nitrite to nitrate [3]. Since then a variety of other organisms have also been tested with LentiKats®.

3.1. ENTRAPMENT OF BACTERIA

While the first experiments with nitrifying organisms were made in artificial media and on lab-scale the current focus is on applying this technique on larger scale under real process conditions. The group of Sievers has reported on their findings when using immobilised nitrifiers on technical scale with feed-stock material from a real waste water treatment plant. Continuous experiments have shown a complete nitrification under these conditions for more than 650 days [7].

The authors have further calculated, that it is possible to ensure complete nitrification by LentiKat® encapsulated nitrifiers with a hydraulic retention time of 0.5 hours, which is approx. 10 times lower compared to the model sewage plant for 100,000 peoples equivalent. Based on a flow rate of 24,500 m^3 per day, a nitrification reactor of 580 m^3 volume compared to common 5,000 to 7,000 m^3 would be sufficient.

Another example for the entrapment of bacteria into LentiKats® are publications about the usage of *Clostridium butyricum* for the valorisation of waste glycerol from biodiesel production sites. Under anaerobic conditions selected strains ferment glycerol to 1,3-propanediol, a bulk-chemical needed in the production of plastics. The complete chemistry for its usage is established since 1,3-propanediol is currently produced chemically in large amounts. Thus, as soon as the biotechnical production is competitive in price the market is available for this compound derived from renewable resources. To decrease the process costs the group around Vorlop and Wittlich pursued two strategies: they screened for highly product tolerant strains which also tolerate the feed of raw glycerol instead of expensive pharmaceutical grade glycerol [8] and they investigated into optimising and stabilising the process conditions by entrapping the clostridia in LentiKats® [9]. Bock reports, that by immobilising the strain *C. butyricum* NRRL B-1024 the productivity could be increased threefold and thus the fermentation time could be shortened by the factor 3.5 [10]. In addition, Schlieker and Wittlich observed that the process was more robust, since a contamination with *Klebsiella pneumonia* could be eliminated by just increasing the flow rate to lower the hydraulic retention time and thereby washing out the non-entrapped microorganisms efficiently (personal communication, unpublished data).

A continuous process to deacidify apple juices and cider was developed by the group of Durieux by entrapping *Oenococcus oeni* in LentiKats®. For a residence time of 0.55 h, malic acid was completely converted into lactic acid when the LentiKats® bioreactor was fed with apple juice at pH between 4.46 and 3.95 [11]. The bioreactor filled with LentiKats® gave better performance than continuous reactor with *Oenococcus oeni* immobilised in alginate beads (specific malic acid consumption increased by a factor of 4.6). The authors explain this improvement with the increase of the ratio external surface to volume, allowing for better mass transfer.

3.2. ENTRAPMENT OF FUNGI

When entrapping fungi the mechanical stability of the matrix is often a problem since the cells are more powerful and break out of the particles. Compared to classical materials like alginate or pectinate the PVA-based LentiKats® show an increased strength and thus can entrap fungi more efficiently. Welter has tested LentiKats® for the bioconversion of glucose and other carbohydrates to itaconic acid, a multifunctional compound used in chemical industries for plastic production, by *Aspergillus terreus* NRRL 1963 [12]. Compared to PVA-hydrogels obtained by cryogelation she found an increased specific activity for LentiKats® and concludes that this is due to the improved shape of the particles and the better mass transport conditions. The immobilised material was tested over a period of six weeks in a stirred tank reactor. During this time span the fermentor was not overgrown by free biomass proving that indeed the biomass was successfully retarded.

For usage in organic solvent Bruß has immobilised the fungus *Beauveria sulfurescens* in LentiKats®. This fungus is capable of selectively hydrolysing epoxides to enantiomerically pure epoxides and diols, respectively [13]. He immobilised whole cells and found, that the immobilised biocatalyst was stable in aqueous and organic solvents for days. Current investigations go into the field of modelling the mass transfer parameters and kinetics for catalysts entrapped in LentiKats® (personal communication).

3.3. ENTRAPMENT OF ENZYMES

For the majority of enzymes the cut-off of hydrogel networks like alginate, pectinate, PVA in general, and LentiKats® in this special case is too large and thus the entrapment of the native enzyme usually fails because the enzyme diffuses readily from the polymer matrix (Figure 6).

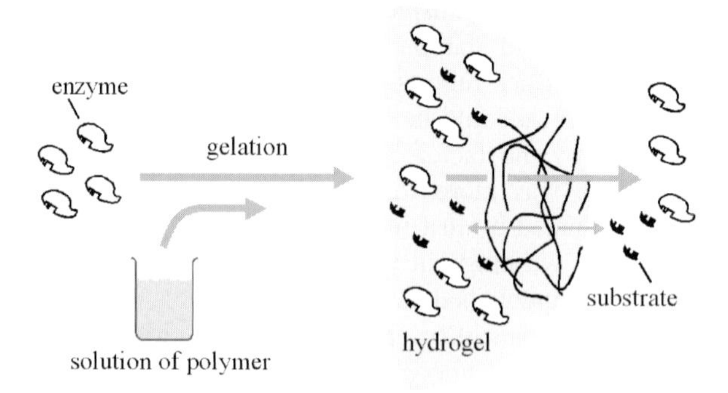

Figure 6. Native enzymes are too small to be retarded by hydrogel matrices.

To counteract this behaviour the molecular weight of the enzyme has to be enlarged. Different techniques have been developed for this. The enzyme molecules can either be cross-linked with each other or co-cross-linked with other polymers. A common chemical to be used is glutardialdehyde which reacts with two amino residues either both on enzyme molecules or on the enzyme and the polymer. A suitable polymer for co-cross-linking can be for example chitosan. The principal effect of this procedure is depicted in Figure 7 and its effectiveness was fine-tuned by Capan for the enzyme (R)-oxynitrilase [14]. This enzyme converts benzaldehyde and hydrogen cyanide to (R)-mandelonitrile, an important chiral building block of the cyanohydrin group. After co-cross-linking the enzyme with chitosan and entrapping in LentiKats® the desired product was obtained in good yields and also with high enantioselectivities of up to >99% ee. Both in long-term tests (>140 h) and after multiple re-usage (20 times) no significant leaching of the catalyst or decrease in activity could be observed.

The group of Guisan has used a comparable procedure to encapsulate penicillin G acylase (PGA). In the first instance, soft enzyme aggregates are either formed by precipitating solely PGA or by precipitating the enzyme in the presence of dextran sulfate and poly-ethyleneimine. The precipitates are then treated with glutardialdehyde and the resulting aggregates, which are very soft and fragile, are encapsulated into LentiKats® to provide mechanical stability. The hydrophilic microenvironment of LentiKats® protects the enzyme from inactivation, allowing higher yields in the

thermodynamically controlled synthesis of cefalosporin G in organic media. An increased stability was observed [15].
Figure 7 also shows a second technique, which was developed by the group of Dautzenberg [16]. Instead of covalently binding the enzyme, the molecules are flocculated by a successive treatment with a polycation and a polyanion. The resulting polyelectrolyte complexes are extraordinarily stable without the risk of loosing enzymatic activity due to chemical reactions in or nearby the active site of the enzyme.

The benefits of this mild procedure were demonstrated using the model enzyme amyloglucosidase [17]. In repeated batch experiments the entrapped enzyme complex showed a constant high level of activity, assayed by hydrolysis of maltose to glucose. The authors conclude that the developed protocol offers a promising way for the entrapment of a variety of enzymes into LentiKats® for industrial applications.

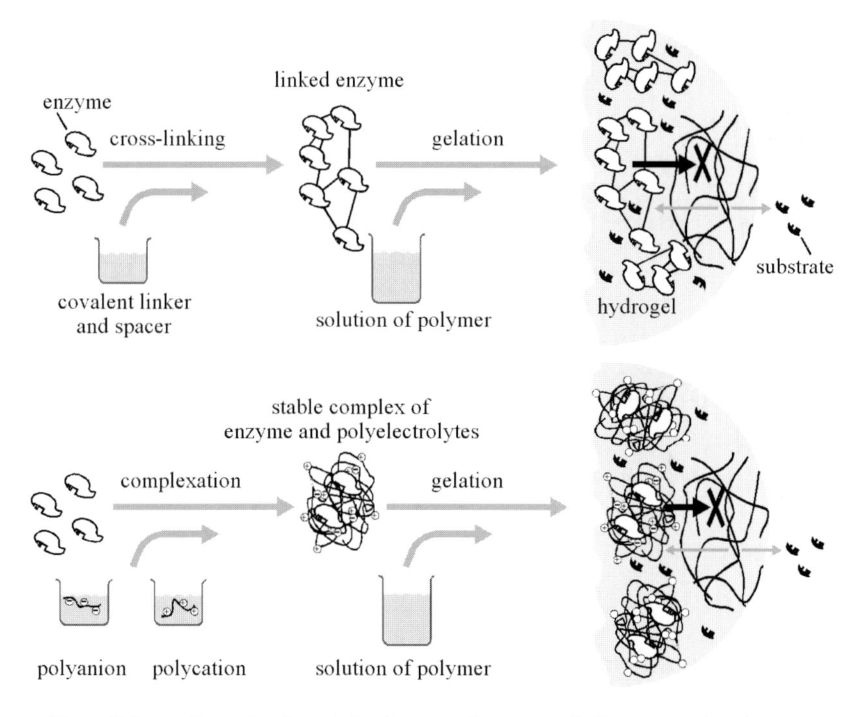

Figure 7. Increasing molecular weight of enzymes by co-cross-linking or complexation.

A universally valid approach for enzyme entrapment with LentiKats® was described also by the group of Vorlop. They prepared LentiKats® with pre-activated particles (Figure 8) and soaked these subsequently with an enzyme solution. The enzyme diffuses into the hydrogel network and is then bonded there by chemical reaction between the enzyme and the reactive particles [18]. The viability of this idea was proven using again the enzyme (R)-oxynitrilase but has still to be optimised for maximum reactivity within the hydrogel. Advancing this technique will allow to produce universal and simple immobilisation kits based on efficient LentiKat® technology for a broad range of enzymes.

4. Conclusions

Compared to established immobilisation carriers like alginate hydrogels LentiKats® are relatively new but all the same various applications for the entrapment of bacteria, fungi and enzymes have been published so far and reviewed above. For all applications the benefits when using LentiKats® were significant. For the process of nitrification in waste water the experiments have reached technical scale and the results are promising enough to continue. The existence of a resulting first application on industrial scale will definitely lead to an increased appreciation of the technology. In addition, the recent work on special pre-activated LentiKats® for the multi-purpose encapsulation of enzymes will further help to establish LentiKats®.

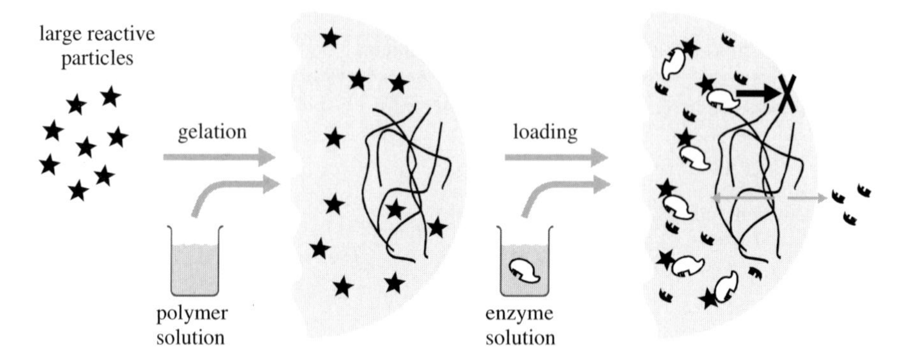

Figure 8. Loading of pre-activated LentiKats® with an enzyme.

References

[1] Lozinsky, V.I. (1998) Cryotropic gelation of poly(vinyl alcohol). Russian Chemical Reviews, English Edition, 67-7: 573-586.

[2] Lozinsky, V.I. and Plieva, F.M. (1998) Poly(vinyl alcohol) crygels employed as matrices for cell immobilization. D. Overview of recent research and developments. Enzyme Microbiol. Tech. 23: 227-242.

[3] Jekel, M.; Buhr, A.; Willke, T. and Vorlop, K.-D. (1998) Immobilization of biocatalysts in LentiKats®. Chem. Eng. Technol. 21: 275-278.

[4] Jahnz, U.; Wittlich, P.; Pruesse, U. and Vorlop, K.-D. (2001) New matrices and bioencapsulation processes. In: Hofmann, M. and Thonart, P. (Eds.) Focus on Biotechnology Vol. 4, Kluwer Academic Publishers, Dordrecht; pp. 293-307.

[5] geniaLab (2002) Tips&Tricks manual, http://www.geniaLab.de/download/tt-english.pdf

[6] Finch, C.A. (1973) Polyvinyl alcohol – properties and applications. John Wiley & Sons, Chichester, UK.

[7] Sievers, M.; Schäfer, S.; Jahnz, U.; Schlieker, M. and Vorlop, K.-D. (2002) Significant reduction of energy consumption for sewage treatment by using LentiKat® encapsulated nitrifying bacteria. Landbauforschung Völkenrode SH 241: 81-86.

[8] Koschik, I.; Bock, R.; Wittlich, P. and Vorlop, K.-D. (2001) Conversion of glycerol to 1,3-propanediol by newly isolated bacterial strains. In: Book of Abstracts of 10th European Congress on Biotechnology, Madrid, Spain; p. 86.
[9] Wittlich, P.; Schlieker, M.; Lutz, J.; Reimann, C.; Willke, T. and Vorlop, K.-D. (1999) Bioconversion of raw glycerol to 1,3-propanediol by LentiKats. Schriftenreihe Nachwachsende Rohstoffe 14: 524-532.
[10] Bock, R.; Koschik, I.; Schlieker, M.; Wittlich, P. and Vorlop, K.-D. (2002) Production of 1,3-propanediol with immobilized microorganisms. In: Conference proceedings, X International BRG Workshop on Bioencapsulation, Prague, Czech Republic; pp. 41-44.
[11] Durieux, A.; Nicolay, X. and Simon, J.P. (2000) Continuous malolactic fermentation by *Oenococcus oeni* entrapped in LentiKats®. Biotechnol. Lett. 22 (21): 1679-1684.
[12] Welter, K. (2000) Biotechnische Produktion von Itaconsäure aus nachwachsenden Rohstoffen mit immobilisierten Zellen. PhD-Thesis. Technical University of Braunschweig.
[13] Bruß, T.; Westermann, B. and Warnecke, H.-J. (2001) Hydrolysis of epoxides with LentiKats®-immobilised microorganisms. In: Book of abstracts of BioTrans 2001 – The 5th International Symposium on Biocatalysis and Biotransformation, Darmstadt; p. 237.
[14] Gröger, H.; Capan, E.; Barthuber, A. and Vorlop, K.-D. (2001) Asymmetric Synthesis of an (R)-cyanohydrin using enzymes entrapped in lens-shaped gels. Org. Lett. 3(13): 1969-1972.
[15] Wilson, L.; Illanes, A.; Abian, O.; Fernandez-Lafuente, R. and Guisan, J.M. (2002) Encapsulation of very soft cross-linked enzyme aggregates (CLEAs) in very rigid LentiKats®. Landbauforschung Völkenrode SH 241: 121-125.
[16] Dautzenberg, H.; Karibyants, H. and Zaitsev, S.Y. (1997) Immobilization of trypsin in polycation-polyanion complexes. Macromol. Rapid Commun. 18: 175-182.
[17] Czichocki, G.; Dautzenberg, H.; Capan, E. and Vorlop, K.-D. (2001) New and effective entrapment of polyelectrolyte-enzyme-compleses in LentiKats. Biotechnol. Lett. 23: 1303-1307.
[18] Capan, E.; Jahnz, U. and Vorlop, K.-D. (2002) Pre-activated LentiKat®-hydrogels for covalent binding of enzymes. Landbauforschung Völkenrode SH 241: 151-153.

STARCHES AS ENCAPSULATION MATERIALS

PIRKKO FORSSELL, KAISA POUTANEN,
TIINA MATTILA-SANDHOLM AND PÄIVI MYLLÄRINEN
VTT Biotechnology, Tietotie 2, Espoo 02044 VTT, Finland –
Fax: +358 9 455 2103 – Email: pirkko.forssell@vtt.fi

1. Introduction

In the food and paper making industries starches have successfully been used as ingredients for a long time [1]. Starches are used in many forms because they offer a large variety of functionalities. In the food and paper production, their main functions are gel formation and binding ability. Most often starches are pregelatinised, partially depolymerised or slightly derivatised before they are applied. In pharmaceutical formulations starches are also widely used as fillers, binders and disintegrant agents in oral solid dosage forms. The other pharmaceutical area is nasal drug delivery, where microspheres prepared of starches by cross-linking starch with epichlorohydrine are used [2].

Chemically modified starches and starch hydrolysates, the latter in the presence of proper emulsifiers, are used as microencapsulation matrices for lipophilic flavours to improve the encapsulant properties, such as easier handling or better stability against evaporation and oxidation. Blending of syrups or sugars with modified or hydrolysed starches have been claimed to lead to optimal encapsulating materials. The increased interest for using bioactive compounds as functional ingredients in formulations of novel foods is a challenging field to apply microencapsulation technologies. Limitations in many available techniques are often their high production cost and lack of food-grade materials [3].

In the past decade, extensive research on starch based biodegradable plastics has been carried out including potentiality of using native starches as raw materials for plastics [4]. Furthermore, starches and other biopolymers have been suggested to be used as edible films in various food applications, *e.g.* for protective coatings and structure improvers [5]. The present chapter deals with functionality of native starch granules and starch polymers related to encapsulation, with the main focus on the microencapsulation of living bacteria.

V. Nedović and R. Willaert (eds.), Fundamentals of Cell Immobilisation Biotechnology, 65-71.

2. Starch granules

2.1. PROPERTIES OF STARCH GRANULES

Starch is unique among carbohydrates because it occurs naturally as discrete particles, called granules. The size of the granules depends on the starch origin ranging from 1 to 100 μm. Starch granules are rather dense and insoluble, and hydrate only slightly in water at room temperature (Figure 1). Starch granules are not easily hydrolysed by amylases, especially granules of tuber starches and of high amylose maize starch are amylase resistant. The granular structure is irreversibly lost when the granules are heated in water to about 80°C, which is visually observed as an increase in the viscosity of the dispersion. If enough heat and mechanical energy is applied, starch granules are totally dissolved and a polymer solution is formed. Depending on the experimental conditions starch polymers are more or less depolymerised during the dissolving process. Commercial starch hydrolysates called dextrins are manufactured by prolonged heating under dry conditions at high temperatures either with or without the addition of an acid. Usually starches used in food and non-food industries are slightly chemically modified in order to match better the needed functionality.

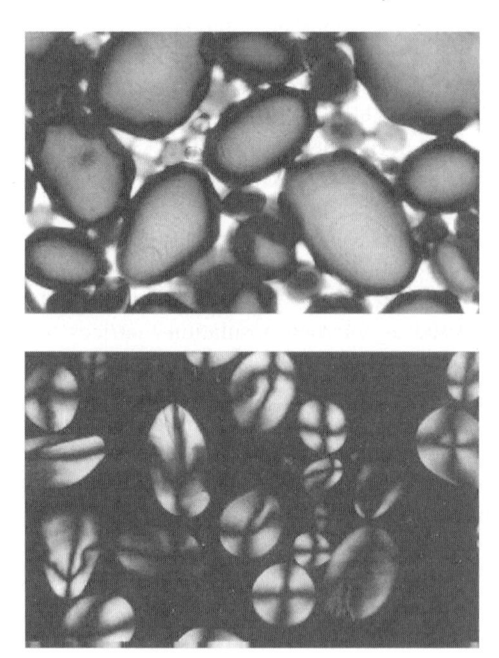

Figure 1. Native potato starch granules under the light microscopy. Iodine staining without (upper) and with polarised light (lower). The polarisation cross (birefringence) is an indication of the ordered, partially crystalline nature of the starch granules.

2.2. USE OF STARCH GRANULES

Normally starches are partially or totally dissolved before they are used in various applications. Traditionally, they are as texture and mouth feeling agents in the food area, as binders in the paper coating industries and as excipients in the pharmaceutical formulations.

Applications based on starch granules are scarce, but recently novel ways to exploit the granular form of starches have been introduced. Partially hydrolysed and cross-linked starch granules were suggested to be suitable carriers for various functional food components [6]. Hydrolysis was performed with the aid of amylases, and corn starch was suggested to be the preferable starch. To improve the absorption of the encapsulant, the surface of the granules was modified by proper agents. Another application – based on enzymatically treated starch granules without performing any cross-links between the granules – was also claimed to be able to function as flavour carrier [7]. In addition to exploit partially hydrolysed starch granules starch granule aggregates were discovered to be formed in a spray-drier, and was suggested to use this as flavour carriers [8]. The aggregate formation occurs for small starch granules when a starch-water dispersion in the presence of small amount of protein or water soluble polysaccharide is dried. In order to modify the release characteristics, the aggregates, which were filled with the active component, could be further coated with a specific polymer.

A few studies describing the encapsulation of living cells in granular starch particles have been reported. Some intestinal bacteria were shown to be able to adhere to starch, and it was suggested that adhesion was sometimes required for the efficient the utilisation of the substrate [9,10]. Recently, a few investigations about the utilisation of starch granules to protect health promoting bacteria were reported. The presence of high amylose maize starch increased the survival of bifidobacteria at a low pH, and during the passage through the intestinal tract of mice [11]. An increased bacterial survival was attributed to the adsorption of the bacteria on starch.

Crittenden and co-workers [12] examined the adhesion of 19 bifidobacteria strains on starch granules of maize, potato, oat and barley (Figure 2, [13]) in order to find out whether adhesion is a common characteristic of bifidobacteria, if there is a link between adhesion and the utilisation of starch, and to evaluate the potential of microencapsulation of these bacteria in starch granules. The results indicated that a strong adhesion of bifidobacteria to granular starch is not characteristic for all strains. Furthermore, the highly adherent strains were observed to be able to hydrolyse the starch granules, but not all amylolytic (amylase producers) strains were adherent. This indicated surprisingly that the adhesion of bifidobacteria is not necessarily needed for substrate utilisation. The binding capacity correlated to the surface area of the granules. Adhesion was assumed to be mediated via surface proteins of the bacterial cells, without any involvement of proteins or peptides associated with the starch granules. The acid- and protease-sensitive nature of the adhesion was demonstrated by performing adhesion measurements under *in vitro* upper gastrointestinal conditions for two adherent strains and for two poorly adhering strains.

The technology to microencapsulate probiotic bacteria by using starch granules as adsorption carriers and high amylose maize starch as the final coating layer, has been developed with the aim of protecting the bacteria during processing, storage and passage through the gastrointestinal tract [14]. By using a proper combination of starch granules, amylase and growth medium, probiotic bacteria were successfully fermented to produce a matrix, in which part of the bacteria were adsorbed on starch granules. The bacteria-carrier matrix was further microencapsulated with high amylose maize starch dissolved in water (Figure 3). After freeze-drying, a white powder was produced (Figure 3). High amylose maize starch was used as the coating material because of its rather high resistivity against gastric juices. The viability of the microencapsulated bacteria was detected to be high and stable under ambient conditions for at least 6 months.

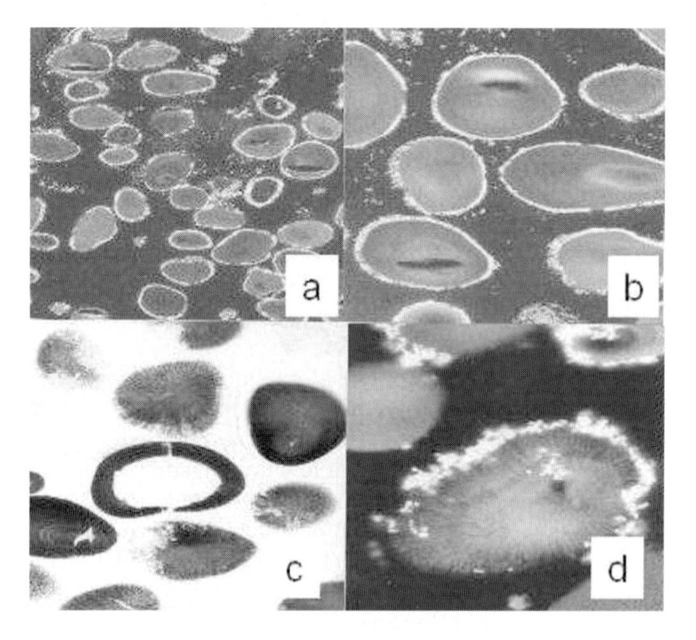

Figure 2. Photomicrographs showing the adhesion of bifidobacteria on the surface of starch granules. Thin cross sections were prepared using DAPI and iodine staining. Magnifications: 290x (a), 725x (b and c) and 1450x (d). [13] (Int. Dairy J. 12: 173-182. Copyright 2002 by Elsevier Science. Reprinted with permission.)

The use of an *in-vitro* GI transit model showed that the microencapsulated bacteria were resistant in the upper intestine. By performing an *in-vitro* fermentation, it was demonstrated that the carrier and coating material was degraded by the colonic microflora [14]. In *in-vivo* tests it was seen that microencapsulated bacteria survived the passage through the human GI-tract. This is the only study reported so far, in which starch granular structure and starch polymers have been used without chemical modifications with the aim to develop microencapsulation technology for living probiotic bacteria.

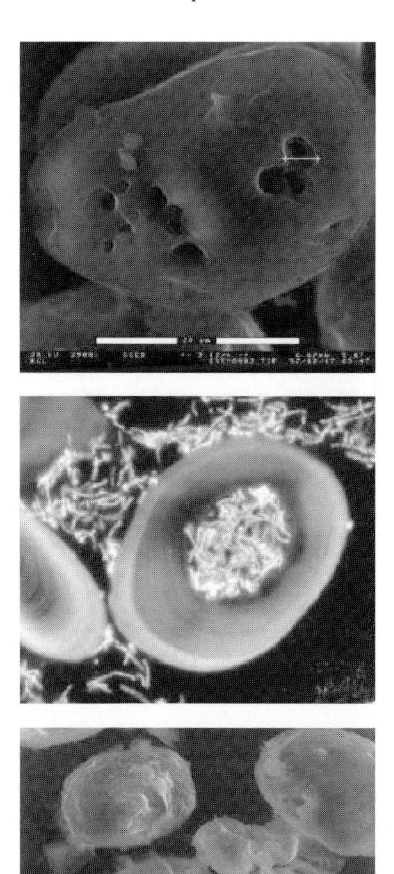

Figure 3. ESEM picture of partially hydrolysed potato starch granules, in which pores are seen on the surface (upper). Potato starch granules filled with probiotic bacteria (middle/light microscopy, DAPI staining, thin cross section). ESEM picture of amylose coated potato starch granule-bacteria matrix (bottom).

3. Starch polymers

3.1. PROPERTIES OF STARCH POLYMERS

Starch granules are usually built up of two glucose polymers called amylopectin and amylose. Amylopectin is a huge and heavily branched polymer composed of α-D-glucopyranosyl units joined by (1→4) linkages. Branching is created by (1→6) linkages. Amylose is essentially a linear chain, which is built up of (1→4)-linked α-D-glucopyranosyl units. When starch granules are heated in water at a proper temperature the granular structure is degraded and a polymer solution is formed. A homogeneous polymer mixture or melt can also be achieved by applying heat and mechanical energy to starch under low water conditions. From both polymer mixtures films can be produced. This property has been known for a long time. Especially investigations of films prepared of amylose have been conducted, and potential applications were suggested already in the early days [15]. The properties of amylose films were claimed to be equivalent to cellulose films. In addition to the good film formation ability, amylose network structures in films and gels are rather resistant to gastric juices [16,17]. As compared with amylopectin, amylose is able to form structures which are mechanically much more stronger and less soluble under an aqueous environment. Due to the hydrophilic nature of both amylose and amylopectin polymers, films prepared from both starch polymers have shown to act as good oxygen barriers under ambient humidity [18].

3.2. ENCAPSULATION APPLICATIONS BASED ON NATIVE STARCH POLYMERS

During the eighties new interest to exploit biodegradable polymers and edible films in food and non-food industries started partly due to an increased demand for new packaging solutions. Only few applications based on native starch polymers have been reported.

An injection moulded starch capsule was developed for mainly pharmaceutical applications [19]. The insolubility of amylose in gastric juices and its degradability by colonic bacteria was exploited in a novel way when colon specific drug delivery formulations were developed [20,21]. Coatings of amylose alone was too porous due to its swelling under aqueous conditions which resulted in drug release. Swelling was, however, successfully controlled by combining amylose with ethyl cellulose. *In vivo* studies demonstrated that glucose was released in the colon from pellets coated with the new polymer blend [22]. It has also been suggested to use the retrogradation property of dissolved starches for the development of microencapsulation technology for lipophilic drug particles [23].

4. Summary

Native starches are safe and readily available, and their processing can be performed in aqueous media. With the aid of enzymes, the properties of starch granules and polymers can be further modified. The versatile granular and polymer properties offer useful functionalities, which can be used in formulations of living bacteria and perhaps also in developing novel drug dosage forms. The safety, environmental and economic benefits can be achieved by applying smart processing methods based on these natural biomaterials.

References

[1] Whistler, R.L.; Miller, J.N. and Paschall, E.F. (Eds.) (1984) Starch: Chemistry and Technology. Academic Press Inc., London.

[2] Pereswetoff-Morath, L. (1998) Microspheres as nasal drug delivery systems. Advanced Drug Del. Rev. 29: 185-194.

[3] Gibbs, B.F.; Kermasha, S.; Alli, I. and Mulligan C.N. (1999) Encapsulation in the food industry: a review. Int. J. Food Sci. and Nutrition. 50: 213-224.

[4] Feil, H. (1995) Biodegradable plastics from vegetable raw materials. Agro-Food-Industry Hi-Tech. July/August: 25-32.

[5] Krochta, J.M.; Baldwin, E.A. and Nisperos-Carriedo (Eds.) (1994) Edible coatings and films to improve food quality. Technomic Publ. Inc., Lancaster, 1994.

[6] Whistler, R.L. and Lammert, S.R. (1989) Microporous granular starch matrix composition. US 4985082.

[7] Kobayashi, S.; Miwa, S. and Tsuzuki W. (1993) Method of preparing modified starch granules. Patent publication EP 0539 910 A1.

[8] Zhao, J. and Whistler, R.L. (1994) Spherical aggregates of starch granules as flavor carriers. Food Technol. 48: 104-105.

[9] Tancula, E.M.J.; Feldhaus, L.A.; Bedzyk, L.A. and Salyers, A.A. (1992) Location and characterization of genes involved in binding of starch to the surface of Bacteroides thetaiotaomicron. J. Bacteriol. 174:5609-5616.

[10] Reeves, A.R.; Wang, G.R. and Salyers, A.A. (1997) Characterization of four aouter membrane proteins that play s role in utilization of starch by Bacteroides thetaiotaomicron. J. Bacteriol. 179: 643-649.

[11] Wang, X.; Brown, I.L.; Evans, A.J. and Conway, P.L. (1999) The protective effects of high amylose maize (amylomaize) starch granules on the survival of Bifidobacterium spp. in the mouse intestinal tract. J. Appl. Microbiol. 87: 631-639.

[12] Crittenden, R.; Laitila, A.; Forssell, P.; Mättö, J.; Saarela, M.; Mattila-Sandholm, T. and Myllärinen, P. (2001) Adhesion of Bifidobacteria to granular starch and its implications in probiotic technologies. Appl. and Environ. Microbiol. 67:3469-3475.

[13] Mattila-Sandholm, T.; Myllärinen, P.; Crittenden, R.; Mogensen, G.; Fondén, R. and Saarela, M. (2002) Technological challenges for future probiotic foods. Int. Dairy J. 12: 173-182.

[14] Myllärinen, P.; Forssell, P.; von Wright, A.; Alander, M.; Mattila-Sandholm, T. and Poutanen, K. (1999) Starch capsules containing microorganisms and/or polypeptides and/or proteins and a process of producing them. World Patent WO9952511A1.

[15] Langlois, D.P. and Wagoner, J.A. (1967) Production and use of amylose. In: Whistler, R.L. and Paschall, E.F. (Eds.) Starch: Chemistry and Technology, Vol II. Academic Press, New York; pp. 451-497.

[16] Ring, S.G.; Gee, J.M.; Whittam, M.; Orford, P. and Johnson, I.T. (1988) Resistant starch: its chemical form in foodstuffs and effect on digestibility in vitro. Food Chem. 28: 97-109.

[17] Botham, R.L.; Morris, V.J.; Noel, T.R.; Ring, S.G.; Englyst, H.N. and Cummings, J.H. (1994) A comparison of the in-vitro and in-vivo digestabilities of retrograded starch. In: Phillips, G.O.; Williams, P.A. and Wedlock, D.J. (Eds.) Gums and Stabilizers for the food industry 7. Oxford Univ. Press, Oxford; pp. 187-195.

[18] Forssell, P.; Lahtinen, R.; Lahelin, M. and Myllärinen, P. (2002) Oxygen permeability of amylose and amylopectin films. Carbohydr. Polym. 47: 125-129.

[19] Wittwer, F.; Tomka, I.; Bodenmann, H.U.; Raible, T. and Gillow, L.S. (1988) Method for forming pharmaceutical capsules from starch compositions. US Patent 4 738 724.

[20] Ring, S.G.; Archer, D.B.; Allwood, M.C. and Newton, J.M. (1991) Delayed release formulations. WO 91/07946.

[21] Milojevic, S.; Newton, J.M.; Cummings, J.H.; Gibson, G.R.; Botham, R.L.; Ring, S.G.; Stockham, M. and Allwood, M.C. (1996) Amylose as a coating for drug delivery to the colon: Preparation and in-vitro evaluation using glucose pellets. J. Control. Release 38: 85-94.

[22] Milojevic, S.; Newton, J.M.; Cummings, J.H.; Gibson, G.R.; Botham, R.L.; Ring, S.G.; Allwood, M.C. and Stockham, M (1994) Amylose, the new perspective in oral drug delivery to the human large intestine. S.T.P. Pharma Sci. 5(1): 47-53.

[23] Rein, H. and Steffens, K.-J. (1997) Surface modification of water-insoluble drug particles with starch. Starch/Stärke 49: 364-371

PROTEINS: VERSATILE MATERIALS FOR ENCAPSULATION

MONIQUE H. VINGERHOEDS AND PAULIEN F.H. HARMSEN
ATO B.V., Department Coating Systems and Active Ingredients,
Bornsesteeg 59, 6708 PD Wageningen, The Netherlands –
Fax: +31 317 475 347 – Email: m.h.vingerhoeds@ato.wag-ur.nl

1. Introduction

Proteins have unique material properties, which make them attractive for several applications including microencapsulation [1-7]. Actually, the first reported microencapsulation process, over 50 years ago, was the manufacture of microcapsules based on gelatin, intended for the development of carbonless copy paper [8]. Relevant protein properties for encapsulation purposes include surface active properties, good film forming and mechanical properties, high gas barrier properties, and high resistance to organic solvents and oils/fats [9]. As proteins can be obtained from many different sources, the variation in intrinsic properties is already very wide. Furthermore, because of the large variation in composition of proteins and their reactive groups, a broad range of modification reactions can be performed (*e.g.* physical, chemical, and enzymatic). Thus, protein properties can be tailored towards specific coating and encapsulation applications.

This chapter deals with proteins as encapsulation materials, and aims to gain insight in the possibilities of the use of different (modified) proteins as carrier material for microcapsules and microspheres. First, general information about proteins is given, with respect to composition, structure and properties, and possibilities for modification, including crosslinking. An overview of several industrial proteins is given and finally different encapsulation methods are discussed where proteins have shown their value, *e.g.* spray drying, complex coacervation and emulsion stabilisation.

2. Composition, structure, and properties

Industrial (bulk) proteins are biopolymers derived from plants or animals. The annual production of protein isolates (> 90% protein) and concentrates (25-80% protein) reaches 1 million metric tonnes world wide with princes ranging from about €1.50/kg to €5/kg [10], with the exception of the more expensive collagen and zein.

Proteins are composed of amino acids linked *via* amide bonds giving chain lengths ranging from about 50 up to more than 100,000 amino acids. The primary structure of

V. Nedović and R. Willaert (eds.), Fundamentals of Cell Immobilisation Biotechnology, 73-102.

proteins is the linear sequence of the linked amino acids. The primary structure folds into the secondary structure, which describes the path that the polypeptide backbone of the protein follows in space. The tertiary structure describes the organisation in three dimensions of all atoms in the polypeptide chain, including the side groups as well as the polypeptide backbone.

Amino acids are amphoteric due to the presence of basic (NH_2) and acidic groups (COOH). There are twenty common amino acids with side chains of different size, shape, charge, and chemical reactivity [11]. The degree of hydrophobicity and hydrophilicity of the amino acids is one of the major determinants of the three-dimensional structure of proteins. The amino acids glycine, alanine, valine, leucine, isoleucine, methionine, and proline have nonpolar aliphatic side chains, while phenylalanine and tryptophan have nonpolar aromatic side groups. These hydrophobic amino acids are generally found in the interior of proteins, forming the so-called hydrophobic core. Other amino acids have ionisable side chains (*i.e.* arginine, aspartic acid, glutamic acid, cysteine, histidine, lysine, and tyrosine). Together with asparagine, glutamine, serine, and threonine, which contain non-ionic polar groups, they are frequently located on the protein surface where they can interact strongly with the aqueous environment.

Proteins show enormous diversity of structures as a result of their ability to generate a huge range of conformations, but certain types of secondary structure resulting from hydrogen bonding between the NH and CO groups of the polypeptide backbone are relatively common. Hydrogen bonding between the backbone CO and NH groups in close proximity (four amino acid residues apart) causes the polypeptide chain to twist into an α-helix, a rod-like structure. A polypeptide chain can also be extended into a sheet-like structure by hydrogen bonding between backbone CO and NH groups of amino acid residues situated at larger distances; the chain may run in the same (parallel β-sheet) or opposite (anti-parallel β-sheet) direction. This β-sheet structure is favoured by the presence of glycine and alanine residues. Each protein has in principle almost indefinite number of conformation possibilities, but in practice, only one or a few conformations are favoured. Proteins can be divided into two general classes based on their tertiary structure:

- Fibrous proteins: elongated structures, long strands; may be α-helix or β-sheet.
- Globular proteins: compact structure, rather irregular tertiary structure often containing many different types of secondary structure.

Protein denaturation is the unfolding of the protein from a structured native state into an (partially) unstructured state with no or little fixed residual structure, resembling a random coil [9]. Denaturation can be induced both by temperature and by denaturants (chemical denaturation). At medium water content (>5-10%) and temperatures higher than 75ºC denaturation occurs. However, it is generally known that the denaturation temperature of proteins may differ due to the protein source, pH, salts, denaturants (*e.g.* urea or guanidinium.HCl), and processing methods. Moreover, denaturation can be reversible, but mostly leads to irreversible protein aggregation.

An important parameter in the development of protein-based encapsulation systems is the film forming ability of proteins. In principle, any protein capable of forming a film is suitable for microencapsulation, provided that the film can also be formed on the surface of the core material, *e.g.* oil droplets [12]. Film formation is based on separation of unfolded proteins from the solvent phase due to for instance:

- Changes in solvent conditions (polarity of pH changes, electrolyte additions).
- Thermal treatments.
- Solvent removal (drying).

Solvents used to prepare protein film-forming solutions are generally based on water, ethanol, and occasionally acetone. However, dispersing proteins in solvents may require addition of reductive agents (*e.g.* mercaptoethanol, sodium sulphite, cysteine, sodium borohydride), pH adjustment, or ionic strength control by electrolyte addition.

The proteins must be in an open or extended form to allow the molecular interactions that are necessary for film formation [12,13]. The extent of these interactions depends on the protein structure (degree of unfolding) and the sequence of hydrophobic and hydrophilic amino acid residues in the protein. Increased molecular interaction results in a film that is strong but less flexible and less permeable. The degree of hydrophilicity of the amino acid residues in a protein controls the influence of moisture on the mass transport properties of the protein film. Most protein films are rather moisture sensitive, but this inherent hydrophilicity makes them excellent barriers to nonpolar substances such as oxygen, some aroma compounds and oils/fats. An increase in crystallinity, density, orientation, molecular weight or crosslinking results in a decrease in permeability. Complicated protein structures make the control of these factors quite challenging.

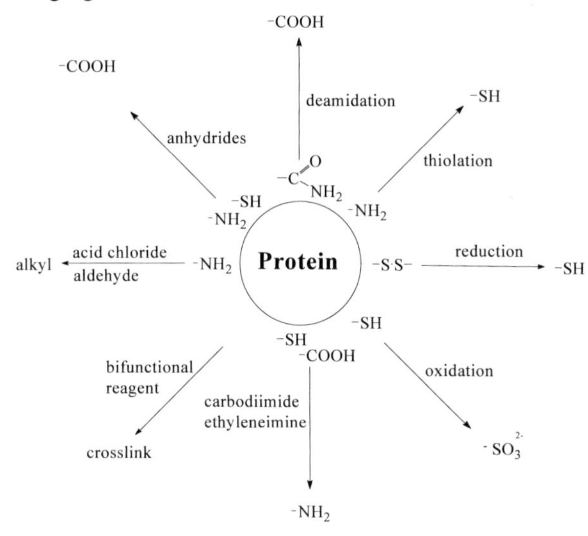

Figure 1. Schematic presentation of some commonly used protein modifications.

3. Protein modification

3.1. INTRODUCTION

Although proteins are versatile materials with interesting properties for microencapsulation, it might be necessary to alter some properties, *e.g.* to improve water resistance, or to increase the water solubility or mechanical strength. A variety of modifications can be carried out with proteins. The different amino acids have side chains of different sizes, shapes, charges, and chemical reactivity. The reactivity of a protein, in terms of its ability to be chemically modified, will be determined largely by its amino acid composition and the sequence location of the individual amino acids in the three-dimensional structure of the molecule. Protein modification can be done by physical means (*e.g.* pressure and temperature), chemical or enzymatic. Some of the possible protein modifications and their expected influence on protein properties are illustrated in Figure 1 and Table 1.

It has to be emphasised that, besides knowledge on protein modification reactions that can be carried out to tailor the functional properties for the desired application, it is also important to gain insight in the effects of environmental factors (*e.g.* salt, pH, temperature) on proteins, as these can play a major role in processing and in the final product [14].

Table 1. Effect of modification on protein properties.

Property	**Hydrophilisation**	**Hydrophobisation**	**Crosslinking**
Solubility in water	Increase	Decrease	Decrease
Film forming	Increase	Unknown	Unknown
Adhesion	Increase	Decrease/increase	Unknown
Strength	Unchanged	Unchanged	Increase
Water resistance	Decrease	Increase	Increase
Gas permeability	Unknown	Unknown	Decrease
Biodegradability	Increase	Unknown	Decrease

3.2. HYDROPHILISATION AND HYDROPHOBISATION

By introducing polar or charged groups, the solubility of proteins in water can be increased. Examples of polar groups are carboxyl, amine, hydroxyl, phosphate, and sulphate. For instance, improved solubility can be achieved by deamidation in which amide residues are being transferred to acid residues. The deamidated protein has a lower isoelectric point and as a result has better solubility in mildly acidic environments [15-17].

Hydrolysis can be done to decrease the molecular weight of a protein, thereby making the protein more water-soluble. Hydrolysis can be done chemically and by utilising enzymes. Important parameters include pH, reaction time, and temperature. The advantages of enzymatic hydrolysis include substrate specificity, low energy costs, and the absence of side reactions [18].

Hydrophobisation makes proteins more lipophilic due to the interaction of apolar groups like alkyl- or aromatic groups. This gives rise to better dispersability in non-solvents (including water in the case of water-insoluble proteins such as wheat gluten), a higher solids content of the dispersions, an increased water resistance, and better adhesion to apolar substances.

3.3. CROSSLINKING

The term 'crosslinking' is commonly used to describe the covalent bonding of a protein to itself or to another protein. As well as causing changes in molecular size and shape, crosslinking often results in important changes in the chemical and functional properties of the proteins. Crosslinking can be used for stabilisation and subsequent tailoring of release properties of protein-based controlled release systems.

Different methods can be used for crosslinking purposes, ranging from physical induced modification, to enzymatic and chemical crosslinking [11,19]. Crosslinking can be controlled by the proper selection of reaction type, reactive groups on protein, and type of crosslinking reagent. The number of reactive groups per protein chain, and the amount, functionality, and spacer length of the crosslinker determine the resulting crosslink density. Table 2 summarises some different crosslinking possibilities.

Several physical treatments, such as irradiation and heat treatment can result in protein crosslinking by rearrangements in the protein. The main disadvantage of heat treatment as crosslinking method is that it is difficult to control the reactions that are taking place. This also applies for the Maillard reaction, which occurs upon heating of proteins in the presence of a source of carbonyl groups, such as reducing sugars [19]. The more reactive the reducing sugar the stronger the gel and the darker its colour.

Protein crosslinking by irradiation was shown amongst others by Brault *et al.* [20], who produced free-standing sterilised edible films based on caseinates. The irradiation of aqueous protein solutions generates upon water radiolysis hydroxyl radicals that produce stable compounds. Sulphur and aromatic amino acids react more readily with free radicals than aliphatic residues. Advantages of the irradiation process are that the method is relatively cheap, and that a sterilised product can be obtained.

Various enzymes have the ability to crosslink proteins. Examples thereof are transglutaminase (TG), disulphide isomerase, peroxidase, lipoxygenase, and catechol oxidase. In a recent overview on enzymatic crosslinking, it was indicated that enzymes available on a sufficient large scale for industrial application include only microbial TG, lactoperoxidase and glucose oxidase [21]. TG is an acyl-transferase that catalyses the introduction of ε-(γ-glutamyl)-lysine crosslinking into proteins, making it an interesting enzyme for use in food grade protein crosslinking. TG has been used to crosslink several proteins [22]. In general, caseins appear to be more susceptible to TG-induced crosslinking than whey-proteins, possible because of the predominantly random structure of caseins, in contrast to the globular structure of whey proteins [19]. Edible films crosslinked with TG appeared to be protease-digestible [23].

A less known type of crosslinking involves the peroxidase catalysed reaction between the side chains of two tyrosines, resulting in a C-C bond between the two carbons in an *ortho* position with respect to the phenol group [24]. This crosslink is a strong, mainly intermolecular produced bond. Within ATO work has been done on the use of co-substrates to enhance the crosslinking efficiency of peroxidases, which reduces the amount of enzyme to be used and therewith increases the cost effectiveness. The co-substrates are phenolic compounds such as mono- and di-hydroxy benzene derivatives (*e.g.* catechol, ferulic acid, and p-hydroxybenzoic acid) and are probably acting as a spacer between the protein molecules.

Table 2. Protein crosslinking (adapted from [11,19] and others).

Physical treatment	
Alkali or heat	β-Elimination of cysteine or phosphoserine residues resulting *via* dehydroalanine residues in lysinoalanine crosslinks with lysine and lanthionine crosslinks with cysteine
Severe heat	Condensation of the ε-amino group of lysine with the amide group of aspartic acid or glutamic acid residues results in the formation of isopeptide bonds
Oxidation	Disulphide crosslinks involving thiol-disulphide interchange reactions
Sugars or aldehydes in combination with heating	Maillard reaction: reversible formation of Schiff base between reducing sugar and amino group of protein resulting in advanced glycation end products
Irradiation	Crosslinking *via* hydroxyl radicals
Lipids	Crosslinking involving *e.g.* free radicals, or malondialdehyde
Enzymatic treatment	
Transglutaminase	Catalysis of acyl-transfer reaction between the γ-carboxyamide group of glutamic acid residues and various primary amines (*e.g.* lysine) resulting in inter- and intramolecular glutamyl-lysine crosslinks
Peroxidase or catalase	Catalysis of reaction between tyrosine side chains
Disulphide isomerase	Catalysis of transient breakage of protein disulphide bonds; exposed cysteine SH-groups form new disulphide bonds
Polyphenol oxidases	Catalysis of hydroxylation of tyrosine residues followed by reaction with amine or thiol groups resulting in intra- or intermolecular bonds
Chemical treatment	
Homobifunctional crosslinking reagents	• Amino group directed, *e.g.* di-isocyanates and dialdehydes • SH-group directed, *e.g.* alkylating agents (*e.g.* aziridines, bis-epoxides), non-covalent crosslinking reagents (*e.g.* avidin and streptavidin) • Hydrogen bond forming agents (tannic acid)
Heterobifunctional crosslinking reagents	• Amino and sulfhydryl group directed • Carboxyl and sulfhydryl or amino group directed • Carbonyl and sulfhydryl group directed
Zero-length crosslinking reagents	Carboxyl group activating reagents, *e.g.* carbodiimides, isoxazolium derivatives, chloroformates, carbonyldiimidazole Reagents for carbohydrate activation or for disulphide formation

Much research has been done on the use of chemical agents in protein crosslinking [11]. The reagents most often used are bifunctional reagents. Bifunctional reagents have two reactive groups, and can be used to introduce inter- and/or intramolecular crosslinks into proteins [19]. These reagents can be classified as homobifunctional or heterobifuntional. Homobifunctional reagents have two identical functional groups, whereas heterobifuntional reagents have two different functional groups. It is difficult to control the reaction of homobifunctional reagents to ensure that crosslinking is exclusively intra- or intermolecular, although some control is possible by choice of the appropriate pH, ionic strength and protein:reagent ratio. A well-known homobifunctional reagent is

glutardialdehyde (GDA). Heterobifunctional reagents can be used in a more discriminating manner; crosslinking can occur in separate sequential steps, and the formation of intermolecular crosslinks can be avoided or promoted.
Tannins are complex polyphenolic substances that can be derived from galls, but also occur in fruit (*e.g.* pomegranate) and tea. Tannic acid (TA) is capable of complexing or crosslinking proteins by the formation of multiple hydrogen bonds [25]. Proteins can hereby be physically crosslinked and rendered more resistant to resolution and to enzymatic degradation. This crosslinking is a partly reversible process. Some metal ions can oxidise TA and proteins, or inhibit the formation of hydrogen bonds.

Summarising, the degree of crosslinking plays a major role in the stability, porosity and release properties of protein-based microspheres or microcapsules, and can be adjusted by proper selection of crosslinking method/agent, time, and concentration. For example, Orienti *et al.* [26] showed a clear difference in release profiles due to different methods of crosslinking. Physical modification induced by heating or irradiation is often difficult to control. However, when a strongly crosslinked matrix is desired and the compound that has to be encapsulated is insensitive to heating or irradiation, these methods can be useful. In case of the Maillard reaction, off-flavour and brown colouring might give extra disadvantages. Enzymes provide an elegant way of protein-crosslinking. Unfortunately, some enzymes are expensive and not available on a large scale. The use of co-factors can reduce the amount of enzyme required and thereby the costs. Especially TG offers great potential as it has been used in food processing and also in stabilisation of microcapsules. Lastly, a large number of chemical crosslinking agents are available. Although a number of them provide a very efficient crosslinking process, most of them are not food approved and are therefore limited to non-food applications.

4. Industrial proteins

4.1. INTRODUCTION

Industrial proteins are proteins that are produced at a kiloton scale per year. Examples of industrial proteins are animal proteins, *e.g.* collagen, gelatin, whey protein, casein, keratin, and plant proteins like wheat gluten, soy protein, pea, and potato protein.

Relevant properties of proteins for microencapsulation include:

- biodegradable, often biocompatible material
- natural emulsifying, dispersing and gelling properties
- gas barrier properties
- variety in source, composition and number of functional groups
- ease of modification
- processability (in melt, aqueous solutions, and dispersions)
- good adhesion and film formation properties
- resistance toward oils and organic solvents
- reasonable pricing while available at large scale
- food grade

Table 3 gives an overview of different industrial proteins, which differ from each other in their chemical and physical structure and thus in functional properties.

4.2. COLLAGEN AND GELATIN

Gelatin and its precursor collagen are very interesting proteins. Based on their special conformation (see below), they exhibit unique reversible gelating properties. Materials based on collagen and gelatin have been developed for medical uses, photographic films, adhesives and for microencapsulation purposes.

Table 3. Protein properties and pricing.

Protein	Source	Mw (kD)	Isoelectric point	Price (€/kg)
		Animal proteins		
Collagen	Skin, bones	285	*	1-50
Gelatin	Skin, bones	15 – 250	4.5-5.5 (type B) 7-9 (type A)	0.3-10
Casein & caseinate	Milk	19-24	4.2-5.3	3.5-5.5
Whey protein - β-lactoglobulin - α-lactalbumin	Milk	 18.6 14.2	 5.3 4.8	6-12
Keratin	Feathers, hair	10	**	<0.5
Albumin - BSA - Ovalbumin	 Bovine serum Egg	 66 45	 4.7 5.5-6.8	 >400*** 7
		Plant proteins		
Gluten - Gliadins - Glutenins	Wheat	 30-80 2000-3000	5-6	< 0.8
Zein	Corn	25-45	**	8-20
Soy protein	Soy beans	150-190 (7S) 320-360 (11S)	4.3-4.8	1-3
Pea protein	Pea	160-350	4.5-5.5	3

*Not relevant; ionisable groups not accessible;
**Cannot be established because of water insolubility;
*** BSA is not regarded as a bulk protein;

Collagen is the most abundant structural protein in vertebrates and is a family of proteins with very high tensile strength [27]. Collagen is a long stretched protein with a Mw of about 285,000 g/mol. The amino acid sequence of collagen is highly distinctive, nearly every third is glycine in a constellation of Gly-X-Y, and collagen is unusually rich in proline, hydroxyproline and hydroxylysine [28-30]. Three polypeptide α-helical chains are twisted around a common axis to give a right-handed triple helix, which is stabilised by intramolecular hydrogen bonds. Due to the presence of intermolecular crosslinks, collagen is insoluble in water [31].

Gelatins are high molecular weight polypeptides derived by partial hydrolysis of collagen and have subsequently a quite similar amino acid composition as collagen. The major sources for gelatin are bones and hides from cows. Type A (pI 7-9) is produced by acid hydrolysis of collagen, whereas type B (pI 4.5-5.5) results from alkali or lime processed collagen [32,33]. Influenced by the isolation process, gelatin has a broad molecular weight distribution (15,000-250,000 g/mol) with an average of about 50,000-70,000 g/mol. Gelatin dissolves in warm water and upon cooling a gel is formed. Gelatin is a potentially useful biomaterial mainly because of its advantageous gelating properties, melting temperature of 37°C, and other protein properties like low price, large availability and biodegradability. Important parameters determining gelatin quality and applicability include Bloom number, which is a measure of gel strength, Mw distribution, viscosity of the gelatin solution and its isoelectric point. Gelatin is considered a versatile encapsulating material [34] and has found use in microencapsulation of herbicides [35], cosmetics, pharmaceutics [36] and in carbonless copy paper [8].

The most common source of collagen is bovine corium or tendon. Since the late nineties there is an ongoing discussion whether collagen and gelatin could be possible contaminated by prions (proteinaceous infection particles) of bovine spongiform encephalopathy (BSE) [37]. This is mainly important for medical and food applications. It is reasoned that there are several ways to minimise the risk for contamination, including careful selection of the herds, use of bovine tendon and skin as these are considered as "materials with no detectable effectivity" [38,39], or the use of alternative gelatin and collagen sources, *e.g.* corium of swine or horse tendon [40], material from marine origin such as shark skin or marine sponge [28,41,42], or recombinant gelatin [43-46].

4.3. MILK PROTEINS (CASEIN AND WHEY)

Milk has a protein content of about 33 g per litre [47,48]. The two major milk protein fractions are caseins (about 80%) and whey proteins (20%). Milk proteins possess numerous functional properties positively influencing film formation and microencapsulation, such as their solubility in water and ability to act as emulsifiers.

Caseins are predominantly phosphoproteins that precipitate at pH 4.6 at 20°C. The amino acid composition is characterised by low levels of cysteine. Consequently, few disulphide crosslinks are present. The molecule has an open, random coil structure [12]. A large fraction of caseins in milk exists as colloidal dispersed micelles ranging in size from 10 to 250 nm [49,50]. Caseinates are produced by treating acid-precipitated caseins with alkali (sodium or calcium hydroxide) at 80-90°C and pH 6.2-6.7 [12]. The caseinates do not reassociate in any manner resembling the original casein micelles, but remain as monomers in an open extended form, that are soluble at pH > 5.5 [12]. Caseins and caseinates form transparent and flexible films from aqueous solutions without further treatment due to their random coil nature and numerous hydrogen bonds. Their amphiphilic nature makes them effective surfactants [51]. It is not surprising that caseins have shown to be useful proteins for microencapsulation, *e.g.* for food and aquaculture applications [52-57].

Whey proteins are characterised by their water solubility at pH 4.6 [50]. Whey protein contains five principal protein types: β-lactoglobulin (BLG), (most abundant protein), α-lactalbumin, bovine serum albumin (BSA), immunoglobulins and proteose-peptones. BLG and α-lactalbumin are the most important proteins for microencapsulation and exist both in a globular form, where α-lactalbumin has a more compact structure. Because of the globular nature of whey proteins, production of films requires heat denaturation to open the globular structure, break existing disulphide bonds and form new disulphide and hydrophobic bonds [12]. Efficient purification procedures for whey protein isolates (WPI) and whey protein concentrates (WPC) have been developed, resulting in protein contents greater than 90% or ranging from 25-80% respectively [58-60]. From our experience, WPI is preferred for encapsulation. A large number of references are available on whey protein as encapsulating material for food applications, especially by the group of Rosenberg [7,56,61-66].

4.4. KERATIN

Keratin is an insoluble fibrous structural protein present in skin, hair, feathers, nails, claws, hoofs and scales. The large amount of cystine present in keratin stabilises the protein network by the formation of many disulphide bonds. These disulphide bonds are mainly responsible for the natural insolubility of keratin and resistance to most proteolytic enzymes [67]. Extensive cleavage of these disulphide linkages is necessary for the extraction and isolation of keratin. A clear advantage of this protein is its low price as it is derived from waste streams. Because of the high tensile strength that can be obtained with keratin and its low degradation rate, keratin is envisaged as useful material in tissue engineering [68]. Keratin usage is also found in textiles and cosmetics. However, no reports were found in literature on encapsulation using keratin.

4.5. ALBUMIN

Serum albumin (BSA) is the most familiar plasma protein. It is a large globular protein with a molecular weight around 66 kD, consisting of a single peptide chain of about 580 amino acids. Serum albumins have a low content of tryptophan and methionine and a high content of cysteine and the charged amino acids aspartic acid, glutamic acid, lysine, arginine [69]. The protein is characterised by high solubility, net negative charge at neutral pH, stability, flexibility and diversity of ligand-binding affinities. The ability to bind free fatty acids, other lipids and flavours can stabilise albumin against thermal denaturation [70]. Egg albumin (ovalbumin) has a lower molecular weight (45 kD) and a somewhat lower hydrophobicity as compared to BSA [71].

Much research has been done on the applicability of albumin, especially BSA, for encapsulation [72-74]. Depending on the application, albumin from different sources (and subsequent different pricing) can be used, *e.g.* serum albumin for medical applications and egg albumin for other applications [75-77]. Egbaria and Friedman compared the binding properties for different dyes of colloidal microspheres based on BSA, HSA (human serum albumin) and ovalbumin [71,78]. Their study suggests that BSA microspheres are most hydrophobic, followed by HSA microspheres. Ovalbumin microspheres are regarded as hydrophilic [71]. Also the porosity of BSA particles was higher than that of ovalbumin microspheres [78]. These studied indicate that BSA cannot be replaced by ovalbumin in microsphere preparation without proper evaluation of the new system.

Most research with serum albumins as carrier materials has been done on imaging [79-82] and drug delivery of cytostatics [83-86], antiviral agents [87], and other drugs

[88-90]. Currently, research interest in albumin as matrix material appears to shift from albumin microspheres to nanoparticles [87,91-96].

4.6. GLUTEN

Gluten from most wheat varieties has more or less the same composition: 55 wt% glutenins (polymeric, elastic) and 45 wt% gliadins (monomeric, viscous). The gliadins and glutenins are the storage proteins of wheat endosperm and they are rich in glutamine and proline. Both fractions consist of water-insoluble proteins. Gliadins are soluble in 70% ethanol, whereas glutenins are insoluble [12].

The gluten complex is believed to be a protein network held together by extensive covalent and non-covalent bonding. The amino acid composition of gluten determines its properties. An important characteristic is its high content of glutamine of about 37%. The elastic and cohesive character of gluten is due to a great extent to the presence of disulphide bonds. The disulphide bonds in gliadin are exclusively intramolecular, while in glutenin they are both intramolecular and intermolecular [97].

Films based on wheat gluten are not-water soluble. They have been used for encapsulation, to improve the quality of cereal products, and to retain antimicrobial or antioxidant additives on food surfaces. Wheat gluten has a remarkable high gas barrier (oxygen and carbon dioxide) and a number of possibilities for chemical modifications. Wheat gluten has found application in adhesives, coatings and cosmetics. Despite the interesting properties and low price of gluten, only a low number of citations was found using gluten in microencapsulation [57,98].

The water-insolubility of gluten is limiting its application possibilities. The solubility can be increased by deamidation, *i.e.* chemical deamidation (acid solubilisation) under acidic conditions and high temperature [99] or enzymatic treatment [100-102]. During this reaction, glutamine is converted to glutamic acid and subsequently the water solubility is increased.

4.7. ZEIN

Zein is the major storage protein of corn and comprises approximately 45-50 wt% of the protein in corn [103]. Zein belongs to the prolamine class of grain proteins and is not soluble in water. This insolubility in water is due to the amino acid composition of zein: a low content of polar amino acids and a high content of non-polar amino acids [12]. Interest in the industrial utilisation of zein is due to the formation of water-resistant films when cast from appropriate solvent systems, usually aqueous aliphatic alcohol like ethanol and isopropanol [12,104]. Zein finds its use in printing inks, floor coatings and greaseproof paper. Films, coatings and controlled release systems based on zein were developed for food (*e.g.* chewing gum) and feed applications [105-110], and for controlled release of pesticides [111], and medically active compounds. Due to its pH dependent behaviour, zein is referred to as a reversed enteric material; *i.e.* solubility at low pH is higher then at increased pH [104,112-114].

4.8. SOY PROTEIN

Soybeans have a protein content of 38-44 wt%, where globulins form the main fraction. Four major fractions of soy protein have been characterised by their sedimentation constants: 2S, 7S (ca. 35%), 11S (ca. 52%) and 15S (S stands for Svedberg units) [115]. The 2S fraction consists of low molecular mass polypeptides (in the range of 8-20 kD) and comprises the soybean trypsin inhibitors. The 7S fraction is highly heterogeneous, consisting of a trimer of β-conglycinin, a sugar containing globulin with a molecular mass in the order of 150-190 kD. The 11S fraction consists of glycinin, the principal protein of soybeans. Glycinin (hexamer) has a molecular mass of 320-360 kD. The polypeptides in native glycinin are tightly folded and stabilised *via* intermolecular disulphide bonds. The 15S protein is probably a dimer of glycinin [1]. Soy protein concentrates and soy protein isolates contain at least 70% and 90% protein on a dry basis, respectively [29,116], with, depending on the isolation method, high soluble protein contents [1].

There is a clear interest in the use of soy protein for a range of applications, including paper coatings and plywood adhesives [1,22,117]. Soy protein has been used for microencapsulation of insecticides, food ingredients and pharmaceutics [118,119]. Recently, Vaz *et al.* demonstrated the usefulness of soy protein in the development of pH sensitive hydrogels for delivery of anti-inflammatory drugs [120].

4.9. PEA PROTEINS

A fairly unfamiliar protein is pea protein. Until now the applicability of peas is rather limited; they are mainly used for animal feed. Pea seed contains 25 wt% of protein. The properties and composition of pea protein are quite similar to those of soy protein. Pea protein is mainly composed of water-soluble proteins, including the globulin (55-65 %) and albumin (18-25 %) fractions [121]. A relatively large variation between cultivars has been observed in the insoluble protein fraction [122]. Pea globulins are composed of two main families, legumin (pI 4.8) and vicillin (pI 5.5), which belong respectively to the 11S and 7S seed storage protein classes [121]. Legumin is the major fraction in peas, the vicillin:legumin ratio varies per cultivar from 1:1.3 to 1:4.2 and is on the average 1:2.

Pea proteins have good emulsifying properties for preparing oil-in-water (O/W) emulsions [123-125]. As pea protein contains a relatively large amount of reactive amino groups, chemical modification reactions such as acetylation or succinylation can be carried out effectively. These reactions appeared to be useful in improving for instance the emulsifying properties of pea proteins [122,126]. Besides our own work [3,127], no references concerning the use of pea proteins as encapsulation material were found, despite the high potential of this renewable resource [3]. However, a small number of articles point to the usefulness of vicillin and legumin in the preparation of nano- and microparticles [128-131].

5. Microencapsulation using proteins

5.1. INTRODUCTION

There is a large number of encapsulating methods available that can be used for the preparation of protein-based controlled release systems, *e.g.* spray drying, coacervation or emulsion stabilisation. For most techniques, a subsequent crosslinking treatment is necessary to harden and stabilise the obtained micro- or nanoparticles. This paragraph discusses the potential of different common and less common technologies for the preparation of protein-based controlled release systems.

5.2. SPRAY DRYING

Spray drying is widely used in the chemical, pharmaceutical, and food industry. During spray drying, a solution containing the product is atomised by compressed air or nitrogen in a desiccation chamber and dried across a current of hot air. This technique can be used to protect sensitive materials against oxidation, *e.g.* fish oil, essential oils, vitamins, and flavours [132-134].

The wall material for encapsulation by spray drying should exhibit high solubility and possess emulsification-, film-forming, and drying properties. In addition, the concentrated solutions should have a low viscosity. Proteins are mostly applied as wall material in spray drying processes for food applications; especially milk proteins (casein and whey) and soy protein. Spray drying of whey proteins, preferably in combination with carbohydrates as lactose and maltodextrin, has been referred to extensively for the successful encapsulation of anhydrous milk fat or volatile materials [7,61-64,135-142].

Whey protein isolate, soy protein isolate, sodium caseinate, and gum arabic were evaluated as encapsulation material for flavours by Kim and co-workers [143,144]. Soy protein was most effective and whey protein least effective in retaining orange oil during spray drying of the orange oil emulsions. The microcapsules based on gum arabic had undergone more shrinkage during drying than the protein microcapsules. Dynamic headspace analysis revealed that the gum arabic particles had the highest volatile release rate and soy protein particles the lowest. Orange oil was best protected against oxidation by soy - and whey protein.

Spray-dried products are often converted to O/W-emulsions again by addition of water. It is usually desirable to obtain a stable emulsion with a small and narrow size distribution of emulsion droplets. Hogan and co-workers prepared spray-dried microparticles of sodium caseinate containing 20-75 wt% soy oil [54]. Mean particle sizes of reconstituted dried emulsions were greater than those of the original emulsions, particularly at high oil/protein ratios (>1.0), suggesting destabilisation of high-oil emulsions during the spray-drying process. Fäldt and Bergenstål have shown that lactose can prevent the coalescence of emulsion droplets, and the subsequent increase in droplet size during spray drying of whey protein/lactose/soybean oil emulsions. Furthermore, lactose improved the wettability of the spray-dried products [145,146].

Figure 2. Microcapsules of pea protein containing sunflower oil prepared by spray drying as revealed by SEM-analysis. Magnification 3240x.

Three different types of casein were used by Keogh *et al.* for the encapsulation of fish oil [56]. Homogenised O/W emulsions containing fish oil, casein, and lactose as filler were spray dried. Shelf life of these products was dependent on type of casein used and homogenisation conditions. Free fat was inversely related to the shelf life.

Pavanetto and co-workers prepared corticosteroid-loaded albumin microspheres by spray drying with a particle size smaller than 10 μm for pharmaceutical applications [147]. By optimising the suitable albumin/drug ratio in the formulation, and coupling the spray drying technique with thermal stabilisation, it was possible to modulate the release rate of drug from the microspheres without adding any other chemical.

Pea proteins have good emulsifying properties that can be useful for the preparation of O/W emulsions [125]. At ATO, pea protein-based microcapsules containing sunflower oil were prepared from O/W emulsions. A scanning electron microscopy (SEM) picture of these microcapsules is presented in Figure 2.

5.3. FLUID BED TECHNOLOGY

The fluidised bed is well known for its granulating and drying efficiency. The system can also be used as a coating device for any type of coating system (*e.g.* solution, suspension, emulsion, latex, hot melt) to be applied to a wide range of particle sizes. Coatings can be applied to fluidised (solid) particles by a variety of techniques, including spraying from the top (top spray), from the bottom (Wurster process), or tangentially (rotating disk) [148].

Primarily, the release rate of a product coated by fluid bed depends on the nature of the coating material. An extensive list of materials successfully applied in the Wurster coating is published by Hall *et al.* [149]. Proteins mentioned here include casein, gelatin, whey, and zein. Susceptibility of proteins to attack by enzymes is mentioned as a possible route for triggered release of the encapsulated components.

An example of sodium caseinate as coating material is described by Dewettinck and Huyghebaert [150]. The effects of several process variables on the coating efficiency of top-spray coatings were evaluated. The functional and compositional properties of the protein concentrate clearly influenced the coating efficiency, and choosing the appropriate processing conditions could probably reduce coating losses drastically.

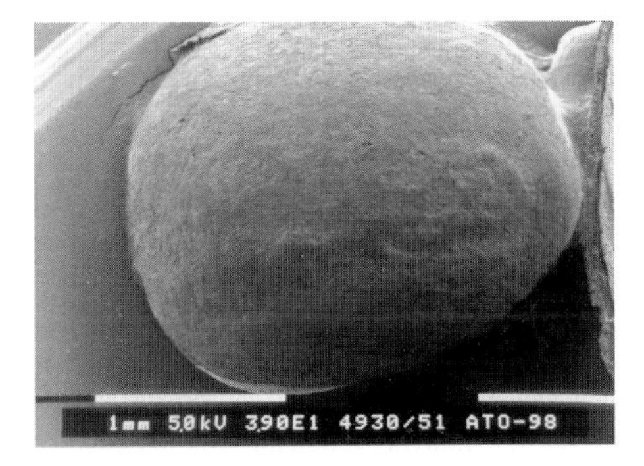

Figure 3. SEM-picture of the surface of cabbage seed coated with soluble wheat gluten. Magnification left picture 39x, right picture 150x.

Applying film coatings on high value vegetable seeds by fluidised bed is common technology nowadays. Care must be taken not to sacrifice germination of the seeds by the various coating treatments. The coating must posses a certain degree of porosity that will allow for the exchange of water, oxygen, carbon dioxide, and other metabolic products between the germinating seed and the environment. It usually contains (more than one) crop protection agent, signal colour, additives, and most importantly the binder. Scheffer and co-workers explored the possibilities of using biodegradable materials for seed coating purposes, to further improve the cost/benefit ratio of seed treatments [151]. Part of this work was done at ATO, where cabbage seeds were coated by fluidised bed using industrial proteins such as whey protein, soy protein, and soluble wheat gluten (Figure 3). Remarkable good adhesive properties and water resistance was found for the seed coating based on soy protein. With soluble wheat gluten, a homogeneous coating was obtained with promising release properties.

5.4. EXTRUSION

The basic idea behind encapsulation using extrusion is to create a molten mass in which the active agents, *i.e.* either liquids or solids, are dispersed [152,153]. Upon cooling, this mass will solidify, thereby entrapping the active components. In most cases, matrix material with a glass transition temperature (*Tg*) above room temperature is chosen. Solidification then yields a good encapsulation material with low rates of diffusion. Extrusion is a continuous and flexible process, using cost-effective and environmentally friendly technology.

The classical extrusion process involves forcing a core material dispersed in a molten carbohydrate mass through a series of dies into a bath of dehydrating liquid [5]. The first patent on flavour encapsulation by this extrusion process was issued in 1957 [154], where an emulsion was extruded into solid pellets. An important development was starting to use the melt-extruder for extrusion-encapsulation. In this apparatus, all separate processing steps (mixing, melting, shaping, heating, and cooling) can be combined and executed in a continuous manner.

To control and modify the properties of the matrix systems, like release characteristics and stability of the system, it is required that control over the dispersed phase morphology is achieved [155]. For starch it was demonstrated that the dispersed phase morphology could be controlled by proper selection of plasticizers and emulsifiers, and processing conditions such as screw speed, presence of die-head, throughput, melt-temperature, and screw configuration [155,156].

Most extrusion-encapsulation processes are dealing with carbohydrates as matrix material. The success of the particular matrix depends on its functionalities (*e.g.* emulsifying properties), on its ability to create a plastic flow, and on interactions that take place between the matrix and the ingredients. Although a lot of research has been done on extrusion of proteins [157-159], only two of the few patents claiming a non-carbohydrate carrier are dealing with proteins as matrix material for extrusion-encapsulation [160]. Sair and Sair used different carbohydrates and/or proteins as carrier material [160]. All components, including 10–40% water were fed to the extruder as a premix. Extrusion temperatures used were in the range of 65-170°C. The extruded polymer glass was dried and ground yielding a stable, microdispersed powder with an active ingredient load of 5% (oil, flavour).

Black *et al.* used whey protein as matrix material [161], resulting in a controlled release product for flavours or other active agents. Whey protein isolate was used solely or mixed with other components, like food protein lipids or different simple or complex carbohydrates. If necessary, a plasticizer (*e.g.* water) was added. All ingredients were premixed and extruded under ambient pressure with no temperature specified. After grinding, a powder with a 3% load was obtained. The patent claimed loadings up to 20%. From our experience, especially milk protein is easily processable by extrusion, probably due to its emulsifying properties. Good encapsulation properties were obtained through an emulsified system of small droplets in a caseinate matrix with loadings up to 30% (unpublished results). Also soy protein thermoplastics proved very versatile materials that can easily be processed by conventional, melt-based technologies. Soy protein thermoplastics are currently explored for encapsulation and the manufacture of pH-dependent matrix systems [162].

5.5. COMPLEX COACERVATION

Complex coacervation is a very common encapsulation method and referred to as true encapsulation. Coacervation is the separation of an aqueous polymeric solution into two miscible liquid phases: a dense coacervate phase and a dilute equilibrium phase. Complex coacervation can result spontaneously on mixing oppositely charged polyelectrolytes in aqueous media. The charges must be sufficiently large to induce interaction, but not large enough to cause precipitation. Complex coacervation is affected by pH, ionic strength, temperature, molecular weight, and concentration [163,164].

Well-known is the complex coacervation that occurs with the neutralisation of two oppositely charged polymers. Often gelatin is used as at least one of the polyelectrolytes. Several combinations can be used, *e.g.* gelatin with gum arabic (acacia), or polyphosphate, but also heparin [165], carrageenan [166], chitosan [167], and sodium dodecyl sulphate (SDS) [168] were reported to form a coacervate with gelatin. Often the coacervates are stabilised by addition of chemical stabilising agents such as glutaraldehyde. However, for food applications this should be avoided, and therefore Schmitt *et al.* studied the formation of stable coacervates between whey protein and acacia gum [169,170]. Another food-approved option is crosslinking complex coacervates using transglutaminase [171,172].

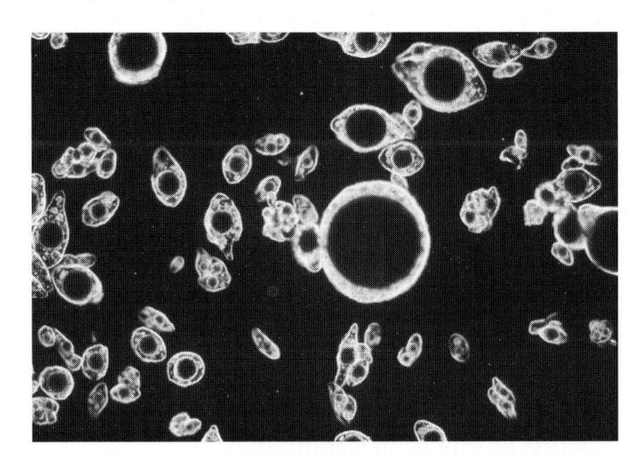

Figure 4. Complex coacervates containing β--carotene shown by light microscopy. Magnification 100x.

The complex coacervation of gelatin and acacia is well defined and described extensively in literature, *e.g.* with respect to the effect of oil phase [173,174], pH [34], or droplet size [175]. Usually commercial gelatins are used, but Cabeza *et al.* studied the possibility of using alkali- or enzyme-extracted gelatins from chrome shavings (produced as waste stream in the leather industry) for encapsulation purposes [176]. No significant differences were found between microcapsules based on acacia and commercial- or extracted gelatins.

Also the combination of gelatin and polyphosphate is common. It was applied for the encapsulation of essential oils by Ribeiro *et al.* [177]. An example of β-carotene encapsulated in gelatin-polyphosphate microcapsules is presented in Figure 4. Gelatin-polyphosphate was also chosen for the encapsulation of bacteria-containing microcapsules, because coacervation between gelatin and polyphosphate leads to very resistant gels [178]. This improved mechanical behaviour of the capsule wall was related to the high gelatin:polyphosphate ratio of 4:1, and it made it easier to separate and dry the particles. Unfortunately, the survival rate of the bacteria in the microcapsules was very poor, probably due to the high oxygen-barrier of the gelatin wall.

The encapsulation of proteins for pharmaceutical applications requires special treatment. Burgess and co-workers developed a method where complex coacervate gel microcapsules were prepared based on albumin and acacia, without the use of crosslinking agents, heat, or organic solvents. This method was suitable for the encapsulation of proteins, but the stability of these particles was questionable, and isolation was problematic due to aggregation [179]. However, spray drying of the albumin-acacia system in combination with polyvinylpyrolidone (PVP) appeared to be useful for the preparation and collection of the microcapsules with minimal loss of the protein activity [180].

5.6. EMULSION STABILISATION

Many encapsulation methods are based on single or double emulsions. Since proteins are generally water-soluble, two systems can be distinguished if proteins are used as wall material. First, single water-in-oil (W/O)-emulsions are employed if the active ingredient is water-soluble. In this case an aqueous protein solution containing the active ingredient is emulsified in a hydrophobic phase like vegetable oil or organic solvent. Second, double oil-water-oil (O/W/O)-emulsions can be used if the active ingredient is hydrophobic. The active ingredient is first emulsified in the aqueous protein phase to form an O/W-emulsion. Then the emulsion is added to a hydrophobic phase to form the double O/W/O-emulsion. Common challenge with emulsion stabilisation is retention of the core; often losses occur during the encapsulation process.

Proteins used for the preparation of microparticles by emulsion stabilisation include gelatin [181-187], albumin [75,76,84,86,188-198], casein [55], and whey protein [65,66,199-203]. Albumin is by far the most investigated natural polymer in microparticle preparation by emulsion stabilisation.

The pH of the protein solution has a major influence on particle formation. This might be related to the differences in protein solubility at various pH-values, as proteins are least soluble around the pI. Generally, proteins are suitable for microsphere preparation if they are (partly) soluble in water. Besides the proteins mentioned earlier, also less water-soluble proteins have shown to be suitable for microsphere preparation under certain conditions, including soluble wheat gluten, zein, soy protein, pea protein [3,4], and potato protein.

Proteins are usually stabilised in order to keep the particles intact, and to prevent the particles from dissolving in water. Different methods can be applied as described in section 3.3, such as chemical, thermal, or enzymatic crosslinking. In search for alternative stabilisation methods, experiments were performed at ATO to evaluate the possibilities of encapsulation by ultra high-pressure treatment (1,000-10,000 bar). This treatment offers advantages over other techniques as it modifies the matrix structure in a more controlled manner under mild conditions. Therefore, it might be a very interesting technique for the encapsulation of (heat) sensitive materials, or for applications where chemical crosslinking is not allowed. Stabilisation or gelation of the protein matrix occurs at >5,000 bar by protein denaturation. Figure 5 presents SEM-pictures of whey protein- and gelatin microspheres formed at 10,000 bar. It appeared that protein solubility and ultra high-pressure enhanced particle formation.

Figure 5. SEM-pictures of whey protein (left) and gelatin (right) microspheres stabilised by ultra high pressure. Magnification 800x.

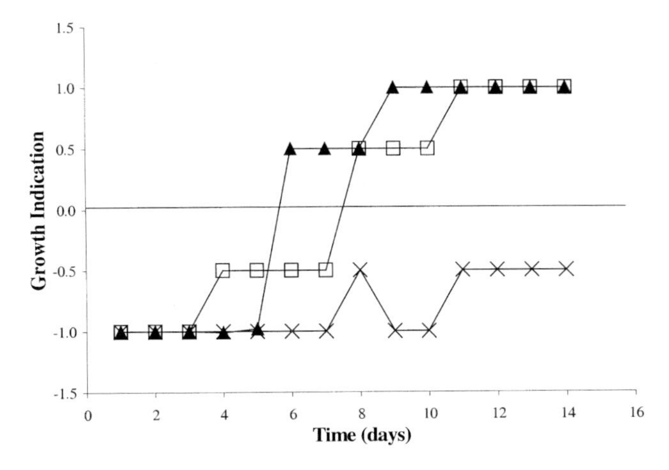

Figure 6. Growth of S. Aureus *(agar bioassay) after incubation with unencapsulated nisin (▲), and nisin-microspheres stabilised by TG treatment (□) or heat treatment* (X).

Providing controlled or slow release of ingredients is one of the aims for microencapsulation. In a recent study, we explored the possibility of encapsulation and subsequent slow release of the antimicrobial peptide nisin to prolong its activity [203, 204]. Nisin was encapsulated in whey protein-based microspheres by emulsion-stabilisation. Heat treatment (by microwave) and enzymatic crosslinking by transglutaminase (TG) were used for stabilisation. The biological activity of the microspheres against *Staphylococcus aureus* is shown in Figure 6. It was demonstrated that especially nisin-containing microspheres stabilised by heat treatment had a prolonged efficacy as compared to unencapsulated nisin or nisin-microspheres stabilised by TG.

5.7. SUPERCRITICAL FLUID TECHNOLOGY

A relative new technology that can be utilised for microencapsulation is supercritical fluid technology. The elegance of this encapsulation technology comes from the mild processing conditions and the lack of required organic solvents. The ingredient to be encapsulated has to be soluble in a supercritical fluid, *e.g.* supercritical CO_2 (properties comparable to hexane). First, microparticles are prepared by *e.g.* extrusion or emulsion stabilisation. Then, the particles and the active ingredient are placed in the reactor. CO_2 at a pressure exceeding 73 bar and a temperature over 31°C results in supercritical CO_2, which dissolves the active ingredient and is believed to cause the microparticles to swell. The swollen microparticles have increased accessibility and are loaded with the active ingredient. Upon reduction of the pressure to atmospheric conditions, the CO_2 is released as a gas and the active ingredient is trapped in the matrix. Within ATO, the potential of this technology was demonstrated for encapsulation of β-carotene in pea protein-based microspheres [3,4,127,205].

5.8. PROTEIN-LIGAND INTERACTIONS

The thus far discussed technologies describe the utilisation of proteins as matrix or wall material for microencapsulation. Active ingredients are entrapped in microparticles in a non-specific manner. A relatively new approach in the development of protein-based controlled release systems is to make use of the ability of proteins to bind ligands (active ingredients) in a specific manner [206]. Most interesting for ligand-specific controlled release systems are applications in which the release of the bound ligand is stimulus-controlled. Suggested stimuli include displacement of the ligand by a related compound, protein denaturation or degradation by *e.g.* proteases, or changes in pH, ionic strength or temperature.

At ATO, β-lactoglobulin (BLG) was studied as matrix for protein-ligand interaction-based controlled release systems [207]. BLG is a relatively small and stable protein that was selected because of its binding capacity for small hydrophobic ligands and the availability of structure information [208,209]. BLG can bind a range of hydrophobic molecules in a fairly specific manner by insertion in a narrow binding pocket, whereby the pocket shields ligands from the surrounding medium, thus providing protection. Molecular modelling analysis was used to predict the binding characteristics of BLG and to suggest possibilities for target-specific modifications of the binding pocket. Binding affinities of flavour compounds and insect repellents were determined by monitoring changes of protein tryptophan fluorescence upon complex formation. Figure 7 presents the displacement of undecanolactone from BLG by a ligand with a higher binding affinity, *i.e.* retinol. The binding experiments demonstrated the potential of protein-based stimulus-induced controlled release systems.

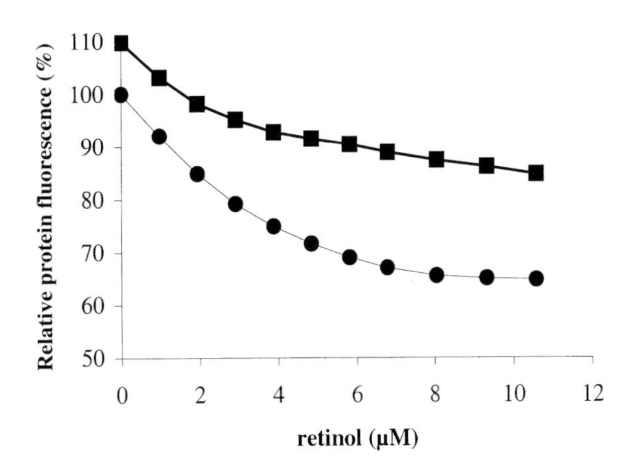

Figure 7. Displacement of the ligand undecanolactone from BLG by titration with retinol, a ligand with higher binding affinity (■). Free protein (●) titration with retinol.

6. Conclusions

This chapter provides an overview of proteins and their possible role in microencapsulation. The large number of references on proteins as encapsulating material already points to the usefulness of these biopolymers. Encapsulation systems described in literature based on proteins often involve relatively expensive proteins like albumin. This can partly be explained by their pharmaceutical application and the added value of the product. The use of industrial proteins such as wheat gluten, soy protein, and milk proteins has not been studied to the greatest extent in encapsulation-technology, although these proteins have the advantages of being abundant, relatively inexpensive, and biodegradable. In addition, they have extremely good emulsifying properties and a lot of functional groups are available for modification.

Table 4 summarises, based on literature and our experience, the different proteins and their general suitability for microencapsulation. Obviously, highly soluble proteins provide the easiest proteins to handle. Examples of these proteins are gelatin, albumin, whey protein, and caseinates. However, also less soluble proteins (mainly plant proteins), like gluten, soy- and pea protein can be utilised for microencapsulation. Adjusting pH, salt concentration or protein modification facilitates the use of these less water-soluble proteins. The use of different protein sources can largely influence processing conditions and the final encapsulated product.

Following the formation of microspheres or microcapsules, the particles have to be stabilised by crosslinking. With respect to particle stabilisation, the number of functional groups, their accessibility and the protein structure are important parameters. Different studies showed that the selected crosslinking method or agent highly influenced the properties of the produced microparticles.

In conclusion, proteins are highly versatile materials, which can be used as cost-effective carriers for microencapsulation. Given the large variety in protein sources, properties, and modification possibilities, tailor-made materials can be designed for a wide range of encapsulation products.

Table 4. Suitability of proteins for microencapsulation

Protein	Suitability
Gelatin	Very suitable for emulsion stabilisation, and also one of the few materials that can be used for complex coacervation
Casein	Especially in the form of caseinates, suitable for microencapsulation by the emulsion stabilisation method
Whey protein	Has been extensively used for encapsulation by spray drying and emulsion stabilisation. Shows also potential in extrusion
Albumin	Mostly used protein for emulsion-stabilisation method. Bovine serum albumin is expensive, egg albumin is most interesting for low-priced applications
Gluten	Only in soluble form (deamidated or hydrolysed) suitable for microencapsulation
Zein	Forms water-resistant films and shows interesting pH behaviour, but can only be processed from organic solvents like ethanol
Soy protein	Suitable for microencapsulation, but dependent on pH. Has shown potential for spray drying, complex coacervation, extrusion and emulsion stabilisation
Pea protein	Suitable for microencapsulation, but highly dependent on pH (solubility) and crosslinking method

Acknowledgements

The authors wish to thank Dr. Ir. L.A. de Graaf, Drs. W.J. Mulder and Dr. J.M. Vereijken for their valuable input.

References

[1] Kumar, R.; Choudhary, V.; Mishra, S.; Varma I.K. and Mattiason B. (2002) Adhesives and plastics based on soy protein products. Ind. Crops Prod. In Press.

[2] de Graaf, L.A. (1998) Non-food applications of cereal proteins. Industrial Proteins 6: 9-11.

[3] de Graaf, L.A.; Harmsen, P.F.H.; Vereijken, J.M. and Mönikes M. (2001) Requirements for non-food applications of pea proteins. Nahrung/Food 6: 408-411.

[4] Harmsen, P.F.H.; Velner, E.M.; de Jonge, H.G.; Harrison, R.M.; Vereijken, J.M. and Vingerhoeds M.H. (2000) Pea protein isolates for microencapsulation purposes. In: Proceedings of the 27th Int. Symp. Control. Rel. Bioact. Mater., Paris (France); p. 556.

[5] Shahidi, F. and Han, X.-Q. (1993) Encapsulation of food ingredients. Crit. Rev. Food Sci. Nutr. 33: 501-547.

[6] Jackson, L.S. and Lee, K. (1991) Microencapsulation and the food industry. Lebensm. Wiss. Technol. 24: 289-297.

[7] Rosenberg, M. and Sheu, T-Y. (1996) Microencapsulation of volatiles by spray-drying in whey protein-based wall systems. Int. Diary Journal 6: 273-284.

[8] Bungenberg de Jong, H.G. (1949) Crystallisation - coacervation - flocculation. In: Kruyt, H.R. (Ed.) Colloid Science. Elsevier, Amsterdam; pp. 232-258.

[9] de Graaf, L.A. (2000) Denaturation of proteins from a non-food perspective. J. Biotechn. 79: 299-306.

[10] Anonymous (2000) Protein Crops. In: Mangan, C. (Ed.) Inieca European Overview report. Brussels (Belgium); pp. 63-65.
[11] Wong, S.S. (1991) Chemistry of protein conjugation and cross-linking. CRC Press, Inc., Boca Raton, Florida
[12] Anker, M. (1996) Edible and biodegradable films and coatings for food packaging - a literature review. Department of food science, Chalmers University of Technology, Sweden.
[13] Krochta, J.M. (1997) Edible protein films and coatings, In: Food Proteins and Their Applications in Foods. Marcel Dekker, New York; pp 529-550.
[14] Liu, L.H. and Hung T.V. (1998) Functional properties of acetylated chickpea proteins. J. Food Sci. 63: 331-337.
[15] Zhang, J.; Lee, T.C. and Ho, C.T. (1993) Kinetics and mechanism of nonenzymatic deamidation of soy protein. J. Food Proc. Pres. 17: 259-268.
[16] Zhang, J.; Lee, T.C. and Ho, C.-H. (1993) Thermal deamidation of proteins in a restricted water environment. J. Agric. Food Chem. 41: 1840-1843.
[17] Shih, F.F. (1991) Effect of anions on the deamidation of soy protein. J. Food Sci. 56: 452-454.
[18] Wu, W.U.; Hettiarachchy, N.S. and Qi, M. (1998) Hydrophobicity, solubility, and emulsifying properties of soy protein peptides prepared by Papain modification and ultrafiltration. J. Am. Oil. Chem. Soc. 75: 845-850.
[19] Singh, H. (1991) Modification of food proteins by covalent crosslinking. Trends Food Sci. Technol. 2: 196-200.
[20] Brault, D.; D'Aprano, G. and Lacroix, M. (1997) Formation of free-standing sterilized edible films from irradiated caseinates. J. Agric. Food Chem. 45: 2964-2969.
[21] Koppelman, S.J. and Wijngaards, G. (1997) Enzymatische verknoping van eiwitten: een literatuurstudie. TNO Zeist, The Netherlands.
[22] Babiker, E.E. (2000) Effect of transglutaminase treatment on the functional properties of native and chymotrypsin-digested soy protein. Food Chem. 70: 139-145.
[23] Yildrim, M. and Hettiarachchy, N.S. (1998) Properties of films produced by cross-linking whey proteins and 11S globulin using transglutaminase. J. Food Sci. 63(2): 248-252.
[24] Michon, T.; Wang, W.; Ferrasson, E. and Guéguen J. (1999) Wheat prolamine crosslinking through dityrosine formation catalyzed by peroxidases: Improvement in the modification of a poorly accessible substrate by "indirect" catalysis. Biotechnol. Bioengineer. 63: 449-458.
[25] Heijmen, F.H.; Pont, J.S. du; Middelkoop, E.; Kreis, R.W. and Hoekstra M.J. (1997) Cross-linking of dermal sheep collagen with tannic acid. Biomaterials 18: 749-754.
[26] Orienti, I. and Zecchi, V. (1993) Progesterone-loaded albumin microparticles. J. Control. Release 27: 1-7.
[27] Rose, P.I. (1988) Gelatin. In: Mark, H.F.; Bikales, N.M.; Overberger, C.G. and Menges, G. (Eds.) Encyclopedia of polymer science and engineering. John Wiley & Sons, New York; pp. 488-513.
[28] Geiger, M. and Friess, W. (2002) Collagen Sponge Implants - Applications, Characteristics and Evaluation: part 1. Pharm. Techn. Eur. 14: 48-56.
[29] Gennadios, A.; McHugh, T.H.; Weller, C.L. and Krochta J.M. (1994) Edible coatings and films based on Proteins (chapter 9). In: Krochta, J.M.; Baldwin, E.A. and Nisperos-Carriedo, M. (Eds.) Edible coatings and Films to Improve Food Quality. Technomic Publishing Company Inc., Lancaster; pp. 201-277.
[30] Stryer, L. (1988) Protein structure and function. In: Stryer, L. (Ed.) Biochemistry. W.H. Freeman and Company, New York; pp. 15-42.
[31] Grobben, A.H. and Visser, A. (1997) Collagen and gelatin: related proteins used in food products. Industrial Proteins 4: 7-8.
[32] Wade, A. and Weller, P.J. (1994) Gelatin. In: Handbook of pharmaceutical excipients, 2nd Edition. APA; The Pharmaceutical Press; pp 199-201.
[33] Mulder, W. (1997) Gelatin in photographic systems. Industrial Proteins 4: 8-12.
[34] Daniels, R. and Mittermaier, E.M. (1995) Influence of pH adjustment on microcapsules obtained from complex coacervation of gelatin and acacia. J. Microencaps. 12: 591-599.
[35] Akin, H. and Hasirci, N. (1994) Effect of loading on the release of 2,4 D from polymeric microspheres. Pol. Preprints 35: 765-766.
[36] Tirkkonen, S.; Turakka, L. and Paronen, P. (1994) Microencapsulation of indomethacin by gelatin-acacia complex coacervation in the presence of surfactants. J. Microencaps. 11: 615-626.
[37] van den Bent, P.M.L.A.; Verzijl, J.M. and van Roon E.N. (1997) Prions: contamination, and decontamination, interfaces with pharmacy. Pharm. Weekblad 132: 143-149.

[38] Anonymous (1999) Note for the guidance for minimising the risk of transmitting animal spongiform encephalopathy agents *via* veterinary medicinal products. EMEA, 7 Westferry Circus, Canary Wharf, London E14 4HB, UK.
[39] Anonymous (2000) European Pharmacopoeia Supplement. EDQM, 226 Avenue de Colmar, BP 907, F-67029, Strassbourg, France.
[40] Anonymous (2001) Rote Liste 2001. Aulendorf, Germany Cantor Verlag.
[41] Yoshimura, K.; Terashima, M.; Hozan, D., Ebato, T.; Nomura, Y.; Ishii, Y. and Shirai, K. (2000) Physical properties of shark gelatin compared with pig gelatin. J. Agric. Food Chem 48: 2023-2027.
[42] Nomura, Y.; Toki, S.; Ishii, Y. and Shirai, K. (2000) The physicochemical property of shark type I collagen gel and membrane. J. Agric. Food Chem 48: 2028-2032.
[43] Werten, M.W.T.; Wisselink, W.H.; Jansen-van den Bosch, T.J.; de Bruin, E.C. and de Wolf, F.A. (2001) Secreted production of a custum-designed, highly hydrophilic gelatin in *Pichia pastoris*. Prot. Engineer. 14: 447-454.
[44] Werten, M.W.T.; van den Bosch, T.J.; Wind, R.J.; Mooibroek, H. and de Wolf, F.A. (1999) High-yield secretion of recombinant gelatins by *Pichia pastoris*. Yeast 15: 1087-1096.
[45] de Wolf, F.A. and Werten, M.W.T. (2001) Gelatins from yeast provide novel possibilities. Industrial Proteins 8: 9-12.
[46] de Bruin, E.C.; Werten, M.W.T.; Laane, C. and de Wolf, F.A. (2002) Endogenous prolyl 4-hydroxylation in Hansenula polymorpha and its use for the production of hydroxylated recombinant gelatin. FEMS Yeast Research, in press.
[47] Dalgleish, D.G. (1989) Milk proteins - chemistry and physics. In: Kinsella, J.E. and Soucie, W.G. (Eds.) Food Proteins. American Oil Chemists Society Champagne, IL; pp. 155-178.
[48] Visser H. and Paulsson M. (2001) Beta-lactoglobulin: a whey protein with unique properties. Industrial Proteins 9: 9-12.
[49] Kinsella, J.E. (1984) Milk proteins: physicochemical and functional properties. CRC Crit. Rev. Food Sci 21: 197-262.
[50] Brunner, J.R. (1977) Milk proteins. In: Whitaker, J.R. and Tannenbaum, S.R. (Eds.) Food Proteins. AVI Publishers, Inc. Westport, CT; pp. 175-208.
[51] Corrigan, O.I. and Heelan, B.A. (2001) Characterization of drug release from diltiazem-loaded polylactide microspheres using sodium caseinate and whey proteins as emulsifying agents. J. Microencaps. 18: 335-345.
[52] Yufera, M.; Pascual, E. and Fernandez-Diaz, C. (1999) A highly efficient microencapsulated food for rearing early larvae of marine fish. Aquaculture 177: 249-256.
[53] Alarcon, F.J.; Moyano, F.J.; Diaz, M.; Fernandez-Diaz, C. and Yufera, M. (1999) Optimization of the protein fraction of microcapsules used in feeding of marine fish larvae using *in vitro* digestibility techniques. Aquac. nutr. 5: 107-113.
[54] Hogan, S.A.; McNamee, B.F.; O'Riordan, E.D. and O'Sullivan, M. (2001) Microencapsulating properties of sodium caseinate. J. Agric. Food Chem. 49: 1934-1938.
[55] Latha, M.S.; Lal, A.V.; Kumary, T.V.; Sreekumar, R. and Jayakrishnan, A. (2000) Progesterone release from glutaraldehyde cross-linked casein microspheres: *In vitro* studies and *in vivo* response in rabbits. Contraception 61: 329-334.
[56] Keogh, M.K.; O'Kennedy, B.T.; Kelly, J.; Auty, M.A.; Kelly, P.M.; Frby, A. and Haahr, A.M. (2001) Stability to oxidation of spray dried fish oil powder microencapsulated using milk ingredients. J. Food Sci. 66: 217-224.
[57] Yu, J.Y. and Lee, W.C. (1997) Microencapsulation of pyrrolnitrin from Pseudomonas cepacia using gluten and casein. J. Ferment. Bioeng. 84: 444-448.
[58] McHugh, T.H. and Krochta, J.M. (1994) Permeability properties of edible films (chapter 7). In: Krochta, J.M.; Baldwin, E.A. and Nisperos-Carriedo, M. (Eds.) Edible coatings and Films to Improve Food Quality. Technomic Publishing Company Inc. Lancaster; pp. 139-187.
[59] De Wit, J.N. and Moulin, J. (2001) Whey protein isolates: manufacture, properties and applications. Industrial Proteins 9: 6-8.
[60] De Wit J.N. (2001) Whey protein concentrates: manufacture, composition and applications. Industrial Proteins 9: 3-5.
[61] Moreau, D.L. and Rosenberg, M. (1993) Microstructure and fat extractability in microcapsules based on whey proteins or mixtures of whey proteins and lactose. Food Structure 12: 457-468.
[62] Moreau, D.L. and Rosenberg, M. (1996) Oxidative stability of anhydrous milk fat microencapsulated in whey proteins. J. Food Sci. 61(1): 39-43.

[63] Sheu, T-Y. and Rosenberg, M. (1995) Microencapsulation by spray drying ethyl caprylate in whey protein and carbohydrate wall systems. J. Food Sci. 60: 98-103.
[64] Rosenberg, M. and Young, S.L. (1993) Whey proteins as microencapsulating agents. Microencapsulation of anhydrous milk fat - structure evaluation. Food Structure 12: 31-41.
[65] Lee, S.J. and Rosenberg, M. (2001) Microencapsulation of theophylline in composite wall system consisting of whey proteins and lipids. J. Microencaps. 18: 309-321.
[66] Lee, S.J. and Rosenberg, M. (2000) Preparation and some properties of water-insoluble, whey protein-based microcapsules. J. Microencaps. 17: 29-44.
[67] Schrooyen, P. (1999) Feather keratins: modification and film formation. Ph.D. Thesis, University of Twente, Enschede, The Netherlands.
[68] Yanauchi, K.; Maniwa,.M. and Mori, T. (1998) Cultivation of fibroblast cells on keratin-coated substrate. J. Biomater. Sci. 9: 259-270.
[69] Friedli, G.L. (1996) Interaction of deamidated soluble wheat protein (SWP) with other food proteins and metals. Ph.D. thesis, University of Surrey, UK.
[70] Damodaran, S. and Kinsella, J.E. (1980) Flavor protein interactions. Binding of carbonyls to bovine serum albumin: thermodynamic and conformational effects. J. Agric. Food Chem. 28: 567-571.
[71] Egbaria, K. and Friedman, M. (1992) Physicochemical properties of albumin microspheres determined by spectroscopic studies. J. Pharm. Sci. 81(2): 186-190.
[72] Luftensteiner, Ch.P.; Horaczek, A.; Maly, P. and Viernstein, H. (1999) Preparation and characterization of hydrophilic and non-aggregating albumin microspheres for intravenous administration. Pharm. Pharmacol. Lett. 9: 44-47.
[73] Larionova, N.V.; Kazanskaya, N.F.; Larionova, N.I.; Ponchel, G. and Duchene, D. (1999) Preparation and characterization of microencapsulated proteinase inhibitor aprotinin. Biochem. 64: 857-862.
[74] Tulsani, N.B.; Kumar, A.; Pasha, Q.; Kumar, H. and Sarma, U.P. (2000) Immobilization of hormones for drug targeting. Art. Cells Blood Subst. Immob. Biotechn. 28: 503-519.
[75] Ishizaka T.; Endo K. and Koishi M. (1981) Preparation of egg albumin microcapsules and microspheres. J. Pharm. Sci. 70(4): 358-363.
[76] Ishizaka, T. and Koishi, M. (1983) *In vitro* drug release from egg albumin microcapsules. J. Pharm. Sci. 72(9): 1057-1061.
[77] Tomlinson, E.; Burger, J.J.; Schoonderwoerd, E.M.A. and McVie, J.G. (1984) Human serum albumin microspheres for intraarterial drug targeting of cytostatic compounds: Pharmaceutical aspects and release characteristics. In: Davis, S.S.; Illum, L.; McVie, J.G. and Tomlinson, E. (Eds.) Microspheres and Drug Therapy: Pharmaceutical, Immunological and Medical Aspects. Elsevier Science Publishers, Amsterdam; pp. 75-89.
[78] Egbaria, K. and Friedman, M. (1992) Adsorption of fluorescein dyes on albumin microspheres. Pharm. Res. 9(5): 629-635.
[79] Sontum, P.C.; Walday, P.; Drystad, K.; Hoff, L.; Frigstad, S. and Chistiansen, C. (1997) Effect of microsphere size distribution on the ultrasonographic contrast efficacy of air-filled albumin microspheres in the left ventricle of dog heart. Invest. Radiol. 32: 627-635.
[80] Killam, A.L.; Mehlhaff, P.M.; Zavorskas, P.A.; Greener, Y.; McFerran, B.A.; Miller, J.J.; Burrascano, C.; Jablonski, E.G.; Anderson, L. and Dittrich, H.C. (1999) Tissue distribution of 125I-labeled albumin in rats, and whole blood and exhaled elimination kinetics of octafluropropane in anesthetized canines, following intravenous administration of OPTISON(R) (FS069). Int. J. Toxicol. 18: 49-63.
[81] Greener, Y.; Killam, A.L.; Cornell, S.T.; Osheroff, M.R. and Wolford, S.T. (1998) Nonclinical safety assessment of intravenous Optison: A perfluoropropane (PFP)-filled albumin microspheres contrast agent for ultrasonography. Int. J. Toxicol. 17: 631-662.
[82] Clark, L.N. and Dittrich, H.C. (2000) Cardiac imaging using Optison. Am. J. Cardiol. 86: 14G-18G.
[83] Truter, E.J.; Santos, A.S. and Els, W.J. (2001) Assessment of the antitumor activity of targeted immunospecific albumin microspheres loaded with cisplatin and 5-fluorouracil: Toxicity against a rodent ovarian carcinoma *in vitro*. Cell Biol. Int. 25: 51-59.
[84] Luftensteiner, C.P.; Schwendenwein, I.; Paul, B.; Eichler, H.G. and Viernstein, H. (1999) Evaluation of mitoxantrone-loaded albumin microspheres following intraperitoneal administration to rats. J. Control. Release 57: 35-44.
[85] Luftensteiner, C.P. and Viernstein, H. (1998) Statistical experimental design based studies on placebo and mitoxantrone-loaded albumin microspheres. Int. J. Pharm. 171: 87-99.
[86] Chen, Y.; McCulloch, R.K. and Gray, B.N. (1994) Synthesis of albumin-dextran sulfate microspheres possessing favourable loading and release characteristics for the anticancer drug doxorubicin. J. Control. Release 31: 49-54.

[87] Merodio, M.; Arnedo, A.; Renedo, M.J. and Irache, J.M. (2001) Ganciclovir-loaded albumin nanoparticles: Characterization and *in vitro* release properties. Eur. J. Pharm. Sci. 12: 251-259.
[88] Ozkan, Y.; Dikmen, N.; Isimer, A.; Gunham, O. and Aboul, E.H.Y. (2000) Clarithromycin targeting to lung: Characterization, size distribution and *in vivo* evaluation of the human serum albumin microspheres. Farmaco-Lausanne 55: 303-307.
[89] Pande, S.; Vyas, S.P. and Dixit, V.K. (1991) Localized rifampicin albumin microspheres. J Microencaps. 8: 87-93.
[90] Bernardo, M.V.; Blanco, M.D.; Gomez, C.; Olmo, R. and Teijon, J.M. (2000) *In vitro* controlled release of bupivacaine from albumin microspheres and a co-matrix formed by microspheres in a poly(lactide-co-glycolide) film. J. Microencaps. 17: 721-731.
[91] Lin, W.; Garnett, M.C.; Schacht, E.; Davis, S.S. and Illum, L. (1999) Preparation and *in vitro* characterization of HSA-mPEG nanoparticles. Int. J. Pharm. 189: 161-170.
[92] Lin, W.; Garnett, M.C.; Davis, S.S.; Schacht, E.; Ferruti, P. and Illum, L. (2001) Preparation and characterisation of rose Bengal-loaded surface-modified albumin nanoparticles. J. Control. Release 71: 117-226.
[93] Roser, M.; Fischer, D. and Kissel, T. (1998) Surface-modified biodegradable albumin nano- and microspheres. II: Effect of surface charges on *in vitro* phagocytosis and biodistribution in rats. Eur. J. Pharm. Biopharm. 46: 255-263.
[94] Weber, C.; Reiss, S. and Langer, K. (2000) Preparation of surface modified protein nanoparticles by introduction of sulfhydryl groups. Int. J. Pharm. 211: 67-78.
[95] Weber, C.; Kreuter, J. and Langer, K. (2000) Desolvation process and surface characteristics of HSA-nanoparticles. Int. J. Pharm. 196: 197-200.
[96] Weber, C.; Coester, C.; Kreuter, J. and Langer, K. (2000) Desolvation process and surface characterisation of protein nanoparticles. Int. J. Pharm. 194: 91-102.
[97] Krull, L.H. and Inglett, G.E. (1971) Industrial Uses of Gluten. Cereal Sci. Today 16: 232-236,261.
[98] Ezpeleta, I.; Irache, J.M.; Stainmesse, S.; Chabenat, C.; Guéguen, J.; Popineau, Y. and Orecchioni, A.M. (1996) Gliadin nanoparticles for the controlled release of all-trans-retinoic acid. Int. J. Pharm. 131: 191-200.
[99] Wu, C.H.; Shuryo, N. and Powrie, W.D. (1976) Preparation and properties of acid solubilized gluten. J. Agric. Food Chem. 24: 504-510.
[100] Kato, A.; Tanaka, A.; Lee, Y.; Matsudomi, N. and Kobayashi, K. (1987) Effects of deamidation with chymotrypsin at pH 10 on the functional properties of proteins. J. Agric. Food Chem. 35: 285-288.
[101] Bollecker, S.; Viroben, G.; Popineau, Y. and Guéguen, J. (1990) Acid deamidation and enzymic modification at pH 10 of wheat gliadins: influence on their functional properties. Sci. Aliments 10: 343-356.
[102] Popineau, Y. and Thebaudin, J.Y. (1990) Functional properties of enzymatically hydrolyzed glutens. In: Bushuk, W. and Tkachuk, R. (Eds.) Gluten Proteins. AACC, St Paul, MN; pp. 277-286.
[103] Shukla, R. and Cheryan, M. (2001) Zein: the industrial protein from corn. Ind. Crop Prod. 13: 171-192.
[104] Mazer, T. (1999) Zein, the versatile reverse enteric. In: Proceedings of the 26th Int. Symp. Control. Rel. Bioact. Mat., Boston, USA, CRS Inc.; pp. 267-268.
[105] Mathiowitz, E.; Bernstein, H.; Morrel, E. and Schwaller, K. (1993) Method for producing protein microspheres. US Patent 5,271,961, Alkermes Controlled Therapeutics, Inc. USA.
[106] Yoshimaru, T.; Takahashi, H. and Matsumoto, K. (2000) Microencapsulation of L-lysine for improving the balance of amino acids in ruminants. J. Faculty Agric., Kyushu Univ. 44: 359-365.
[107] Zibell, S.E. (1989) Chewing gum containing zein coated high-potency sweetener and method. US Patent 4,863,745, Wm. Wrigley Jr., Chicago, IL, USA.
[108] Zibell, S.E.; Yatka, R.J. and Tyrpin, H.T. (1992) Aqueous zein coated sweeteners and other ingredients for chewing gum. US Patent 5,112,625, Wm. Wrigley Jr., Chicago, IL,USA.
[109] Courtright, S.B. and Barrett, K.F. (1990) Chewing gum containing high-potency sweetener particles with modified zein coating. US Patent 4,931,295. Wm. Wrigley Jr., Chicago, IL, USA.
[110] Campbell, A.A. and Zibell, S.E. (1992) Zein/shellac encapsulation of high intensity sweeteners in chewing gum. US Patent 5,164,210, Wm. Wrigley Jr. Company, USA.
[111] Demchak, R.J. and Dybas, R.A. (1997) Photostability of abamectin/zein microspheres. J. Agric. Food Chem. 45: 260-262.
[112] Mazer, T.B.; Meyer, G.A.; Hwang, S.-M.; Candler, E.L.; Drayer, L.R. and Daab-Krzykowski, A. (1992) System for delivering an active substance for sustained release. US Patent 5,160,742, Abbott Laboratories, USA.
[113] Ardaillon, P. and Bourrain, P. (1991) Granules for feeding ruminants with an enzymatically degradable coating. US Patent 4,983,403, Rhone-Poulenc Sante, France.

[114] Autant, P.; Cartillier, A. and Pigeon, R. (1989) Compositions for coating feeding stuff additives intended for ruminants and feeding stuff additives thus coated. US Patent 4,876,097, Rhone-Poulenc Sante, France.
[115] Kinsella, J.E. (1979) Functional properties of soy proteins. J. Am. Oil Chem. Soc. 56: 242-258.
[116] McHugh, T.H. (1993) Effects of chemical properties and physical structure on mass transfer in whey protein-based edible films system. Ph.D. Thesis, University of California, Davis, USA.
[117] Rhim, J.W.; Gennadios, A.; Weller, C.L. and Hanna, M.A. (2002) Sodium dodecyl sulfate treatment improves properties of cast films from soy protein isolate. Ind. Crops Prod. 15: 199-205.
[118] Atterholt, C.A.; Delwiche, M.J.; Rice, R.E. and Krochta, J.M. (1998) Study of biopolymers and paraffin as potential controlled-release carriers for insect pheromones. J. Agric. Food Chem. 46: 4429-4434.
[119] Wilson, W.W.; Polemenakos, S.C.; Potter, J.L.; Mangold, D.J.; Harlowe, W.W. and Schlameus, H.W. (1989) Microencapsulated and bait. US Patent 4,874,611, The Dow Chemical Company, USA.
[120] Vaz, C.M.; de Graaf, L.A.; Reis, R.L. and Cunha, A.M. (2002) Soy protein-based systems for different tissue regeneration applications. In: Reis, R.L. and Cohn, D. (Eds.) Polymer based systems on tissue engineering, replacement and regenration. Kluwer Academic Publ., Dordrecht, The Netherlands; (in press)
[121] Guéguen, J. and Cerletti, P. (1994) Legume seed proteins. In: Hudson, B.J.F. (Ed.) New and developing sources of food proteins. Chapman and Hall, London, UK; p. 145.
[122] Guéguen, J. (2000) Pea proteins: new and promising protein ingredients. Industrial Proteins 8: 6-8.
[123] Franco, J.M.; Partal, P.; Ruiz-Marquez, D.; Conde, B. and Gallegos, C. (2000) Influence of pH and protein thermal treatment on the rheology of pea protein-stabilized oil-in-water emulsions. J. Am. Oils Chem. Soc. 77: 975-983.
[124] Lu, B.Y.; Quillien, L. and Popineau, Y. (2000) Foaming and emulsifying properties of pea albumin fractions and partial characterisation of surface-active components. J. Sci. Food Agric. 80: 1964-1972.
[125] Sijtsma, L.; Tezera, D.; Hustinx, J. and Vereijken, J.M. (1998) Improvement of pea protein quality by enzymatic modification. Nahrung/Food 42: 215-216.
[126] Legrand, J.; Guéguen, J.; Berot, S.; Popineau, Y. and Nouri, L. (1997) Acetylation of pea isolate in a torus microreactor. Biotechnol. Bioengin. 53: 409-414.
[127] Harmsen, P.F.H.; Vingerhoeds, M.H.; Berendsen, L.B.J.M.; Harrison, R.M. and Vereijken, J.M. (2001) Microencapsulation of β-carotene by supercritical CO_2 technology. IFSCC Magazine 4: 34-36.
[128] Irache, J.M.; Bergounoux, L.; Ezpeleta, I.; Guéguen, J. and Orecchioni, A.M. (1995) Optimization and *in vitro* stability of legumin nanoparticles obtained by a coacervation method. Int. J. Pharm. 126: 103-109.
[129] Ezpeleta, I.; Irache, J.M.; Stainmesse, S.; Chabenat, C.; Guéguen, J. and Orecchioni, A.-M. (1996) Preparation of lectin-vicilin nanoparticle conjugates using the carbodiimide coupling technique. Int. J. Pharm. 142: 227-233.
[130] Mirshahi, T.; Irache, J.M.; Guéguen, J. and Orecchioni, A.M. (1996) Development of drug delivery systems from vegetal proteins: legumin nanoparticles. Drug Dev. Ind. Pharm. 22: 841-846.
[131] Ezpeleta, I.; Irache, J.M.; Guéguen, J. and Orecchioni, A.M. (1997) Properties of glutaraldehyde cross-linked vicilin nano- and microparticles. J. Microencaps. 14(5): 557-565.
[132] Benoit, J-P.; Marchais, H.; Rolland, H. and van de Velde, V. (1996) Biodegradable microspheres: advances in production technology. In: Benita, S. (Ed.) Microencapsulation: methods and industrial applications. Marcel Dekker Inc., New York; pp. 36-72.
[133] Re, M.I. (1998) Microencapsulation by spray drying. Drying Techn. 16: 1195-1236.
[134] Giunchedi, P. and Conte, U. (1995) Spray-drying as preparation method of microparticulate drug delivery systems: an overview. S.T.P. Pharma Sciences 5: 276-290.
[135] Keogh, M.K. and O'Kennedy, B.T. (1999) Milk fat encapsulation using whey proteins. Int. Diary J. 9: 657-663.
[136] Moreau, D.L. and Rosenberg, M. (1998) Porosity of whey protein-based microcapsules containing anhydrous milkfat measured by gas displacement pycnometry. J. Food Sci. 63: 819-823.
[137] Rosenberg, M. (1997) Milk derived whey protein-based microencapsulating agents and a method of use. US Patent 5,601,760, University of California, USA.
[138] Young, S.L.; Sarda, X. and Rosenberg, M. (1993) Microencapsulating properties of whey proteins. 1. Microencapsulation of anhydrous milk fat. J. Dairy Sci. 76: 2868-2877.
[139] Moreau, D.L. and Rosenberg, M. (1999) Porosity of microcapsules with wall systems consisting of whey proteins and lactose measured by gas displacement pycnometry. J. Food Sci. 64: 405-409.
[140] Sheu, T.-Y. and Rosenberg, M. (1998) Microstructure of microcapsules consisting of whey proteins and carbohydrates. J. Food Sci. 63: 491-494.
[141] Young, S.L.; Sarda, X. and Rosenberg, M. (1993) Microencapsulating properties of whey proteins. 2. Combination of whey proteins with carbohydrates. J. Dairy Sci. 76: 2878-2885.

[142] O'Brien, C.M.; Grau, H.; Neville, D.P.; Keogh, M.K. and Arendt, E.K. (2000) Effects of microencapsulated high-fat powders on the emperical and fundamental rheology properties of wheat flour doughs. Cereal Chem. 77: 111-114.

[143] Kim, Y.D. and Morr, C.V. (1996) Microencapsulation properties of gum arabic and several food proteins: spray-dried orange oil emulsion particles. J. Agric. Food Chem. 44: 1314-1320.

[144] Kim, Y.D.; Morr, C.V. and Schenz, T.W. (1996) Microencapsulation properties of gum arabic and several food proteins: liquid orange oil emulsion particles. J. Agric. Food Chem. 44: 1308-1313.

[145] Faldt, P. and Bergenstahl, B. (1996) Spray-dried whey-protein/lactose/soybean oil emulsions 1. Surface composition and particle structure. Food hydrocolloids 10: 421-429.

[146] Faldt, P. and Bergenstahl, B. (1996) Spray-dried whey-protein/lactose/soybean oil emulsions 2. Redispersability, wettability and particle structure. Food hydrocolloids 10: 431-439.

[147] Pavanetto, F.; Genta, I.; Guinchedi, P.; Conti, B. and Conte, U. (1994) Spray-dried albumin microspheres for the intra-articular delivery of dexamethasone. J. Microencaps. 11: 445-454.

[148] Jones, D.M. (1988) Air suspension coating. Pharm. Technol. Enc. January 1988: 2-27.

[149] Hall, H.S. and Pondell, R.E. (1980) The Wurster process. In: Kydonieus, A.F. (Ed.) Controlled release technologies: Methods, theory and applications. CRC Press, New York; pp. 133-155.

[150] Dewettinck, K. and Huyghebaert, A. (1998) Top-spray fluidized bed coating: effect of process variables on coating efficiency. Food Sci. Techn. 31: 568-575.

[151] Scheffer, R.J.; Harmsen, P.F.H. and Drift, van der E. (2000) Release of agrochemicals in relation to seed treatments. In: Proceedings of the 27th Int. Symp. Control. Rel. Bioact. Mater., Paris (France); p. 551.

[152] Yuryev, V.P.; Zasypkin, D.V. and Tolstoguzov, V.B. (1990) Structure of protein texturates obtained by thermoplastic extrusion. Die Nahrung 34: 607-613.

[153] Tolstoguzov, V.B. (1993) Thermoplastic extrusion - the mechanism of the formation of extrudate structure and properties. JAOCS 70: 417-424.

[154] Swisher, H.E. (1957) Solid essential-oil containing components. US Patent 2,809,895.

[155] Yilmaz, G.; Jongboom, R.O.J.; van Soest, J.J.G. and Feil, H. (1999) Effect of glycerol on the morphology of starch-sunflower oil composites. Carboh. Polym. 38: 33-39.

[156] Yilmaz, G.; Jongboom, R.O.J.; Feil, H. and Hennink, W.E. (2001) Encapsulation of sunflower oil in starch matrices *via* extrusion: effect of the interfacial properties and processing conditions on the formation of dispersed phase morphologies. Carboh. Polym. 45: 403-410.

[157] Camire, M.E. (1991) Protein functionality modification by extrusion cooking. JAOCS 68: 200-205.

[158] Ledward, D.A. and Tester, R.F. (1994) Molecular transformations of proteinaceous foods during extrusion processing. Trends Food Sci. Techn. 5: 117-120.

[159] Prudencio-Ferreira, S.H. and Areas, J.A.G. (1993) Protein-protein interactions in the extrusion of soya at various temperatures and moisture contents. J. Food Sci 58: 378-381.

[160] Sair, L. and Sair, R.A. (1980) Food supplement concentrate in a dense glasseous extrudate. US Patent 4,232,047, Griffith Laboratories, USA.

[161] Black, M.; Popplewell, L.M. and Porzio, M.A. (1998) Controlled release encapsulation compositions. US Patent 5,756,136, McCormick & Company, USA.

[162] Vaz, C.M.; van Doeveren, P.F.N.M.; Yilmaz, G.; de Graaf, L.A.; Reis, R.L. and Cunha, A.M. (submitted for publication) Processing and characterization of biodegradable soy thermoplastics: Effect of crosslinking with glyoxal and thermal treatment. ATO, Wageningen, The Netherlands.

[163] Arshady, R. (1990) Microspheres and microcapsules: a survey of manufacturing techniques. Part 2: Coacervation. Polym. Eng. Sci. 30: 905-914.

[164] Burgess, D.J. (1990) Practical analysis of complex coacervate systems. J. Colloid Interf. Sci. 140: 227-238.

[165] Tsung, M. and Burgess, D.J. (1997) Preparation and stabilization of heparin/gelatin complex coacervate microcapsules. J. Pharm. Sci. 86: 603-607.

[166] Michon, C.; Cuvelier, G.; Launay, B.; Parker, A. and Takerkart, G. (1995) Study of the compatibility/incompatibility of gelatin/iota-carrageenan/water mixtures. Carboh. Polym. 28: 333-336.

[167] Remuñán-Lopez, C. and Bodmeier, R. (1996) Effect of formulation and process variables on the formation of chitosan-gelatin coacervates. Int. J. Pharm. 135: 63-72.

[168] Vinetsky, Y. and Magdassi, S. (1997) Formation of surface properties of microcapsules based on gelatin-sodium dodecyl sulphate interactions. Colloids and surfaces A: Physicochem. Eng. Aspects 122: 227-235.

[169] Schmitt, C.; Sanchez, C.; Thomas, F. and Hardy, J. (1999) Complex coacervation between β-lactoglobulin and acacia gum in aqueous medium. Food Hydrocolloids 13: 483-496.

[170] Schmitt, C.; Sanchez, C.; Despond, S.; Renard, D.; Thomas, F. and Hardy, J. (2000) Effect of protein aggregates on the complex coacervation between β-lactoglobulin and acacia gum at pH 4.2. Food Hydrocolloids 14: 403-413.
[171] Soper, J.C. and Thomas, M.T. (2001) Enzymatically protein encapsulating oil particles by complex coacervation. US Patent 6,325,951, Givaudan Roure Flavors Corporation, USA.
[172] Soper, J.C. and Thomas, M.T. (2000) Enzymatically protein encapsulating oil particles by complex coacervation. US Patent 6,039,901, Givaudan Roure Flavors Corporation, USA.
[173] Rabiskova, M.; Song, J.; Opawale, F.O. and Burgess, D.J. (1994) The influence of surface properties on uptake of oil into complex coacervate microcapsules. J. Pharm. Pharmacol. 46: 631-635.
[174] Rabiskova, M. and Valaskova, J. (1998) The influence of HLB on the encapsulation of oils by complex coacervation. J. Microencaps. 15: 747-751.
[175] Övez, B.; Çitak, B.; Öztemel, D.; Balbas, A.; Peker, S. and Çakir, S. (1997) Variation of droplet sizes during the formation of microcapsules from emulsions. J. Microencaps. 14: 489-499.
[176] Cabeza, L.F.; Taylor, M.M.; Brown, E.M. and Marmer, W.N. (1999) Potential applications for gelatin isolated from chromium-containing solid tannery waste: microencapsulation. JALCA 94: 182-189.
[177] Ribeiro, A.; Arnaud, P.; Frazao, S.; Venâncio, F. and Chaemeil, J.C. (1997) Development of vegetable extracts by microencapsulation. J. Microencaps. 14: 735-742.
[178] Amiet-Charpentier, C.; Benoit, J.P.; Gadille, P. and Richard, J. (1998) Preparation of rhizobacteria-containing polymer microparticles using a complex coacervation method. Colloids Surfaces A: Physicochem. Eng. Aspects 144: 179-190.
[179] Burgess, D.J. and Singh, O.N. (1993) Spontaneous formation of small sized albumin/acacia coacervate particles. J. Pharm. Pharmacol. 43: 586-591.
[180] Burgess, D.J. and Ponsart, S. (1998) B-Glucuronidase activity following complex coacervation and spray drying microencapsulation. J. Microencaps. 15: 569-579.
[181] Yoshioka, T.; Hashida, M.; Murinashi, S. and Sezaki, H. (1981) Specific delivery of mitomycin C to the liver, spleen and lung: nano- and microspherical carriers of gelatin. Int. J. Pharm. 81: 131-141.
[182] Chemtob, C.; Assimacopoulos, T. and Chaumeil, J.C. (1988) Preparation and characteristics of gelatin microspheres. Drug Dev. Ind. Pharm. 14: 1359-1374.
[183] Esposito, E.; Cortesi, R. and Nastruzzi, C. (1996) Gelatin Microspheres: influence of preparation parameters and thermal treatment on chemico-physical and biopharmaceutical properties. Biomaterials 17: 2009-2020.
[184] Leucuta, S.E.; Ponchell, G. and Duchene, D. (1997) Dynamic swelling behaviour of gelatin/poly(acrylic acid) bioadhesive microspheres loaded with oxprenolol. J. Microencapsulation 14: 201-510.
[185] Leucuta, S.E.; Ponchell, G. and Duchene, D. (1997) Oxprenolol relase from bioadhesive gelatin/poly(acrylic acid) microspheres. J. Microencapsulation 14: 511-522.
[186] Cortesi, R.; Nastruzzi, C. and Davis, S.S. (1998) Sugar cross-linked gelatin for controlled release microspheres and disks. Biomaterials 1998: 1641-1649.
[187] Shu, X.Z. and Zhu, K.J. (2001) Chitosan/gelatin microspheres prepared by modified emulsification and ionotropic gelation. J. Microencapsulation 18: 237-245.
[188] Cremers, H.F.M. (1993) Biodegradable ion-exchange microspheres for site specific delivery of adriamycin, Ph.D. Thesis, Department of Chemical Technology, University of Twente, Enschede, The Netherlands.
[189] Jong, A. de (1994) Albumin microspheres: Preparation, characteristics and applications. Graduation thesis, University of Utrecht, The Netherlands.
[190] Gallo, J.M.; Hung, C.T. and Perrier, D.G. (1984) Analysis of albumin microsphere preparation. Int. J. Pharm. 22: 63-74.
[191] Pasqualine, R.; Plassio, G. and Sosi, S. (1969) The preparation of albumin microspheres. J. Biol. Nucl. Med. 13: 80-84.
[192] Scheffel, U.; Buck, A.; Rhodes, T.K. and Wagner, H.N. (1972) Albumin microspheres for study of the reticuloendothelial system. J. Nucl. Med. 13: 498-503.
[193] Rubino, O.P.; Kowalsky, R. and Swarbrick, J. (1993) Albumin microsphers as a drug delivery system: relation among turbidity ratio, degree of cross-linking and drug release. Pharm. Res. 10: 1059-1065.
[194] Burger, J.J.; Tomlinson, E.; Mulder, E.M.A. and McVie, J.G. (1985) Albumin microspheres for intra-arterial tumour targeting. I. Pharmaceutical aspects. Int. J. Pharm. 23: 333-344.
[195] Torrado, J.J.; Illum, L.; Cadorniga, R. and Davis, S.S. (1990) Egg albumin microspheres containing paracetamol for oral administration. II. *In vivo* investigation. J. Microencapsulation 7(4): 471-477.
[196] Torrado, J.J.; Illum, L.; Cadorniga, R. and Davis, S.S. (1990) Egg albumin microspheres containing paracetamol for oral administration.I. *In vitro* characterization. J. Microencapsulation 7(4): 463-470.

[197] Orienti, I.; Coppola, A.; Gianasi, E. and Zecchi, V. (1994) Influence of physico-chemical parameters on the release of hydrocortisone acetate from albumin microspheres. J. Contr. Rel. 31: 61-71.
[198] Kramer, P.A. (1974) Albumin microspheres as vehicles for achieving specificity in drug delivery. J. Pharm. Sci. 63: 1646-1647.
[199] Heelan, B.A. and Corrigan, O.I. (1998) Preparation and evaluation of microspheres prepard from whey protein isolate. J. Microencaps. 15: 93-105.
[200] Lee, S.J. and Rosenberg, M. (1999) Preparation and properties of glutaraldehyde crosslinked whey protein-based microcapsules containing theophylline. J. Control. Rel. 61: 123-136.
[201] Lee, S.J. and Rosenberg, M. (2000) Whey protein microcapsules prepared by double emulsification and heat gelation. Lebensm. Wiss. Technol 23: 80-88.
[202] Lee, S.J. and Rosenberg, M. (2000) Microencapsulation of theophylline in whey proteins: Effects of core-to-wall ratio. Int. J. Pharm. 205: 147-158.
[203] Spillane, S.M.; Vingerhoeds, M.H.; van der Bent, A. and van Amerongen, A. (2000) Antimicrobial effect of nisin-containing microspheres. In: Proceedings of the 27th Int. Symp. Control. Rel. Bioact. Mat., CRS Inc.; pp. 1373-1374.
[204] Vingerhoeds, M.H. and Harmsen, P.F.H. (2000) Proteins: versatile materials for microencapsulation. In: Proceedings of the COST 840 meeting (Bioencapsulation innovation and technologies) "Structure-function properties of biopolymers in relation to bioencapsulation", Helsinki (Finland); pp. 1-4.
[205] Langelaan, H.C.; Litjens, M.J.J.; Poortier, E.; Harmsen, P.F.H. and Vingerhoeds, M.H. (2001) Microencapsulation for food applications using supercritical fluid technology. In: Proceedings of the 13th Int. Symp. Microencapsulation, Angers (France); C-VI.
[206] de Wolf, F.A. and Brett, G.M. (2000) Ligand-binding proteins: their potential for application in systems for controlled delivery and uptake of ligands. Pharmaco. Rev. 52: 207-236.
[207] Muresan, S.; Vingerhoeds, M.H.; van der Bent, A. and de Wolf, F.A. (1998) Fluorescence and equilibrium analysis studies on the binding of small ligands to beta-lactoglobulin. In: Proceedings of the 25th Int. Symp. Control. Rel. Bioact. Mat., Las Vegas, USA, CRS Inc.; pp. 332-333.
[208] Brownlow, S.; Morais Cabral, J.H.; Cooper, R.; Flower, D.R.; Yewdall, S.J.; Polikarpov, I.; North, A.C. and Sawyer, L. (1997) Bovine beta-lactoglobulin at 1.8 A resolution--still an enigmatic lipocalin. Structure 5: 481-495.
[209] Oliveira, K.M.; Valente-Mesquita, V.L.; Botelho, M.M.; Sawyer, L.; Ferreira, S.T. and Polikarpov, I. (2001) Crystal structures of bovine beta-lactoglobulin in the orthorhombic space group C222(1). Structural differences between genetic variants A and B and features of the Tanford transition. Eur. J. Biochem. 268: 477-483.

POLYELECTROLYTE COMPLEXES FOR MICROCAPSULE FORMATION

IGOR LACÍK
Department of Special Polymers and Biopolymers, Polymer Institute of the Slovak Academy of Sciences, Dúbravská cesta 9, 842 36 Bratislava, Slovak Republic – Fax: +421 2 5477 5923 – Email: upollaci@savba.sk

1. Introduction

Polyelectrolyte complexes are formed by reaction of oppositely charged polymers containing covalently bound either anionic (polyanions) or cationic (polycations) groups. They represent an attractive class of polymer-based materials finding an irreplaceable role in many areas of the everyday life used for preparation of membranes, (micro)capsules and various types of controlled release devices [1,2]. Polyelectrolyte complexes have been considered as a challenging scientific field overlapping the areas of (i) the general macromolecular chemistry, (ii) the specific features of polyelectrolytes, (iii) the aspects of hydrogels and physically crosslinked networks and, (iv) the membranes.

The necessity to consider each of these areas of polymer chemistry becomes important in designing the ultimate properties of the polyelectrolyte complex stemming from the complexity of polyelectrolytes [1,3] and process of the complex formation [1,4-6]. Numerous factors affect the properties of the polyelectrolyte complex placing the stringent requirements on the selected polyelectrolytes (molecular weight, molecular weight distribution, number, type and distribution of the charged groups, chain flexibility, presence of hydrophobic groups) as well as on the preparation conditions (concentration, reaction time, temperature, ionic strength, pH, presence of other polyelectrolytes). From the point of view of polymer hydrogels, the polyelectrolyte complexes belong to the category of the physically crosslinked gels [7,8] with the crosslinks of small but finite energy and/or of finite lifetime [7]. Unlike in the covalently crosslinked gels, the number and position of crosslinks in the physical gels fluctuate with time and temperature and their nature, involving Coulombic, dipole-dipole, van der Waals, charge transfer, hydrophobic and hydrogen bonding interactions, is not known unambiguously. The crosslinks are usually not at the point of the polymer chain but they rather form more extended junction zones with the possibility for a subsequent lateral aggregation of chains after the initial contact.

The aim of this review is to highlight the recent achievements in the polyelectrolyte complex-based systems, where both the process conditions and the selection of polyelectrolytes are specifically carried out in a way to provide the well-defined

V. Nedović and R. Willaert (eds.), Fundamentals of Cell Immobilisation Biotechnology, 103-120.

spherical membrane in the form of a microcapsule of required properties serving the desired functions, *e.g.* immunoprotection, controlled release, specific environment for the encapsulated substance, *etc.* Predominantly the most recent original papers have been selected as primary information to a reader for the further studies of a given capsule type. A major attention is paid to the polyelectrolyte complex-based microcapsules for encapsulation of the islets of Langerhans because the microencapsulation is considered as the most promising approach for the encapsulation of living cells [9] and, hence, there is a strong development in this area. Consequently, since the capsule formation in the presence of living matter has to be carried out under the most rigorous criteria [9-12], any other microcapsule and its membrane formed under the less stringent conditions, *e.g.*, the controlled release materials used in biotechnology, pharmacology or agriculture, can be derived by following the principles and using the materials originally designed for the encapsulation of living cell. The final part of the review attempts to summarise and recommend the main principles to be followed in the process of capsules formation by polyelectrolyte complexation.

2. Sodium alginate based capsules

2.1. ALGINATE – POLY-L-LYSINE – ALGINATE CAPSULE

The literature contains exhausting information on sodium alginate as the basic material for the capsule formation in cell and enzyme immobilisation [13] and it certainly is not a purpose of this paper to repeat it. Nevertheless, the classical work of Lim and Sun [14] cannot be overlooked because the utilisation of the alginate - poly-L-Lysine – alginate (APA) capsule for the encapsulation of living cells initiated an intense interest in this area of science and application. This, at a glance, a rather simple procedure has been implemented in many laboratories virtually without the need for a special training in the field of polymer chemistry. The initial enthusiasm regarding this capsule leading to the first clinical trial [15] declined after realising that the APA membrane exhibits poor mechanical stability and biocompatibility. It was recognised that the capsule, membrane and gel quality and performance critically depend on the alginate microstructure (concentration and distribution of guluronic and mannuronic acid units, molecular weight, purity, supplier, batch number), molecular weight of poly-L-Lysine and the preparation conditions. This was the reason why the cell transplantation data reported by different groups for the APA capsule were sometimes irreproducible or contradictory. The variation in capsule properties as well as the complex stability between various types of alginate and poly-L-Lysine were studied in detail by Thu and co-workers providing the quantitative data on the interactions of alginate with poly-L-Lysine for the membrane formation as well as for the stability of alginate coat [16,17]. Based on these data, the group performing the first clinical trial [15] critically discussed the previous work on APA capsule [18-20]. A number of capsule modifications were focused on improving the capsule and the membrane mechanical stability, the absence of the positive charge from poly-L-Lysine on the capsular surface, and decreased diffusion of entrapped material out of the capsule. The recommendations for transplantation of the APA capsules to the higher animals were postulated involving (i) using the guluronic acid enriched alginate in both core and coat, (ii) avoiding the degelling step, (iii) supporting the anisotropy in distribution of alginate in the core, (iv) using the combination of calcium and barium cations, (v) and entrapment of the standard APA microcapsules in the alginate matrix forming the macrocapsule. The newcomers to the field of microcapsule formation in any application area, who often start to work with the APA microcapsule, should realise the complexity hidden behind the alginate gelling protocol [16-20] before selecting the starting materials to meet the goal of the encapsulation.

The small size of the APA capsule (< 350 μm) made with an electrostatic pulse system [21] has been reported as a feasible way to overcome the poor mechanical stability [22] of the standard size (1 mm) capsules. Quantitative evidence was provided revealing that increased reaction time and poly-L-Lysine concentration together with using a higher guluronic content alginate positively contributed to the mechanical strength and biocompatibility [23] as a consequence of mechanical stability, surface smoothness and the absence of physical imperfections. The safety of the intrahaepatic implantation of these small capsules [24] is determined by the microcapsule size and requires the strict control over the capsule swelling.

The recently published data on APA capsules concentrate on a detailed identification of the possible reasons for the transplant failure [9]. An improvement of the membrane mechanically stability was obtained by optimising the capsule formation conditions in terms of the poly-L-Lysine reaction step and avoiding the degelling step. The alginate purification appears to be crucial, however, even then a small fraction of capsules (up to 10 %) transplanted to the peritoneal cavity of rats exhibited the signs of fibrosis overgrowth. The assumption was made that the small fraction of capsules is not biocompatible because of their physical imperfections due to an inadequate encapsulation of individual islets, and thereby an inadequate immune protection as well as insufficient biocompatibility. The comparison between smaller and larger capsules of diameter 500 and 800 μm, respectively, revealed that the number of retrieved capsules was significantly smaller in case of the smaller capsules than in case of the larger ones. This contradictory observation in respect to the data discussed previously [23] are explained by a higher extent of the inadequately encapsulated islets in the smaller capsules due to swelling and shrinking of the bead and capsule during the capsule preparation steps.

Development and implementation of the new and sophisticated experimental techniques opened the possibility for a thorough investigation of the biocompatibility issues, which has recently been devoted to the APA capsule [25]. Concentration of poly-L-Lysine and time of reaction with alginate bead were shown to play the crucial role in initiating the immune reaction. Experimental evidence is provided suggesting a detailed mechanism of stimulation of the immune system by the presence of poly-L-Lysine leading to the fibrosis overgrowth of the capsular surface. Even when the capsules were prepared under the mild conditions regarding the presence of poly-L-Lysine and show only a minor fibrosis overgrowth, the reported amount of retrieved capsules from the peritoneal cavity was below 60 %. Poly-L-Lysine was identified as a strong inducer of proinflamatory cytokine TNF production with more pronounced effect in case of a higher molecular weight. The data demonstrate that typically used alginate coating layer is insufficient from shielding the poly-L-Lysine from the immune response. It was suggested that when alginate is the only capsule gel-forming material, the other ways of their stabilisation against swelling should be used including barium ions, inhomogeneous gel formation, more efficient coating layer than that of alginate and using alginates with controlled composition and sequential distribution of monomers for improved gelling properties. The outcome of this work is that the amount of poly-L-Lysine used for the capsule preparation should be minimised which corresponds with the requirements for chemical and mechanical stability of membrane to avoid leaching of poly-L-Lysine from the polyelectrolyte complex.

The versatility of the APA capsules was also tested for the encapsulation of the recombinant cells (epithelial cells, myoblasts and fibroblasts) [26]. Three types of capsules were used including calcium-gelled alginate with the solubilised core and alginate-poly-L-Lysine membrane, barium alginate beads and the barium alginate beads with poly-L-Lysine membrane. Some differences were observed but generally all cells proliferated in the capsules and their products in the molecular weight range from 20 to 300 kDa were released from each type of capsule at the similar rates. This paper was selected for this review also because it provides the information on collection time in the formation of alginate beads, which, unfortunately, is omitted in almost all the papers dealing with the bead and capsule preparation. According to the described encapsulation procedure, the alginate drops were collected for 90 seconds and immediately washed. It follows that the gelling time for the entire batch ranges from zero (the last drop) to 90 seconds (the first drop). Although the beads prepared within the batch may look macroscopically similar, there must be a large distribution among the beads in the gel properties in terms of the degree of crosslinking, thickness of the gelled zone and availability of the anionic sites for the successive reaction with polycation. This dilemma of setting collection and reaction times is discussed in part 5.

As stated by deVos and co-workers [9], the highly biocompatible and purified alginate is apparently not a sufficient requirement for the ideal performance of the APA membrane. A minor modification of the encapsulation procedure could have a crucial impact on the capsule biocompatibility and on the graft survival. Obviously, these differences contribute to the enormous variation in reported success rates of encapsulation allo- and xenografts. In concluding this part it should be emphasised that the APA has got a long history and thereby it has been tested in the most detailed and critical way. The issues highlighted here on biocompatibility, mechanical and chemical stability are obviously valid for other microcapsules with different chemistries. The similar approaches as recently carried out with the APA capsule [25] should be applied under a standard protocol to the other microcapsules types in order to have a direct comparison among the different encapsulation systems.

2.2. ALGINATE – POLY-L-ORNITHINE – ALGINATE CAPSULE

Calafiore and his co-workers uniquely use poly-L-ornithine instead of poly-L-Lysine with encouraging results. This polycation differs from poly-L-Lysine by a hydrophobic group (pendant group shorter by one CH_2 group) and in combination with a high M alginate (61 % mannuronic acid) provides a tighter permeability and apparently a more stable complex than in case of poly-L-Lysine in APA capsule. Alginate-poly-L-ornithine-alginate membrane has been used in formulation of coherent and medium size microcapsules [27-29]. Coherent microcapsules, prepared by the double emulsification technique [27], are thin, skin-like hydrogel films enveloping each individual islet with little dead space between islet and the artificial coating membrane. The *in vivo* transplantation of canine islets encapsulated in coherent microcapsules into portal vein of liver of adult pigs showed a thick layer of dense inflammatory cell tissue encasing the microcapsules [29]. Since the islets were not protruding through the membrane and the membrane was impermeable to IgG, it was assumed that the reason for the transplant failure was either in the membrane rupture or in the insufficient immunoprotection for xenotransplanted cells because of the very thin membrane. The next approach of this group was to test the capsules with a better mechanical resistance and a larger idle space by using the so called medium size microcapsules [29] of the size in the range from 300 to 400 μm prepared by the spray-gelling technique. The studies with the canine islets transplanted intraperitoneally to three spontaneously diabetic dogs showed the positive data in two dogs. This work led to the conclusions that the peritoneal cavity can be considered as a suitable transplantation site also for the large animals, and possibly also for humans. It should also be mentioned that this type of capsules has been applied to the human volunteers in the xenotransplantation studies [30].

2.3. ALGINATE – BASED CAPSULES: MISCELLANEOUS SYSTEMS

The following text comprises some of the recently announced polyelectrolyte encapsulation systems, which use alginate as the primary material for either encapsulation of living cells (instead of the APA capsule) or for the other applications.

An interesting approach for augmenting the mechanical and chemical stability as well as resistance to swelling was used in formulation of alginate-aminopropyl-silicate-alginate microcapsule tested as a bioartificial pancreas utilising the combination of the polyelectrolyte complexation and the sol-gel process [31]. The method involves formation of Ca-alginate beads, which are dispersed in n-hexane containing silicon alkoxides (inorganic- or aminopropyl-silicate or their combination). Silicon alkoxides penetrate into the Ca-alginate network and the membrane is formed via the sol-gel process of silicate by water contained in the bead. When amino-silicate is used, also the membrane is formed with alginate via the electrostatic interactions. The residual positive charges are neutralised by applying the alginate coat thus leading to the sandwiched aminopropyl-silicate layer between the core alginate gel and outer alginate layer. The type and selection of silicates together with the application of the alginate outer layer are crucial for the surface biocompatibility and partitioning of glucose and proteins selected in the molar mass range up to 157 kDa.

The optimisation of the membrane design for encapsulation of pancreatic cells was performed by formation of the multilayer capsule [32]. The membrane is built on the barium-alginate bead by several successive steps using a number of commercially available polyanions and polycations. Each layer is formed of a polymer of the opposite charge regarding the preceding layer. A different selection of the layer-forming polymers, their source (natural *vs.* synthetic) and molecular weight influence differently the capsule properties quantified in terms of mechanical resistance, permeability and membrane thickness. The membrane composition did not show any effect on insulin secretion and biocompatibility compared to the barium beads without any membrane.

In case of hepatocyte encapsulation, the APA capsule was made in the presence of galactosylated alginate in the core forming the 3D structure [33]. The galactose moieties in the galactosylated alginate provide the anchorage sites for hepatocytes. Consequently, hepatocytes interact with the galactosylated alginate twenty-times more intensively than with alginate, which promotes their organisation to the three-dimensional architecture and improved viability.

In the context of the encapsulation of the islets of Langerhans it is important to include also the very simple encapsulation system used for allotransplantation of pancreatic islets to NOD and BALB/c mice [34]. The simple barium-alginate membrane provided a sufficient allo- and autoimmune protection for over one year. Although the functionality of this type of membrane in case of xenografts remains to be answered, the protection of islets in this material points out that the polyelectrolyte complex-based membrane with a tight control over the mechanical strength, permeability and other parameters might not be strictly required, at least in the case of allotransplantation. The requirements on the entrapping materials were limited to using alginates of high purity and checking the membrane stability by determination of the fraction of retrieved, broken and cell-overgrown capsules. The most stable membrane was formed in case of alginate of high mannuronic acid content gelled by $BaCl_2$, which was superior over both membrane formed in the presence of $CaCl_2$ or PLL membrane and capsule formed of high-guluronic acid content alginate. It would be appropriate to have information on the permeability of these beads in order to compare them with the other encapsulation systems.

Generally, alginates are used very commonly for the sustained drug release [13]. The control over the release rate can be achieved by various principles. The retarded release of the drug from the porous matrix of calcium alginate was achieved by dispersing the drug in the waxy matrix of glyceryl behenate before its entrapment in the calcium alginate matrix [35]. A controlled lag time for duration of the capsular membrane followed by burst release of a drug according to the requirements of organism was achieved by using the calcium-alginate beads coated with carboxy-n-propylacrylamide copolymers [36]. The release profile is governed by the dissociation of calcium ions from complex with alginate upon the diffusion of non-gelling sodium cations into the gel bead core. The fast destabilisation of the alginate bead and an uncontrolled drug release is prevented by the presence of coating layer of carboxy-n-propylacrylamide, which resists the osmotic pressure and controls the permeability. Above a certain limit, the osmotic pressure inside the core induces the breakage of the coating copolymer layer followed by an accelerated drug release. The time lag can be triggered by the thickness of the coating layer of copolymer. Mixing of the beads coated under different conditions provides the system in which each fraction of beads bursts at different time lag achieving the pulsatile drug release.

The emulsification process represents an alternative way to formation of the beads by extruding the polyanion solution through a nozzle. For example, the emulsification technique was used to encapsulate the probiotic bacteria [37]. This encapsulation technique involves emulsification of alginate solution containing bacteria in oil with surfactant. The addition of $CaCl_2$ solution leads to formation of beads, although, as reported, their size and shape remain to be improved. Incorporation of starch to alginate solution as a prebiotic results in a prolonged activity of encapsulated cultures compared to free cells. Other examples include using the emulsification technique for preparation of the calcium alginate microspheres in the presence of poly-vinylpyrrolidone or ethylcellulose for the tablet formulations of various flow properties, aggregation and drug release [38] and testing the presence of different polymers on stability and drug release characteristics of microspheres in different environments [39].

Alginate has also been a common material for the controlled release formulations used in the agriculture for preventing an uncontrolled leaching of active compounds – herbicides, pesticides, fertilisers – causing the pollution of environment. The presence of natural bentonite in the alginate-based controlled release formulation was found to be beneficial for a reduced release rate of herbicide [40]. The calcium alginate-bentonite beads were stable providing the controlled release properties for herbicide tested in the greenhouse soil.

3. Capsule formulations using poly(methylene-co-guanidine)

3.1. ALGINATE – CELLULOSE SULPHATE – POLY(METHYLENE-CO-GUANIDINE) CAPSULE

The low molar mass polycation poly(methylene-co-guanidine), PMCG, originally used in the capsule formulation [41,42] at Vanderbilt University in Nashville (TN), has recently found a frequent use in the polyelectrolyte complex-based encapsulation technologies. Detailed studies [42] resulted in the capsule made of a multi-reactant composition of five active ingredients involved in the capsule formation process: high viscosity sodium alginate (SA-HV), sodium cellulose sulphate (CS), poly(methylene-co-guanidine) hydrochloride (PMCG), calcium chloride and sodium chloride. The polyelectrolyte complexation of these compounds exhibits the unique features in terms of the cell encapsulation technology: (1) solution of SA-HV with CS represents a physical mixture of two entangled polyanions that provide both pH-sensitive (carboxylic) and permanently charged (sulphate) groups, where the CS-PMCG interactions are responsible for the chemical stability of the polyelectrolyte complex while the presence of SA-HV provides the possibility to control the extent of swelling of the membrane by hydrophilic regions (free carboxylate groups) as well as the spherical shape of the capsule in gelling reaction with Ca^{2+}, (2) presence of $CaCl_2$ in the cation solution ensures not only the formation of the gelled bead after the drop of polyanion is immersed to the cation solution but also an increase in the local polyanion concentration at the bead surface, (3) character of the polycation (PMCG), *i.e.* low molecular weight and high charged density, combines both high mobility and high reactivity, (4) presence of PMCG together with $CaCl_2$ in the cation gives rise to the competitive binding of these two reactive cations based on their diffusion and reactivity towards the anion groups, and (5) NaCl provides the anti-gelling sodium ions that affect the competitive reaction of $CaCl_2$ and PMCG with the polyanion matrix.

The advantage of this microcapsule relates to the possibility to vary the capsule parameters essentially independently of each other. The permeability can be adjusted in the range from 3 to over 200 kDa towards the dextran standards by using several principles [43]. It can be modified by (i) polyanion concentration, (ii) composition of polycation mixture, where the concentration ratio of gelling/anti-gelling small cations (Na^{+}/Ca^{2+}) plays an important role, and (iii) additional treatment with the secondary polycation demonstrated by using poly-L-Lysine and polyvinylamine. The latter principle was the most efficient way to alter the capsule permeability with the minimum potential stress on cells. In addition, techniques to assess the capsule permeability have been thoroughly investigated by (i) the inverse size exclusion chromatography [44] with capsules tested as a SEC column packing predominately by the dextran standards of known molecular weight providing the membrane pore size distribution, and (ii) the method to measure efflux/influx of the biologically relevant proteins [43]. The mechanical strength of this capsule can be adjusted in the range of at least an order of magnitude higher than that of APA by the membrane thickness usually formed in the time-scale of a few tens of seconds [41,42].

The extent of surface roughness depending on the experimental conditions was quantified by using the atomic force microscopy [45]. It was demonstrated that the rate of capsule formation in the polycation after the polyanion drop penetrates the air/polycation interface is, together with the reasonably high polyanion viscosity, a crucial parameter in controlling the surface roughness (wrinkles) and capsular shape [46]. Slowing down the rate of freezing the capsular shape by increasing the anti-gelling/gelling cation ratio resulted in an improved shape recovery and smoother surface.

A unique feature associated with the development of this capsule is represented by the newly developed multiloop reactor for a continuous production of highly uniform capsules [47] with a precise control of the reaction time. In addition, this reactor provides a great flexibility, when, for example, two reactors connected in series allowed for changing the composition of the cation solution during the capsule formation process to provide the conditions for smooth membrane surface as well as mechanically and chemically stable membrane [46].

The encapsulated rat islets into this capsule were tested both *in vitro* and *in vivo* [41,42]. The *in vitro* perfusion data showed that the insulin secretion of encapsulated cells was comparable to that from free cells. The *in vivo* experiments of transplantation of encapsulated rat islets in cases of the chemically induced diabetes in C57/BL6 mice and spontaneously developed diabetes in NOD mice, respectively, normalised the blood glucose levels for the time period of over 6 months in both groups of animals. Recently the new data from animal studies have been announced revealing a long-term survival of porcine islets in this type of capsules after transplantation to NOD mice and dogs [48].

These data on one-step formulation of SA-HV – CS – PMCG capsule have been reconfirmed in the follow-up work [49] investigating also the possibility to replace cellulose sulphate with iota-carrageenan and providing the further view on the membrane permeability [50]. This capsule was successfully tested in the encapsulation and cryopreservation of hepatocytes [51]. The same laboratory recently described a detailed evaluation of the effect of polyvinylamine for the control of permeability and mechanical resistance in case of the one-step SA–CS– PMCG capsule as well as in the two-step procedure, where firstly the calcium-polyanion beads were formed followed by the membrane formation in the polycation solution containing either PMCG alone or the mixture of PMCG and polyvinylamine [52]. The effect of polyvinylamine in a strong reduction of the membrane permeability was similar as reported previously [43].

3.2. POLY(METHYLENE-CO-GUANIDINE) BASED CAPSULES: MISCELLANEOUS SYSTEMS

The membrane formed of the polyelectrolyte complex of alginate and either poly(methylene-co-guanidine) or poly(hexamethylene-co-guanidine), PHMG, fulfilled the requirements for biocompatibility, membrane resistance to permeation of hydrolytic enzyme DNAase nuclease of molecular weight 31 kDa, and allowing the free access of the low molecular weight carcinogens to encapsulated DNA [53]. The encapsulated DNA mimics the DNA of living cells and forms adducts with carcinogenic or mutagenic agents (*e.g.* ethidium bromide). The time-scale for membrane formation in this type of encapsulation, *i.e.* 3 hours for alginate gelling in Ca^{2+} solution and 10 – 60 minutes for reaction with co-guanidine, is orders of magnitude higher than that in case of the SA-CS-PMCG capsule [42] and apparently this encapsulation protocol is not feasible for the encapsulation of living cells due to the long exposure to PMCG. The same type of capsule was used for encapsulation of urease [54]. Increased PMCG concentration and longer exposure times resulted in ridges on the surface and shrunken capsules indicating the intense reaction between PMCG and calcium alginate under the given conditions.

4. Capsule formulations using chitosan

4.1. CHITOSAN AS THE OUTER POLYMER

Alginate-chitosan microcapsules with alginate as the core material were thoroughly investigated as the bioadhesive drug delivery system in order to prolong the residence time of a drug carrier in the gastrointestinal tract [55]. The one- and two-stage processes were used based on the presence or absence of Ca^{2+} in the receiving chitosan solution set to pH equal to 5. The beads were prepared in a way to differ in the level of homogeneity of the alginate concentration gradient through the cross-section of the bead [56] by addition of sodium chloride to the calcium chloride solution. The number average molecular weight of chitosan was in the range from 1.5 to 210 kDa. All chitosan coated capsules exhibited enhanced adhesive properties compared to the uncoated beads. In addition, the homogeneous capsules were more adhesive than the inhomogeneous, explained by either a higher amount of bound chitosan or a lower degree of negative charges from alginate at the surface, independently of the type of chitosan and capsule generation process. Capsule mechanically strength and permeability strongly depend on the process of capsule preparation [57,58]. The stability of chitosan-alginate complex in the pH ranging from 2 to 5 and in the presence of salt significantly exceeds the stability of the complex alginate-poly-L-Lysine [16] as revealed by determining the amount of released radiolabelled chitosan. This is an important feature in lowering the probability of the immune reaction due to the leakage of a polycation from the membrane. In the one-stage procedure (in the absence of Ca^{2+}

in chitosan solution), chitosan is located only at the interface as a thin alginate-chitosan membrane with a weak mechanical resistance. The capsules were much stronger when the two-stage protocol was used. Stable gelled core remained after treatment with the calcium sequestrant indicating the presence of chitosan throughout the capsule. This difference between two protocols of capsule formation is due to the ability of chitosan to penetrate through the membrane. The capsules were made impermeable to IgG after building the additional multi-layers of alginate and chitosan.

Bartkowiak and Hunkeler used the low molecular weight oligochitosans with the molar mass in the range from 2 to 3 kDa [59,60]. This chitosan is soluble at the physiological pH and allows for a large variation in the degree of ionisation and, hence, the reactivity. The alginate-chitosan capsules were formed for 20 min in the absence of the gelling step of alginate by calcium. The kinetics of membrane formation and the capsule parameters (thickness, permeability and mechanical strength) depend on the concentration of components, molar masses of both alginate and chitosan, reaction time, pH and ionic strength. No data are provided on the long term stability under the physiological conditions. This work represents an excellent example demonstrating how a minor modification of the capsule formation protocol modifies the final properties of the capsule and emphasises an enormous variation in the capsule formation conditions in case of just two components involved in the capsule formation step.

4.2. CHITOSAN AS THE INNER POLYMER

Capsules made of the high molar mass chitosan (1 600 kDa) as the core polymer and oligophosphate (hexamethaphosphate) of the molar mass 3 kDa as the outside polymer were formed under various conditions [61]. The capsule stability was sensitive to the presence of salt. Increased concentrations of both NaCl and oligophosphate resulted in a significant decrease of bursting force indicating a strong influence of the ionic strength on the membrane formation rate. The membrane stability depends also on pH of solutions and reaction times. Due to the intense reaction between polycation and oligopolyanion, the reaction time (in range of 2 minutes) had to be precisely controlled in order to keep the capsule swollen. This capsule is characterised by a relatively narrow window of the capsule formation conditions, which also depend on the capsule size. All these parameters were shown to be critical for the stability under the physiological NaCl concentration. These factors are correlated with both mechanical and diffusion properties of the membrane with the molecular weight cut-off between 15 and 40 kDa towards the dextran standards [62].

Alginate-reinforced chitosan beads were formed as a potential drug delivery system [63]. Chitosan was dropped to the alginate and tripolyphosphate solution, i.e., also this system represents a combination of a complex coacervation and ionotropic gelation. The encapsulation efficiency of bovine serum albumin was increased from 50 to over 90% by addition of pectin to the polyanion solution likely as a consequence of the reduction in the membrane permeability.

Another similar one-stage system consists of chitosan-alginate bead reinforced by genipin used as the naturally occurring anionic crosslinking agent [64]. Genipin can spontaneously react with the amino acids and proteins and is not toxic compared to glutaraldehyde. Chitosan-alginate membrane located at the bead surface is supported by the covalently crosslinked gel of chitosan-genipin via nucleophilic attack of chitosan amine group on the dihydropyran ring of genipin. Curing of the inner core by genipin takes a relatively long time (12 hours). These beads were tested as a possible drug carrier system showing also a good cellular biocompatibility.

5. Miscellaneous encapsulation systems

A new polyelectrolyte complex-based material for encapsulation of living cells was recently patented under the name "Biodritin heteropolysaccharide" [65]. It consists of covalently bound chodroitin sulphate and sodium alginate preferably in the ratio 1:3 or 1:4 linked by divinyl sulphone. Biodritin represents a logical combination of groups providing gels with the calcium cations (alginate) as well as strong interactions with the cationic polymers like poly-L-Lysine or poly-L-ornithine by means of the sulphogroups (chondroitin sulphate) and in this respect it resembles the principles controlling the SA-CS-PMCG capsule formation [42]. The Biodritin gels are chemically and mechanically stable and therefore suitable for the capsule formation. Empty capsules implanted to mice showed no inflammation reaction. Biodritin material has been used for the islet encapsulation stressing the importance of the extracellular matrix for the viability of the islets [66].

Two critical requirements for the therapeutic use of the encapsulated islets of Langerhans is to decrease the volume of transplant at the high islet number and to decrease the membrane thickness in order to obtain a fast mass transfer through the membrane. The thinnest possible coating of islets reaching 10 μm was achieved by the interfacial photopolymerization by using ionic and lipophilic dyes of Eosin type adsorbed on the islet surface [67]. Binding of the ionically interacting photoinitiator dye to the cell is the key event of this encapsulation process. While the lipophilic dye (Eosin DHPE) has the ability to be inserted to the cell membrane, the anionic dye (Eosin Y) binds by its carboxylic group to the cationic and basic surface domains. The dyes serve as the photoinitiators for starting polymerisation of the pre-polymer formed of diacrylated poly(ethylene glycol), triethanolamine and 1-vinyl-2-pyrrolidone forming the conformal coating around the islets. The photopolymerization takes only a few tens of seconds and the cell viability was not reduced by this process.

Cellulose sulphate-poly(dimethyldiallylammonium chloride) capsules originally developed by Dautzenberg and co-workers [68] were recently used for the encapsulation and transplantation of hybridoma cells for delivering the monoclonal antibodies into the blood stream [69]. Capsules implanted into the intraperitoneal cavity remain mobile and non-vascularised whereas capsules implanted under the skin started to be vascularised after 3 days with the complete vascularization after 2 to 3 weeks. In the latter case the higher blood concentration of antibody was observed and capsules functioned at the implantation site as long as 10 months. Cellulose sulphate capsules are claimed to offer several advantages over other capsule types – larger pore sizes to release also the molecules as large as antibodies, excellent mechanical properties due to strongly interacting sulphate groups and a simple encapsulation procedure enabling a large scale production.

6. Summary and recommendations for capsule formation

The above overview of the current microencapsulation systems made by the polyelectrolyte complexation reveals the various principles taking place to govern the polyelectrolyte complex formation process in a direction to obtain the spherical capsule with the defined properties. The final properties of microcapsules are dictated by their applications. For the biomedical applications in the cell encapsulation field, the capsules have to be prepared under the mild and physiological conditions, have to exhibit biocompatibility, mechanical and chemical stability, a proper permeability cut-off and proper geometry without protruded cells, whereas in the case of other types of the controlled release devices the major factor is to physically or chemically capture the active substance and to control its release to the environment.

This part emphasises the major factors influencing the process of capsule formation and attempts to provide the guidelines, which can be fundamental either in optimising the already developed microcapsules or in designing the new ones. Obviously, these guidelines cannot cover all possibilities, since the major message from the above overview is that all the complexity due to working with polymers, and specifically with polyelectrolyte complexes, have to be individually addressed to the specific systems and the specific applications.

6.1. SELECTION OF SUITABLE POLYELECTROLYTES

The primary task is to select the polyelectrolytes capable of forming the polyelectrolyte complex in the shape of the microcapsule. The tendency from about 5 to 10 years ago to identify a number of potentially useable polyelectrolytes perhaps lacking in a deeper investigation of each component and process conditions has been changed. Instead of trying to provide a number of possible candidates [10], the trend is to understand and optimise the individual polyelectrolyte complex from many points of view. Chitosan-alginate system represents a very good example of such an approach [55,57-60]. In principle, the primary structure of any two oppositely charged polyelectrolytes can be tailored in the way that they eventually form the microcapsule. In this respect, synthetic polymers possess an advantage over the natural ones, because their properties can be tailored in an easier way. On the other hand, the clear advantage of the polyelectrolytes from the natural sources (polysaccharides) is their diversity and ability to form hydrogels.

6.2. MOLECULAR WEIGHT AND MOLECULAR WEIGHT DISTRIBUTION

It becomes more obvious, that polyelectrolytes should be considered not only as the polymers characterised by a certain type, number and distribution of charges, but also by their molecular weight and molecular weight distribution. The latter parameter, characterised so far rarely in the microcapsule formation process [16,39], governs the solution rheology, complex stability and the mutual diffusion of the gelling zones and may have a significant effect on the capsule properties.

Regarding the molecular weight of the outer polymer, apart from the molecular weight starting from a few tens of thousand Daltons, the oligomeric polyelectrolytes have been successfully introduced including poly(methylene-co-guanidine) [41,42], chitosan [59,60] and hexamethaphosphate [61,62]. The advantage of the oligomers is their high diffusivity through the membrane made upon the first contact of the polyelectrolyte solutions and fast membrane development. On the other hand, their reactivity should be sufficient to counterbalance the lower number of interactions of the individual polymer chains, where the molecular weight should not be lower than the critical chain length to form the stable complex in a given matrix-oligomer system [5].

6.3. VISCOSITY OF SOLUTIONS

The viscosity of polymer solutions is crucial for both the inner and outer polymer. The inner solution has to be viscous in order to provide the proper conditions for the spherical capsule shape and surface quality. It has been reported by several authors [9,59] that the higher viscosity of the inner polymer solution leads to more spherical capsules. This phenomenon has been explained as a consequence of the drop shape relaxation by viscous forces after the capsule penetrates the air-cation solution interface [46], where the time required for the drop shape relaxation is inversely proportional to the viscosity of the inner solution and proportional to the drop size. The viscosity can be increased either by increasing the polymer concentration or by using the polymer of higher molecular weight or by applying the viscosity modifying agent. Each of this steps alters the conditions for drop formation of an inner solution as well as the capsule and membrane properties. For example, using a higher concentration of low molecular weight polymer causes a higher osmotic pressure and lower extent of the chain entanglements with an impact on the mechanical properties. Another variable playing the role in the capsule shape recovery is the distance between the nozzle and receiving solution containing the outer polymer. The viscosity and, simultaneously, the density of the receiving solution should be carefully selected to allow for an easy drop penetration of the interface without creating the surface imperfections.

6.4. COMPLEX STABILITY

The selected polymers should form the highly stable complex providing the long-term mechanical and chemical stability. The mechanical stability has been routinely tested depending on the application and processing conditions. However, this is not a frequent case for evaluation of the chemical stability, which represents a crucial property for membrane biocompatibility, mechanical stability and durability under the conditions where the capsule is intended to be used. The examples of testing the complex stability were shown for APA [16] as well as chitosan-alginate [37] capsules by using the radiolabelled polyelectrolytes, and for SA-CS-PMCG capsules the substitution of PMCG by polyvinylamine was assessed by means of the UV spectroscopy [52].

Swelling of the beads and capsules represent the factor relating to the complex chemical stability and lateral aggregation of the polymer molecules reaching the osmotic equilibrium in a given environment. Swelling has been an issue followed in detail for example in case of the APA capsule [9,24,31] since it may be responsible for the failure of the graft. In general, swelling should be suppressed either by increasing the intensity of the interactions of polyelectrolytes and the membrane strength or by decreasing the osmotic pressure of internal polymer by avoiding the degelling step and using the high molar mass polymers.

6.5. IONIC STRENGTH AND pH OF POLYELECTROLYTE SOLUTIONS

The reactivity of the ionic groups and the chain conformation and solubility of polyelectrolytes naturally depend on ionic strength and pH of the solution. The flexibility to trigger both the complex stability and the membrane properties by often only a slight altering of these two parameters has been discussed in parts devoted to the individual capsules. The ionic strength of receiving bath and combination of gelling and non-gelling cations affects the homogeneity of distribution of the core polymer throughout the bead [16,56], diffusion of the outer polymer and membrane thickness [55,59] as well as the rate of competing reaction between the ionic and cationic groups [42] affecting the membrane permeability [43], surface roughness and capsule shape [45,46]. It is obvious but yet worth mentioning that the pH and ionic strength of the outer polyelectrolyte solution will be valid also for the inner solution as soon as the drops are immersed to the inner solution, since the small ions are the fastest diffusing species in the system. Thereby the chain conformation and charge of the inner polyelectrolyte will be determined by the conditions set for the outer polymer, e.g., in case of weak polyelectrolytes, decreased pH of receiving bath containing polycation results in the increased degree of ionisation and reactivity of the polycation but decreased degree of ionisation and reactivity of the polyanion.

6.6. DROP COLLECTION AND CAPSULE REACTION TIMES

As for any polymer crosslinked network, the character of the gel formed by polyelectrolyte and/or ionotropic gelation depends upon many conditions. For constant thermodynamic conditions and type and composition of reactants, the character of a gel depends on the time of reaction, which determines the gel density (number of crosslinks, effective molecular weight between crosslinks) and the thickness of the gelled zone. Translated to the process of capsule formation, reaction time governs properties of the formed semipermeable membrane from the viewpoints of permeability, mechanical resistance, complex chemical stability and mass transfer.

The capsular membrane formed by polyelectrolyte complexation is almost solely prepared discontinuously in two steps. In the first step, the drops of one polyelectrolyte are collected in the solution of a polyelectrolyte of the opposite charge or a small ionotropically gelling ion for a given time (collection time), until the drop collection is ceased. The crosslinking reaction continues in the solution used for collection within the additional time period (reaction time) to form the capsules of required properties. It logically follows that unless the collection time is much shorter than the reaction time, the capsules within one batch are heterogeneous in the gel character simply because the collection time contributes to the reaction time differently for each individual capsule. This may lead to the heterogeneous character and a broad distribution of the final capsule properties observed as heterogeneous membrane thickness, different extent of swelling, biocompatibility, etc. Importantly, the entire batch of capsule may consist of the capsule fraction failing to serve the goal of encapsulation thus leading to the false conclusion that the used polymers are inappropriate and, hence, they may be discarded from the list of the polymer candidates.

The majority of papers on microcapsules formed by polyelectrolyte complexation provide the reaction time only very rarely, while the information on the collection time is almost always missing. The reasons behind not revealing these times is unknown and may be due to (i) not paying attention to this factor, (ii) the awareness and, at the same time, the inability to solve this problem, and, possibly, (iii) keeping the process conditions confidential. Thus the impossibility to reproduce the process can be an additional source of irreproducibility among the different groups. In the alginate-based systems, where the bead collection in $CaCl_2$ usually takes a few minutes and reaction proceeds for additional few minutes, the heterogeneity is not directly visible, yet the qualitative difference in the gel properties are present. The collection and reaction times become to be more critical in the polyelectrolyte systems exhibiting the higher reactivity, where, for example, the heterogeneity in membrane thickness [42] or membrane shrinkage [61] are the visible signs leading to the requirement for the strict control over collection and reaction times.

The dilemma between collection and reaction times led to the invention of the continuous capsule-producing reactor [47]. It continuously generates uniform capsules at a high production rate with a little monitoring by eliminating the step of capsule collection. The basic component of the reactor is a single loop, when the number of loops can be increased depending on the required reaction time. The steady stream of the cation solution introduced to the reactor flows out at the exit at the same rate. The polyanion drops directed into the reactor are drawn by the cation stream, flow through the tube and react forming the capsule, and exit at the other end, where the reaction is quenched by a buffer medium. The residence time of each capsule in the cation solution is the same leading to high batch homogeneity.

7. Conclusion

This review describes the main features of the currently used microencapsulation systems formed by the polyelectrolyte complexation. Many different encapsulation systems have been developed for a number of diverse fields, where a high complexity involved in the polyelectrolyte complexation should be taken as an advantage and utilised for triggering the capsule properties of interest. However, the performance of the capsules depends on a number of factors, which have to be recognised and controlled taking into account the selection of materials and optimising each step in the capsule formation process.

Acknowledgement

The author wishes to express his thanks to Prof. Taylor G. Wang from Vanderbilt University, Nashville, TN, for valuable comments to the final form of this paper.

References

[1] Dautzenberg, H.; Jaeger, W.; Kötz, J.; Philipp, B.; Seidel, Ch. and Stscherbina, D. (1994) Polyelectrolytes: Formation, characterization and application. Hanser Publishers, Munich, Vienna, New York, ISBN 3-446-17127-4.

[2] Park, J.K. and Chang, H.N. (2000) Microencapsulation of microbial cells. Biotechnol. Adv. 18: 303-319.

[3] Kötz, J.; Kosmella, S. and Beitz, T. (2001) Self-assembled polyelectrolyte systems. Prog. Polym. Sci. 26: 1199-1232.
[4] Decher, G. (1996) Layered nanoarchitectures via directed assembly of anionic and cationic molecules. In: Sauvage, J. P. and Hosseini, M. W. (Eds.) Comprehensive supramolecular chemistry. Vol. 9: Templating, self-assembly and self-organization. Pergamon Press, Oxford; pp. 507-528.
[5] Karibyants, N. and Dautzenberg, H. (1998) Preferential binding with regard to chain length and chemical structure in the reactions of formation of quasi-soluble polyelectrolyte complexes. Langmuir 14: 4427-4434.
[6] Tsuchida, E. and Abe, K. (1986) Polyelectrolyte complexes. In: Wilson, A D and Prosser, H J (Eds.) Development in ionic polymers-2. Elsevier Applied Science Publishers, London and New York, Chapter 5; pp. 191-263.
[7] Ross-Murphy, S.B. (1991) Physical gelation of synthetic and biological macromolecules. In: DeRossi, D.; Kajiwara, K.; Osada, Y. and Yamauchi A. (Eds.) Polymer gels: fundamentals and biomedical applications, Plenum Press New York and London, ISBN 0-306-43805-4; pp. 21-39.
[8] Hoffman, A.S. (2001) Hydrogels for biomedical applications. Adv. Drug. Deliv. Rev. 54: 3-12.
[9] de Vos, P.; Hamel, A.F. and Tatarkiewicz, K. (2002) Considerations for successful transplantation of encapsulated pancreatic islets. Diabetologia 45: 159-173.
[10] Hunkeler, D.; Prokop, A.; Powers, A.; Haralson, M.; DiMari, S. and Wang, T. (1997) A screening of polymers as biomaterials for cell encapsulation. Polym. News 22: 232-240.
[11] Hou, Q.P. and Bae, Y.H. (1999) Biohybrid artificial pancreas based on macrocapsule device. Adv. Drug Deliv. Rev. 35: 271-287.
[12] Zimmermann, U.; Mimietz, S.; Zimmermann, H.; Hillgärtner, M.; Schneider, H.; Ludwig, J.; Hasse, C.; Haase, A.; Rothmund, M. and Fuhr, G. (2000) Hydrogel-based non-autologous cell and tissue therapy. BioTechniques 29: 564-581.
[13] Gombotz, W.R. and Wee, S.F. (1998) Protein release from alginate matrices. Adv. Drug Deliv. Rev. 31: 267-285.
[14] Lim, F. and Sun, A.M. (1980) Microencapsulated islets as bioartificial pancreas. Science 210: 908-910.
[15] Soon-Shiong, P.; Heintz, R.E.; Merideth, N.; Yao, Q.X.; Yao, Z.; Zheng, T.; Murphy, M.; Moloney, M.K.; Schmehl, M.; Harris, M.; Mendez, R. and Sandford, P.A. (1994) Insulin independence in a type I diabetic patient after encapsulated islets transplantation. Lancet 343: 950-951.
[16] Thu, B.; Bruheim, P.; Espevik, T.; Smidsrød, O.; Soon-Shiong, P. and Skjåk-Bræk, G. (1996) Alginate polycation microcapsules I. Interaction between alginate and polycation. Biomaterials 17: 1031-1040.
[17] Thu, B.; Bruheim, P.; Espevik, T.; Smidsrød, O.; Soon-Shiong, P. and Skjåk-Bræk, G. (1996) Alginate polycation microcapsules I. Interaction between alginate and polycation. Biomaterials 17: 1069-1079.
[18] Soon-Shiong, P.; Heintz, R.E. and Skjåk-Bræk, G. (1998) Microencapsulation of cells. US Patent 5,762,959.
[19] Soon-Shiong, P.; Desai, N.P. and Heintz, R.E. (1998) Method for making cytoprotective, biocompatible, retrievable microcapsule containment system. US Patent 5,788,988.
[20] Soon-Shiong, P.; Desai, N.P. and Heintz, R.E. (1999) Method of treating patients with diabetes. US Patent 5,879,709.
[21] Hallé, J.-P.; Leblond, F.A.; Pariseau, J.-F.; Jutras, P.; Brabant, M.J. and Lepage, Y. (1994) Studies on small (<300 μm) microcapsules: II. Parameters governing the production of alginate bead by high voltage electrostatic pulses. Cell Transplant. 3: 365-372.
[22] Leblond, F.A.; Tessier, J. and Hallé, J.-P. (1996) Quantitative method for the evaluation of biomicrocapsule resistance to mechanical stress. Biomaterials 17: 2097-2102.
[23] Robitaille, R.; Pariseau, J.-F.; Leblond, F.A.; Lamoureux, M.; Lepage, Y. and Hallé, J.-P. (1999) Studies on small (<350 μm) alginate-poly-L-Lysine microcapsules. III. Biocompatibility of smaller versus standard microcapsules. J. Biomed. Mater. Res. 44: 116-120.
[24] Leblond, F.A.; Simard, G.; Henley, N.; Rocheleau, B.; Huet, P.-M. and Hallé, J.-P. (1999) Studies of smaller (~315 μm) microcapsules: IV. Feasibility and safety of intrahepatic implantations of small alginate poly-L-Lysine microcapsules. Cell Transplant. 8: 327-337.
[25] Strand, B.L.; Ryan, L.; In't Velt, P.; Kulseng, B.; Rokstad, A.M.; Skjåk-Bræk, G. and Espevik, T. (2001) Poly-L-Lysine induces fibrosis on alginate microcapsules *via* the induction of cytokines. Cell Transplant. 10: 263-275.
[26] Peirone, M.; Ross, C.J.D.; Hortelano, G.; Brash, J.L. and Chang, P. (1998) Encapsulation of various recombinant mammalian cell types in different alginate microcapsules. Biomaterials 42: 587-596.
[27] Calafiore, R.; Basta, G.; Osticioli, L.; Luca, G.T.; Tortoioli, C. and Brunetti, P. (1995) Coherent microcapsules for pancreatic islet transplantation: a new potential approach for bioartificial pancreas. Transplant. Proc. 28: 822-823.

[28] Calafiore, R.; Basta, G.; Luca, G.; Boselli, C.; Bufalari, A.; Giustozzi, G.M.; Gialletti, R.; Moriconi, F. and Brunetti, P. (1998) Transplantation of allogeneic/xenogeneic pancreatic islets containing coherent microcapsules in adult pigs. Transplant. Proc. 30: 482-483.
[29] Calafiore, R.; Basta, G.; Luca, G.; Boselli, C.; Bufalari, A.; Bufalari, A.; Cassarani, M.P.; Giustozzi, G.M. and Brunetti, P. (1999) Transplantation of minimal volume microcapsules in diabetic high mammalians. Ann. NY Acad. Sci. 875: 219-232.
[30] Anonymous, www.diatranz.co.nz
[31] Sakai, S.; Ono, T.; Ijima, H. and Kawakami, K. (2001) Synthesis and transport characterisation of alginate/aminopropyl-silicate/alginate microcapsule: application to bioartificial pancreas. Biomaterials 22: 2827-2834.
[32] Schneider, S.; Feilen, P.J.; Slotty, V.; Kampfner, D.; Preuss, S.; Berger, S.; Beyer, J. and Pommersheim, R. (2001) Multilayer capsules: a promising microencapsulation system for transplantation of pancreatic islets. Biomaterials 22: 1961-1970.
[33] Yang, J.; Goto, M.; Ise, H.; Cho, C.-S. and Akaike, T. (2002) Galactosylated alginate as a scaffold for hepatocytes entrapment. Biomaterials 23: 471-479.
[34] Duvivier-Kali, V.F.; Omer, A.; Parent, R.J.; O'Neil, J.J. and Weir, G.C. (2001) Complete protection of islets against allorejection and autoimmunity by a simple barium-alginate membrane. Diabetes 50: 1698-1705.
[35] Mirghani, A.; Idkaidek, N.M.; Salem, M.S. and Najib, N.M. (2000) Formulation and release behaviour of diclofenac sodium in Compritol 888 matrix beads encapsulated in alginate. Drug Dev. Ind. Pharm. 26: 791-795.
[36] Iskakov, R.M.; Kikuchi, A. and Okano, T. (2002) Time-programmed pulsative release of dextran from calcium-alginate gel beads coated with carboxy-n-propylacrylamide copolymers. J. Control. Release 80: 57-68.
[37] Sultana, K.; Godward, G.; Reynolds, N.; Arumugaswamy, R.; Peiris, P. and Kailasapathy, K. (2000) Encapsulation of probiotic bacteria with alginate-starch and evaluation of survival in simulated gastrointestinal conditions and in yoghurt. Int. J. Food. Microbiol. 62: 47-55.
[38] Chan, L.W. and Heng, P.W.S. (1998) Effects of poly(vinylpyrrolidone) and ethylcellulose on alginate microspheres prepared by emulsification. J. Microencapsulation 15: 409-420.
[39] Gürsoy, A.; Karakus, D. and Okar, I. (1999) Polymers for sustained release formulations of dipyridamol-alginate microspheres and tabletted microspheres. J. Microencapsulation 16: 439-452.
[40] Fernández-Pérez, M.; González-Pradas, E.; Villafranca-Sánchez, M. and Flores-Céspedes (2000) Mobility of isoproturon from an alginate-bentonite controlled release formulation in layered soil. Chemosphere 41: 1495-1501.
[41] Wang, T.G.; Lacík, I.; Brissová, M.; Anilkumar, A.V.; Prokop, A.; Hunkeler, D.; Green, R.; Shahrokhi, K. and Powers, A.C. (1997) An encapsulation system for the immunoizolation of pancreatic islets. Nature Biotechnol. 15: 358-362.
[42] Lacík, I.; Brissová, M.; Anilkumar, A.V.; Powers, A.C. and Wang, T.G. (1998) New capsule with tailored properties for the encapsulation of living cells. J. Biomed. Mater. Res. 39: 52-60.
[43] Brissová, M.; Lacík, I.; Anilkumar, A.V.; Powers, A.C. and Wang, T.G. (1998) Control and measurement of permeability for design of microcapsule cell delivery system. J. Biomed. Mater. Res. 39: 61-70.
[44] Brissová, M.; Petro, M.; Lacík, I.; Powers, A.C. and Wang, T.G. (1996) Evaluation of microcapsule permeability via inverse size exclusion chromatography. Anal. Biochem. 242: 104-111.
[45] Xu, K.; Hercules, D.; Lacík, I. and Wang, T.G. (1998) Atomic force microscopy used for the surface characterization of microcapsule immunoisolation devices. J. Biomed. Mater. Res. 41: 461-467.
[46] Lacík, I.; Anilkumar, A.V. and Wang, T.G. (2001) A two-step process for controlling the surface smoothness of polyelectrolyte-based microcapsule. J. Microencapsulation 18: 479-490.
[47] Anilkumar, A.V.; Lacík, I. and Wang, T.G. (2001) A novel reactor for making uniform capsules. Biotechnol. Bioeng. 75: 581-589.
[48] Wang T.G. (2002) Microencapsulation methods: PMCG capsules. In: Atala, A. and Lanza, R. (Eds.) Methods of Tissue Engineering. Academic Press, Vol. 75; pp. 841-857.
[49] Bartkowiak, A.; Canaple, L.; Ceuasoglu, I.; Nurdin, N.; Renken, A.; Rindisbacher, L.; Wandrey, Ch.; Desvergne, B. and Hunkeler, D. (1999) New multicomponent capsule for immunoisolation. Ann. NY Acad. Sci. 875: 135-145.
[50] Nurdin, N.; Canaple, L.; Bartkowiak, A.; Desvergne, B. and Hunkeler, D. (2000) Capsule permeability *via* polymer and protein ingress/egress. J. Appl. Polym. Sci. 75: 1165-1175.

[51] Canaple, L.; Nurdin, N.; Angelova, N.; Saugy, D.; Hunkeler, D. and Desvergne, B. (2001) Maintenance of primary murine hepatocyte functions in multicomponent polymer capsules – in vitro cryopreservation studies. J. Hepatology 34: 11-18.
[52] Grigorescu, G.; Rehor, A. and Hunkeler, D. (2002) Polyvinylamine hydrochloride-based microcapsules: polymer synthesis, permeability and mechanical properties. J. Microencapsulation 19: 245-259.
[53] Quong, D. and Neufeld, R.J. (1999) DNA encapsulated within co-guanidine membrane coated alginate beads and protection from extracapsular nuclease. J. Microencapsulation 16: 573-585.
[54] Hearn, E. and Neufeld, R.J. (2000) Poly(methylene-co-guanidine) coated alginate as an encapsulation matrix for urease. Process Biochem. 35: 1253-1260.
[55] Gåserød, O.; Jolliffe, I.G.; Hampson, F.C.; Dettmar P.W. and Skjåk-Bræk, G. (1998) The enhancement of the bioadhesive properties of calcium alginate beads by coating with chitosan. Int. J. Pharm. 175: 237-246.
[56] Skjåk-Bræk, G.; Grasdalen, H. and Smidsrød, O. (1989) Inhomogeneous polysaccharide ionic gels. Carbohydr. Polym. 10: 31-54.
[57] Gåserød, O.; Smidsrød O. and Skjåk-Bræk, G. (1998) Microcapsules of alginate-chitosan. I. A quantitative study of the interaction between alginate and chitosan. Biomaterials 19: 1815-1825.
[58] Gåserød, O.; Sannes, A. and Skjåk-Bræk, G. (1999) Microcapsules of alginate-chitosan. II. A study of capsule stability and permeability. Biomaterials 20: 773-783.
[59] Bartkowiak, A. and Hunkeler, D. (1999) Alginate-oligochitosan microcapsules: A mechanistic study relating membrane and capsule properties to reaction conditions. Chem. Mater. 11: 2486-2492.
[60] Bartkowiak, A. and Hunkeler, D. (2000) Alginate-oligochitosan microcapsules: II. Control of mechanical resistance and permeability of the membrane. Chem. Mater. 12: 206-212.
[61] Angelova, N. and Hunkeler, D. (2001) Stability assessment of chitosan-sodium hexametaphosphate capsules. J. Biomater. Sci. Polym. Edn. 12: 1207-1225.
[62] Angelova, N. and Hunkeler, D. (2001) Effect of preparation conditions on properties and permeability of chitosan-sodium hexametaphosphate capsules. J. Biomater Sci. Polym. Edn. 12: 1317-1337.
[63] Aral, C. and Akbuga, J. (1998) Alternative approach to the preparation of chitosan beads. Int. J. Pharm. 168: 9-15.
[64] Mi, F.L.; Sung, H.W. and Shyu, S.S. (2002) Drug release from chitosan-alginate complex beads reinforced by a naturally occurring cross-linking agent. Carbohydr. Polym. 48: 61-72.
[65] Mares-Guia, M. and Ricordi, C. (2001) Hetero-polysaccharide conjugate and methods of making and using the same. US Patent 6,281,341
[66] Maria-Engler, S.S.; Mares-Guia, M.; Correa, M.L.C.; Oliveira, E.M.C.; Aita, C.A.M.; Krogh, K.; Genzini, T.; Miranda, M.P.; Ribeiro, M.; Vilela, L.; Noronha, I.L.; Eliaschewitz, F.G. and Sogayar, M.C. (2001) Microencapsulation and tissue engineering as an alternative treatment of diabetes. Braz. J. Med. Biol. Res. 34: 691-697.
[67] Desmangles, A.-I.; Jordan, O. and Marquis-Weible, F. (2001) Interfacial photopolymerization of β-cell clusters: approaches to reduce coating thickness using ionic and lipophilic dyes. Biotechnol. Bioeng. 72: 634-641.
[68] Dautzenberg, H.; Loth, F.; Fechner, K.; Mehlis, B. and Pommerening, K. (1985) Preparation and performance of symplex capsules. Makromol. Chem. Suppl. 9: 203-210.
[69] Pelegrin, M.; Marin, M.; Noël, D.; Del Rio, M.; Saller, R.; Stange, J.; Mitzner, S.; Günzburg, W.H. and Pechaczyk, M. (1998) Systemic long-term delivery of antibodies in immunocompetent animals using cellulose sulphate containing antibody-producing cells. Gene Ther. 5: 828-834.

PRE-FORMED CARRIERS FOR CELL IMMOBILISATION

FERDA MAVITUNA
Chemical Engineering Department, UMIST, Sackville Street, PO Box 88, Manchester, M60 1QD, United Kingdom – Fax: 44 0161 200 4399 – Email: f.mavituna@umist.ac.uk

1. Introduction

Cell immobilisation in pre-formed carriers involves passive/natural immobilisation usually *in situ* in the bioreactors or in the culture environment. Numerous inorganic and organic materials have been used as pre-formed carriers: reticulated polyurethane and polyvinyl formal foam, other polymers, plastics, stainless steel, ceramic, glass, synthetic (ion exchange) resins, activated charcoal, aluminium oxide, diatomaceous earth, sand, cellulose, lignocellulose, cellulose acetate, and others. In this chapter however, only polyurethane foam, stainless steel knitted mesh, ceramic, glass and cellulose will be considered.

Most of these carriers are porous with a wide range of pore sizes to suit immobilisation of various bacterial, yeast, fungal, plant and animal cells and tissues. For passive immobilisation, cells, flocs, mycelia, cell aggregates or spores are inoculated into the sterilised medium containing empty pre-formed carriers. Depending on the cell and the carrier type, immobilisation then takes place in a combination of filtration, adsorption, growth and colonisation processes. Furthermore, surfaces of the carriers can be modified by various pre-treatments to enhance immobilisation efficiency.

Reticulated polyurethane foam matrices and stainless steel knitted mesh spheres were some of the earliest porous pre-formed carriers to be used [1]. In fact, CAPTOR® [2,3] and LINPOR® [4] processes which were developed in the mid-1970s and commercially introduced during the early 1980s, and subsequent processes such as KALDNES [5] are based on the use of porous polyurethane or other polymeric materials as the pre-formed carriers for activated sludge microbial consortia in the secondary treatment of wastewater. Biological wastewater treatment is still the largest scale application of immobilised cell technology. The fact that these innovative processes rely on immobilisation in pre-formed carriers is a testimony to the ease, robustness, and cost-effectiveness of these cell immobilisation materials and methods.

V. Nedović and R. Willaert (eds.), Fundamentals of Cell Immobilisation Biotechnology, 121-139.

2. Mechanisms of passive cell immobilisation in pre-formed carriers

The main factors that determine the mechanisms of passive cell immobilisation in or on pre-formed carriers are the cell type and the carrier type, especially whether it is porous or not. Although, the following is a summary of various factors affecting cell immobilisation in porous pre-formed carriers, most of them are valid for non-porous pre-formed carriers as well:

- Biological characteristics of the cell and its surface (e.g., requirement of some mammalian cells for biological extracellular matrix components)
- Biocompatibility/cytotoxicity of the carrier
- Immunocompatibility of the carrier (for artificial organ/tissue implants)
- Pore size, distribution and porosity of the carrier
- Cell, floc, aggregate size and distribution
- Surface chemistry of the carrier (hydrophobicity, charge, ligand-complexes)
- Hydrodynamic/shear forces
- Specific gravity of the carrier
- Carrier hold-up
- Carrier size and shape
- Mechanical strength and flexibility of the carrier
- Biodegradability of the carrier
- Inoculum type (e.g., spore or vegetative mycelia) and concentration
- Space available for growth, cell migration and colonisation
- 3-D structure inside the carrier for cell scaffolding
- Surface smoothness
- Effects of cell products, media components on the carrier (shrinkage, degradation, gas evolution, mineralisation, precipitate formation).

Depending on the factors above, cells are immobilised through filtration and/or adhesion/adsorption, growth and colonisation in the pores of the carrier. Of course, if a suitable bioreactor configuration, design and operation mode is chosen, then the immobilisation can be very efficient.

2.1. INTERACTIONS OF CELLS WITH SURFACES AND BIOFILM FORMATION

In nature, the majority of microorganisms grow in association with surfaces. Even in natural aquatic ecosystems, surface-associated microorganisms vastly outnumber organisms in suspension [6]. This implies some strong survival and/or selective advantages for the surface-associated state compared to the planktonic. A further implication therefore, is the evolutionary existence of genetic-level control and regulation, signal transduction and cell-to-cell and cell-surface interactions for microorganisms to recognise when they become in contact with surfaces [7]. Recent genetic and molecular approaches have identified that cells develop surface-sensing responses following cell-support contact [8,9]. Microorganisms interact with surfaces through many specialised structures on their walls (*e.g.*, pili and pilus-associated adhesions), exopolymers (glycocalyx in the case of bacteria) and complex ligand interactions involving signalling molecules. They have quorum sensing mechanisms as

well. Therefore sensing either a biotic or an abiotic surface triggers genetic switches that can change cells' metabolism, phenotype and morphology [10,11].
Surface-associated growth of microorganisms in nature usually leads to the formation of highly structured biofilms, which are in fact naturally immobilised cells [7,12]. In the passive immobilisation of microbial cells in pre-formed carriers therefore, the natural tendency of microbial cells to form biofilms is exploited by providing the surface and the hydrodynamic conditions conducive to biofilm formation. Immobilisation of plant and animal cells in the pre-formed carriers on the other hand, is only a natural and in most cases a necessary extension of *in vitro* suspension cultures in order to simulate some aspects of these cells' natural environment in multicellular, differentiated and organised tissues/organs [13,14]. Any observed changes in their physiology compared to freely suspended cells can therefore be explained usually on the basis of their multicellular origins. Compared to immobilised microbial cells, the interactions between the plant or animal cells and surfaces of immobilisation matrices are studied using more sophisticated approaches involving cell signalling and extracellular matrix interactions [15].

Cell surface changes with physiological state and the environmental conditions. Critical physicochemical parameters that govern interactions of cells with surfaces can be quantified *via* several techniques: electrostatic properties by zeta potentials and ion exchange chromatography, cell hydrophobicity using contact angle measurements, hydrophobic interaction chromatography and partitioning assay. The types of forces involved with cell-immobilisation surface interactions can involve Lewis acid/base (hydrophobic), Lifshitz-van der Waals (electrodynamic) and Coulombic (electrostatic). Carrier surfaces can therefore be modified in order to affect cell immobilisation.

2.2. TREATMENT OF CARRIERS BEFORE IMMOBILISATION

A routine treatment of pre-formed carriers, especially polymer matrices, before cell immobilisation involves immersing them in distilled water and autoclaving for 15-30 min in order to remove any toxic monomers or fillers which may otherwise leach out into the culture medium. They may then be washed several times with distilled water and then sterilised by autoclaving usually immersed in the culture medium at 121°C for 15 min. They will then be ready for inoculation with the culture to be immobilised [16]. The polyurethane foam matrix can be coated with activated carbon [17] or FeS [18] to facilitate immobilisation of anaerobic bacteria.

The surface of most inorganic supports is mainly composed of oxide and hydroxyl groups, such as silanol groups in glass, which provide a mildly reactive surface for activation and binding of cell-surface proteins. Although the derivatization of a relatively inert inorganic carrier is difficult, it can be accomplished by an inorganic and/or a cross-linked organic coating. For example, silanisation method involves the use of trialkoxy silane derivatives containing an organic functional group. Coupling of these reagents to the carrier is assumed to take place by displacement of the alkoxy residues on the silane by hydroxyl groups or the oxidised surface of the inorganic support to form a metal-O-Si linkage.

Stainless steel wire spheres can be pre-treated by burning (oxidation) in the ethanol flame zone for 10 min, or modified chemically with $TiCl_4$ or γ-AS [19], ceramic monoliths are pre-treated for biocompatibility by boiling with 10% nitric acid and rinsing with deionised water and phosphate-buffered saline [20]. Similarly, glass beads can be soaked in chromosulphuric acid, thoroughly washed with phosphate-buffered saline and deionised water and dried at 105°C [21]. Glass surface can be silanised by two min exposure to 3-aminopropyltrimethoxy silane (2% v/v) dissolved in methanol:water (24:1) followed by washing with methanol and drying [22]. Treatments may involve chemical or physical attachment of chemical groups/complexes to carrier surface in order to enhance cell adhesion/adsorption, covalent bond formation with cell surface chemical groups.

For immobilisation of animal cells, carrier surface can be modified to mimic biological extracellular matrix (ECM) proteins such as fibronectin, collagen, vitronectin, elastin, laminin; growth factors such as EGF, PDGF; surface smoothness, roughness, and micropatterning can be altered by etching using lasers, and photolithography [15].

2.3. PASSIVE IMMOBILISATION

When the freely suspended cell inoculum is introduced aseptically into the sterile liquid medium containing the pre-formed carriers in shake flasks or bioreactors, the hydrodynamic conditions should be such that the cells are carried from the bulk liquid into the porous carriers by the liquid elements flowing through the open pore structure of the initially empty carrier. This is best achieved by agitating the bulk liquid or making it flow through the carriers. If the specific gravity of the carrier is close to that of the liquid medium, as is the case with reticulated polyurethane foam particles, then holding the carrier stationary in the agitated/flowing liquid helps increase the efficiency and speed of the initial entrapment of cells. Various bioreactor configurations such as packed bed, circulating bed, sheet, rotating disk, have been developed to achieve this [16,17,23]. Carriers with specific gravity higher than the liquid medium can be fluidised or used in stirred-tank bioreactors if they are robust enough.

The size distribution of the cells or cell aggregates compared to that of the carrier and the relative movement between the cells and the carrier are the most important factors affecting the initial physical entrapment, which is by filtration [24]. If the size distribution of cell aggregates does not match that of the pores of the carrier, the fine aggregates and individual cells will pass through the carrier pores and the large aggregates will be excluded from the pores. If very fine cell suspensions made up of mostly single cells, or spores are used, then either a carrier with suitably fine pore size should be chosen or the culture should not be agitated, at least not vigorously, so that the initial passive immobilisation should be through cell adhesion/adsorption onto carrier (pore) surface. Immobilisation due to cell adhesion is concomitant with filtration, in any case.

After the initial immobilisation mechanisms due to filtration and adhesion, cells can form biofilms and grow in the pores and on the reticulate surfaces of the carrier matrix depending on the available space and diffusional limitations. In most porous pre-formed carriers, there is a degree of convectional mass transfer in addition to molecular diffusion across the carrier particle [16].

Cell aggregate size may be altered to suit the pore size of the carrier [25], cells can be made to flocculate or adhere to carrier surface more readily by changing the physico-chemical conditions or by genetic modification of cells, or co-immobilisation with other microorganisms in order to facilitate the initial immobilisation stages. These do not alter the passive nature of this method of immobilisation in pre-formed carriers.

2.4. CELL GROWTH

One of the most important interactions between the immobilised cells and their immobilisation matrices is due to cell growth. Growth in carriers usually result in heterogeneous microcolonies, outgrowth on the carrier particles, cell leakage into the bulk medium, changes in pore size distribution, mechanical stresses imposed on and by microbial cells, distortion of cell shape and the reduction in cell size. Pre-formed carriers generally accommodate cell growth well compared to gels. In fact, in most cases cell growth further consolidates the passive immobilisation process. There can be freely suspended cells in the bulk liquid to varying levels because of some cell leakage but mainly because of outgrowth and attrition. For instance, reticulated polyurethane foam carriers are best suited for mycelial or flocculant microorganisms, plant and some types of animal cell cultures because when used with bacterial or single cell cultures, there can be plenty of freely suspended cells in the bulk liquid.

3. Reticulated polyurethane and polyvinyl foam carriers

Polyurethanes (PU) have been widely used as cell carriers due to their mechanically strong and biochemically inert characteristics. Polyurethanes have an open cell/pore structure as a result of condensation of polycyanates (R-CNO) and polyols (R-OH). Upon polymerisation carbon dioxide escapes from the matrix leaving pore spaces behind. In the manufacture of polyurethane foams (PUF), surfactants and fillers are also used which can be toxic [26].

Polyurethane foam can be produced form a water-based prepolymer of polyurethane (HYPOL® 2000) which consists of a proprietary prepolymer at 97% (w/w) and toluene diisocyanate at 3% (w/w) [27]. 5 ml of saline (0.85% NaCl) is mixed with 0.5 g of HYPOL® 2000 in a 5 ml sterile test tube. The mixture is then stirred 100 strokes with a sterile wood stick, which is then formed into a cylindrical-shaped sponge with an average dimension of 10 mm (diameter) x 50 mm (length). The foam then can be cut into pieces. However, it is easier and more reliable to purchase polyurethane carriers from one of the commercial suppliers given later in this chapter. The pore size of polyurethane foam matrices is usually expressed as the number of pores per linear inch (ppi). In industry, polyurethanes are used mainly as filters, fillers, upholstery material and packing material.

Some examples of cell immobilisation in polyurethane foam carriers are given in Table 1.

Table 1. Some examples of cell immobilisation in polyurethane foam carriers.

Organism	Product/Process	Comments	Ref
Saccharomyces diastaticus	Ethanol	6-mm cubes of PUF, repeated fed-batch, 2 l airlift bioreactor	[28, 29, 30]
Nitrifier bacteria	Nitrification of high strength ammonia wastewaters	6-mm cubes of PUF (Linpor), Polyethylene "pasta" shapes of 1 cm diameter (Kaldnes), 25% particle holdup in 4 l CSTR	[31]
Pseudomonas sp.	Naphthalene degradation	Shake flasks and continuous packed-bed bioreactor, re-use of immobilised cells over 90-120 days	[32]
Aspergillus niger, *Penicillium variabile*	Organic acid production for solubilisation of inorganic (rock) phosphates in fermentation and soil	Five repeated batch cycles	[33]
Co-culture of *Ralstonia* sp. and *Pseudomonas putida*	Degradation of aromatic compounds in the effluents of anaerobic digestors treating olive mill wastewaters	10- mm cubes of PUF in 700 ml packed bed bioreactor	[34]
Rhizopus oryzae	Methanolysis of soybean oil for biodiesel fuel production from plant oils	6-mm cubes of PUF, 50 ppi, 150 particles per 500 ml flask containing 100 ml medium	[35]
Taxus spp. (yew tree)	Growth, taxanes	Eight PUF sheets of 3x15x0.5 cm with 45 ppi, arranged vertically like baffles in a batch 4 l stirred tank bioreactor with 3.5 l working volume	[36, 37]
Baculovirus infected Sf9 (insect) cells	Recombinant protein (β-galactosidase)	2-mm cubes of PVF with 60 μm mean pore diameter in shake flasks, $5x10^7$ cells/cm^3 of PVF particle	[38]
Aspergillus niger	Citric acid	PUF attached to each side of plastic disks mounted on a horizontal shaft in a 2 l, batch, rotating biological contactor with 1 l working volume	[39]
Bacillus sp.	Polygalacturonase	Shake flasks	[40]
Rhizopus arrhizus	Lipase	55x20x8 mm PUF slab with 15 ppi in 250 ml flask containing 50 ml medium	[41]
Rat hepatocytes	Tissue engineering for bioartificial liver	Carrier was a polyurethane foam disk of 30 mm diameter, 6 mm thickness with micropore size <100 μm (mean pore size 35 μm, but more than 40% of micropores had pore size of 50-100 μm), macropore size > 100 μm, and porosity of 90%.	[42]
Streptomyces sp.	Xylanase	10-mm cube PUF particles with pore size of 100-500 μm in 250 ml shake flasks	[43]

Table 1. Some examples of cell immobilisation in polyurethane foam carriers (continued).

Organism	Product/Process	Comments	Ref
Oryza sativa (rice) callus	Cultivation for somatic embryogenesis	3-mm cube PUF (ester type) particles with average pore size of 1.3 mm, in 600 ml working volume turbine blade bioreactor	[44, 45]
Prototheca zopfii	Degradation of n-alkanes	8-mm cube PUF particles in a bubble-column bioreactor	[46]
Rhizopus arrhizus	Biosorption of copper ions from dilute solutions	12 PUF sheets of 14x6x1 cm with 20 ppi arranged vertically like baffles in a batch stirred tank bioreactor with 6 l working volume	[47, 48]
Rhodococcus fascians	Limonin degradation	2-cm cubes of PUF with pore size of 0.75 mm, in 2 l continuous bioreactor with 1 l working volume, 4.4 mg cells per particle	[49]
Penicillium janthinellum	Chitinolytic enzymes	PUF cubes with an average volume of 0.03 cm^3 were used in shake flasks and 1 l (working volume) bubble column bioreactor.	[50]
Mirabilis jalapa (Marvel of Peru/four o'clock flower)	Protease	Five 1-cm cubes of PUF in 250 ml flasks	[51]
E. coli	*L*(-)-carnithine	PUF particles with 0.75 mm pore size in tubular, continuous packed-bed bioreactors	[52]
Phanerochaete chrysosporium (white-rot basidiomycete)	Peroxidase	2x2x1-cm PUF particles with 35-50 ppi, 7-14 particles in 250 ml flask	[53]
Aspergillus niger	Solubilisation of rock phosphate by acidification	1-cm and 0.5-cm cubes of PUF with pore size of 0.6-0.8 mm in 250 ml flasks	[54]
Thiobacillus ferrooxidans	Oxidation of ferrous iron	6-mm cube PUF particles with and without coating with activated carbon. Porosity of the particles was 97%, with an average pore size of 0.3 mm (80 ppi).	[17, 55, 56]
Anaerobic sludge	Wastewater treatment	5-mm cubes of PUF in horizontal flow, continuous bioreactor	[57]
Chlorella vulgaris, Chlorella kessleri, Scenedesmus quadricauda (microalgae)	Swage treatment, Manure treatment	1-cm cubes of PUF and polystyrene in 2-8 Plexiglas columns of 1 l volume, each reactor packed with 474 particles.	[58]
Aerobic co-culture of *Pseudomonas* sp And *Alcaligenes* sp	Mineralisation of low-chlorinated biphenyls (PCBs)	PUF in 0.5 l glass fixed bed bioreactor, batch with medium recycle	[59]
Catharanthus roseus	Biomass growth	PUF particles in 500 ml flasks	[60]
Scenedesmus obliquus (green algae)	Nitrate removal from water	4-5-mm cubes of PUF and PVF in 250 ml flasks and 25 ml glass column photobioreactor	[61]

Table 1. Some examples of cell immobilisation in polyurethane foam carriers.(continued)

Organism	Product/Process	Comments	Ref
Citrobacter N14	Uranium recovery by biosorption	Fifty five 1-cm cubes or hundred and sixty 3 mmx1 cmx1 cm slices of PUF in plug flow methacrylate column with 3.5 cm diameter, 15 cm height	[62]
Streptomyces coelicolor	Actinorhodin	11-13 cm^3 slab of PUF with 60 ppi in 500 ml flasks	[63, 64]
Mouse myeloma MPC-11 cells	Growth	3-mm cubes of PVF with mean pore diameter of 60 μm, in shake flask and CSTR	[65, 66, 67]
Aspergillus awamori	Glucoamylase	6-mm cubes of PVF with 97% porosity were used in shake flasks.	[68]
Capsicum frutescens, Catharanthus roseus, Daucus carota, Mentha spicata, Mucuna pruriens	Growth, secondary metabolite production, biotransformation	Five 1-cm cubes of PUF in 250 ml flasks	[69]
Capsicum frutescens	Growth, capsaicin production	Five 1-cm cubes of PUF in 250 ml flasks, 4 l bioreactor containing 600 1-cm cubes of PUF, integrated with liquid-liquid extraction column for the product	[13, 16, 24, 70, 71]
Trichoderma reesei	Cellulase	10x25x70 mm PUF sheets with 10 ppi in 500 ml flasks	[72]

4. Stainless steel knitted mesh carriers

These are produced by knitting stainless steel wires usually in stocking-stitch and then by giving the knitted sheet a form by rolling into spheres, doughnut-shaped pads and cylindrical cartridges. They are used in industry mainly as filters, coalescent components, grinders and abrasives (industrial and domestic). Manufacturers characterise them by quoting the number of stitches per linear centimetre, wire thickness, specific gravity and porosity.

Some examples of cell immobilisation in stainless steel knitted mesh carriers are given in Table 2.

5. Ceramic carriers

Ceramic pre-formed carriers are available as monoliths with cylindrical or square/rectangular prism form having parallel open channels/macro-pores of various size running along the longitudinal axis. The length can also vary. Additionally, the ceramic walls of the channels may be porous or non-porous throughout the monolith or in alternating channels creating various medium/product perfusion alternatives.

Ceramic carriers are also available as porous spheres of various sizes, Raschig rings or granules. Table 3 gives some examples of ceramic carriers used for cell immobilisation.

Table 6. Some examples of commercial suppliers of Pre-formed carriers

Pre-formed carrier	Supplier	Ref.
Polyurethane foam	Recticel Declon, Corby, UK.	[16]
	PPL Polyurethane Products Ltd, Retford, UK.	
	British Vita Plc, Manchester, UK.	
	Caligen Foam Ltd, Accrington, UK.	
	Polymer Ltd, Cardiff, UK.	
	Recticel, Wetteren, Belgium.	[53,62]
	LINPOR ®, Linde AG, Germany.	[31]
	Calther, Salamanca, Spain.	[62]
	ILPO, Bologna, Italy.	[34,59]
	Polyol International BV, Switzerland.	[49.52]
	Kirin-Bridgestone Co., Ltd., Osaka, Japan.	[35,81]
	Terumo Co., Tokyo, Japan.	[42]
	INOAC, Nagoya, Japan.	[44,45]
Polyvinyl formal foam	Kanebo Kasei Co., Japan.	[67]
Polyurethane prepolymer	(HYPOL) Hampshire Chemical Corp., Boston, MA, USA	[112]
Other plastics	(KALDNES) Anox AB, Sweden, Kaldnes Co., Sweden.	[31]
Stainless Steel Mesh	KnitMesh Limited, Surrey, England.	[1,76]
	Scotch Bride 3M, Spain.	[75]
Glass	(Sintered, Raschig rings or SIRAN spheres) Schott Glaswerke, Mainz, Germany.	[81,88,92]
	Fisons, UK.	
	(Borosilicate glass cover slips) Fisher Scientific, Albany, NY, USA.	[90]
	Merck, Dietikon, Switzerland.	[21]
	Roth, Karlsruhe, Germany.	[21]
	Huber, Reinach, Switzerland.	[21]
Ceramic	(Monolith) Corning Inc. Corning, NY, USA.	
	(OptiCell®) USA Scientific Inc., USA.	[85]
	(Biolite® beads) Degremont, Bilbao, Spain.	[78]
	(Beads) Gallenkamp, UK.	[62]
	(Raschig rings, Berl saddles) Norton, USA and UK.	[62]
Cellulose	VWR Scientific, Canada.	[99]
	(Granular DEAE-cellulose) Sigma Chemical Company	[102]
	(AQUACEL®) Acordis, Spondon, UK.	[107]
	(AQUACEL®) ConvaTec (Bristol-Myers Squibb Co.), USA.	
Cellulose	Biomaterial Co. Ltd, Japan.	[104]
	Sakai Eng. Co. Ltd., Fukui, Japan.	[106
	(Polystyrene coated with DEAE-cellulose) Cultor® Ltd, Finnsugar Bioproducts, Helsinki, Finland.	[103]

9. Conclusions

Cell immobilisation in pre-formed carriers is easy, natural and therefore harmless to cells. These carriers are mostly inert materials with good mechanical strength. They can be autoclaved and since immobilisation is carried out *in situ*, the risk of contamination during immobilisation is the lowest compared to other methods and materials of immobilisation. They can be used repeatedly with immobilised cells. Immobilised cells in these carriers can be stored in the medium or saline solution in the refrigerator or several weeks. Bacteria, yeast and most animal cells can be removed from these carriers easily. It is however, more difficult to remove mycelial and plant cells. As with most immobilisation materials, cells can leak from these carriers, but they support good cell growth.

Monitoring and quantification of immobilised cells in these carriers is possible using wet/dry weight measurements, although some carriers may loose weight slightly during high temperature drying. Of course, microscopy such as SEM, various staining techniques (viability), indirect quantification based on measurable cell activity (*e.g.*, respiration), and NMR are some of the other techniques used for quantification and characterisation of cells immobilised in pre-formed carriers.

As mentioned in the introduction, the largest industrial scale application of immobilised cells, biological wastewater treatment and vinegar production, rely on the pre-formed carriers.

References

[1] Atkinson, B.; Black, G.; Lewis, P. and Pinches, A. (1979) Biological particles of given size, shape and density for use in biological reactors. Biotechnol. Bioeng. 21: 193-200.

[2] Cooper, P.F.; Walker, I.; Crabtree, H.E. and Aldred, R.P. (1986) Evaluation of the CAPTOR® process for uprating an overloaded sewage works. In: Webb, C.; Black, G.M. and Atkinson, B. (Eds.), Process Engineering Aspects of Immobilised Cell Systems, The Institution of Chemical Engineers, Rugby, UK, ISBN 0 85295 196 5, pp. 205-217.

[3] Golla, P.S.; Reddy, M.P.; Simms, M.K. and Laken, T.J. (1994) Three years of full-scale Captor® process operating at Moundsville WWTP, Water Sci. Technol. 29: 175-181.

[4] http://www.linde-anlagenbau.de/en

[5] http://www.kaldnes.com

[6] Dunne, W.M. (2002) Bacterial adhesion: seen any good biofilms lately? Clinical Microbiol. Rev. 15: 155-166.

[7] O'Toole, G.; Kaplan, H.B. and Kolter, R.. (2000) Biofilm formation as microbial development. Annu. Rev. Microbiol. 54: 49-79.

[8] Sauer, K. and Camper, A.K. (2001) Characterization of phenotypic changes in *P. putida* in response to surface-associated growth. J. Bacteriol. 183: 6579-6589.

[9] Wen, Z.T. and Burne, R.A. (2002) Functional genomics approach to identifying genes required for biofilm development by *S. mutans*, Appl. Env. Microbiol. 68:1196-1203.

[10] Eberl, L.; Molin, S. and Givskov, M. (1999) Surface motility of *S. liquefaciens* MG1. J. Bacteriol. 181: 1703-1712.

[11] Loo, C.Y.; Corliss, D.A. and Ganeshkumar, N. (2000) *S. gordonii* biofilm formation: identification of genes that code for biofilm phenotypes. J. Bacteriol. 182: 1374-1382.

[12] Shapiro, J.A. and Dworkin, M. (Eds.) (1997) Bacteria as Multicellular Organisms. Oxford Univ. Press, New York, USA.

[13] Williams, P.D. and Mavituna, F. (1992) Immobilised plant cells. In: Plant biotechnology, Comprehensive biotechnology, second supplement, Fowler, M.W. and Warren, G.S. (Eds.), Moo-Young, M. (Ed-in-Chief), Pergamon Press, Oxford, UK, pp. 63-78.

[14] Liu, D. and Dixit, V. (Eds.) (1997) Porous materials for tissue engineering. Materials Sci. Forum, Vol. 250, TTP, Switzerland.

[15] Chapekar, M.S. (2000) Tissue engineering: challenges and opportunities. J. Biomed. Mater. Res. (Appl. Biomater.) 53: 617-620.

[16] Mavituna, F.; Park, J.M.; Williams, P.D. and Wilkinson, A.K. (1987) Characteristics of immobilised plant cell reactors. In: Plant and Animal Cells: Process Possibilities, Webb, C. and Mavituna, F. (Eds.), Ellis Horwood Ltd, Chichester, UK. ISBN 0 7458 0145 5, pp. 92-115.

[17] Nemati, M. and Webb, C. (1999) Combined biological and chemical oxidation of ferrous sulfate using immobilised *Thiobacillus ferrooxidans*. J. Chem. Technol. Biotechnol. 74: 562-570.

[18] Armstrong, J.L.; Mavituna, F. and Stephens, G.M. (1997) Novel immobilisation supports for anaerobic bacteria. The 1997 Jubilee Research Event, Vol. 2, IChemE, Rugby. UK, ISBN 0 85295 389 5, pp. 969-972.

[19] Bekers, M.; Ventina, E.; Karsakevich, A.; Vina, I.; Rapoport, A.; Upite, D.; Kaminska, E. and Linde, R. (1999) Attachment of yeast to modified stainless steel wire spheres, growth of cells and ethanol production. Process Biochem. 35: 523-530.

[20] Grampp, G.E.; Applegate, M.A. and Stephanopoulos, G. (1996) Cyclic operation of ceramic-matrix animal cell bioreactors for controlled secretion of an endocrine hormone. A comparison of single-pass and recycle modes of operation. Biotechnol Prog. 12: 837-846.

[21] Simoni, S.F.; Schäfer, A.; Harms, H. and Zehnder, A.J.B. (2001) Factors affecting mass transfer limited biodegradation in saturated porous media. J. Contam. Hydrol. 50: 99-120.

[22] Jerabkova, H.; Kralova, B. and Nahlik, J. (1999) Biofilm of *Pseudomonas* C12B on glass support as catalytic agent for continuous SDS removal. Intern. Biodeterior. Biodegrad. 44: 233-241.

[23] Ileri, R. and Mavituna, F. (1998) A theoretical study of biosorption by immobilized dead biomass in a batch sheet bioreactor. Trans. IChemE 76 (B3): 249-258.

[24] Park, J.M. and Mavituna, F. (1986) Factors affecting the immobilisation of plant cells in biomass support particles. In: Process Engineering Aspects of Immobilised Cell Systems, Webb, C.; Black, G.M. and Atkinson, B. (Eds.), Pergamon Press, Oxford, pp. 295-303.

[25] Williams, P D, Wilkinson, A K, Lewis, J A, Black, G M and Mavituna, F (1988) A method for the rapid production of fine plant cell suspension cultures, Plant Cell Reports, 7, 459-462.

[26] Wirpsza, Z. (1993) Polyurethanes: chemistry, technology and applications. Ellis Horwood, Chichester, UK.

[27] Bang, S S.; Galinat, J.K. and Ramakrishnan, V. (2001) Calcite precipitation induced by polyurethane-immobilized *Bacillus pasteurii*. Enzyme Microb. Technol. 28: 404-409.

[28] Kishimoto, M.; Beluso, M.; Omasa, T.; Katakura, Y.; Fukuda, H. and Suga, K. (2002) Construction of a fuzzy control system for a bioreactor using biomass support particles. J. Mol. Catalysis (B: Enzymatic) 17: 207-213.

[29] Liu, Y.; Kondo, A.; Ohkawa, H.; Shiota, N. and Fukuda, H. (1998) Bioconversion using immobilized recombinant flocculent yeast cells carrying a fused enzyme gene in an `intelligent' bioreactor. Biochem. Eng. J. 2: 229-235.

[30] Furuta, H.; Arai, T.; Hama, H.; Shiomi, N.; Kondo, A. and Fukuda, H. (1997) Production of glucoamylase by passively immobilised cells of a flocculant yeast *Saccharomyces diastaticus*. J. Ferment. Bioeng. 84: 169-171.

[31] Rostron, W.M.; Stuckey, D.C. and Young, A.A. (2001) Nitrification of high strength ammonia wastewaters: comparative study of immobilisation media. Water Res. 35: 1169-1178.

[32] Manohar, S.; Kim, C.K. and Karegoudar, T.B. (2001) Enchanced degradation of naphthalene by immobilisation of *Pseudomonas* sp. strain NGK1 in polyurethane foam. Appl. Microbiol. Biotechnol. 55: 311-316.

[33] Vassilev, N.; Vassileva, M.; Fenice, M. and Federici, F. (2001) Immobilised cell technology applied in solubilisation of insoluble inorganic (rock) phosphates and P plant acquisition. Bioresource Technol. 79: 263-271.

[34] Bertin, L.; Majone, M.; Gioia, D.D. and Fava, F. (2001) An aerobic fixed-phase biofilm reactor system for the degradation of the low-molecular weight aromatic compounds occurring in the effluents of anaerobic digestors treating olive mill wastewaters. J. Biotechnol. 87: 161-177.

[35] Ban, K.; Kaieda, M.; Matsumoto, T.; Kondo, A. and Fukuda, H. (2001) Whole cell biocatalyst for biodiesel fuel production utilising *Rhizopus oryzae* cells immobilised within biomass support particles. Biochem. Eng. J. 8: 39-43.

[36].Tang, C.W.; Zalat, E. and Mavituna, F. (2001) Initiation, growth and immobilisation of cell cultures of *Taxus* spp. for paclitaxel production. In: Focus in Biotechnology: Engineering and Manufacturing for Biotechnology, Hofman, M. and Thonart, P. (Eds.), Volume IV, Anne, J. and Hofman M, (Ser. Eds.), Kluwer Publishers, Dordrecht, The Netherlands, pp. 429-448.

[37] Tang, C.W. and Mavituna, F. (2001) Cell immobilisation of *Taxus media*. In: Novel Frontiers in the Production of Compounds for Biomedical Use, Van Broekhoven, A.; Shapiro, F. and Anne, J. (Vol. Eds.) Volume I, Part 6: Antitumor Compounds, Anne, J. and Hofman M. (Ser. Eds.), Kluwer Publishers, Dordrecht, The Netherlands, pp. 401- 407.

[38] Yamaji, H.; Tagai, S.I.; Sakai, K.; Izumoto, E. and Fukuda, H. (2000) Production of recombinant protein by Baculovirus-infected insect cells in immobilised culture using porous biomass support particles. J. Bioscience Bioeng. 89: 12-17.

[39] Jianlong, W. (2000) Production of citric acid by immobilized *Aspergillus niger* using a rotating biological contactor (RBC). Bioresource Technol. 75: 245-247.
[40] Kapoor, M.; Beg, Q.K.; Bhushan, B.; Dadhich, K. S. and. Hoondal, G. S (2000) Production and partial purification and characterization of a thermo-alkali stable polygalacturonase from *Bacillus* sp. MG-cp-2. Process Biochem. 36: 467-473.
[41] Elibol, M. and Özer, D. (2000) Lipase production by immobilised *Rhizopus arrhizus*. Process Biochem. 36: 219-223.
[42] Kurosawa, H.; Yasumoto, K.; Kimura, T. and Amano, Y. (2000) Polyurethane membrane as an efficient immobilisation carrier for high-density culture of rat hepatocytes in the fixed-bed reactor. Biotechnol. Bioeng. 70: 160-166.
[43] Beg, Q.K.; Bhushan, B.; Kapoor, M. and Hoondal, G.S. (2000) Enhanced production of a thermostable xylanase from *Streptomyces* sp. QG-11-3 and its application in biobleaching of eucalyptus kraft pulp. Enzyme Microb. Technol. 27: 459-466.
[44] Liu, C.; Moon, K.; Honda, H. and Kobayashi, T. (2000) Immobilization of rice (*Oryza sativa* L.) callus in polyurethane foam using a turbine blade reactor. Biochem. Eng. J. 4:169-175.
[45] Moon, K.H.; Honda, H. and Kobayashi, T. (1999) Development of a bioreactor suitable for embryogenic rice callus culture. J. Bioscience Bioeng. 87: 661-665.
[46] Yamaguchi, T.; Ishida, M. and Suzuki, T. (1999) An immobilized cell system in polyurethane foam for the lipophilic micro-alga *Prototheca zopfii*. Process Biochem. 34: 167-172.
[47] Ileri, R. and Mavituna, F. (1998) A theoretical study of biosorption by immobilized dead biomass in a batch sheet bioreactor. Trans. IChemE, 76 (B3), 249-258.
[48] Ileri, R.; Mavituna, F.; Parkinson, M. and Turker, M. (1990) The use of biosorption for the uptake of low level contaminants by immobilised cells. Proc. 5th European Congress on Biotechnology, Christiansen, C.; Munck, L. and Villadsen, J. (Eds.), Munksgaard International Publisher, Copenhagen, Vol. II, pp. 663-666.
[49] Cánovas, M.; García-Cases, L. and Iborra, J.L. (1998) Limonin consumption at acidic pH values and absence of aeration by *Rhodococcus fascians* cells in batch and immobilized continuous systems. Enzyme Microb. Technol. 22: 111-116.
[50] Fenice, M.; Di Giambattista, R.; Raetz, E.; Leuba, J.L. and Federici, F. (1998) Repeated-batch and continuous production of chitinolytic enzymes by *Penicillium janthinellum* immobilised on chemically-modified macroporous cellulose. J. Biotechnol. 62: 119-131.
[51] Tamer, I.M. and Mavituna, F. (1997) Protease from freely suspended and immobilised *Mirabilis jalapa*. Process Biochem. 32: 195-200.
[52] Obón, J.S.; Maiquez, J.R.; Canovas, M.; Kleber, H.P. and Iborra, J.L. (1997) L(-)-Carnitine production with immobilized *Escherichia coli* cells in continuous reactors. Enzyme Microb. Technol. 21: 531-536.
[53] Gerin, P.A.; Asther, M. and Rouxhet, P.G. (1997) Peroxidase production by the filamentous fungus *Phanerochaete chrysosporium* in relation to immobilization in "filtering" carriers. Enzyme Microb. Technol. 20: 294-300.
[54] Vassilev, N.; Vassileva, M. and Azcon, R. (1997) Solubilization of rock phosphate by immobilized *Aspergillus niger*. Bioresource Technol. 59 (1), 1-4.
[55] Nemati, M. and Webb, C. (1996) Effect of ferrous iron concentration on the catalytic activity of immobilised cells of *Thiobacillus ferooxidans*. Appl. Microbiol. Biotechnol. 46: 250-255.
[56] Jensen, A.B. and Webb, C. (1994) A trickle bed reactor for ferrous sulphate oxidation using *Thiobacillus ferroxidans*. Biotechnol. Techn. 8: 87-92.
[57] Zaiat, M.; Cabral, A.K.A. and Foresti, E. (1996) Cell wash-out and external mass transfer resistance in horizontal-flow anaerobic immobilized sludge reactor. Water Res. 30: 2435-2439.
[58] Travieso, L.; Benitez, F.; Weiland, P.; Sánchez, E.; Dupeyrón, A. and Dominguez, A.R. (1996) Experiments on immobilization of microalgae for nutrient removal in wastewater treatments. Bioresource Technol. 55: 181-186.
[59] Fava, F.; Baldoni, F.; Marchetti, L. and Quattroni, G. (1996) A bioreactor system for the mineralization of low-chlorinated biphenyls. Process Biochem. 31: 659-667.
[60] Zong, D.H. and Ying, J.Y. (1995) Fuzzy growth kinetics of immobilized *C. roseus* cells in polyurethane foams. Chem. Eng. Sci. 50: 3297-3301.
[61] Urrutia, I.; Serra, J.L. and Llama, M.J. (1995) Nitrate removal from water by *Scenedesmus obliquus* immobilized in polymeric foams. Enzyme Microb. Technol. 17: 200-205.
[62] Roig, M.G.; Manzano, T.; Díaz, M.; Pascual, M.J.; Paterson, M. and Kennedy, J.F. (1995) Enzymically-enhanced extraction of uranium from biologically leached solutions. Int. Biodeterior. Biodegrad. 35: 93-127.

[63] Ozergin-Ulgen, K. and Mavituna, F. (1994) Production of actinorhodin by immobilised and freely suspended *Streptomyces coelicolor*. Progr. Biotechnol. 9: 497-500.

[64] Ozergin-Ulgen, K. and Mavituna, F. (1994) Comparison of the activity of immobilised and freely suspended *Streptomyces coelicolor* A3(2). Appl. Microbiol. Biotechnol. 41: 197-202.

[65] Yamaji, H. and Fukuda, H. (1994) Growth kinetics of animal cells immobilised within porous support paticles in a perfusion culture. Appl. Microbiol. Biotechnol. 42: 531-535.

[66] Yamaji, H. and Fukuda, H. (1992) Growth and death behaviour of anchorage-independent animal cells immobilised within porous support matrices. Appl. Microbiol. Biotechnol. 37: 244-251.

[67] Yamaji, H.; Fukuda, H.; Nojima, Y. and Webb, C. (1989) immobilisation of anchorage-independent animal cells using reticulated polyvinyl formal resin biomass support particles. Appl. Microbiol. Biotechnol. 30: 609-613.

[68] Bon, E. and Webb, C. (1989) Passive immobilisation of *Aspergillus awamori* spores for subsequent glucoamylase production. Enzyme Microb. Technol. 11: 495-499.

[69] Wilkinson, A.K.; Park, J.M.; Williams, P.D. and Mavituna, F. (1990) Immobilisation of plant cells and bioreactor design. Proc. APBioChE '90, Korea, pp. 173-176.

[70] Mavituna, F. and Park, M. (1985) Growth of immobilised plant cells in reticulate polyurethane foam matrices. Biotechol. Lett. 7: 637-640.

[71] Mavituna, F.; Williams, P.D.; Wilkinson, A.K. and Park, J.M. (1987) Bioreactor performance for the production of secondary metabolites by immobilised plant cells. Proc. of the 4th European Congress on Biotechnology, Amsterdam, June 14-20,2, pp. 385-387.

[72] Turker, M. and Mavituna, F. (1987) Production of cellulase by freely suspended and immobilised cells of *Trichoderma reesei*. Enzyme Microb. Technol. 9: 739-743.

[73] Bekers, M.; Laukevics, J.; Karsakevich, A.; Ventina, E.; Kaminska, E.; Upite, D.; Vina, I.; Linde, R. and Scherbaka, R. (2001) Levan-ethanol biosynthesis using *Zymomonas mobilis* cells immobilized by attachment and entrapment. Process Biochem. 36: 979-986.

[74] Bekers, M.; Ventina, E.; Laukevics, J.; Kaminska, E.; Upite, D. and Vigants, A. (1997) Levan production by *Zymomonas mobilis* cells attached to plaited spheres. Acta biotechnol, 17: 265-275.

[75] Gomez, J.M.; Cantero, D. and Webb C. (2000) Immobilisation of *Thiobacillus ferrooxidans* cells on nickel alloy fibre for ferrous sulfate oxidation. Appl. Microbiol. Biotechnol. 54: 335-340.

[76] Cross, P.A. and Mavituna, F. (1987) Yeast retention fermenters for beer production Proc. of the 4th European Congress on Biotechnology, Amsterdam, June 14-20 Vol. l, pp. 199-200.

[77] Yongming, Z.; Liping, H.; Jianlong, W.; Juntang, Y.; Hanchang, S. and Yi, Q. (2002) An internal airlift loop bioreactor with *Burkholderia pickttii* immobilized onto ceramic honeycomb support for degradation of quinoline. Biochem. Eng. J. 11: 149-157.

[78] Prieto, M.B.; Hidalgo, A.; Serra, J.L. and Llama, M.J. (2002) Degradation of phenol by *Rhodococcus erythropolis* UPV-1 immobilized on Biolite® in a packed-bed reactor. J. Biotechnol. 97: 1-11.

[79] Resende, M.M.; Ratusznei, S.M.; Suazo, C.A.T. and Giordano, R.C. (2002) Simulating a ceramic membrane bioreactor for the production of penicillin: an example of the importance of consistent initialization for solving DAE systems. Process Biochem. 37: 1297-1305.

[80] Martin, M.; Mengs, G.; Plaza, E.; Garbi, C.; Sanchez, M.; Gibello, A.; Gutierrez, F. and Ferrer, E. (2000) Propachlor removal by *Pseudomonas* strain GCH1 in an immobilised-cell system. Appl. Environ. Microbiol. 66: 1190-1194.

[81] Christov, P.; Spassov, G. and Pramatarova, V. (1999) Effect of matrix on (S)-p-chlorodiphenylmethanol production by immobilized *Debaryomyces marama*. Process Biochem. 34: 231-237.

[82] Ohashi, R.; Kamoshita, Y.; Kishimoto, M. and Suzuki, T. (1998) Continuous production and separation of ethanol without effluence of wastewater using a distiller integrated scm-reactor system. J. Ferment. Bioeng. 86: 220-225.

[83] Martin-Montalvo, D.; Mengs, G.; Ferrer, E.; Allende, J.L.; Alonso, R. and Martin, M. (1997) Simazine degradation by immobilized and suspended soil bacterium. Int. Biodeterior. Biodegredat. 40: 93-99.

[84] Salter, G.J.; Kell, D.B.; Ash, L.A.; Adams, J.M.; Brown, A.J. and James, R. (1990) Hydrodynamic deposotion: a novel method of cell immobilisation. Enzyme Microbial. Technol. 12: 419-430.

[85] Bodeker, B.G.; Hubner, G.E.; Hewlett, G. and Schlumberger, H.D. (1987) Production of human monoclonal antibodies from immobilised cells in the Opticell culture system. Dev. Biol. Stand. 66: 473-479.

[86] Barbucci, R.; Magnani, A.; Lamponi, S.; Pasqui, D. and Bryan, S. (2003) The use of hyaluronan and its sulphated derivative patterned with micrometric scale on glass substrate in melanocyte cell behaviour. Biomaterials 24: 915-926.

[87] Barbucci, R.; Lamponi, S.; Magnani, A. and Pasqui, D. (2002) Micropatterned surfaces for the control of endothelial cell behaviour. Biomol. Eng. 19: 161-170.
[88] Kuncova, G.; Triska, J.; Vrchotova, N. and Podrazky, O. (2002) The influence of immobilization of *Pseudomonas* sp. 2 on optical detection of polychlorinated biphenyls. Mat. Sci. Eng. (C) 21: 195-201.
[89] Cornish, T.; Branch, D.W.; Wheeler, B.C. and Campanelli, J.T. (2002) Microcontact printing: a versatile technique for the study of synaptogenic molecules. Molec. Cell. Neurosci. 20: 140-153.
[90] Kam, L.; Shain, W.; Turner, J.N. and Bizios, R. (2002) Selective adhesion of astrocytes to surfaces modified with immobilized peptides. Biomaterials 23: 511-515.
[91] Premkumar, J.R.; Lev, O.; Marks, R.S.; Polyak, B.; Rosen, R. and Belkin, S. (2001) Antibody-based immobilization of bioluminescent bacterial sensor cells. Talanta 55: 1029-1038.
[92] Bonin, P.; Rontani, J.F. and Bordenave, L. (2001) Metabolic differences between attached and free-living marine bacteria: inadequacy of liquid cultures for describing *in situ* bacterial activity. FEMS Microbiol. Lett. 194: 111-119.
[93] Pogliani, C. and Donati, E. (2000) Immobilisation of *Thiobacillus ferrooxidans*: importance of jarosite precipitation. Process Biochem. 35: 997-1004.
[94] Noll, T. and Biselli, M. (1998) Dielectric spectroscopy in the cultivation of suspended and immobilised hybridoma cells. J. Biotechnol. 63: 187-198.
[95] Yokoi, H.; Tokushige, T.; Hirose, J.; Hayashi, S. and Takasaki, Y. (1997) Hydrogen production by immobilised cells of aciduric *Enterobacter aerogenes* strain HO-39. J. Ferment. Bioeng. 83: 481-484.
[96] Edgehill, R.E. (1996) Degradation of pentachlorophenol (PCP) by *Arthrobacter* strain ATCC 33790 in biofilm culture. Water Res. 30: 357-363.
[97] Truck, H.U.; Chmiel, H.; Hammes, W.P. and Trosch, W. (1990) A study of N- and P-dependence of nikkomycin production in continuous culture with immobilised cells. Appl. Microbiol. Biotechnol. 33: 139-144.
[98] Porto, A.L.M.; Cassiola, F.; Dias, S.L.P.; Joekes, I.; Gushikem, Y.; Rodrigues, J.A.R.; Moran, P.J.S.; Manfio, G.P. and Marsaioli, A.J. (2002) *Aspergillus terreus* CCT 3320 immobilized on chrysotile or cellulose/TiO_2 for sulfide oxidation. J. Mol. Catalysis (B: Enzymatic) 19-20: 327-334.
[99] Dahiya, J.; Singh, D. and Nigam, P. (2001) Decolourisation of molasses wastewater by cells of *Pseudomonas fluorescens* immobilised on porous cellulose carrier. Bioresource Technol. 78: 111-114.
[100] Kumar, N. and Das, D. (2001) Continuous hydrogen production by immobilised *Enterobacter cloacae* IIT-BT 08 using lignocellulosic materials as solid matrices. Enzyme Microb. Technol. 29: 280-287.
[101] Fujii, N.; Oki, T.; Sakurai, A.; Suye, S. and Sakakibara, M. (2001) Ethanol production from starch by immobilised *Aspergillus awamori* and *Saccharomyces partorianus* using cellulose carriers. J. Indust. Microbiol. Biotechnol. 27: 52-57.
[102] Smogrovicova, D. and Domeny, Z. (1999) Beer volatile by-product formation at different fermentation temperature using immobilised yeasts. Process Biochem. 34: 785-794.
[103] van Iersel, M.F.M.; van Dieren, B.; Rombouts, F.M. and Abee, T. (1999) Flavour formation and cell physiology during the production of alcohol-free beer with immobilized *Saccharomyces cerevisiae*, Enzyme Microb. Technol. 24: 407-411.
[104] Fenice, M.; Di Giambattista, R.; Raetz, E.; Leuba, J.L. and Federici, F. (1998) Repeated-batch and continuous production of chitinolytic enzymes by *Penicillium janthinellum* immobilised on chemically-modified macroporous cellulose. J. Biotechnol. 62: 119-131.
[105] Liu, Y.K.; Seki, M.; Tanaka, H. and Furusaki, S. (1998) Characteristics of Loofa (*Luffa cylindrica*) sponge as a carrier for plant cell immobilisation. J. Ferment. Bioeng. 85: 416-421.
[106] Chen, J.P. and Wang, J.B. (1997) Wax ester synthesis by lipase-catalyzed esterification with fungal cells immobilized on cellulose biomass support particles. Enzyme Microb. Technol. 20: 615-622.
[107] Matsumura, M.; Yamamoto, T. Wang, P.C.; Shinabe, K. and Yasuda, K. (1997) Rapid nitrification with immobilized cell using macro-porous cellulose carrier. Water Res. 31: 1027-1034.
[108] Catalan-Sakairi, M.A.; Wang, P.C. and Matsumura, M. (1997) High-rate seawater denitrification utilizing a macro-porous cellulose carrier. J. Ferment. Bioeng. 83: 102-108.
[109] Matsumura, M.; Tsubota, H.; Ito, O.; Wang, P.C. and Yasuda, K. (1997) Development of bioreactors for denitrification with immobilized cells. J. Ferment. Bioeng. 84: 144-150.
[110] Ogbonna, J.C.; Tomiyama, S. and Tanaka, H. (1996) Development of a method for immobilisation of non-flocculating cells in loofa (*Luffa cylindrica*) sponge. Process Biochem. 31: 737-744.
[111] Kumakura, M.; Yoshida, M. and Asano, M. (1992) Preparation of immobilized yeast cells with porous substrates. Process Biochem. 27: 225-229.

[112] Bang, S.S.; Galinat, J.K. and Ramakrishnan, V. (2001) Calcite precipitation induced by polyurethane-immobilised *Bacillus pasteurii*. Enzyme Microb. Technol. 28: 404-409

.

MICROCARRIERS FOR ANIMAL CELL CULTURE

ELENA MARKVICHEVA[1] AND CHRISTIAN GRANDFILS[2]
[1]*Shemyakin&Ovchinnikov Institute of Bioorganic Chemistry, Russian Academy of Sciences, Miklukho Maklaya Str, 16/10, 117997, Moscow, Russia – Fax: 007(095)3351011 – Email: lemark@ibch.ru*
[2]*Centre Interfacultaire des Biomatériaux (CEIB), University of Liège, Chemistry Institute, B6c, 4000 Liège (Sart-Tilman), Belgium – Fax: 32(0)43663623 – Email: C.Grandfils@ulg.ac.be*

1. Introduction

Due to a great recent progress in animal cell biotechnology in particular in gene engineering, large-scale production of important biological materials, such as viral vaccines, viral vectors, as well as various cell products has become an industrial reality. The list of these products includes recombinant proteins and peptides, lymphokines and cytokines, enzymes, monoclonal antibodies, hormones, growth factors (*e.g.* platelet-derived growth factor (PDFG), epithelial growth factor (EGF), colony-stimulating factor (CSF), tissue plasminogen activator, serum proteins (*e.g.* factor VIII, factor IX), tumour necrosis factor (TNF), erythropoietin (EPO), nucleic acids *etc.* Some of these products are available on the market as human and animal diagnostics and therapeutics. Many others are currently undergoing evaluation in a developmental process or in clinical trials.

As for recombinant DNA technology, although earlier bacteria, yeast and fungi have been considered as the most appropriate and the cheapest sources to obtain recombinant biological products, later it has become obvious that these expression systems could not perform post-translational modifications (*i.e.* amidification, glycosylation, phosphorylation, carboxylation, *etc.*). Additionally, it was realized that these systems were frequently unable to excrete the bioproducts of interest. Animal cells possess all these potentials, and that is why they became unavoidable for the controlled production of recombinant proteins and vaccines.

Animal cell lines can be cultivated *in vitro* in two basically different modes: as anchorage-independent cells by growing in suspension or as anchorage-dependent cells which require attachment to a solid substrate for their spreading and proliferation.

Naturally, most mammalian cells require a solid substrate for spreading and further proliferation like it occurs *in vivo*. Although within two last decades, several continuous established cell lines have been adopted to grow as free-suspended cultures, their

V. Nedović and R. Willaert (eds.), Fundamentals of Cell Immobilisation Biotechnology, 141-161.

application for production of human and veterinary pharmaceuticals is still questionable for several reasons:

- their tumourigenic potential
- varying antigenicity of viral particles with time (which does not occur in the case of microcarrier cell cultures)
- often worse cell proliferation and productivity of these free-suspended cell lines compared to microcarrier cultures.

In the latter case, several recent reports have suggested that proliferation and production of rCHO free-suspended cells were inferior to those ones obtained for the same cells cultured on microcarriers. This different cell behaviour has been correlated to the activation of cell inhibitors, such as the p27 molecule, delaying the transition from G1 to S phase [1]. Others have observed that cell adhesion to adequate substrates has resulted in a higher expression level of various cyclines (cyclines G1, D and E), which represent key molecules of the cellular cycle [2].

Suspended cells in large reactors are additionally subjected to physical/mechanical aggressions, which can alter their productivity, quality of bioproducts as well as reproducibility of processes [3-4]. Anyway, it should be noted that the fact mentioned above (namely, that most animal cells do need to be attached to a solid substrate in order to grow *in vitro*), was a reason of the delay for the use of standard microbiological homogeneous cultivation techniques for the propagation of animal cells in bioreactors.

In an attempt to provide systems that offer large accessible surfaces for cell growth in small-culture volume, a number of techniques have been proposed: the roller bottle system, the stack plates propagator, the spiral film bottles, the packed-bed system, the plate exchanger system *etc*. All these systems are non-homogeneous and suffer from limited potential for scale-up, difficulties to take samples, limited potential to measure and to control cultivation parameters [5].

To overcome these limitations van Wezel, who was called the "father of the microcarrier" by van der Velden-de Groot [6] developed a concept of the microcarrier culture system. In fact this system combines features of monolayer and suspension cultures. The idea of the method is very simple: van Wezel proposed to grow cells as monolayers on the surface of small beads (microspheres) which are suspended in culture medium by gentle stirring [7]. He suggested the ion exchange gel DEAE-Sephadex A-50 beads as microcarriers and demonstrated that a homogeneous microcarrier system could be used for the large scale culture of anchorage-dependent cells. Later it has been shown that the microcarrier technique could be scaled-up for the variety of large-scale production processes [8]. DEAE-Sephadex A-50 provided a charged culture surface with a large surface area/volume ratio, a beaded form, good optical properties and a suitable density. Since the yield of anchorage-dependent cells depends on the surface area available for growth, it was believed that the maximum cell density in microcarrier cultures would depend on the microcarrier surface area [9]. However, DEAE-Sephadex A-50 was toxic for cells at concentrations more than 1 mg/ml. Levine *et al.* [10-13] suggested lowering the level of DEAE substitution, and these new developments have led to the commercial production of Cytodex microcarriers (launched in 1981) as a result of cooperation between van Wezel and Pharmacia Biotech AB.

The exciting history of the microcarrier is perfectly described by van der Velden-de Groot [6]. Since 1967 ("a birthday of microcarrier") a lot of companies developed a wide spectrum of microcarriers based on different natural and synthetic materials. Hence classical microporous microcarriers (for instance Cytodex, Amersham Biosciences) as well as later developed macroporous microcarriers which allow cell growth inside the beads, thereby increasing cell density and protecting cells against mechanical stress at agitation (Cultispher-G, -S, -GL, Percell Biolytica) can be found on the market. Choosing the correct microcarrier one has to keep in mind the type of cells being cultured and the purpose of the culture.

Now we could say that microcarrier culture technique has a lot of advantages for the commercial manufacturer. It operates in batch or perfusion modes and is well suited to efficient process development and smooth scale-up. Additionally, the bioreactors can be modified to grow other organisms.

Microcarrier culture technique is not only limited to the large-scale production of cells, viruses or cell products. It offers other possibilities, one of them is fundamental studies on cell function, metabolism and differentiation, as will be reported later on.

Finally, microcarrier culture technique opens up a whole range of possibilities for therapeutic applications. Multicellular systems are becoming promising tools which open the way to the development of patient specific cellular therapies and tissue engineering where biocompatible tri-dimensional polymer micromatrices could serve as temporary scaffolds to support and to promote cell colonisation.

Several excellent papers and reviews on microcarriers have been published [5, 14] in which the most important key-aspects, such as economical and technological advantages of microcarrier cell culture for industries have been reported as well as the description of microcarrier properties (mechanical stability, optical properties, specific density, sedimentation velocity, surface specificity, porosity, *etc.*), in more than practical considerations useful to optimise cell culture in bioreactors.

The aim of this chapter is to review the development of microcarriers for the last five-seven years both in terms of their evolution on the market (number, diversity) and their scientific analysis in the literature. Special attention will be paid to the control and characterisation of microcarrier surface properties. As well known from cell biologists and from biomaterial engineers, these surface properties are indeed crucial to control cell adhesion and spreading on a specific substrate. Finally, new promising applications of microcarriers in medicine and biotechnology will be considered.

2. Microcarrier design

2.1. MATERIALS

The choice of material for the microcarrier preparation is a crucial point, because it determines several essential features, such as physico-chemical, mechanical and optical properties. As for microcarrier density, biocompatibility, porosity, shape, surface characteristics, their level of adjustment also depends on the nature of the material. Moreover, the stability of these various specifications of the microcarriers will also be directly influenced by the material nature.

As mentioned above, the first microcarriers proposed in the late sixties were hydrophilic ion-exchange chromatographic beads based on cross-linked dextran. Later observations showed that cells can adhere and spread on a variety of substrates, and a wide range of materials has been used to produce microcarriers. Among them we would like to mention some inorganic materials (glass, silica and ceramics), but mainly organic ones, both natural polymers (collagen and gelatin, cellulose and its derivatives) and synthetic ones, such as polystyrene, polyacrylamide, polyacrylates, polyethylene, polypropylene, polyvinylchloride, polycarbonate, polyethyleneterephtalate, polyvinyldifluoroethylene, polydimethylsiloxane, *etc*.

2.2. SURFACE PROPERTIES AND CHARGE

It is not surprising that many compounds have been proposed as materials for the preparation of microcarriers. Indeed, if cells can adhere to a large variety of substrates, it appears that cell behaviour on the microcarrier is mostly influenced not by the properties of the raw material, but first of all by the microcarrier surface properties. Therefore, from a physical, chemical and mechanical point of view surfaces can be designed in such a way that their properties can be completely different from the ones inside the matrix.

As for surface tailoring, several general aspects are important:

- First of all, one has to realize a thickness of the material layer – which really contacts the surrounding liquid media – within the nanometre range. This corresponds to several (macro-) molecule layers, which are chemically or physically bonded to the internal matrix core. The chemical nature of these molecules, their mobility (*i.e.* local movements in the medium, conformational changes, diffusion in/out of the matrix), their spatial orientation and distribution, their interaction with others as well as their electric field, have to be considered as quite original and different compared to those they would have when being located inside the matrix. The different behaviour of the molecules results from the fact that the solid/liquid interphase has both its own chemical composition and characteristics in terms of intra- and intermolecular interactions (short and long ranges).
- The interface interacts with cell culture system on two different scales, in particular on a molecular level by facing ions, low molecular weight organic molecules and proteins, and on a microscopic level when contacting cells, it is necessary to envisage microcarrier surface homogeneity/heterogeneity on both scales.
- A compromise for tailoring these microcarrier surface properties, on one hand, should consist of providing good conditions for the cells (in terms of adhesion, spreading, proliferation and production status), and, on the other hand, in avoiding non-specific adsorption of cell products as well as cultivation medium components.

The surface design is rather difficult for several reasons:

- It has to be realised for small structures (micrometer range).
- It has to modulate these surface properties in function of biotechnological process phase. Thus, during the cell attachment phase the surface should promote extracellular matrix protein adsorption and cell adhesion, while at the

end of the cultivation process, it has to provide easy cell detachment from a solid support.

- It has to be taken into account that cells can not "really see" a native microcarrier surface. The original interphase is indeed subjected to a rapid screening when a solid substrate is placed in contact with the cell culture media. This is the case when the microcarrier surface is initially positively charged. The resulting surface electrokinetic potential has more propensity to adsorb proteins due to a favourable ionic interaction with a majority of negatively charged proteins of cultivation medium at physiological pH. However, entropic as well as other enthalpic contributions are thermodynamically preferable for a rapid non-specific adsorption process on a surface of a solid substrate [15].

Considering surface characteristics of microcarriers, one should also take into account their roughness properties. In particular, a homogeneous spreading and growth of cells requires the microcarrier surface to have an even continuous contour.

Although the history of microcarriers started with positively charged microcarriers, later it became evidently that negatively charged materials as well as amphoteric materials (*e.g.* proteins or amino acids covalently attached to microcarrier surface) can also be used.

2.3. POROSITY

Microcarriers can be non porous (non permeable even for low molecular compounds), microporous (allowing easy diffusion of macromolecules up to 100 kDa) or macroporous (the average pore size of different types is being within the range of 30-400 μm). The latter represents a second generation of microcarriers on the market. The porosity of macroporous carriers is defined as the percentage volume of pores compared to the total microcarrier volume and usually is 60-99%. Their continuous and large pores allow cells to grow in a 3D environment at high densities. The microcarrier structure protects the cells from shear forces at stirring. The process advantage of macroporous microcarriers is that high perfusion rates can be used, resulting in a homogeneous environment, which ensures sufficient nutrient supply and removal of toxic metabolites. Since the use of macroporous microcarriers provides a stabilisation of cell population and decreases the need for external growth factors, it is easy to use low serum, serum-free or protein-free media that definitely reduces costs of yield products. Macroporous microcarriers can be used both for anchorage-dependent and free suspension cultures, for instance hybridoma cells.

2.4. BUOYANT DENSITY

Buoyant density should be slightly higher compared to that of the culture medium, thus facilitating easy separation of cells and medium. On the other hand, it should be sufficiently low to allow complete suspension of the microcarriers using only gentle stirring. The optimum density for microporous microcarriers is within the range of 1.030-1.045. Macroporous microcarriers have specific densities 1.04-2.5. However, sedimentation velocity is a more suitable parameter, because it reveals the kinematics of the microcarriers in a bioreactor, considering the microcarrier size and shape, medium viscosity as well as the concentration of microcarriers. A velocity of 150-250 cm/min has been defined as an optimal value to prevent microcarrier agglomeration and simultaneously to provide efficient circulation of cell culture media and oxygen and nutrient supply in the bioreactor [14].

2.5. MEAN SIZE AND SIZE DISTRIBUTION

The mean size and size distribution should be narrow so that a homogeneous suspension of all microcarriers is achieved and that confluence is reached at approximately the same stage on each microcarrier. The preferable size, which can provide the best cell growth, lies within diameter limits of 100-230 μm for microporous microcarriers. This size range corresponds to a compromise between the maximisation of the surface, the stability of the dispersion (both factors are in favour of decreasing the microcarrier size) and a surface unit per microbead, which is needed to provide adhesion and spreading of several hundred cells. Larger diameters up to several mm are evidently required for macroporous microcarriers with macropores of 20-400 μm whilst preserving their mechanical stability.

2.6. OPTICAL PROPERTIES

The optical properties should provide a routine observation of cells on microcarriers using standard microscopy techniques. This criterion is of special importance for vaccine production when one needs to carefully control the cell morphology on microcarriers, in order to identify the correct moment to infect cells with the virus or to harvest the virus. In some cases, transparency can be enhanced by transferring microcarriers with growing cells after dying to a liquid with higher refraction index, for instance glycerol. Unfortunately, because of the size, 3D structure and non-transparent material of some microcarriers, for instance polystyrene-based microcarriers, cell observation is very difficult using a light microscope. For macroporous microcarriers, confocal laser scanning microscopy is a nice tool to observe cells within pores [16].

2.7. CYTOCOMPATIBILITY

Non-toxic microcarriers are required not only for cell survival and good growth but also to satisfy requirements in veterinary or clinical applications.

2.8. MECHANICAL PROPERTIES

The rigidity of the microcarrier, although not quantitatively estimated in the literature concerning the mechanical properties, has to be equilibrated in order to avoid cell damage during microcarrier collisions in stirred tanks. But on the other hand, some rigidity of the solid surface has been reported to be essential to promote the cell spreading.

2.9. SHAPE

It is well known that a sphere provides the most favourable thermodynamic situation with a corresponding reduced surface/volume ratio, and thereby it can be explained why various emulsion/dispersion precipitation or polymerisation techniques allow to produce round microcarriers. However, this ideal shape does not maximise the surface of the material. That is why non-spherical microcarriers, such as disks made of polyester fibres [17] or the cylindrical shape, for instance DEAE-cellulose microcarriers, are also used. Additionally, some cell types have demonstrated a better affinity towards these anisotropic microcarriers [18].

3. Evolution of microcarriers on the market and in scientific publications

The first microcarriers proposed by van Wezel in the late sixties were hydrophilic ion-exchange chromatographic beads, namely DEAE-Sephadex A-50 carriers based on cross-linked dextran. Many scientists contributed to the research and development of other microporous microcarriers. The next important step in microcarrier design elaboration was the development of gelatin macroporous microcarriers [19].

Commercially available microcarriers as well as these reported in the scientific literature for the last two decades are listed in Tables 1 and 2, respectively. Additional criteria for classification matrix nature and surface properties of the microcarriers have been used. Obviously polymer microcarriers are preferred over inorganic supports due to their flexibility in adjusting the macromolecular properties to the requirements of the microcarrier design. Reticulated hydrogel beads, such as dextran, gelatin or acrylamide, have been adopted as hydrophilic matrix microcarriers; hydrophobic materials (polydimethylsiloxane, polyethylene, polystyrene) can also be used. The latter microcarriers are especially of interest in order to enhance the mechanical stability of macroporous supports.

In order to assess the evolution of commercial microcarriers during the last 5-7 years, we indicated in Table 1 (a and b) products which were available in 1998 and compared them to those which are available today. Although we have tried to carry out an exhaustive market research, the list is not exhaustive. From Tables 1a and 1b, it is obvious that, in spite of a rather short period during which we scanned the market of microcarriers, several products (for example about half of microporous microcarriers) disappeared from a trade place. The same tendency is observed for macroporous microcarriers, although a decrease of diversity is not so pronounced and concerns only a few products. There are at least 2 possible reasons, which can explain this evolution:

- small companies involved initially in the production and commercialisation of some microcarriers disappeared or merged with larger ones having prevailing position on the market.
- alternative cell culture technologies, *i.e.* hollow fibres or cell encapsulation methods appeared, decreasing an interest to microcarriers.

Table 1a. Commercially available microporous microcarriers.

Raw material	Type	Trade Name	Chemical functionalities	Company 1998	Company 2002
Natural polymer	Reticulated dextran	Cytodex-1	DEAE	Pharmacia	Amersham Biosciences
		Cytodex-2	Quaternary amine coated dextran		
		Cytodex-3	Surface layer of collagen chemically bonded to dextran		Amersham Biosciences
		Superbeads	DEAE	Flow Labs	
		Microdex	DEAE	Dextran Products	
		Dormacell	DEAE-dimers	Pfeifer and Langer	
	Cross-linked gelatin	Gelibeads	N.D.	KC Biologicals/ Hazelton Labs	
		Ventragel	N.D.	Ventrex	
	Cellulose	DE-52/53	DEAE	Whatman	Whatman
		DEAE, QAE or TEAE	DEAE, QAE or TEAE	Sigma	Sigma
Synthetic polymers	Polyacrylamide	Biocarrier	Dimethyl-aminopropyl	Bio-Rad	Bio-Rad
	Polystyrene	Biosilon	Negative charge	Nunc	Nunc
		2D Microhex	N.D.		
		Cytospheres	Negative charge	Lux	
		Bioplas	Collagen coating	Solo Hill Labs Inc.	SoloHill Engineering Inc.
		Plastic Plus	N.D.		
		Pronectin ® F	Coating with Pronectin ® F, a recombinant protein		
	Based on trimethylamine core	Hillex	N.D.		
Glass	Coating of glass on Polymer beads	Glass coated	N.D.	Solo Hill Engs	
		Ventreglas	N.D.	Ventrea	

N.D. no data available

Table 1b. Commercially available macroporous microcarriers.

Type	Type	Trade Name	Company 1998	Company 2002
Natural polymer	Reticulated gelatin	CultiSphere-G, S,GL	Percell Biolytica	Percell Biolytica
		Informatrix	Biomat. Corp.	
		Microsphere	Cellex	
	Cellulose	Cytopore 1,2	Pharmacia	Amersham Biosciences
Synthetic polymer	Polyethylene	Cytoline 1,2	Pharmacia	Amersham Biosciences
	Polydimethylsiloxane	Immobasil	Ashby Scientific Ltd	Ashby Scientific Ltd
Glass	Glass	Siran	Schott Glasswerke	Schott Glasswerke

Table 2. Non-commercial microporous microcarriers.

Material class	Matrix composition	Chemical functionality	Ref
Synthetic polymer	Styrene copolymers	Covalently linked trimethylamine	[20]
	Polyhydroxyethylmethacrylate (PHEMA)	Fibronectin-coated	[21]
	PHEMA and polystyrene (beads prepared by suspension polymerisation)	HEMA plus other acrylic monomers (*i.e.* MMA, EGDMA, DMAEMA) Polystyrene beads coated with various alkylamine monomers	[22]
Natural polymers	Gelatin	Fibronectin-coated	[23]
	Dextran	Chitosan-coated. Fructose covalently bounded.	[24]
Inorganic	Porous mineral (sepiolite, clay, pozzolana and foam glass-Poraver)		[25]
	Hollow biocompatible ceramic		[26]
Degradable synthetic polymers	Poly(D,L-lactide-co-glycolide)		[27]

Knowing that the main application of microcarrier cell culture concerns human pharmaceuticals, it is not surprised that only medium size or big companies are in a position to offer reliable products corresponding to all pharmaceutical criteria (*i.e.* production capacity, batch to batch reliability, GMP facilities, availability of Regulatory Support File and Drug Master Files *etc.*).

In order to evaluate the level of success of microcarriers in cell culture technology within the scientific community, we have followed the evolution of the publication number and used either "microcarrier" or "animal cell culture" as key words for a search in the Chemical Abstract database (Figure 1 A, B). Comparing these 2 figures, this is not of great interest to analyse the difference in absolute number of publications per year knowing that the "animal cell culture" keyword encloses all the aspects of animal cell culture technology. This is the time of evolution which is quite interesting in both cases and which has to be considered.

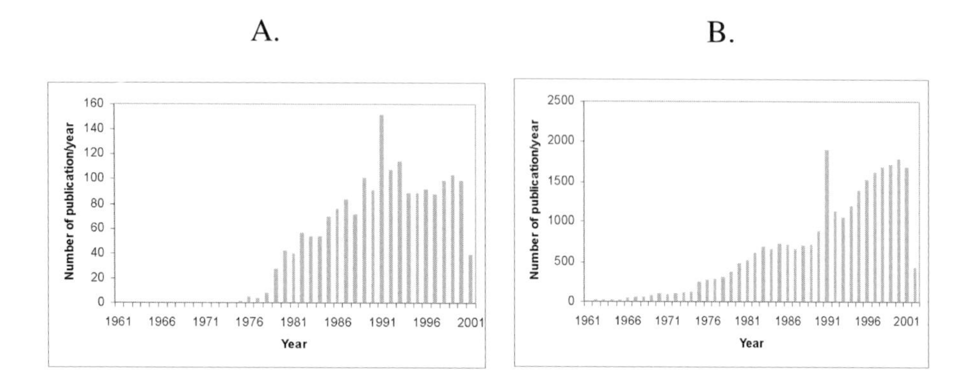

Figure 1. Publication rate of (A) "microcarrier" and (B) "Animal cell culture" (Source, Chemical abstract, SciFinder).

Contrary to a significant progress in a number of publications on animal cell culture which proceeded in two phases with a stagnation period in 1985-1990, the level of microcarrier citations first increased in 1975-1992, but then slightly decreased and remains quite stable since this "pioneer" period. It's important to emphasize that our search using "microcarrier" as a key word includes not only publications on the optimisation of microcarriers themselves, but also papers on their use in cell biology research. Thus, with reference to the evolution of the animal cell culture field, this bibliography analysis states a decrease of the interest towards microcarriers in the research community. This conclusion needs to be discussed at least by two remarks. The first and the main important one stems from the fact that microcarriers have been engineered initially for large-scale applications, and therefore had to be used mostly in this field. Thus, it is not surprisingly that academic works did not pay too much attention to materials, which were not specially directed, to them. The second remark, which could explain our observations, is that after the optimisation of the microcarriers (a classical research/development task for academic centres), efforts have been mostly focused on industrial applications using satisfactory supports, but not on additional improvements of microcarrier design.

4. Interfacial aspects of microcarriers

Working in the field of microcarrier cell culture, one has to realize that this is really a borderland between cell culture technology and biomaterial engineering. Although a lot of efforts have been focused on the optimization of soluble components in cell culture, resulting in development of many commercially available culture media and supplements, rather little research has been done to improve non soluble phases [28]. Obviously it is simpler to adjust the concentration of soluble compounds than to design microcarrier surfaces, which can mimic an extracellular matrix. Moreover, even if molecular reactions accompanying processes of cell adhesion, spreading, migration, proliferation and differentiation are now better understood, they still remain complex. They are multiparametric in nature and are occurring at interphases which are evolving rapidly both from the side of the cell membrane and the cell support. Additionally, without considering the dynamic dimension of microcarrier surface, their characterization (from a physico-chemical and topological point of view) is really challenging because of their sensitivity to contamination and the need to analyze them in environmental conditions (with respect to hydration level, ionic strength, pH, temperature, *etc*.).

Cell interaction with a foreign surface consists of cell migration to the surface, cell attachment through new externalised adhesion proteins, production and secretion of extracellular matrix proteins, reorganisation of cytoskeletal components allowing cell spreading, proliferation and/or differentiation. Protein adsorption is a key event to control cell adhesion to the surface [29-30]. It determines the nature of proteins presented at the interface, their conformation status, their supramolecular level of organisation, their relative concentration as well as their adsorption kinetics.

As this is the case for protein nature and protein concentration, if an interphase represents a favourable thermodynamic position of proteins, it has to be kept in mind that the total area developed by the interphase, remains limited compared to the total protein amount present in solution. Therefore, protein adsorption is characterized by an adsorption isotherm. In the case of a protein mixture, there is a competition for the interphase among proteins. Their diffusion rate, ability to alter their conformation status, hydration state, their redistribution of charged groups compared to the surface, the nature and amplitude of their electrokinetics potential *versus* the one which predominates at the surface, affect the final composition of the protein layer generated non-specifically at the interphase [15]. Therefore, it is obviously difficult to foresee the final composition starting from a crude mixture of proteins encountered in a cell culture medium. Moreover, presently in biomaterials engineering there is no approach, which completely prevents this non-specific protein adsorption, although some surfaces are recognised to be more resistant than others [31-33].

To highlight the importance of the contribution of 5 parameters responsible for protein adsorption, we can cite that:

- The preferential adsorption of proteins on a surface providing a stable and steric repulsion barrier against the adsorption of proteins of the extracellular matrix has been demonstrated to impede cell adhesion both *in vitro* and *in vivo* [34-36]. This competition between adhesion proteins, albumin and synthetic polymers with amphiphilic properties, such as polyethylene moieties, has been described by Dewez *et al.* [37]. These authors have shown that the steric repulsion efficiency of Pluronic F68 to prevent cell adhesion was not only related to the substrate surface energy but was also dependent on a sequence of added copolymers to the adhesion proteins.

- The concentration gradient of adhesion proteins on a substrate is of great importance. Ruoslahti and Pierschbacher demonstrated that cells tended to migrate preferentially to zones of a higher adhesion protein concentration. [38].
- As for protein adsorption kinetics, the first proteins which adsorb, are those, which are the most relatively abundant in a solution and with having a higher diffusivity (for instance albumin). But according to a mutual repulsion/exclusion phenomenon, called the "Vroman effect", first adsorbed proteins can be displaced by others, less abundant, with more affinity towards an interface, such as fibrinogen. This competition is evidently also influenced by the surface nature. Due to a higher entropic contribution in the thermodynamical balance afforded by largely hydrophobic surfaces, their binding with albumin is more stable compared to that for hydrophilic surfaces [39-41].
- According to thermodynamic considerations, the adsorption of proteins to a substrate is in favour of conformational changes in order to reduce especially structured water in contact with hydrophobic domains. Those configuration modifications influence cell behaviour *in vitro*. For example, Garcia *et al.* have reported that conformational changes of fibronectin adsorbed to the surface could influence a number of fibronectin-integrin bounds, and thereby modulate cell proliferation and differentiation [42]. Similarly, the thermal denaturation of collagen triple helix of collagen has been shown to induce an increase of the adhesion of mouse hepatocytes [43]. This observation has been explained by a better accessibility of the ubiquitous RGD peptide sequence in the open state of collagen molecules compared to its native form.
- Taking into account that the adhesion protein layer does not interact with cells continuously, but by an intermediate of focal plates, which determine lateral movement of transmembrane adhesion proteins, it is evidently that the supramolecular organisation of cell receptors could influence a cell answer. By supramolecular organisation should be understood the spatial organisation of the binding sites both in the x-, y- and z-directions, while keeping in mind that we deal with an interphase. Grinnel *et al.* showed that the behaviour of human fibroblasts depended upon the topographical collagen organisation [44]. A reversible change of morphology, cell proliferation rate and collagen metabolism have also been disclosed for hepatic stellate cells in function of the extracellular matrix organisation [45]. Being a function of this parameter, the morphology of these cell lines changes from star-shaped stellate cells to the fibroblast type. Such a modification in cell phenotype has been explained by these authors by a variation in integrin binding to the extracellular matrix and by consequences of the cytoskeleton assembly reorientation, mediated by signal transduction.

A contribution of a spatial organisation of extracelllular matrix proteins has also been reported, stressing the importance of this parameter for tissue engineering purposes. [46]. However, it should not be forgotten that cells not only "feel" their environment in a "chemical way", but they need to be supported by a matrix having mechanical properties corresponding to their needs (mechanotransduction). The importance of this contribution explains the approach recently used by Roeder *et al.* [47]. These authors have investigated the tensile properties of collagen I matrices according to the aggregation conditions of this protein.

Last but not least among the surface parameters controlling the cell behaviour is the matrix surface topology. If the importance of the contribution of this parameter to cell behaviour could be weighted against the chemical nature of a surface, the former would act predominantly [48]. Thus, surface properties of materials are obviously critical to control the cell behaviour both *in vitro* and *in vivo*. Therefore, the creation of micro- or nanopatterned heterogeneous surfaces, either on chemical or topological levels, has to be adjusted, in order to orientate cell migration on a matrix or to promote cell co-culture [49-50].

Trying to increase the concentration of ion-exchange chromatographic beads, which were initially adopted as microcarriers, van Wezel and some other researches immediately faced the cell toxicity of the supports [51]. Levine *et al.* [13] have reported that an optimal charge density was required to avoid toxicity and to promote good cell adhesion and proliferation. This optimal value corresponds to so called a discrete range of surface charges which is for negatively charged surfaces within the range of 2 - 10 charges /nm^2 [52] and to 0.6 mEq/g for Cytodex II (about 600 charges/nm^2). At a lower surface potential, the density of the cells attached to the surface remains limited and after some days of culture cells finally desorb. While at an higher charge density, a toxic effect is observed which has been interpreted as an inhibition of transmembrane protein movements resulting from multiple adhering bonds [53]. In general, it should be mentioned that a smaller charge density is preferred because it provides a reduction of non-specific protein sorption. In this case serum proteins can be removed from the culture medium by simple washing at the end of the cultivation process. The charges should be distributed regularly on the microcarrier surface in order to provide an homogeneous inoculum distribution.

Cell adhesion behaviour depends on the type of the charge on the microcarrier surface. On negatively charged surfaces cells form discontinuous focal contacts mediated with filopodia, while in the case of positive charges, one can observe continuous cell contacts with substrates [54-57]. Knowing that cell adhesion is controlled by specific recognition at the substrate surface, a difference in cell behaviour should be interpreted not directly in terms of surface/cell repulsion or attraction, but on the basis of surface/protein/cell interphase.

Despite the fact that protein adsorption on a microcarrier surface is of obvious value both for cell culture control and for avoiding non-specific sorption of yield protein product, very few publications have been published on this subject. This lack could be explained by the difficulty to get access to the interphase curve using classical analysis techniques (XPS, ellipsometry, contact angle, *etc.*). However, some specific methods have been reported to study protein adsorption kinetics on microcarriers [58] as well as to analyse their conformation changes [59]. Although initially charge density has been considered only in terms of optimising microcarrier surface properties, more recently different strategies have been proposed to add more specific functionalities. The latter could allow:

- to use serum free culture media, microcarrier surface being able to promote good cell adhesion, proliferation and product yield,

- to stimulate a specific biological response of cells when interacting with specific ligands grafted on the microcarrier surface.

The use of serum free media allows not only to simplify and to reduce the costs for make preparing the various cell products (for instance recombinant proteins), but becomes imperative for pharmaceutical applications where additives of animal origin are avoided for security reasons. Table 3 gives several examples of microcarriers with modified surface properties developed for the last ten years.

Table 3. Microcarriers with modified surfaces.

Microcarrier	Type of surface modification	Ref
Physical adsorption		
Cytodex 1	Immobilized insulin	[60]
Dextran microcarrier	Fructose-chitosan	[61]
DEAE-dextran	Physical adsorption of laminin (10 micrograms/cm^2)	[62]
Poly(D,L-lactide-co-glycotide) microcarriers	Collagen	[27]
Polystyrene microcarriers	Pronectin-F and poly-L-lysine, either alone or in combination.	[63]
Chemical modification of the surface		
Cytodex III	Type 1 Collagen	[64-65]
Beads having a styrene copolymer core	Trimethylamine exterior shell	[20]
PHEMA and polystyrene beads	Coated with different alkylamine monomers (*i.e.* EDA, ALAM, TEA)	[22]
Polystyrene-based microcarriers	Surface modification (triethylamine, maltamine or N-methylglucosamine)	[66]
Polystyrene sodium sulphonate or Sephadex derivatives	Composite ligand site made of a combination of different chemical functional groups	[67]
	Cell-growth-factor proteins and cell-adhesion-factor proteins	[68]
PHEMA microcarriers	Collagen or fibronectin	[69]

As can be seen in Table 3, the simplest strategy relies upon physical adsorption of an adhesion protein. Adsorbed collagen, fibronectin and laminin on microcarrier surfaces are the most popular proteins presently used to stimulate cell adhesion. To enhance the stability of a coating, some authors proposed chemical grafting of the proteins to the microcarrier surface. Recently to avoid any component of animal origin, more specific biochemical moieties have been evaluated. For example, Varani *et al.* suggested that recombinant Pronectin-F alone or in conjunction with a cationic polymer could be used instead of gelatin or collagen to modify microcarrier surfaces [63]. Some authors demonstrated the stimulation of a specific cell response using adequate grafting of specific biochemical signal molecules. For example, Yagi *et al.* reported grafting fructose-modified chitosan on a microcarrier surface to promote hepatocyte attachment [61]. Cell adhesion was significantly enhanced without altering hepatocyte morphology and liver-specific functions, such as urea synthesis and drug metabolism. On the other hand, hepatocytes adsorbed on a collagen-coated surface underwent large morphological changes with concomitant loss of their drug metabolisation potential.

In all these biosignalling and specific approaches, careful attention should be paid to avoid a non-specific protein adsorption contribution at functional interphases. Otherwise all efforts made for surface engineering could rapidly vanished because proteins can mask the ligands grafted to microcarrier surfaces [70].

5. Advantages and disadvantages of the microcarrier cell culture technology

The microcarrier approach evidently marked the beginning of a new era, since this technology resulted in a homogeneous culture system that is most easily scale-able. Starting with the first product obtained in microcarrier cultures on industrial scale at economically reasonable costs, namely Inactivated Polio Vaccine and leading to large-scale production of numerous recombinant proteins as a result of developing recombinant DNA technology, this technology really has demonstrated its great potential. The main advantages of the microcarrier approach can be summarized as follows:

- it provides an essential surface for cell adhesion within a limited bioreactor volume, while keeping relatively homogeneous and controlled environmental culture conditions. The relative homogeneous conditions of the cell culture is in favour of the scaling-up as well as controlling the various cultivation parameters (such as pH, pO_2 and pCO_2 concentration), as well as the analysis of amino acids, glucose or lactic acid [71];
- it is a cost-effective method compared to other cell culture technologies;
- it facilitates the purification of biological products due to the easy separation of media/microcarriers;
- it allows to culture both anchorage dependent and anchorage independent cells;
- it permits to obtain a high cell density in long-term cultures;
- it can be used in different bioreactor systems (stirred tank reactor, packed bed reactor, fluidized bed reactor).

As for the implantation for cell therapy purposes, the dispersion state provided by microcarriers allows easy manipulation, such as observations using electronic microscopy, or transferring cells for post-treatments after culturing.

It should be noted that macroporous microcarriers have additional benefits compared to microporous ones, in particular an increase of surface/volume ratio providing a higher plating efficiency, an inoculum reduction, an enhancement of mechanical protection due to cell localization within the macropores of microcarriers. Macroporous microcarriers allow cells to grow in a 3D-microenvironment at high densities, and thereby enhance the cell stability, productivity and longevity of the culture. A stabilised cell population decreases the need for external growth factors, which permits to use low serum, serum free, and sometimes even protein-free media [72-73].

An alternative strategy to avoid the use of serum-containing media, and thereby to reduce production costs, relies upon the surface fonctionnalisation of the microcarrier. For example, adsorption of insulin on the surface of Cytodex 1 microcarriers allowed to get a growth rate of CHO cells in serum free medium comparable to the one using untreated microcarriers in low serum medium (2% of serum) [60].

Mechanical shear forces generated by the stirrer, spin filter and air/oxygen sparging, is a limiting factor in cell proliferation. Based on a numerical analysis of particle dynamics, Qiu *et al.* [26] have showed that a limited coverage of microcarriers may be attributed both to a high shear stress imparted to a particle surface and to collisions between microcarriers and a bioreactor wall. Recently, the methods available for quantifying a magnitude of specific forces experienced by cells on microcarriers as well as resistance threshold in function of cell type have been reported [74]. However, in some cases shear forces on cells encountered on microcarrier surfaces are quite beneficial in order to enhance an expression of a definite cell phenotype. Frondoza *et al.* compared the behaviour of chondrocytes cultivated on microcarriers and onto flat surfaces [75]. They have discovered that shear forces stimulated chondrocytes to reacquire a rounded shape and to produce cartilage matrix components, while in monolayer chondrocytes looked like "fibroblastoid" cells. This change was characterized by a shift from high- to low molecular weight proteoglycans and from collagen type II production to collagen type I.

Microcarrier cell culture technology can provide benefits not only in the field of large scale production of biochemicals. For example, recently Hodder *et al.* [76] have demonstrated some merits of the microcarrier approach in the field of drug discovery research. The combination cell-microparticles was proposed as an interesting *in vitro* model to assess pharmacological activity of new chemical entities. The authors have used microcarriers with attached cells as a disposable and renewable surface for directly performed pharmacological assay, instead of using several assays on the same group of cells. This original approach allowed to avoid problems associated with a decreasing pharmacological response in common repetitive assays on the same group of cells.

Stephens *et al.* [77] used merits of microcarrier technology to simplify the manipulation of cells and to largely increase cell interaction area for the characterisation of procoagulant and inhibitory activities of different normal and transformed cells (haemopoietic, endothelial, muscle and connective tissue phenotypes). Moreover, for this specific application microcarriers were used as an indicator of cell biological activity.

An elegant and more sensitive analysis of ion-flux uptake adopting astrocytomas precultured on microcarrier surface was also reported by Braham *et al.* [78]. Pawlowski *et al.* [79] used Cytodex to study the differentiation of chick embryo skeletal muscle cells. Smith and Vale [80-81] have developed a superperfusion column technique to study rat anterior pituitary cells and the modulation of pituitary secretions by gonadotrophins and cocarcinogens. Vosbeck and Roth [82] used a microcarrier culture to investigate the effects of different treatments on intercellular adhesion. Confluent cell monolayers were cultivated on microcarriers and intercellular adhesion was examined by studying the binding of ^{32}P-labelled cells to the monolayer.

Cell microcarrier technology has also some drawbacks, obviously the scale-up of cells on microcarriers is more complex than for suspended free cells (in this case it is possible to adopt a cell line for cultivation in suspension). The microcarrier approach suffers from complicated washing procedures. Counting cells as well as harvesting is not complete, and it is even more difficult in the case of macroporous microcarriers because of the high cell densities. It is also difficult to infect all cells simultaneously, especially when using non-lytic viruses. Finally, large microcarriers with small pores may limit the diffusion of nutrients to some cells, especially during long term cultivation when high cell densities are reached.

6. Future applications of microcarriers

In our opinion, rather new and promising biomedical applications are currently being evaluated. Microcarriers have been proposed to treat burns using the transplantation of keratinocytes grown on gelatin microcarriers [83]. Recently, Del Guerra *et al.* [84] reported that gelatin CultiSpher-S microcarriers can be used for prolonged survival and function of dispersed islet cells prepared from large mammal (bovine) pancreatic islets. The use of dispersed cells could conceivably facilitate oxygen and nutrient supply to the cells. Encapsulation of these cell-containing microcarriers in sodium alginate-poly-L-lysine microcapsules can provide immunoisolation of the microcarriers for the potential use in transplantation therapies to treat diabetes. Finally, an interesting approach to dose control of therapeutic factors using hollow fibre devices with a discrete numbers of cellulose macroporous microcarriers CytoPorel (Amersham Biosciences) with growing cells has been recently described [85]. Encapsulation of rapidly proliferating cell lines on microcarriers, namely PC-2 and C2C12 mouse myoblasts delivering neurotransmitters and neurotrophic factors (CNTF), respectively, allow to control cell proliferation both *in vitro* and *in vivo*. The technique with encapsulated microcarriers allowing dose control of therapeutics has been proposed for the treatment of neurodegenerative diseases. Growing cells on degradable microcarriers and using the entire cell/carrier complex has been also discussed in the report of Schugens *et al.* [86]. Considering all these rather new interesting applications, we suggest that the attention of biomaterial engineers in the nearest future should be focused on developing new biocompatible biodegradable microcarriers for biomedical purposes.

In conclusion, we would like to cite van Velden-de Groot [6] who noted that the microcarrier technology developed by van Wezel with the aim to improve human health care by the preparation of better vaccines has completely fulfilled his ideals, since the applications have extended from the production of biologicals towards the direct use for the treatment of patients.

Acknowledgements

C. Grandfils thanks Dr. I. Knott (GlaxoSmithKline, Belgium) for the fruitful discussion on microcarrier benefits for the pharmaceutical industry.

References

[1] Nishijima, K.; Fujiki, T.; Kojima, H. and Iijima, S. (2000) The effects of cell adhesion on the growth and protein productivity of animal cells. Cytotechnology 33 (1-3): 147-155.

[2] Hulleman, E.; Bijvelt, J.J.M.; Verkleij; A.J.; Verrips, C.T. and Boonstra J. (1999) Integrin signalling at the M/G1 transition induces expression of cyclin E. Exp. Cell Res. 253(2): 422-431.

[3] Hu, W.S. and Aunins, J.G. (1997) Large-scale mammalian cell culture. Curr. Opin Biotechnol. 8: 148-153.

[4] Papoutsakis, E.T. (1991) Fluid-mechanical damage of animal cells in bioreactors. Trends Biotechnol. 9(12): 427-437.

[5] Reuveny, S. (1990) Microcarrier culture systems. Bioprocess Technol. 10: 271-341.

[6] van Der Velden-De Groot, C.A.M. (1995) Microcarrier technology, present status and perspective. Cytotechnology 18 (1/2): 51-56.

[7] van Wezel, A.L. (1967) Growth of cell-strains and primary cells on microcarriers in homogeneous culture. Nature 216: 64-65.
[8] van Wezel, A.L. and van der Velden-de Groot, C A. (1978) Large scale cultivation of animal cells in microcarrier culture. Process Biochem. 13: 6-8.
[9] van Hemert, P.; Kilburn, D.G. and van Wezel, A.L. (1969) Homogeneous cultivation of animal cells for the production of virus and virus products. Biotechnol. Bioeng. 11: 875-885.
[10] Levine, D.W. (1979) Production of anchorage-dependent cells on microcarriers. PhD thesis, MIT, Cambridge.
[11] Levine, D.W.; Wong, J.S.; Wang D.I.C. and Thilly W.G. (1977) Microcarrier cell culture : New methods for research scale application. Somatic Cell Genet. 3: 149-155.
[12] Levine, D.W.; Thilly, W.G. and Wang, D.I.C. (1979) Parameters affecting cell growth on reduced charge microcarriers. Dev. Biol. Standard 42: 159-164.
[13] Levine, D.W.; Wang, D.I.C. and Thilly, W.G. (1979) Optimization of growth surface parameters in microcarrier cell culture. Biotechnol. Bioeng. 21: 821-845.
[14] Lundgren, B. and Bluml, G. (1998) Microcarriers in cell culture production. In: Subramanian, G (Ed) Bioseparation and Bioprocessing. "Wiley-VCH Verlag GmbH, Weinheim, Germany"; Vol 2; pp 165-222.
[15] Norde, W. and Lyklema J. (1991) Why proteins prefer interfaces. J. Biomater. Sci. Polymer Edn. 2: 183-202.
[16] Bancel, S. and Hu, W-Sh. (1996) Confocal laser scanning microscopy examination of cell distribution in macroporous microcarriers. Biotechnol. Prog. 12(3): 398-402.
[17] Bohak, Z.; Kadouri, A.; Sussman, M.V. and Feldman, A.F. (1987) Novel anchorage matrices for suspension culture of mammalian cells. Biopolymer 26: 205-213.
[18] Lazar, A.; Reuveny, S.; Geva, J.; Marcus, D.; Silberstein, L.; Ariel, N.; Epstein, N.; Altbaum, Z.; Sinai, J. and Mizrahi; A. (1987) Production of carcinoembrionic antigen from a human colon adenocarcinoma cell line. I. Large-scale cultivation of carcinoembryonic antigen-producing cells on cylindric cellulose-based microcarriers. Dev. Biol. Standard. 66: 423-428.
[19] Nilsson, K.; Birnbaum, S. and Mosbach, K. (1987) Growth of anchorage-dependent cells on macroporous microcarriers. In: Spier R. and Griffiths J.B. (Eds) Modern Approaches to Animal Technology. Butterworths; pp 492-503
[20] Hillegas, W. J.; Solomon, D.E. and Wuttke, G.H. (2001) Microcarrier beads having a styrene copolymer core and a covalently linked tri-methylamine exterior. Solohill Engineering, Inc. Assignee; US Patent, 6,214,618, April 10.
[21] Dixit, V.; Piskin, E.; Arthur, M.; Denizli, A.; Tuncel, S.A.; Denkbas, E. and Gitnick, G. (1992) Hepatocyte immobilization on PHEMA microcarriers and its biologically modified forms. Cell Transplantation 1(6): 391-399.
[22] Kiremitci, M. and Piskin, E. (1990) Cell adhesion to the surfaces of polymeric beads. Biomaterials, Artificial cells, and Artificial organs 18(5): 599-603.
[23] Haselton, F.R.; Dworska, E.; Evans, S.S.; Hoffman, L.H. and Alexander, J.S. (1996) Modulation of retinal endothelial barrier in an *in vitro* model of the retinal microvasculature. Exp. Eye Res. 63(2): 211-222.
[24] Yagi, K.; Michibayashi, N.; Kurikawa, N.; Nakashima, Y.; Mizoguchi, T.; Harada, A.; Higashiyama, S.; Muranaka, H. and Kawase, M. (1997) Effectiveness of fructose-modified chitosan as a scaffold for hepatocyte attachment. Biological & Pharmaceutical Bulletin 20(12): 1290-1294.
[25] Pereira, M.A.; Alves, M.M.; Azeredo, J.; Mota, M. and Oliveira, R. (2000) Influence of physico-chemical properties of porous microcarriers on the adhesion of an anaerobic consortium. Journal of Industrial Microbiology & Biotechnology 24(3): 181-186.
[26] Qiu, Q-Q.; Ducheyne, P. and Ayyaswamy, P.S. (1998) Growth and differentiation of osteoblasts on hollow biocompatible ceramic microcarriers under microgravity conditions. 362 (Advances in Heat and Mass Transfer in Biotechnology): 49-53
[27] Xu, A.S.L. and Reid, L.M.(2001) Soft, porous poly(D,L-lactide-co-glycotide) microcarriers designed for ex vivo studies and for transplantation of adherent cell types including progenitors. Annals of the New York Academy of Sciences 944 (Bioartificial Organs III): 144-159.
[28] Von Recum, H.; Kikuchi, A.; Yamato, M.; Sakurai, Y.; Okano, T. and Kim, S.W. (1999) Growth factor and matrix molecules preserve cell function on thermally responsive culture surfaces. Tissue Engineering 5(3): 251-265.
[29] Ishihara, K.; Ishikawa, E.; Iwasaki, Y. and Nakabayashi, N.J. (1999) Inhibition of fibroblast cell adhesion on substrate by coating with 2-methacryloyloxyethyl phosphorylcholine polymers. Biomater. Sci. Polymer Ed. 10: 1047–1061.

[30] Cox, E.A. and Huttenlocher, A. (1998) Regulation of integrin-mediated adhesion during cell migration. Microscopy Research and Technique 43: 412-419.
[31] Horbett, T. (1982) Plasma adsorption on biomaterials. In: Cooper S.L and Peppas, N.A (Eds) Biomaterials : Interfacial Phenomena and Applications. Washington, DC; ACS; pp 233-244.
[32] Lee, J.; Martic, P.A. and Tan, J.S. (1989) Protein adsorption on pluronic copolymer-coated polystyrene particles. J. Coll. Interf. Sci. 131: 252-266.
[33] Ishihara, K.; Aragaki, R.; Ueda, T.; Watanabe, A. and Nakabayashi, N. (1990) Reduced thrombogenicity of polymers having phospholipid polar groups. J. Biomed. Mat. Res. 20: 1069-1077.
[34] Marchal, Th.G.; Verfaillie, G.; Legras, R.; Trouet, A.B. and Rouxhet, P.G. (1998) Heterogeneous polymer surfaces used as biomaterials: protein adsorption and cell adhesion 63(4a): 1109-1116.
[35] Williams, R. L.; Hunt, J. A. and Tengvall, P. (1995) Fibroblast adhesion onto methyl-silica gradients with and without preadsorbed protein. J. Biomed. Mater. Res. 29(12): 1545-1555.
[36] Tamada, Y. and Ikada, Y. (1993) Effect of preadsorbed proteins on cell adhesion to polymer surfaces. J. Colloid Interface Sci. 155(2): 334-339.
[37] Dewez, JL.; Doren, A.; Schneider, Y.J. and Rouxhet, P.G. (1999) Competitive adsorption of proteins: key of the relationship between substratum surface properties and adhesion of epithelial cells. Biomaterials 20(6): 547-559.
[38] Ruoslahti, E. and Pierschbacher, M.D. (1987) New perspectives in cell adhesion: RGD and integrins. Sciences 238(4826): 491-497.
[39] Elgerstma, A.V.; Zsom, R.I.J.; Norde, W. and Lyklema, J. (1991) The Adsorption of Different Types of Monoclonal Immunoglobulin on Positively and Negatively Charged Polystyrene Lattices. Colloid Surface 54: 89-101.
[40] Norman, M.E.; Williams, P. and Illum, L. (1993) *In vivo* evaluation of protein adsorption to sterically stabilized colloidal carriers. J. Biomed. Mat. Res. 27: 861-866.
[41] Pizzoferrato, A.; Arcioal, C.R.; Cenni,E.; Ciapetti, G. and Sassi, S. (1995) *In vitro* biocompatibility of a polyurethane catheter after deposition of fluorinated film. Biomaterials 16: 361-367.
[42] Garcia, A.J.; Vega, M.D. and Boettiger, D. (1999) Modulation of cell proliferation and differentiation through substrate-dependent changes in fibronectin conformation. Molecular Biology of the Cell 10: 785-798.
[43] Berman, A.; Morozevich, G.; Karmansky, I.; Gleiberman, A. and Bychlova, V. (1993) Adhesion of mouse hepatocytes to type I Collagen, Role of supramolecular forms and effect of proteolytic degradation. Biochemical and Biophysical Research Communications 194: 351-357.
[44] Grinnell, F.; Nakagawa, S. and Ho, C.H. (1989) The collagen recognition sequence for fibroblasts depends on collagen topography. Dep. Cell Biol. Anat., Exp. Cell Res 182 (2): 668-672.
[45] Senoo, H.; Imai, K.; Matano, Y. and Sato, M. (1998) Parenchymal and Mesenchymal Cell Interaction in the Liver. Journal of Gastroenterology and Hepatology 13 (Suppl.): S19-S32.
[46] Griffiths, L. G (2000) Polymeric biomaterials. Acta Materialia 48(1): 263-277.
[47] Roeder, B.A.; Kokini, K.; Sturgis, J.E.; Robinson, J.P. and Voytik-Harbin, S.L. (2002) Tensile mechanical properties of three-dimensional type I collagen extracellular matrices with varied microstructure. Journal of Biomechanical Engineering 124 (2): 214-222.
[48] Curtis, A. and Wilkinson, C. (2001) Nanotechniques and approaches in biotechnology. Trends in Biotechnology 19(3): 97-101.
[49] Detrait, E.; Lhoest, J.-B.; Knoops, B.; Bertrand, P. and van den Bosch de Aguilar, Ph. (1998) Orientation of cell adhesion and growth on patterned heterogeneous polystyrene surface. Journal of Neuroscience Methods 84(1-2): 193-204.
[50] Clark, P.; Connolly, P. and Moores, G.R. (1992) Cell guidance by micropatterned adhesiveness *in vitro*. Journal of Cell Science 103 (Pt 1): 287-292.
[51] van Wezel A.L. (1973) Microcarrier culture of animal cells. In: Kruse P.E. and Patterson M.K. (Eds) Tissue Culture: Methods and Applications. Academic Press, New York; pp 372-377
[52] Maroudas, N.G. (1977) Sulfonated polystyrene as an optimal substratum for the adhesion and spreading of mesenchymal cells in monovalent and divalent saline solution. J. Cellular Physiology 90(3): 511-520.
[53] Reuveny, S. (1983) Research and development of animal cell microcarrier cultures. PhD. thesis, The Hebrew University Jerusalem.
[54] Davies, J.E. (1998) The importance and measurement of surface charge species in cell behavior at the biomaterial interface. In : Ratner, B.D. (Ed) Surface characterisation of biomaterials,. Elsevier, Amsterdam.; pp 219-234
[55] Sheltoon, R.M.; Rasmussen, A.C. and Davies J.E. (1988) Protein adsorption at the interface between charged polymer substrata and migrating osteoblasts. Biomaterials 9: 219-234.

[56] Varani, J.; Bendelow, M.J.; Chun, J.H. and. Hillegas, W.A. (1986) Cell growth on microcarriers. Comparison of proliferation on and recovery from various substrates. J. Biol. Standard 14: 331-336.
[57] Varani, J.; Dame, M.; Beads, T.F. and Wass, J.A. (1983) Growth of three established cell lines on glass microcarriers. Biotechnol. Bioeng. 25: 1359-1372.
[58] Sammons, R.L.; Sharpe, J. and Marquis, P.M. (1994) Use of enhanced chemiluminescence to quantify protein adsorption to calcium phosphate materials and microcarrier beads. Biomaterials 15(10): 542-527.
[59] Wolff, C. and Lai, C.S. (1989) Fluorescence energy transfer detects changes in fibronectin structure upon surface binding. Archives of Biochemistry and Biophysics 268 (2): 536-545.
[60] Wang, Y.-C.; Kuo, C.-H., and Hsieh, H-J. (2001) Enhancement of cell growth on microcarriers immobilized with insulin. Journal of the Chinese Institute of Chemical Engineers 32 (2): 125-133.
[61] Yagi, K.; Michibayashi, N.; Kurikawa, N.; Nakashima, Y.; Mizoguchi, T.; Harada, A.; Higashiyama, S.; Muranaka, H. and Kawase, M. (1997) Effectiveness of fructose-modified chitosan as a scaffold for hepatocyte attachment. Biol. Pharm. Bull. 20 (12): 1290-1294.
[62] Varani, J.; Fligiel, S.E.; Inman D.R.; Beals, T. F. and Hillegas, W.J. (1995) Modulation of adhesive properties of DEAE-dextran with laminin. Journal of Biomedical Materials Research 29(8): 993-997.
[63] Varani, J.; Inman, D.R.; Fligiel, S.E. and Hillegas, W.J. (1993) Use of recombinant and synthetic peptides as attachment factors for cells on microcarriers. Cytotechnology 13(2): 89-98.
[64] Budak, V.; Herak-Perkovic, V. and Weber, M. (2000) Cultivation of MPK (minipig kidney) cells on cytodex microcarriers. Current Studies of Biotechnology 1 (Biomedicine): 89-92.
[65] Lira, R; Rosales-Encina, J.L. and Arguello, C. (1997) Leishmania mexicana: binding of promastigotes to type I collagen. Experimental parasitology 85(2):149-157.
[66] Roder, B.; Zuhlke, A.; Widdecke, H. and Klein J. (1993) Synthesis and application of new microcarriers for animal cell culture. Part II: Application of polystyrene microcarriers. Journal of Biomaterials Science, Polymer Edition 5(1-2): 79-88.
[67] Oturan, N.; Serne, H.; Reach, G. and Jozefowicz, M. (1993) RINm5F cell culture on Sephadex derivatives. Journal of Biomedical Materials Research 27(6): 705-715.
[68] Ito, Y. and Imanishi, Y. (1994) A biomaterial as a strong biosignal. Polym News 19(7): 198-202.
[69] Denizli, A.; Piskin, E.; Dixit, V.; Arthur, M. and Gitnick, G. (1995) Collagen and fibronectin immobilization on PHEMA microcarriers for hepatocyte attachment. International journal of Artificial Organs 18(2): 90-95.
[70] Williams, D. (2002) Reassessing Bioactive Surfaces. Medical Device Technology 13: 8-9.
[71] Griffiths, B. (2001) Scale-up of suspension and anchorage-dependent animal cells. Molecular Biotechnology 17(3): 225-238.
[72] Kawada, M.; Nagamori, S.; Aizaki, H.; Fukaya, K.; Niiya, M.; Matsuura, T.; Sujino, H.; Hasumura, S.; Yashida, H.; Mizutani, S. and Ikenaga, H. (1998) Massive culture of human liver cancer cells in a newly developed radial flow bioreactor system: ultrafine structure of functionally enhanced hepatocarcinoma cell lines. *In Vitro* Cell. Dev. Biol.: Animal 34(2): 109-115.
[73] Molnar, G.; Schroedl, N.A.; Gonda, S.R. and Hartzell, C.R. (1997) Skeletal muscle satellite cells cultured in simulated microgravity. *In Vitro* Cellular and Developmental Biology: Animal 33(5): 386-391.
[74] Chisti, Y. (2001) Hydrodynamic damage to animal cells. Critical Reviews in Biotechnology 21(2): 67-110.
[75] Frondoza, C.; Sohrabi, A. and Hungerford, D. (1996) Human chondrocytes proliferate and produce matrix components in microcarrier suspension culture. Biomaterials 17(9): 879-888.
[76] Hodder, P.S. and Ruzicka, J. (1999) A flow injection renewable surface technique for cell-based drug discovery functional assays. Analytical Chemistry 71(6): 1160-1166.
[77] Stephens, R. W.; Orning, L.; Stormorken, H.; Hamers, M.J.; Petersen, L.B. and Sakariassen, K.S (1996) Characterisation of cell-surface procoagulant activities using a microcarrier model. Thrombosis Research 84(6): 453-461.
[78] Bramham, J.; Carter, A.N. and Riddell, F.G. (1996) The uptake of Li+ into human 1321 N1 astrocytomas using 7Li NMR spectroscopy. Journal of Inorganic Biochemistry 61(4): 273-84.
[79] Pavlovski, R.; Krajcik, R.; Loyd, R. and Przybyllski, R. (1979) Skeletal muscle development in culture on beaded microcarriers Cytodex 1. J. Cell Biol 115 A: 112-117.
[80] Smith, M.A. and Vale, W.W. (1980) Superfusion of rat anterior pituitary cells attached to Cytodex beads: validation of a technique. Endocrinology 107: 1425-1431.
[81] Smith, M.A. and Vale, W.W. (1981) Desentization to gonadotropin-releasing hormone observed in superfused pituitary cells on Cytodex beads. Endocrinology 108: 752-759.
[82] Vosbeck, K. and Roth, S. (1976) Assay of intracellular adhesiveness using cell-coated Sephadex beads as collecting particles. J. Cell Sci 22: 657-670.

[83] Malakhov, S.F.; Paramonov, B.A.; Emel'ianov, A.V. and Terskikh, V.V. (1997) New approaches to the treatment of severe burns: the transplantation of keratinicytes grown in culture. Voenno-Meditsinskii zhurnal 318(9): 16-23.

[84] Del Guerra, S.; Bracci, C.; Nilsson, K.; Belcourt, A.; Kessier, L.; Lupi, R.; Marselli, L.; De Vos P. and Marchetti, P. (2001) Entrapment of dispersed pancreatic islet cells in CultiSpher-S macroporous gelatin microcarriers: preparation, *in vitro* characterisation and microencapsulation, Biotechnology and Bioengineering 76(6): 741-744.

[85] Li, R.H.; Scott, W.; White, M. and Rein, D. (1999) Dose control with cell lines used for encapsulated cell therapy. Tissue Engineering 5(5): 453-465.

[86] Schugens, C.; Grandfils, C; Jérôme, R.; Teyssié, Ph.; Delree, P.; Martin, D.; Malgrange, B. and Moonen; G. (1995) Preparation of a macroporous biodegradable polylactide implant for neuronal transplantation. J. Biomed. Mat. Res. 29: 1349-1362.

PART 2

METHODS AND TECHNOLOGIES FOR CELL IMMOBILISATION/ENCAPSULATION

MICROCAPSULE FORMULATION AND FORMATION

BERIT L. STRAND[1], GUDMUND SKJÅK-BRÆK[1] AND OLAV GÅSERØD[2]
[1]Department of Biotechnology, Norwegian University of Science and Technology, Trondheim, Norway – Fax: +47 73591283 – Emails: blstrand@stamme.chembio.ntnu.no; gudmund@chembio.ntnu.no
[2]FMC Biopolymer, P.O. Box 494 Brakerøya, N-3002 Drammen, Norway

1. Microencapsulation – materials and techniques

Microcapsules are referred to capsules of spherical shape with a diameter of about 100 to 1000 μm. Besides traditional capsules with a well defined shell and core structure, encapsulation in microbeads without a distinct membrane has also been successful in certain applications. For immobilisation of living cells such as bacteria, yeast or mammalian cells, the encapsulation needs to be performed under relative mild conditions, depending on the cell type. The capsules or beads should be semipermeable to allow diffusion of oxygen and nutrients into the cells inside the capsules and waste products out of the capsules. For immunoisolation purposes, the capsule must be impermeable to host cells and soluble components of the immune system.

The possible uses for encapsulated cells in industry, medicine, and agriculture are numerous, ranging from nitrification of domestic waste water [1], ethanol and lactic acid production [2,3,4], production of monoclonal antibodies by hybridoma cells [5] to production of artificial seed by entrapment of plant embryos [6]. The demand for specific properties of the capsules may vary very much depending on the system where it is to be used. Proliferating cells may need a stronger capsule membrane than non-proliferating cells, but the former may tolerate a tougher encapsulation procedure as viable cells will grow and replace the dead cells in the capsules. The criteria for wide industrial acceptance of immobilisation technology may be different than for biomedical applications when it comes to safety, simplicity, long-term immobilisation, activity and price. Also, the freedom to choose stabilising surrounding solution may vary with the use of capsules, *e.g.* in continuous reactors and in implantation.

Microcapsules are almost exclusively produced from water-soluble polymers. As a consequence of the hydrophilic nature of the material, there is a low or zero interfacial tension with surrounding fluids and tissues, which minimises the protein adsorption and cell adhesion. Furthermore, the soft and pliable features of the gel reduce the mechanical or frictional irritations to surrounding tissue. And, finally, they provide a high degree of permeability for low molecular weight nutrients and metabolites, which

V. Nedović and R. Willaert (eds.), Fundamentals of Cell Immobilisation Biotechnology, 165-183.

is required for the optimal functioning of living cells [7,8]. Water-insoluble polymers include the use of an organic solvent, which usually interferes with cellular function [7].

Many different materials and techniques for microencapsulation have been studied [9-14]. Most semipermeable membranes for cell encapsulation have been formulated by phase inversion, polyelectrolyte coacervation, interfacial precipitation [10] or by uncoated gel spheres (Figure 1). Phase-inversion techniques have been used to generate both ultra- and microfiltration membranes in which pore sizes ranges from 2 to 50 and 0.1 to 1 μm, respectively. The process involves the induction of phase separation in a previously homogeneous polymer solution either by a temperature change or by exposing the solution to a nonsolvent component either in a bath (wet process) or in a saturated atmosphere (dry process). Polymer precipitation time, polymer-diluent compatibility and diluent concentration all influence phase separation and hence, membrane porosity [12]. Different membranes have been produced as flat sheets and hollow fibres, but also microcapsules may be produced in this manner. In a series of investigations, a variety of cell types have been encapsulated in semipermeable polyacrylate capsules [11]. Membranes produced from poly(hydroxyethylmethacrylate-co-methyl methacrylate) are hydrolytically stable and mechanically durable. However, heterogeneity in capsule permeability and membrane thickness can occur [12].

In the polyelectrolyte coacervation process, a hydrogel membrane is formed by the complexation of oppositely charged polymers to yield an interpenetrating network. Mass transport characteristics can be modulated by osmotic conditions, diluents, and the molecular weight distribution of the polyionic species [9, 10, 12]. Many examples of binary polymer blends exist, including alginate with either protamine or spermine, cellulose sulphate with poly (diallyldimethyl ammonium chloride), and carboxymethyl cellulose and chitosan or diethylaminoethyl dextran [12,15]. Addition of one or more coating steps with oppositely charged polymer has been performed to further modify biocompatibility, reduce membrane permeability or improve mechanical properties. Also the use of multicomponent polyelectrolyte blends with mixtures of both polyanionic and polycationic species have been suggested. Capsules with molecular weight cut off between 40 000 and 230 000 have been produced with an alginate-cellulose sulphate-calcium chloride-poly(methylene-co-guanidine) system [9,12,16,17].

Alginate-calcium chloride systems are based on interfacial precipitation [12]. The original work by Lim and Sun [18] involved the gelation of alginate-cell suspension in a calcium chloride bath, coating the outer membrane surface with poly-L-lysine (PLL), and the subsequent dissolving of the Ca-alginate gel with sodium citrate. Many modifications of the original procedure of making alginate-PLL capsules have been investigated, including coating of the polycation by an additional layer of alginate or another polyanion [19,20], formation of double-membrane capsule [21] or multilayer capsule [22-24] and modify or exchange PLL with other polycations [25,26]. Excluding the step of liquefying the core, formation of an inhomogeneous core structure as well as variability of capsule parameters with composition of the alginate in the core and PLL molecular weight and exposure have been evaluated [20,27,28]. Some recent reports suggest that simple calcium or barium alginate gels, without polycation membrane could function in both allo- and xenotransplantation models [29-33].

Polyanions and polycations may form a gel by covalent or ionic cross-linking. For alginate, covalent cross-linking may include direct cross-linking of the carboxyl groups

[34] or covalent grafting of alginate with synthetic polymers [35]. Multivalent cations ionically cross-link alginate chains and will be discussed in more detail later. Combination of covalent and ionic cross-linking may also be used [36]. However, covalent cross-linking involves reactive chemicals and their applications are mainly limited to immobilisation of enzymes, fast growing or dead cells. Hydrogels of other natural polysaccharides applied for encapsulation of living cells have consisted of agarose [37-39], κ-carrageenan, chitosan [40], pectin, pectates and gellan [40, 41]. Agarose, κ-carrageenan and pectin form gels by thermal gelation, whereas pectate and chitosan form gels by ionic cross-linking. Gellan gels are formed by a combination of thermal gelation and ionic cross-linking [40].

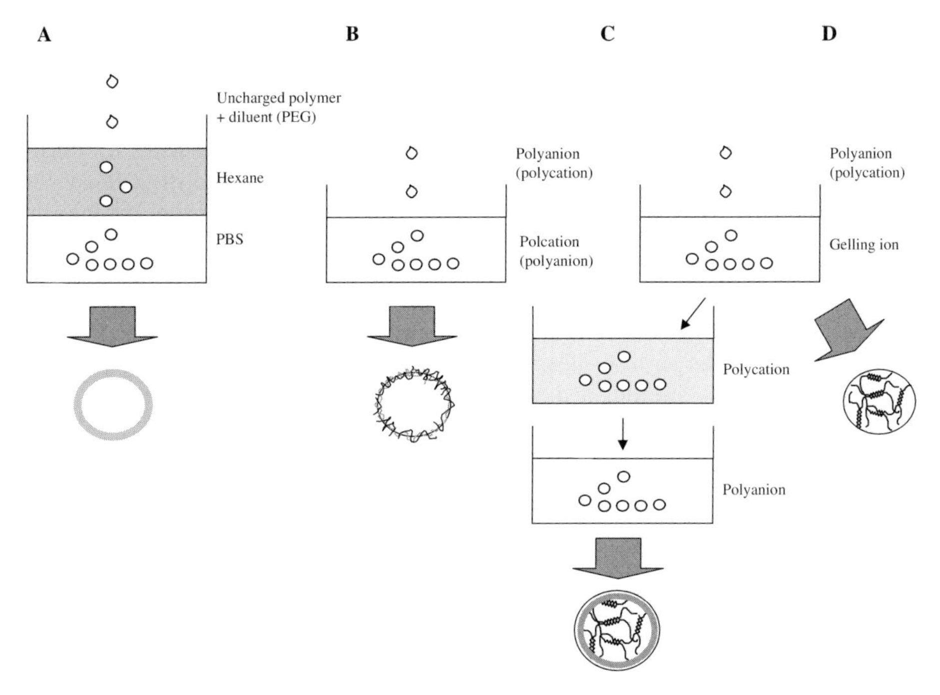

Figure 1. Illustration of four principals of microcapsule formation: A: Phase inversion, B: Polyelectrolyte coacervation, C: Interfacial precipitation and D: Un-coated gel spheres.

Hence, the number of materials, different systems and possibilities for cell immobilisation by microencapsulation are numerous. A great number of different materials for cell encapsulation by polyelectrolyte coacervation have been evaluated by Prokop and co-workers [9]. However, Hunkeler [10] has estimated that more than 85% of all articles dealing with cell encapsulation published since Lim and Suns paper in 1980 [18] have involved modifications of the alginate-calcium chloride-PLL system. In the rest of this chapter we will therefore focus mainly on alginate and PLL. Also chitosan will be discussed since this polycation has been regarded as more biocompatible than PLL.

Methods for gel formation include emulsification techniques [42, 43] and dripping methods where the core polymer is dripped into a solution containing polymer for coacervation or a cross-linking agent. Dripping methods ensure a narrow size distribution of the microcapsules. Different techniques for droplet formation have been established [44]. This includes the established technique based on laminar airflow placed coaxial to the needle feeding the polymer solution. The size of the polymer droplet can also be reduced by breaking up extruded polymer solution by a vibration device or a cutting device [45,46]. Other techniques are based on electrostatic forces to pull the droplets off the needle by creating an electrostatic potential between the gelling bath and the needle. The potential can be static or applied in pulses [47,48]. With the traditional encapsulation technique based on the air jet system, it is difficult to obtain evenly sized capsules smaller than 500 μm in diameter, but by a high voltage electrostatic system, capsules with diameters < 200 μm can be prepared [49-51].

Reduction in capsule size has been emphasised to enhance mass transfer of both nutrients into the encapsulated cells and products from the encapsulated cells out of the capsule. Colton [52] indicates anoxia in the centre of spheres of 250 μm in diameter. It has been shown that the response time of encapsulated Langerhans islets to glucose increase with capsule size [53]. Moreover, reduction in size lowers the shear forces [54] may increase their long-time stability. For implantation purposes, smaller capsules are by themselves shown to be more biocompatible than bigger capsules [55] and by reducing the implanted volume more freedom in the choice of site for implantation is gained [56]. However, by reducing the size of the polymer droplet and capsule more of the capsule is exposed to the surface as small capsules have a larger surface to volume ratio compared to larger capsules. This results in a capsule more vulnerable to destabilisation both by swelling and by exposure to oppositely charged ions [50].

2. Alginate gels

Alginate is the far most used polymer for cell encapsulation as it allows rapid gelling under physiological conditions. Cells can be encapsulated in alginate gel beads simply by dripping a mixture of cells and alginate solution into a solution of multivalent cations (Figure 2). Alginate is a family of unbranched binary copolymers of 1→4 linked ß-D-mannuronic acid (M) and α-L-guluronic acid (G) (Figure 3). The monomers are arranged in a block-wise pattern along the chain with homopolymeric regions of M and G termed M- and G-blocks respectively, interspaced with regions of alternating structure (MG-blocks) [57]. The composition and sequential structure of alginate is widely varying depending on the organism and tissue it is isolated from [58, see also Melvik and Dornish in this book]. In algal alginates, the fraction of single G residues (F_G) varies typically in the range of 0.3 to 0.7 [58]. Alginates with more extreme compositions can be isolated from the bacterium *A. vinelandii*, which in contrast to *Pseudomonas* species produces polymers containing G-blocks [59]. Alginates with specific structures can be made by enzymatic modification. The last step in the biosynthesis of alginate is the conversion of M residues in the mannuronan chain into G residues. This remarkable reaction is catalysed by enzymes called mannuronan C-5 epimerases (Figure 4) [60]. These enzymes determine the properties of the alginate by controlling the polymer's chemical composition and the sequential distribution of the monomers along the chain. Seven epimerases from *A. vinelandii* have been sequenced and cloned and expressed recombinantly. They have different product specificity as some make G-blocks, some make MG-blocks and some make a mixture [61,62]. By applying these enzymes *in vitro*, novel alginates with more defined structures can be made.

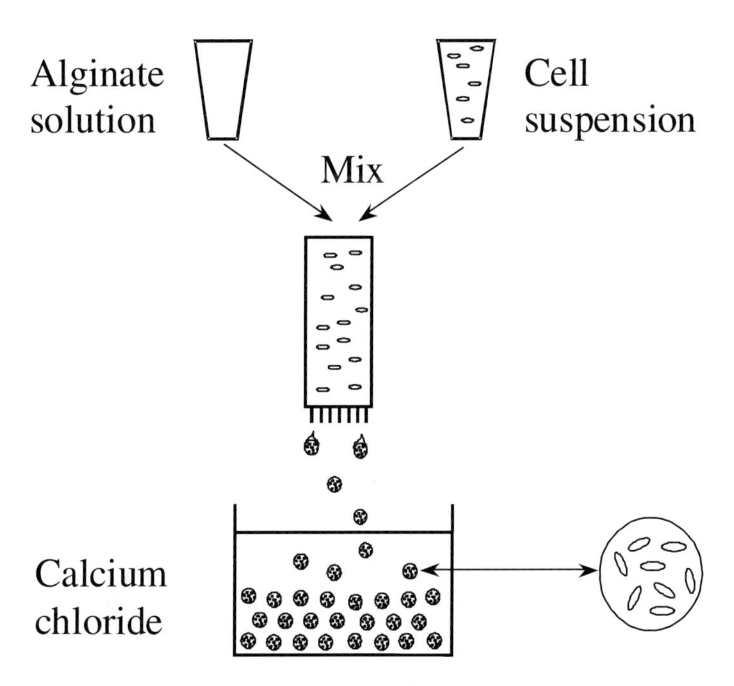

Figure 2. Illustration of the encapsulation of cells in Ca-alginate bead by mixing Na-alginate and cell suspension and dripping the mixture into $CaCl_2$ solution. All solutions are at physiological pH, which makes the immobilisation of cells very mild.

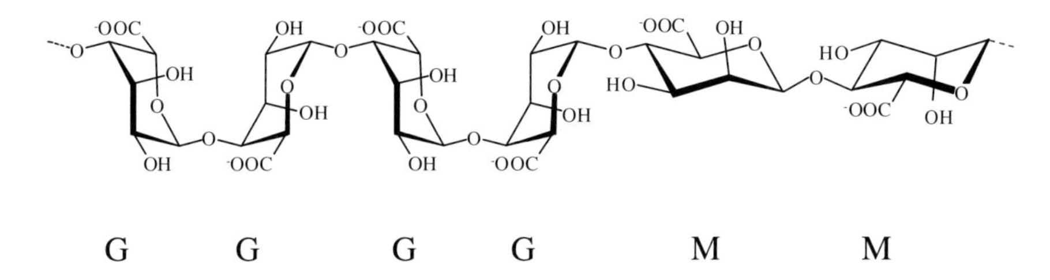

Figure 3. Structure of alginate: 1→4 linked β-D-mannuronic acid (M) and α-L-guluronic acid (G).

Table 1. Ca-alginate gel properties as immobilisation material as a function of alginate composition. High M alginate: about 40 % G, High G alginate: about 70 % G.

Alginate gel properties	High M	High G
Initial size	+	-
Gel strength	-	+
Stability	-	+
Permeability	+	-
Binding of polycation	+	-

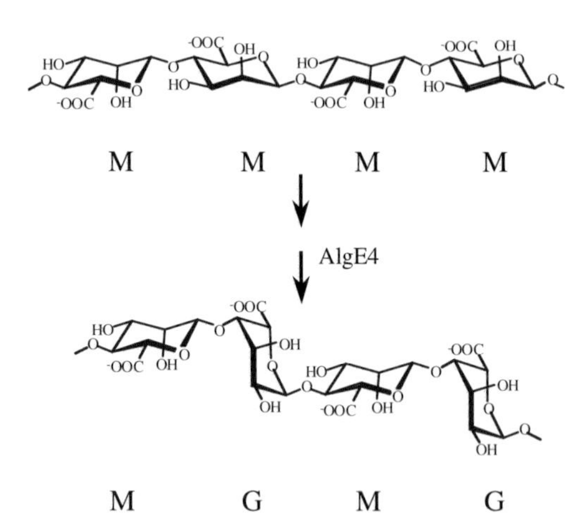

Figure 4. Conversion of M to G in the alginate chain by mannuronan C-5 epimerases: AlgE4 converts M-blocks to alternating structure of M and G (Adapted from [91]).

Alginate gel-forming properties depend to a high extent on composition and sequence [63]. Table 1 summarises some Ca-alginate gel bead properties as a function of the content of M and G that will be discussed in the continuing of this chapter. The selective binding of cations to alginate [58,63, see also Melvik and Dornish in this book] accounts for its capacity to form ionotropic gels. The di-axially linked G-residues form cavities acting as binding sites for ions, and sequences of such sites form bonds to similar sequences in other polymer chains giving rise to the junction zones in the gel network as demonstrated in Figure 5 [64]. Calcium, strontium and barium bind preferentially to the G-blocks in a highly co-operative manner. The size of the co-operative units for Ca-binding is reported to be between approximately 8 and 20 monomers [65,66]. Thus, in general, beads with the highest mechanical strength are made from alginate with a content of G higher than 70 % and an average length of G-blocks ($N_{G>1}$) of about 15 [58,67]. The mechanical strength of the alginate gel beads also depends on the molecular size of the alginate molecules. Above certain molecular weights, the mechanical strength of alginate gels is determined mainly by the chemical composition and block structure and is independent of the molecular weight of the polymer. Below a certain critical molecular weight, *e.g.* 300 kDa for 1% high-G alginate, the gel-forming properties of alginates are reduced [63,67].

The gelling ion may in principle be any of the multivalent cations. Due to their high toxicity, the use of most of the ions, in particular Pb, Cu and Cd, is strictly limited due to their high toxicity and only stabilisation with strontium and low concentrations of barium can be used for entrapment of living cells. Ca^{2+} and Ba^{2+} are mostly used because of their selective and co-operative binding to the G-blocks [63]. Ca^{2+} is non-toxic and has so far been the most used ion for cell immobilisation purposes. Ba^{2+} forms a stronger gel with alginate than Ca^{2+}, but using high concentrations of Ba^{2+} in the gelling process may lead to leakage of toxic ions. However, extensively washing and keeping the Ba-concentration low will reduce the leakage of Ba-ions to a minimum [20]. The rigidity of ionic alginate gels increases generally with the affinity of the ion, except for the earth alkaline metals that have higher rigidity due to the co-operative binding. Hence, the rigidity of ionic alginate gels decreases in the order: Ba > Sr > Ca > Pb > Cu > Ni > Cd > Zn > Co > Mn [58,63,68]. More rigid alginate gels may be made by gelling first against Ca-ions and then extensively against ions of higher affinity [68].

The major limitation of the use of calcium alginate as cell immobilisation matrix is its sensitivity towards chelating compounds, such as phosphate, citrate, carbonates and lactate or non-gelling cations such as sodium or magnesium ions [69]. Hence, the gel will be destabilised, swell and dissolve in their presence. The simplest way to overcome this is to keep the gel beads in a medium containing a few millimoles per litre of free calcium and to keep the sodium calcium ratio less than 25:1 for high-G and 3:1 for high-M alginates. However, replacing calcium ions with other divalent or even trivalent cations having a higher affinity for alginate can also stabilise Ca-alginate gels [20,70,71]. As alginates with a high content of G bind the divalent ions better than alginates with a low content of G, the gels of high-G alginate are more stable to chelating compounds and exchange of monovalent against divalent ions than gels of high-M alginate [67,70]. Other approaches for stabilising alginate gels are covalent cross-linking and by forming an outer polyanion-polycation complex membrane as will be discussed later. It should also be noted that calcium alginate gels, in contrast to other hydrogels, are stabile in a range of organic solvents, and as such has a potential use as immobilisation matrix for enzymes in non-aqueous systems.

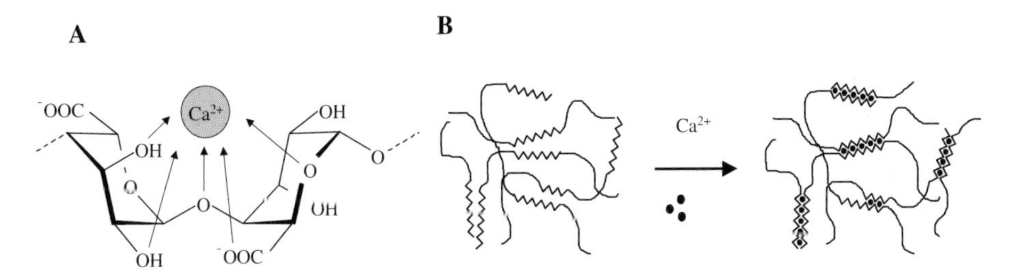

Figure 5. Probable Ca-binding site in a GG-sequence (A) and ionic cross-linking of two homopolymeric blocks of G-residues by the egg-box model (B) [64].

For all uses of microcapsules with living cells, the ingress of oxygen and nutrients and the egress of cell products are of importance. Hence, the diffusion characteristics are essential for the use of alginate gels as immobilisation matrix. Ca-alginate gels have a network with a high range of pore sizes between 50 and 1500 Å [72]. Self-diffusion of small molecules seems to be very little affected by the alginate gel matrix. The self-diffusion of glucose has been reported to be as high as about 90% of the diffusion rate in water [73]. For larger molecules higher resistance of diffusion occurs, and large proteins will leak out of the gel beads with a rate depending on their molecular size [74-76] as well as the shape and charge [77]. Low ionic strengths will improve the electrostatic interactions between charged proteins and the ionic gel network as a result of providing less shielding. This leads to a reduced ability for the protein to move inside the gel. Capsule diffusion rates for blue dextran and cytochrome C are considerably reduced by measuring in pure water relative to salt containing solutions. Hence, the permeability will depend on the ionic strength as well as the network density and the molecule of interest [78-80]. The smallest of the immunoglobulins, IgG with M_W 150 kDa, is found to be on the limit of being excluded from entering alginate gel beads (Ca/Ba = 50/1) of various compositions at physiological conditions. However, a polycation membrane is needed to exclude IgG completely [28,50,77,78]. The highest diffusion rates of proteins, indicating the most open pore structure, are found in beads made from high-G alginate and increasing alginate concentration reduces the permeability of the gel [73,75]. The Ca^{2+} concentration in the gelling solution will also influence the permeability of the Ca-alginate gel. At high concentrations of calcium, more than two G-block sequences take part in the junction zones. This will leave more space in between the junction zones and thus make a more porous gel [81]. As alginates are heterogeneous both regarding the composition and molecular weight, the alginate gel will contain a fraction of alginate that will not form stable junction zones. These polymers will leak out of the gel. The sol fraction is depleted in G and enriched in M, and comprises the low molecular weight tail of the alginate sample [65,82].

The alginate gel bead can also be made more stable and less permeable by the formation of an inhomogeneous alginate gel core. For alginate gels in general formed by diffusing of cross-linking ions into an alginate solution, gels with varying degrees of anisotropy with respect to polymer concentration can be formed by controlling the kinetics of the gel-formation (Figure 6A) [83-85]. This inhomogeneity is governed by the relative rate of diffusion of calcium ions and the rate of diffusion of the sodium alginate molecules towards an inward moving gelling zone. Simply by adjusting the concentration of alginate and the crosslinking ions, the distribution of polymer in the gel can be controlled, and alginate beads with a capsular structure (with a polymer concentration of about 10% on the surface and less than 0.2% in the core) have been made without adding polycations or any other non-gelling polymer [83]. Also the composition of alginate influences the bead structure as high-G alginates form more homogeneous gels than high-M alginates. When gels are formed in the presence of non-gelling cations such as sodium or magnesium, the coupled diffusion between alginate and counter-ions is impaired and more homogeneous gels are obtained. The polymer distribution in microcapsules can easily be easily visualised in the confocal scanning laser microscope by using fluorescent labelled polymers (Figure 6B) [86].

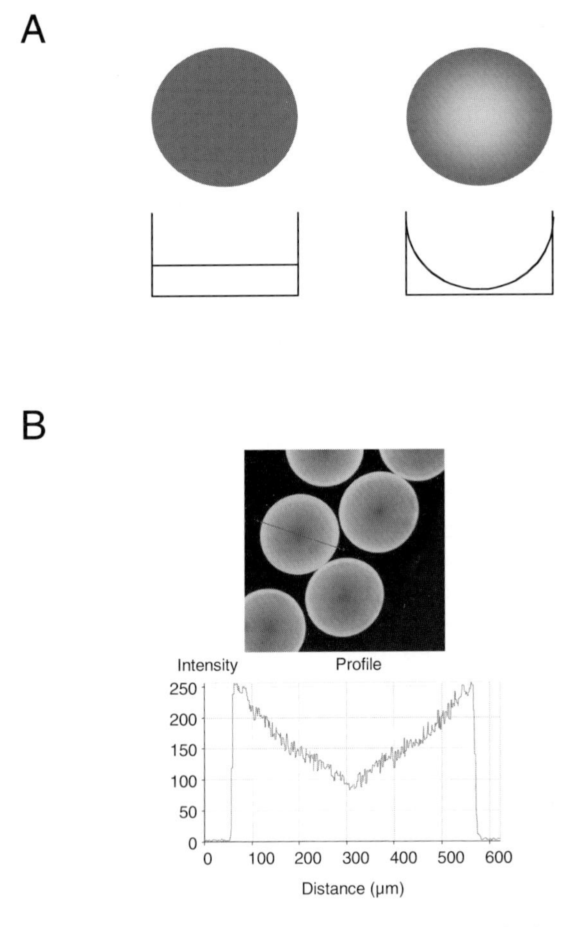

Figure 6. Polymer distribution in alginate gel beads. A: Illustration of polymer gradients in homogeneous and inhomogeneous alginate gel beads (Adapted from [20]). B: Confocal scanning microscopy pictures of alginate gel beads showing an optical section of the beads taken through the capsule equator giving the intensity of the fluorescent light (fluorescent labelled alginate) across the capsule diameter [86].

Homogeneous alginate gels can also be made by internal release of calcium from Ca:EDTA or from Ca:Citrate in the presence of a slow acidifier like glucono-δ-lactone (GDL) [87,88]. The alginate molecules are then "locked" in a mixed H^+ Ca^{2+} gel that can be converted into the complete calcium form by dialysis against a calcium chloride solution [87]. By applying this technique for immobilising cells, alginate gels with various shapes can be made [89]. However, due to the acidifier and the low pKa values of the calcium complexing agents, the pH in the gel would have to be between 3-4 to obtain gel-formation. These pH values are too low for many living cells. A neutral calcium alginate gel can be formed by internal gelation using GDL and $CaCO_3$ [90].

Enzymatic modification of the alginate by exchanging the M-blocks with MG-blocks reduces the initial size of the alginate gel [91]. As the flexibility of the alginate chain depend on the composition and increases in the order GG<MM<MG, the conversion of M-blocks to alternating MG-blocks increases the flexibility of the polymer [92]. The elastic chain segments does not contribute in the cross-links of divalent ions, but since stronger gels are formed by the increase in flexibility, the number of cross-links has increased. Hence, more G-blocks are allowed to come together in cross-links, either two or more together. This also leads to increased stability of Ca-alginate gels and alginate-PLL capsules to destabilisation against osmotic pressure and exchange of cross-linking ions against monovalent ions. Finally, the permeability of both alginate beads and alginate-PLL microcapsules is reduced by this conversion [91].

3. Alginate-polycation microcapsules

After the formation of an ionically cross-linked alginate core, a polycation coating is often added to stabilise the gel against osmotic swelling and to reduce the permeability. An alginate gel bead or capsule can be viewed as an osmotic swelling system. Here, the main contribution to the swelling pressure comes from the difference in concentration of mobile ions between the inside and the outside of the gel. At equilibrium the volume of the gel is stabilised both by the elastic forces of the gel network and the polyanion-polycation complex membrane [93-95]. Alginate and the polycation form strong complexes due to cooperative electrostatic forces between the opposite charges. Chitosan [96,97], polypeptides (*e.g.* polylysine [18,27], polyornithine [98]) or synthetic polymers (*e.g.* polyethyleneimine [99,100], poly (diallylmethylammonium chloride) (PDADMAC) [101] are the most commonly used polycations. Their interaction can be controlled to some extent by shielding the charges by elevated ionic strengths, but once formed the complexes can rarely be dissolved. It is important to control this reaction carefully since an insufficient reaction creates a thin, often rigid shell that makes the capsule break if the capsule core swells.

Capsules can be made with more or less distinct core and shell structure. A crucial point here is whether interpenetrating networks are formed or if a thin, low porous membrane is formed preventing further diffusion of polycation into the alginate core. The density or porosity of the alginate matrix and the molecular weight, or even more important, the molecular extension of the polycation is crucial for how it diffuses into the alginate phase. This principal is valid for both the processes of interfacial precipitation and complex coacervation. Molecular extension is closely linked to both the length (M_w) and the flexibility of the polymer chains. For instance a PLL and a chitosan having approximately the same average molecular weight shows tenfold difference in molecular extension measured by the hydrodynamic volume, due to large differences in chain flexibility (Table 2) [96]. This will result in very different molecular weight "cut-off" for the diffusion of different materials into the same alginate gel bead. Chitosans with M_n around 20 000 diffuses into an alginate gel bead but at a low binding rate needing several hours to form a good membrane [96]. PLL at the same molecular weight needs 5-10 minutes to do the same [27]. As the polycation enters the network several factors affect the binding rate. For instance chitosan binds better at higher levels of calcium chloride, increasing pH and higher fraction of acetyl groups, all factors either increasing the pore size of the gel bead or decreasing the molecular extension of the chitosan [96]. As the polycation molecular size is significantly less than the average gel pore size other factors than restricted diffusion becomes important as for PLL.

Table 2. Experimental and calculated data related to the polymer chain extension for chitosan and poly-L-lysine (Adapted from [96])

Polymer	DP_n	M_n, (g/mol)	[η], ml/g	V_h, (ml/mol)
Chitosan, F_A 0.09	121	24000	171	1640 000
Poly-L-lysine	127[a]	21000[a]	19	160 000

[a] Given by Sigma chemicals co. (Determined by LALLS).

PLL has become the most commonly used polycation. It is positively charged at neutral pH and is a very flexible molecule because of the flexible peptide bounds that links the lysine residues. Increasing exposure to PLL (time and concentration) increases the binding to the alginate gel bead and the strength of the capsules towards osmotic pressure [20,27,28,102]. The molecular weight of PLL is of importance as PLL of 18 kDa forms a more stable membrane than PLL of 55 kDa and PLL of 4 kDa [27,103,104]. PLL binds to a higher extent to alginate beads with a high content of M compared to alginate beads with a high content of G, but the binding to high-G alginate is more stable than the binding to high-M alginate [27,105]. More PLL is also bound to inhomogeneous alginate beads compared to homogeneous beads which yields a more stable capsule [20,27]. The polyanion-polycation complex membrane controls the permeability of alginate-polylysine-alginate microcapsules towards immunoglobulins and cytokines [28,77,102,106,107]. Increasing molecular weights of PLL results in alginate-PLL capsules with higher permeability and increasing exposure of PLL reduces the permeability of the capsules [77,104]. Poly-D-lysine (PDL) gives less permeable capsules than PLL for the same exposure. However, PDL increases the stability towards osmotic pressure to a lesser extent than PLL [20,91], but also increasing concentration of PDL result in increasing stability for both high-G and high-M alginate capsules [20].

Binding of PLL to the alginate gel is mainly governed by the amount of negative charges on the bead surface and the binding is enhanced in inhomogeneous beads due to the higher charge density on the surface [27]. By reducing the alginate gel bead size relatively more of the alginate gel is exposed to the surface. Hence, the gel is more easily destabilised by non-gelling ions or by exposure to polycation. Shrinking or even collapse is observed by neutralising the alginate network with PLL. This collapse may be avoided by washing the beads with non-charged osmolyte mannitol instead of saline. Washing with saline will partly dissolve the gel by exchange the calcium ions bound to mannuronic acid residues as well as to the short G-blocks. This would subsequently lead to an enhanced binding of PLL with a concomitant destabilisation of the gels [50].

When capsules are made by complex coacervation adding a polycation solution to an alginate solution or visa versa, other conditions has to be set than in the interphacial precipitation procedure. For instance when a 2 % alginate solution is dropped into a 0.3 % solution of chitosan, 100 times less chitosan was found in the resulting capsule compared to the interphacial precipitation method and also weak capsules. Also no chitosan could diffuse into the alginate matrix after the initially formed membrane [96].

However, others have succeeded in making strong capsules by using significantly lower polycation molecular weight (2-10 kDa) and preferably lower alginate concentrations [108]. The same factors have also been highlighted as important in work on the same capsule formation principal but with other materials, for instance with cellulose sulphate and PDADMAC [101].

Most chitosans are soluble only in an acidic pH range. This is of course a major disadvantage when working cells and tissues requiring physiological pH, but there are some ways to avoid this problem. Using chitosans of very low average molecular weight (< 6 kDa) leads to a shift in the pKa allowing solubility and membrane formation at neutral pH [108]. Also it is shown that by increasing the fraction of acetyl groups (F_A<0.4) complete solubility is shown at both neutral and alkaline pH values [109]. However most chitosans are soluble up to pH 6-6.5, and for encapsulation of less sensible biocatalysts such as bacteria and yeast this will often not represent a problem.

4. Biocompatibility

For biomedical use, the microcapsules should evoke no or only minimal fibrous tissue reaction, macrophage activation, and cytokine and cytotoxic agent release [110-112]. Although alginate fulfils the requirements for additives in food and pharmaceuticals, standard alginates contain small amounts of poly-phenols that might be harmful to sensitive cells [113]. In connection with transplantation, the alginate must also be free from pyrogens and immunogenic materials such as proteins and complex carbohydrates [114,115]. It is previously shown that alginates with a high content of M (poly-M, F_G<0.10) evoke an inflammatory response by stimulating monocytes to produce proinflammatory cytokines such as tumour necrosis factor (TNF), interleukin (IL)-1 and IL-6 when presented both in soluble and in gel state. Poly-M stimulates to TNF production via binding to CD14 while this is not seen for alginates with a high content of G [116,117]. The cytokine-inducing effect increases with the molecular weight and depends on the form of presentation as the potential to stimulate monocytes to TNF production increases when the alginate is linked to the surface of a particle [118,119]. Recent discovery shows that toll-like receptor (TLR)-2 and TLR-4 are important signal transduction molecules in the stimulation of monocytes by poly-M [120]. Hence, poly-M has been suggested as a general immunostimulator for therapeutic purposes as it protects mice from lethal radiation when given prophylactic and stimulates murine haematopoiesis [121]. Injected alginate is cleared from the bloodstream of mice within 48 hours whereas alginate given orally is not detected in the blood stream [122]. A discussion has been going on whether high-G or high-M alginates are most biocompatible. Antibodies are found to alginates of transplanted capsules of high-M alginates (F_G=0.4), but no antibodies are found to high-G capsules [123]. As mentioned above an increasing M-content of the alginate will increase the potential of stimulating the immune system and trigger inflammatory responses. On the other hand high-G alginates has the potential of lacking the ability to initiate such responses but also to actively inhibit them. Isolated G-blocks are shown to reduce the production of inflammatory cytokines when monocytes are stimulated either with high-M alginates or with bacterial lipopolysaccharides, most probably by competitive binding to the receptors [118]. Zimmermann and co-workers have argued that also high-M alginates can be purified to lower degrees of bioactivity [124]. However the procedure includes a step where the alginate is extensively washed in a Ba^{2+} cross-linked state. Since all alginate samples are heterogeneous with respect to structure, the fraction containing the highest content of M and highest bioactivity will, due to the lack of Ba^{2+}-binding G-blocks, most probably be washed out [63,65]. It is previously shown that alginate leaching from gels have higher M content than in the original gel material [20]. Thus, an alginate purified with this method may lack the alginate fraction representing the highest bioactivity.

For alginate-polycation microcapsules the polycation is apparently the main problem regarding biocompatibility. As polycations bind to negatively charged surfaces and matrices it is not surprising that these materials interact with cells and biological surfaces, which generally both are negatively charged. Several polycations have been screened and found to be more or less cytotoxic [9]. A direct comparison showed that the toxic effects of PLL were significantly higher than for chitosan [125]. PLL causes necrotic cell death in higher concentrations, whereas it in lower concentrations stimulates monocytes to TNF production [126]. TNF has potent proinflammatory activities and increases the proliferation of fibroblasts, which is believed to be the link to the observed fibrotic overgrowth of implanted microcapsules [19,126-130]. The cytotoxicity seems to be connected to the charge density of PLL, as the cytotoxicity of PLL has been reduced by grafting the polymer by adding carbamyl groups to the peptide side group [131]. It is shown that the alginate-PLL capsules themselves activate complement and macrophages to IL-1 production [132,133]. Grafting of the PLL with monomethoxy poly (ethylene glycol) decreases the adsorption of complement compounds and increases the biocompatibility of the capsule [25]. However, no antibodies to PLL are found after transplantation of alginate-PLL-alginate capsules to rats [123]. It is also important to consider the stability of the final complex since *in vivo* release of free polycation may trigger foreign body reactions. Compared to PLL, chitosans are for instance released from an alginate polycation capsules at a much lower level [27,134]. PLL is shown to bind to a higher extent to high-M alginate gels than to high-G gels [27]. This may be the explanation for why better biocompatibility is observed for alginate-PLL-alginate capsules with high-M core alginate compared to capsules with a high-G alginate core [125,128,129,135,136]. Coating the polycation with an outer layer of polyanion or by a poly (ethylene glycol) (PEG)-based hydrogel increases the biocompatibility of the capsules [19,25,137]. The mostly used coating material has been alginate. However, binding of alginate to the alginate-PLL microcapsule is dependent on both molecular weight and composition, as low molecular weight alginates with alternating structure of M and G binds better and increases the biocompatibility of alginate-PLL-alginate microcapsules than alginates of other structures [27,91,137].

Finally, the biocompatibility of the capsules may also depend on their surface roughness, the site of implantation and the host species [111,21]. Fibrosis has also been attributed to leaching cells or cell products, or cells protruding through the capsular membrane and is depending on the phylogenetic disparity between the graft and the host [19,135,138].

References

[1] Vogelsang, C.; Husby, A. and Østgaard, K. (1997) Functional stability of temperature-compensated nitrification in domestic wastewater treatment obtained with PVA-SBQ/alginate gel entrapment. Water Res. 31: 1659-1664.

[2] King, V.A. and Zall, R.R. (1983) Ethanol fermentation of whey using calcium alginate entrapped yeasts. Process Biochem. 12: 17-30.

[3] Klinkenberg, G.; Lystad, K.Q.; Levine, D.W. and Dyrset, N. (2001) pH-controlled cell release and biomass distribution of alginate-immobilized *Lactococcus lactis* subsp *lactis*. Journal of Appl. Microbiol. 91: 705-714.

[4] Larisch, B.C.; Poncelet, D.; Champagne, C.P. and Neufeld, R.J. (1994) Microencapsulation of *Lactococcus-lactis* subsp *cremoris*. J. Microencapsul. 11: 189-195.

[5] Jarvis, A.P. and Grdima, T.A. (1983) Production of biologicals (interferon) from microencapsulated living cells. Biotechn. 1: 24-27.

[6] Brodelius, P. and Mosbach, K. (1979) Immobilization of plant cells. Adv. Appl. Microbiol. 28: 1-26.

[7] de Vos, R. and van Schilfgaarde, R. (1999) Biocompatibility issues. In: Kühtreiber, W.M.; Lanza, R.P. and Chick, W.L. (Eds.) Cell encapsulation technology and therapeutics. Birkhäuser, Boston; pp. 63-78.

[8] DeVos, P.; DeHaan, B.J. and VanSchilfgaarde, R. (1998) Is it possible to use the standard alginate-PLL procedure for production of small capsules? Transplant. Proc. 30: 492-493.

[9] Prokop, A.; Hunkeler, D.; DiMari, S.; Haralson, M.A. and Wang, T.G. (1998) Water soluble polymers for immunoisolation I: Complex coacervation and cytotoxicity. Advances in Polymer Science 136: 1-51.

[10] Hunkeler, D. (1997) Polymers for bioartificial organs. Trends Polymer Sci. 5: 286-293.

[11] Babensee, J.E.; Anderson, J.M.; McIntire, L.V. and Mikos, A.G. (1998) Host response to tissue engineered devices. Adv. Drug Deliv. Rev. 33: 111-139.

[12] Chaikof, E.L. (1999) Engineering and material considerations in islet cell transplantation. Annual Review of Biomedical Engineering 1: 103-127.

[13] Li, R.H. (1998) Materials for immunoisolated cell transplantation. Advanced Drug Delivery Reviews 33: 87-109.

[14] Uludag, H.; DeVos, P. and Tresco, P.A. (2000) Technology of mammalian cell encapsulation. Adv. Drug Deliv. Rev. 42: 29-64.

[15] Dautzenberg, H.; Schuldt, U.; Grasnick, G.; Karle, P.; Muller, P.; Lohr, M.; Pelegrin, M.; Piechaczyk, M.; Rombs, K.V.; Gunzburg, W.H.; Salmons, B. and Saller, R.M. (1999) Development of cellulose sulfate-based polyelectrolyte complex microcapsules for medical applications. Ann. N.Y. Acad. Sci. 875: 46-63.

[16] Lacik, I.; Brissova, M.; Anilkumar, A.V.; Powers, A.C. and Wang, T. (1998) New capsule with tailored properties for the encapsulation of living cells. J. Biomed. Mater. Res. 39: 52-60.

[17] Wang, T.; Lacik, I.; Brissova, M.; Anilkumar, A.V.; Prokop, A.; Hunkeler, D.; Green, R.; Shahrokhi, K. and Powers, A.C. (1997) An encapsulation system for the immunoisolation of pancreatic islets. Nature Biotechnol. 15: 358-362.

[18] Lim, F. and Sun, A.M. (1980) Microencapsulated islets as a bioartificial endocrine pancreas. Science 210: 908-910.

[19] Vandenbossche, G.M.R.; Bracke, M.E.; Cuvelier, C.A.; Bortier, H.E.; Mareel, M.M. and Remon, J.P. (1993) Host reaction against empty alginate-polylysine microcapsules. Influence of preparation procedure. J. Pharm. Pharmacol. 45: 115-120.

[20] Thu, B.; Bruheim, P.; Espevik, T.; Smidsrød, O.; Soon-Shiong, P. and Skjåk-Bræk, G. (1996) Alginate polycation microcapsules. II. Some functional properties. Biomaterials 17: 1069-1079.

[21] Weber, C.J.; Kapp, J.A.; Hagler, M.K.; Safley, S.; Chryssochoos, J.T. and Chaikof, E.L. (1999) Long-term survival of poly-L-lysine-algiante microencapsulated xenografts in spontaneous diabetic NOD mice. In: Kühtreiber, W.M.; Lanza, R.P. and Chick, W.L. (Eds.) Cell encapsulation technology and therapeutics. Birkhäuser, New York/Boston; pp. 117-137.

[22] Pommersheim, R.; Schrezenmeir, J. and Vogt, W. (1994) Immobilization of enzymes by multilayer microcapsules. Macromolecular Chemistry and Physics 195: 1557-1567.

[23] Schneider, S.; Feilen, P.J.; Slotty, V.; Kampfner, D.; Preuss, S.; Berger, S.; Beyer, J. and Pommersheim, R. (2001) Multilayer capsules: a promising microencapsulation system for transplantation of pancreatic islets. Biomaterials 22: 1961-1970.

[24] Sakai, S.; Ono, T.; Ijima, H. and Kawakami, K. (2000) Control of molecular weight cut-off for immunoisolation by multilayering glycol chitosan-alginate polyion complex on alginate-based microcapsules. J. Microencapsul. 17: 691-699.

[25] Sawhney, A.S. and Hubbell, J.A. (1992) Poly(ethylene oxide)-graft-poly(L-lysine) copolymers to enhance the biocompatibility of poly(L-lysine)-alginate microcapsule membranes. Biomaterials 13: 863-870.
[26] Calafiore, R.; Basta, G.; Luca, G.; Boselli, C.; Bufalari, A.; Bufalari, A.; Cassarani, M.P.; Giustozzi, G.M. and Brunetti, P. (1999) Transplantation of pancreatic islets contained in minimal volume microcapsules in diabetic high mammalians. Ann. N.Y. Acad. Sci. 875: 219-232.
[27] Thu, B.; Bruheim, P.; Espevik, T.; Smidsrød, O.; Soon-Shiong, P. and Skjåk-Bræk, G. (1996) Alginate polycation microcapsules. I. Interaction between alginate and polycation. Biomaterials 17: 1031-1040.
[28] Vandenbossche, G.M.R.; Van Oostveld, P.; Deemester, J. and Remon, J.P. (1993) The molecular-weight cutoff of microcapsules is determined by the reaction between alginate and polylysine. Biotechnol. Bioeng. 42: 381-386.
[29] Lanza, R.P.; Kuhtreiber, W.M.; Ecker, D.; Staruk, J.E. and Chick, W.L. (1995) Xenotransplantation of porcine and bovine islets without immunosupressing using uncoated alginate microspheres. Transplantation 59: 1377-1384.
[30] Zekorn, T.D.C.; Horcher, A.; Siebers, U.; Schnettler, R.; Hering, B.; Zimmermann, U.; Bretzel, R.G. and Federlin, K. (1992) Barium-cross-linked alginate beads: A simple one-step-method for successful immuno siolated transplantation of islets of Langerhans. Acta Diabetol. 29: 99-106.
[31] Lanza, R.P.; Ecker, D.M.; Kuhtreiber, W.M.; Marsh, J.P.; Ringeling, J. and Chick, W.L. (1999) Transplantation of islets using microencapsulation: studies in diabetic rodents and dogs. J. Mol. Med. 77: 206-210.
[32] Lanza, R.P.; Kuhtreiber, W.M.; Ecker, D.M.; Marsh, J.P.; Staruk, J.E. and Chick, W.L. (1996) A simple method for xenotransplanting cells and tissues into rats using uncoated alginate microreactors. Transplant Proc. 28: 835-835.
[33] Duvivier-Kali, V.F.; Abulkadir, O.; Parent, R.J.; O´Neil, J.J. and Weir, G.C. (2001) Complete protection of islets against allorejection and autoimmunity by a simple barium-alginate membrane. Diabetes 50: 1698-1704.
[34] Birnbaum, S.; Pendelton, R.; Larsson, P.O. and Mosbach, K. (1981) Covalent stabilisation of algiante gel for the entrapment of living whole cells. Biotechnol. Lett. 3: 393-400.
[35] Soon-Shiong, P.; Desai, N.P.; Sanford, P.A.; Heitz, R. and Sojomihardjo, S. (1993) Crosslinkable polysaccharides, polycations and lipids useful for encapsulation and drug release. Patent PCT/US92/09364. World International Property Organization; pp. 1-52.
[36] Hertzberg, S.; Moen, E.; Vogelsang, C. and Østgaard, K. (1995) Mixed photo-cross-linked polyvinyl-alcohol and calcium-alginate gels for cell entrapment. Appl. Microbiol. Biotechnol. 43: 10-17.
[37] Iwata, H. and Ikada, Y. (1999) Agarose. In: Kühtreiber, W.M.; Lanza, R.P. and Chick, W.L. (Eds.) Cell encapsulation technology and therapeutics. Birkhäuser, New York/Boston; pp. 97-107.
[38] Tun, T.; Inoue, K.; Hayashi, H.; Aung, T.; Gu, Y.J.; Doi, R.; Kaji, H.; Echigo, Y.; Wang, W.J.; Setoyama, H.; Imamura, M.; Maetani, S.; Morikawa, N.; Iwata, H. and Ikada, Y. (1996) A newly developed three-layer agarose microcapsule for a promising biohybrid artificial pancreas: Rat to mouse xenotransplantation. Cell Transplantation 5: S59-S63.
[39] Shoichet, M.S.; Li, R.H.; White, M.L. and Winn, S.R. (1996) Stability of hydrogels used in cell encapsulation: An in vitro comparison of alginate and agarose. Biotechnol. Bioeng. 50: 374-381.
[40] Murano, E. (2000) Natural gelling polysaccharides: indispensable partners in bioencapsulation technology. Minerva Biotec. 12: 213-222.
[41] Kurillova, L.; Gemeiner, P.; Ilavsky, M.; Stefuca, V.; Polakovic, M.; Welwardova, A. and Toth, D. (1992) Calcium pectate gel beads for cell entrapment. 4. Properties of stabilized and hardened calcium pectate gel beads with and without cells. Biotechnol. Appl. Biochem. 16: 236-251.
[42] Poncelet, D.; Desmet, B.P.; Beaulieu, C.; Huguet, M.L.; Fournier, A. and Neufeld, R.J. (1995) Production of alginate beads by emulsification internal gelation. 2. Physicochemistry. App. Microbiol. Biotechnol. 43: 644-650.
[43] Poncelet, D.; Lencki, R.; Beaulieu, C.; Halle, J.P.; Neufeld, R.J. and Fournier, A. (1992) Production of alginate beads by emulsification internal gelation. 1. Methodology. Appl. Microbiol. Biotechnol. 38: 39-45.
[44] Dulieu, C.; Poncelet, D. and Neufeld, R. (1999) Encapsulation and immobilization techniques. In: Kühtreiber, W.M.; Lanza, R.P. and Chick, W.L. (Eds.) Cell encapsulation technology and therapeutics. Birkhäuser, New York/Boston; pp. 3-17.
[45] Serp, D.; Cantana, E.; Heinzen, C.; von Stockar, U. and Marison, I.W. (2000) Caracterization of an encapsulation device for the production of monodisperse alginate beads for cell immobilization. Biotechnol. Bioeng. 70: 41-53.

[46] Prüsse, U.; Dalluhn, J.; Breford, J. and Vorlop, K.D. (2000) Production of spherical beads bu JetCutting. Chem. Eng. Technol. 23: 1105-1110.
[47] Goosen, M.F.A.; Al-Ghafri, A.S.; El Mardi, O.; Al-Belushi, M.I.J.; Al-Hajri, H.A.; Mahmoud, E.S.E. and Consolacion, E.C. (1997) Electrostatic droplet generation for encapsulation of somatic tissue: Assesment of high-voltage power supply. Biotechnol. Prog. 13: 497-502.
[48] Halle, J.P.; Leblond, F.A.; Pariseau, J.F.; Jutras, P.; Brabant, M.J. and Lepage, Y. (1994) Studies on small (<300 μm) microcapsules: II - Parameters governing the production of algiante beads by high voltage electrostatic pulses. Cell Transplant. 3: 365-372.
[49] Pjanovic, R.; Goosen, M.F.A.; Nedovic, V. and Bugarski, B. (2000) Immobilization/encapsualtion of cells using electrostatic droplet generator. Minerva Biotec. 12: 241-248.
[50] Strand, B.L.; Gåserød, O.; Kulseng, B.; Espevik, T. and Skjåk-Bræk, G. (2002) Alginate-polylysine-alginate microcapsules – effect of size reduction on capsule properties. J. Microencapsul. 19: 615-630.
[51] Klokk, T.I. and Melvik, J.E. (2002) Controlling the size of alginate gel beads by use of a high electrostatic potential. J. Microencapsul. 19: 415-424.
[52] Colton, C.K. (1995) Implantable biohybrid artificial organs. Cell Transplant. 4: 415-436.
[53] Chicheportiche, D. and Reach, G. (1988) *In vitro* kinetics of insulin release by microencapsulated rat islets: effect of the size of the microcapsules. Diabetologia 31: 54-57.
[54] Poncelet, D. and Neufeld, R.J. (1989) Shear breakage of nylon membrane microcapsules in a turbine reactor. Biotechnol. Bioeng. 33: 95-103.
[55] Robitaille, R.; Pariseau, J.F.; Leblond, F.; Lamoureux, M.; Lepage, Y. and Halle, J.P. (1999) Studies on small (<350 μm) alginate-poly-L-lysine microcapsules. III. Biocompatibility of smaller versus standard microcapsules. J. Biomed. Mater. Res. 44: 116-120.
[56] Leblond, F.A.; Simard, G.; Henley, N.; Rocheleau, B.; Huet, P.M. and Halle, J.P. (1999) Studies on smaller (similar to 315 mu M) microcapsules: IV. Feasibility and safety of intrahepatic implantations of small alginate poly-L-lysine microcapsules. Cell Transplant. 8: 327-337.
[57] Haug, A.; Larsen, B. and Smidsrød, O. (1966) A study of the constitution of alginic acid by partial hydrolysis. Acta Chem. Scand. 20: 183-190.
[58] Smidsrød, O. and Skjåk-Bræk, G. (1990) Alginate as immobilization matrix for cells. Trends Biotechnol. 8: 71-78.
[59] Govan, J.R.W.; Fyfe, J.A.M. and Jarman, T.R. (1981) Isolation of alginate-producing mutants of *Pseudomonas fluorescens*, *Pseudomonas putida* and *Pseudomonas mendonica*. J. Gen. Microbiol. 125: 217-220.
[60] Skjåk-Bræk, G. and Larsen, B. (1985) Biosynthesis of alginate: Purification and characterisation of mannuronan C-5-epimerase from *Azotobacter vinelandii*. Carbohydr. Res. 139: 273-283.
[61] Ertesvåg, H.; Høidal, H.K.; Hals, I.K.; Rian, A.; Doseth, B. and Valla, S. (1995) A family of modular type mannuronan c-5-epimerase genes controls alginate structure in azotobacter-vinelandii. Mol. Microbiol. 16: 719-731.
[62] Ertesvåg, H.; Høidal, H.K.; Schjerven, H.; Svanem, B.I.G. and Valla, S. (1999) Mannuronan C-5-Epimerases and their application for *in Vitro* and *in Vivo* design of new alginates useful in biotechnology. Metabol. Eng. 1: 262-269.
[63] Smidsrød, O. (1974) Molecular basis for some physical properties of alginates in the gel state. J. Chem. Soc. Farad. Transact. 57: 263-274.
[64] Grant, G.T.; Morris, E.R.; Rees, D.A.; Smith, P.J.C. and Thom, D. (1973) Biological interactions between polysaccharides and divalent cations: The egg-box model. FEBS Lett. 32: 195-198.
[65] Stokke, B.T.; Smidsrød, O.; Zanetti, F.; Strand, W. and Skjåk-Bræk, G. (1993) Distribution of uronate residues in alginate chains in relation to alginate gelling properties. 2. Enrichment of beta-D-mannuronic acid and depletion of alpha-L-guluronic acid in sol fraction. Carbohydr. Polymers 21: 39-46.
[66] Kohn, R. and Larsen, B. (1972) Preparation of water-soluble polyuronic acids and their calcium salts, and the determination of calcium ion activity in relation to the degree of polymerization. Acta Chem. Scand. 26: 2455-2468.
[67] Martinsen, A.; Skjåk-Bræk, G. and Smidsrød, O. (1989) Alginate as immobilization material: I. Correlation between chemical and physical properties of alginate gel beads. Biotechnol. Bioeng. 33: 79-89.
[68] Smidsrød, O. (1973) Some physical properties of alginates in solution and in the gel state. Thesis. NTNF, Trondheim,.
[69] Schlemmer, U. (1989) Studies of the Binding of copper, zinc and calcium to pectin alginate carrageenan and gum guar in HCO_3^- - CO_2 buffer. Food Chem. 32: 223-234.

[70] Skjåk-Bræk, G. and Martinsen, A. (1991) Applications of some algal polysaccharides in biotechnology. In: Guiry, M.D. and Blunden, G. (Eds.) Seaweed Resources in Europe: Uses and Potential. John Wiley & Sons Ltd.; pp. 219-257.
[71] Rochefort, W.E.; Regh, T. and Chau, P.C. (1986) Trivalent cation stabilization of alginate gel for cell immobilization. Biotechnol. Lett. 8: 115-120.
[72] Andresen, I.; Skipnes, O.; Smidsrød, O.; Østgaard, K. and Hemmer, P.C. (1977) Some biological functions of matrix components in benthic algae in relation to their chemistry and the composition of seawater. ACS Symp. Series 48: 361-381.
[73] Øyaas, J.; Storrø, I.; Svendsen, H. and Levine, D.W. (1995) The effective diffusion-coefficient and the distribution constant for small molecules in calcium-alginate gel beads. Biotechnol. Bioeng. 47: 492-500.
[74] Tanaka, H.; Matsumura, M. and Veliky, I.A. (1984) Diffusion characteristics in Ca-alginate gel beads. Biotechnol. Bioeng. 26: 53-58.
[75] Martinsen, A.; Storrø, I. and Skjåk-Bræk, G. (1992) Alginate as immobilization material. 3. Diffusional properties. Biotechnol. Bioeng. 39: 186-194.
[76] Gombotz, W.R. and Wee, S.F. (1998) Protein release from alginate matrices. Adv. Drug Deliv. Rev. 31: 267-285.
[77] Kulseng, B.; Thu, B.; Espevik, T. and Skjåk-Bræk, G. (1997) Alginate polylysine capsules as an immune barrier: Permeability of cytokines and immunoglobulins over the capsule membrane. Cell Transplant. 6: 387-394.
[78] Gåserød, O.; Sannes, A. and Skjåk-Bræk, G. (1999) Microcapsules of alginate-chitosan II. A study of a capsule stability and permeability. Biomaterials 20: 773-783.
[79] Rilling, P.; Walter, T.; Pommersheim, R. and Vogt, W. (1997) Encapsulation of cytochrome C by multilayer microcapsules. A model for improved enzyme immobilization. J. Membrane Sci. 129: 283-287.
[80] Huguet, M.L.; Neufeld, R.J. and Dellacherie, E. (1996) Calcium-alginate beads coated with polycationic polymers: Comparison of chitosan and DEAE-dextran. Process Biochem. 39: 347-353.
[81] Stokke, B.T.; Draget, K.I.; Smidsrod, O.; Yuguchi, Y.; Urakawa, H. and Kajiwara, K. (2000) Small-angle X-ray scattering and rheological characterization of alginate gels. 1. Ca-alginate gels. Macromolecules 33: 1853-1863.
[82] Stokke, B.T.; Smidsrød, O.; Bruheim, P. and Skjåk-Bræk, G. (1991) Distribution of uronate residues in algiante chains in relation to algiante gelling properties. Macromolecules 24: 4637-4645.
[83] Thu, B.; Gåserød, O.; Paus, D.; Mikkelsen, A.; Skjåk-Bræk, G.; Toffanin, R.; Vittur, F. and Rizzo, R. (2000) Inhomogenious alginate gel spheres: An assesment of the polymer gradients by synchrotron radiation-induced X-ray emmision, magnetic resonance microimaging, and mathematic modeling. Biopolymers 53: 60-71.
[84] Skjåk-Bræk, G.; Grasdalen, H. and Smidsrød, O. (1989) Inhomogeneous polysaccharide ionic gels. Carbohydr. Polymers 10: 31-54.
[85] Hills, B.P.; Godward, J.; Debatty, M.; Barras, L.; Saturio, C.P. and Ouwerx, C. (2000) NMR studies of calcium induced alginate gelation. Part II. The internal bead structure. Magn. Res. Chem. 38: 719-728.
[86] Strand, B.L.; Mørch, Y.A.; Espevik, T. and Skjåk-Bræk, G. (2003) Visualisation of alginate-polylysine-alginate microcapsules by confocal laser scanning microscopy. Biotechnol. Bioeng. (In press).
[87] Skjåk-Bræk, G.; Smidsrød, O. and Larsen, B. (1986) Tailoring of alginates by enzymatic modification *in vitro*. Int. J. Biol. Macromol. 8: 330-336.
[88] Pealez, C. and Karel, M. (1981) Improved method for preparation of fruit-simulating alginate gels. J. Food Process. Preserv. 5: 63-81.
[89] Flink, J.M. and Johansen, A. (1985) A novel method for immobilizing yeast cells in alginates of various shapes by internal liberation of Ca-ions. Biotechnol. Lett. 7: 765-768.
[90] Draget, K.I.; Østgaard, K. and Smidsrød, O. (1990) Homogeneous alginate gels - a technical approach. Carbohydr. Polymers 14: 159-178.
[91] Strand, B.L.; Mørch, Y.A.; Syvertsen, K.R.; Espevik, T. and Skjåk-Bræk, G. (2003) Microcapsules made by enzymatically tailored alginate. J. Biomed. Mater. Res. (In press).
[92] Smidsrød, O.; Glover, R.M. and Whittington, S.G. (1973) The relative extension of alginates having different chemical composition. Carbohydr. Res. 27: 107-118.
[93] Flory, P.J. (Ed.) (1953) Principles of polymer chemistry. Oxford University, Ithaca.
[94] Tanaka, T. (1979) Phase transitions in gels and a single polymer. Polymer 20: 1404-1412.
[95] Moe, S.T.; Draget, K.I.; Skjåk-Bræk, G. and Smidsrød, O. (1995) Alginates. In: Stephen, A.M. (Ed.) Food polysaccharides and their applications. Marcel Decker Inc., New York, U.S.
[96] Gåserød, O.; Smidsrød, O. and Skjåk-Bræk, G. (1998) Microcapsules of alginate-chitosan-I: A quantitative study of the interaction between alginate and chitosan. Biomaterials 19: 1815-1825.

[97] Rha, C.K. (1984) Chitosan as biomaterial. In: Colwell, R.R.; Pariser, E.R. and Sinskey, A.J. (Eds.) Biotechnology in the marine sciences. Wiley, New York; pp. 177-189.
[98] Calafiore, R.; Basta, G.; Falorni, A.; Picchio, M.L.; Gambelunghe, G.; Del Sindaco, P. and Brunetti, P. (1992) Fabrication of high performance microcapsules for pancreatic islet transplantation. Diabetes, Nutrition and Metabolism 5: 173-176.
[99] Tanaka, H.; Kurosawa, H.; Kokufuta, E. and Veliky, I.A. (1984) Preparation of immobilized glucoamylase using Ca-alginate gel coated with partially quartenized poly(ethyleneimine). Biotechnology and Bioengineering 26: 1393-1394.
[100] Veliky, I.A. and Williams, R.E. (1981) The production of ethanol by *Saccharomyces cerevisiae* immobilized in polycation stabililzed calcium alginate gels. Biotechnol. Lett. 3: 275-280.
[101] Dautzenberg, H.; Schuldt, U.; Grasnick, G.; Karle, P.; Müller, P.; Löhr, M.; Pelegrin, M.; Piechaczyk, M.; Rombs, K.V.; Günzburg, W.H.; Salmons, B. and Saller, R.M. (1999) Development of cellulose sulfate-beased polyelectrolyte complex microcapsules for medical applications. In: Hunkeler, D.; Prokop, A.; Cherrington, A.; Rajotte, R. and Sefton, M. (Eds.) Bioartificial organs II: Technology, medicine and materials. Ann. N.Y. Acad. Sci. 875: 46-63.
[102] King, G.; Daugulis, A.; Faulkner, P. and Goosen, M. (1987) Alginate-polylysine microcapsules of controlled membrane molecular weight cutoff for mammalian cell culture engineering. Biotechnol. Progress 3: 231-240.
[103] Ma, X.J.; Vacek, I. and Sun, A. (1994) Generation of alginate-poly-l-lysine-alginate (apa) biomicroscopies -the relationship between the membrane strength and the reaction conditions. Artif. Cells Blood Substit. Immobil. Biotechnol. 22: 43-69.
[104] Goosen, M.F.A.; O´Shea, G.M.; Gharapetian, H.M.; Chou, S. and Sun, A.M. (1985) Optimization of microencapsulation parameters: Semipermeable microcapsules as a bioartificial pancreas. Biotechnol. Bioeng. 27: 146-150.
[105] Dupuy, B.; Arien, A. and Minnot, A.P. (1994) FT-IR of membranes made with alginate/polylysine complexes. Variations with the mannuronic or guluronic content of the polysaccharides. Artif. Cells Blood Substit. Immobil. Biotechnol. 22: 71-82.
[106] Halle, J.P.; Bourassa, S.; Leblond, F.A.; Chevalier, S.; Beaudry, M.; Chapdelaine, A.; Cousineau, S.; Saintonge, J. and Yale, J.F. (1993) Protection of islets of langerhans from antibodies by microencapsulation with alginate-poly-l-lysine membranes. Transplantation 55: 350-354.
[107] Tai, I.T.; Vacek, I. and Sun, A.M. (1995) The alginate-poly-L-lysine-alginate membrane: Evidence of a protective effect on microencapsulated islets of Langerhans following exposure to cytokines. Xenotransplantation 2: 37-45.
[108] Bartkowiak, A. (2001) Optimal conditions of transplantable binary polyelectrolyte microcapsules. In: Hunkeler, D.; Cherrington, A.; Prokop, A. and Rajotte, R. (Eds.) Bioartificial Organs III: Tissue sourcing, Immunoisolation and clinical trials. Ann. N.Y. Acad. Sci. 944: 120-134.
[109] Vårum, K.M.; Ottøy, M. and Smidsrød, O. (1994) Water solubility of partially *N*-acetylated chitosans as a function of pH: effect of chemical composition and depolymerisation. Carbohydr. Polymers 25: 65-70.
[110] Rihova, B. (2000) Immunocompatibility and biocompatibility of cell delivery systems. Adv. Drug Deliv. Rev. 42: 65-80.
[111] Gin, H.; Dupuy, B.; Bonnemaisonbourignon, D.; Bordenave, L.; Bareille, R.; Latapie, M.J.; Baquey, C.; Bezian, J.H. and Ducassou, D. (1990) Biocompatibility of polyacrylamide microcapsules implanted in peritoneal-cavity or spleen of the rat - effect on various inflammatory reactions invitro. Biomater. Artif. Cells Artif. Organs 18: 25-42.
[112] Miller, K.M. and Anderson, J.M. (1988) Human monocyte macrophage activation and interleukin-1 generation by biomedical polymers. J. Biomed. Mater. Res. 22: 713-731.
[113] Skjåk-Bræk, G.; Murano, E. and Paoletti, S. (1989) Alginate as immobilization material. 2. Determination of polyphenol contaminants by fluorescence spectroscopy, and evaluation of methods for their removal. Biotechnol. Bioeng. 33: 90-94.
[114] De Vos, P.; De Haan, B.J.; Wolters, G.H.J.; Strubbe, J.H. and Van Schilfgaarde, R. (1997) Improved biocompatibility but limited graft survival after purification of alginate for microencapsulation of pancreatic islets. Diabetologia 40: 262-270.
[115] Zimmermann, U.; Klock, G.; Federlin, K.; Hannig, K.; Kowalski, M.; Bretzel, R.G.; Horcher, A.; Entenmann, H.; Sieber, U. and Zekorn, T. (1992) Production of mitogen-contamination free alginates with variable ratios of mannuronic acid to guluronic acid by free-flow electrophoresis. Electrophoresis 13: 269-274.

[116] Otterlei, M.; Østgaard, K.; Skjåk-Bræk, G.; Smidsrød, O.; Soon-Shiong, P. and Espevik, T. (1991) Induction of cytokine production from human monocytes stimulated with alginate. J. Immunother. 10: 286-291.
[117] Espevik, T.; Ottrerlei, M.; Skjåk-Bræk, G.; Ryan, L.; Wright, S.D. and Sundan, A. (1993) The involvment of CD14 in stimulation of cytokine production by uronic acid polymers. Eur. J. Immunol. 23: 255-261.
[118] Otterlei, M.; Sundan, A.; Skjåk-Bræk, G.; Ryan, L.; Smidsrød, O. and Espevik, T. (1993) Similar mechanisms of action of defined polysaccharides and lipopolysaccharides - characterization of binding and tumor-necrosis-factor-alpha induction. Infect. Immun. 61: 1917-1925.
[119] Berntzen, G.; Flo, T.H.; Medvedev, A.; Kilaas, L.; Skjåk-Bræk, G.; Sundan, A. and Espevik, T. (1998) The tumor necrosis factor-inducing potency of lipopolysaccharide and uronic acid polymers is increased when they are covalently linked to particles. Clin. Diagn. Lab. Immunol. 5: 355-361.
[120] Flo, T.H.; Ryna, L.; Latz, E.; Takeuchi, O.; Monks, B.G.; Lien, E.; Halaas, Ø.; Akira, S.; Skjåk-Bræk, G.; Golenbock, D.T. and Espevik, T. (2002) Involvement of toll-like receptor (TLR) 2 and TLR4 in cell activation by mannuronic acid polymers. J. Biol. Chem. 277: 35489-35495.
[121] Halaas, O.; Olsen, W.M.; Veiby, O.P.; Lovhaug, D.; Skjåk-Bræk, G.; Vik, R. and Espevik, T. (1997) Mannuronan enhances survival of lethally irradiated mice and stimulates murine haematopoiesis *in vitro*. Scand. J. Immunol. 46: 358-365.
[122] Skaugerud, Ø.; Hagen, A.; Borgersen, B. and Dornish, M. (1999) Biomedical and pharmaceutical applications of alginate and chitosan. Biotechnol. Genet. Eng. Rev. 16: 23-40.
[123] Kulseng, B.; Skjåk-Bræk, G.; Ryan, L.; Anderson, A.; King, A.; Faxvaag, A. and Espevik, T. (1999) Antibodies against alginates and encapsulated porcine islet-like cell clusters. Transplantation 67: 978-984.
[124] Klöck, G.; Frank, H.; Houben, R.; Zekorn, T.; Horcher, A.; Siebers, U.; Wöhrle, M.; Federlin, K. and Zimmermann, U. (1994) Production of purified alginates suitable for use in immunoisolated transplantation. Appl. Microbiol. Biotechnol. 40: 638-643.
[125] Carreno-Gómez, B. and Duncan, R. (1997) Evaluation of the biological properties of soluble chitosan and chitosan microspheres. Int. J. Pharmac. 148: 231-240.
[126] Strand, B.L.; Ryan, L.; In`t Veld, P.; Kulseng, B.; Rokstad, A.M.; Skjåk-Bræk, G. and Espevik, T. (2001) Poly-L-lysine induces fibrosis on alginate microcapsuels *via* the induction of cytokines. Cell Transplantation 10: 263-275.
[127] De Vos, P.; De Haan, B.J. and Van Schilfgaarde, R. (1996) The effect of alginate composition on the biocompatibility of alginate-polylysine microcapsules. Biomaterials 18(3): 273-278.
[128] Vandenbossche, G.M.R.; Van Oostveld, P. and Remon, J.P. (1993) Host reactions against alginate-polylysine microcapsules containing living cells. J. Pharm. Pharmacol. 45: 121-125.
[129] Clayton, H.A.; London, N.J.M.; Colloby, P.S.; Bell, P.R.F. and James, R.F.L. (1991) The effect of capsule composition on the biocompatibility of alginate-poly-l-lysine capsule. J. Microencapsul. 8: 221-233.
[130] Clayton, H.A.; London, N.J.M.; Bell, P.R.F. and James, R.F.L. (1992) The transplantation of encapsulated islets of Langerhans into the peritoneal cavity of the BiobBreeding rat. Transplantation 54: 558-559.
[131] Ekrami, H.M. and Sheng, W.-C. (1995) Carbamylation decreases the cytotoxicity but not the drug-carrier properties of polylysines. J. Drug Target 2: 469-475.
[132] Darquy, S.; Pueyo, M.E.; Capron, F. and Reach, G. (1994) Complement activation by alginate-polylysine microcapsules used for islet transplantation. Artif. Organs 18: 898-903.
[133] Pueyo, M.E.; Darquy, S.; Capron, F. and Reach, G. (1993) *In vitro* activation of human macrophages by alginate-polylysine microcapsules. J. Biomater. Sci. Polym. Ed. 5: 197-203.
[134] Gåserød, O.; Sannes, A. and Skjåk-Bræk, G. (1999) Microcapsules of alginate-chitosan II. A study of a capsule stability and permeability. Biomaterials 20: 773-783.
[135] Gill, R.G. and Wolf, L. (1995) Immunobiology of cellular transplantation. Cell Transplant. 4: 361-370.
[136] de Vos, P.; Hamel, A.F. and Tatarkiewicz, K. (2002) Considerations for successful transplantation of encapsulated pancreatic islets. Diabetologia 45: 159-173.
[137] King, A.; Strand, B.L.; Rokstad, A.M.; Kulseng, B.; Andersson, A.; Skjåk-Bræk, G. and Sandler, S. (2003) Improvement of the biocompatibility of alginate/poly-L-lysine/alginate microcapsules by the use of epimerised alginate as coating. J. Biomed. Mater. Res. (In press).
[138] Cole, D.R.; Waterfall, M.; McIntyre, M. and Baird, J.D. (1992) Microencapsulated islet grafts in the BB/E rat: a possible role for cytokines in graft failure. Diabetologia 35: 231-237.

LIQUID CORE CASPULES FOR APPLICATIONS IN BIOTECHNOLOGY

IAN MARISON[1], ANNE PETERS[1] AND CHRISTOPH HEINZEN[2]
[1]Laboratory for Chemical and Biochemical Engineering, Swiss Federal Institute of Technology, CH-1015 Lausanne, Switzerland – Fax: +41 21 693 31 94 – Email: ian.marison@epfl.ch
[2]Inotech Encapsulation AG, Kirchstrasse 1, CH-5605 Dottikon, Switzerland - Fax: +41 56 624 29 88 – Email: heinzen@inotech.ch

1. Introduction

Encapsulation is the process by which a gaseous, liquid or solid encapsulant, is surrounded by a continuous film or coating. As a result the core of the capsule contains the encapsulant, which is prevented from contact with the surroundings by the capsule wall or membrane. Capsules are frequently classified with respect to size: macro-, micro-, and nano-capsules, core material or wall polymer.

The encapsulation process requires a technique for producing capsules of the desired size and properties from the required wall polymer(s). The choice of encapsulation process is dependent on the nature of the core material and wall polymer, which themselves are governed by the application of the resulting capsules. Thus stability, diffusion properties, temperature and pH resistance are just some of the characteristics which define the choice of polymer and encapsulation procedure.

While encapsulation has been used for many years in the food [1,2], cosmetic and pharmaceutical industries [3,4], interest is rapidly growing in the fields of biotechnology and medicine for the encapsulation of whole cells and enzymes for use in cell culture, high throughput screening, implants and bioconversion processes [5-7]. These fields place important emphasis on the ability to produce small (< 700 μm diameter) monodisperse capsules using low shear procedures involving non-cytotoxic polymers and materials at between 20 to 40°C and physiological pH. While immobilisation of cells and enzymes offers many advantages over suspensions, in both cases a direct contact exists with the environment and cell/enzyme leaching can occur. Encapsulation offers distinct advantages in that a solid or liquid matrix or core containing the cells or enzymes is surrounded by a semi-permeable membrane or wall which completely protects the encapsulant from the environment and allows selective diffusion of compounds in either direction. A typical example involves cell implantation [8], such as pancreatic islets and hepatocytes [9,10], in which small molecules and proteins such as insulin can freely diffuse into and out of the capsules, while larger

V. Nedović and R. Willaert (eds.), Fundamentals of Cell Immobilisation Biotechnology, 185-204.

proteins such as antibodies cannot. In this way the transplanted cells (xenotransplants) are hidden from the host immune system and may survive and function over extended periods.

The focus of the present paper will be to describe techniques suitable for the formation of liquid core capsules, particularly for use in biotechnological applications. For techniques used in other fields the reader is referred to the excellent review articles of King [2] and Risch [11].

2. Encapsulation methods

2.1. EMULSION-BASED TECHNIQUES

An emulsion is a dispersion of two immiscible liquids in the presence of a stabilizing compound or emulsifier. When the core phase is aqueous this is termed a water-in-oil emulsion (w/o) while a hydrophobic core phase is termed an oil-in-water emulsion (o/w). Emulsions are simply produced by the addition of the core phase to a vigorously stirred excess of the second phase that contains the emulsifier. Although readily scalable, the technique produces capsules with an extremely large size distribution and is not suitable for the encapsulation of volatile compounds.

The double emulsion technique, such as water-in-oil-in-water (w/o/w), is a modification of the basic technique in which an emulsion is made in of an aqueous solution in a hydrophobic wall polymer such as polyester. This emulsion is then poured, with vigorous agitation, into an aqueous solution containing stabilizer [12]. The loading capacity of the hydrophobic core is limited by the solubility and diffusion to the stabilizer solution. This technique has found particular application in the encapsulation of peptides, proteins and hydrophilic pharmaceutical compounds [13].

2.2. MELT ENCAPSULATION

This technique involves dispersion of the core material into an agitated molten wall compound, followed by cooling and even grinding to form a powder. This powder or wax is then added to a suitable medium with heating to obtain the desired capsule size, followed by cooling to solidify the wall material. The method is easy to scale-up and has a high encapsulation efficiency. It is not however suitable for the encapsulation of compounds, which are thermally labile, or have low boiling points [14].

2.3. INTERNAL GELATION

This is a form of emulsion technique that usually involves charged polysaccharides, such as alginates. In this case an alginate solution is dispersed in an oil dispersion containing an insoluble calcium citrate complex. Addition of an oil soluble acid, such as glacial acetic acid, initiates release of the calcium ions which complex with alginate to form a gel. Choice of a suitable reactor, which provides low turbulence and homogeneous shear, is essential, yet cannot prevent a large size distribution of the resulting capsules [15,16].

2.4. PHASE SEPARATION (COACERVATION)

In this technique the core material is suspended in the wall polymer, and the polymer solution induced to separate by a variety of techniques, termed simple, complex or salt coacervation [12]. The only constraint is that the core material must be compatible with the polymer coating solution and insoluble or weakly soluble in the coacervation medium. An example is the production of polyvinyl alcohol (PVA) microcapsules (5-20 μm diameter). In this case the core (silicone oil) is dispersed in the wall polymer (aqueous PVA solution). A phase separation inducer is then added to generate a new, viscous, polymer-rich phase [17]. In the case of simple coacervation this may be achieved by the addition of a water immiscible non-solvent such as ethanol or, in the case of salt coacervation, by the addition of an electrolyte such as sodium sulphate. The polymer absorbs on to the surface of the hydrophobic oil droplets to form the capsule wall. The use of salts as phase separation inducer makes control of capsule size difficult with the tendency for capsules to aggregate. The resulting capsules may be stabilized by the addition of cross-linking agents or through variation of temperature and/or pH.

Complex coacervation occurs through the interaction of two oppositely charged colloids in an aqueous solution, such as cationic gelatin with anionic gum arabic [17]. The morphology of the capsules may be changed from mononuclear to polynuclear by varying the core material to wall polymer ratio, the size of the core droplet, concentration of stabilizer, agitation speed, viscosity, temperature and contact time [18].

2.5. INTERFACIAL POLYMERIZATION

This technique may be applied to the encapsulation of aqueous solutions, water immiscible liquids and solids, including carbon-less paper ink, pesticides and herbicides [12,19]. For water-immiscible cores, a multi-functional monomer, such as an isocyanate, is dissolved in the liquid core material [20]. The solution is dispersed in an aqueous phase containing the dispersed agent to achieve the desired droplet size. A co-reactant, such as a multi-functional amine, is then added and polymerization occurs. The inverse sequence of events is performed if the core material is an aqueous solution. The main disadvantage of this method is that the reagent is dissolved in the core material with which it may chemically react.

Many factors influence the capsule properties, including composition of the capsule wall, degree of cross-linking (multi-functional monomers are used for higher degrees of cross-linking), the capsule thickness, which is itself a function of the monomer concentration and the capsule size, which is determined by the amount of emulsifying agent and level of agitation. This method is ideal for liquid encapsulation, although the wall thickness is small compared with the relatively thick gelatinous walls characteristic of coacervation techniques.

2.6. PRODUCTION OF NANOCAPULES

Nanocapsules have been successfully produced using polyalkylcyanoacrylate (PACA) using interfacial polymerization [21]. In this case an organic phase, such as an ethanolic solution of the monomer containing lipophilic drug, is mixed with the oil based core material (miglyol, benzylbenzoate, ethyl oleate or lipiodol). The organic phase is passed slowly through a micropipette into an agitated aqueous phase. The latter is usually a non-ionic surfactant, such as Pluronic F-68 (0.5%) at pH 4-10. The ethanol is then removed by evaporation to yield capsules of 200-300 nm diameter. The exact capsule size is dependent on the nature and volume of the oil used. In order to obtain nanocapsules, rather than nanospheres, a dynamic process must be achieved in which the monomer is transferred to the oil-water interface, by the use of a co-solvent. The latter, such as ethanol, must be a solvent for both the oil and monomer and at the same time miscible with the aqueous phase [22].

2.7. *IN SITU* POLYMERIZATION

This technique is mainly used to produce small microcapsules of 3-6 μm in diameter. Monomers are added directly to the encapsulation reactor, although no reactive reagents to the core material. Polymerization occurs exclusively in the continuous phase and on the continuous phase-side of the interface formed by the dispersed core material with the continuous phase. As a result a prepolymer is deposited onto the surface of the dispersed core. An example of this technique is the encapsulation of water-immiscible liquids surrounded by walls formed by the reaction of urea with formaldehyde at acid pH [19].

2.8. SPRAY DRYING

This technique is mainly used for the encapsulation of flavour compounds [23,24]. The first step involves the dispersion of the core (usually an oil) in a concentrated aqueous solution of the wall polymer, which is usually a water soluble compound such as gum arabic, to form an emulsion with droplets of 1-3 μm diameter. The emulsion is then fed as droplets to the heated chamber of the spray drier where they are dehydrated to form capsules with diameters between 10 to 300 μm. The main disadvantage is that the capsules are irregular in size and shape and tend to form aggregates, while low boiling point and polar active agents, such as ethyl acetate, are difficult to encapsulate in this way. The main advantage is that as the polymer is water soluble, it will dissolve in contact with water to release the core material. Depending on the temperature during the spray drying process and the boiling point of the core material, it may be possible to produce liquid-core capsules, however in general this process is used for generating dry capsules.

The formation of an aerosol in the spray dryer is termed nebulization. Two methods exist to create these aerosols. The first, or turbine method involves the use of compressed air to turn a disc with calibrated orifices at high speed. This provokes shearing of the solution as it passes through the orifices and thus can be used for viscous solutions. In the second method an atomizer of 0.5 to 1 μm diameter is used. Compressed air or nitrogen is mixed directly with the core material and polymer solution and sprayed through the atomizer. The gas pressure, viscosity and flow rate of the solution can be used to vary the capsule diameter.

2.9. SPRAY CHILLING

This is a variation of the spray drying method in which the capsule wall (wax) is solidified at room temperature or below. The active agent is dispersed into the molten carrier and atomized through a heated nozzle into a cooling chamber where the wall material solidifies. Capsules produced in this way are generally not soluble in water and is frequently used to encapsulate flavour oils or aqueous flavours which are thermally unstable in the spray drying method [25].

2.10. SPRAY FREEZING

This is similar to spray chilling except that the core material and wall polymer are frozen into a rigid capsule. The solvent is then extracted by solvent exchange or vacuum drying. As a result it is suitable for heat labile compounds and has a high encapsulation efficiency [26].

2.11. FLUIDIZED-BED COATING: THE WURST PROCESS

This process is used to encapsulate tablets, granules, crystals and powders, but is not suitable for liquid core capsules [27]. The solid material is suspended in an airstream in a coating chamber, such that the air stream causes a cyclical flow of the particles. A nozzle at the bottom of the chamber sprays the coating material onto the particles. The particles are then carried away from the nozzle to the top of the chamber where the coating material solidifies, and the particles fall back again to receive a second coating. This process is repeated until the desired coating thickness is achieved (>150 μm).

2.12. PAN COATING

This process, which is widely used in the pharmaceutical industry, is not suitable for forming liquid-core capsules. Solid particles are tumbled in a rotating pan and a coating material slowly added using a carefully controlled temperature profile. The process is slow and difficult to control and thus gives a low yield and low quality of the capsule wall [28].

2.13. TURBOTAK ATOMIZATION

A Turbotak atomizer is used to spray sodium alginate solution into a calcium chloride solution using nitrogen as atomization gas. The resulting beads are then coated with poly-*L*-lysine and alginate to produce alginate-polylysine-alginate microcapsules [29]. The Turbotak device is a hollow stainless steel cylinder that contains internal and external chambers. The sodium alginate solution is fed into the internal chamber, using pressurized gas, from the top and flows into the external chamber via an orifice, falling into the calcium chloride solution. Small particles (5-15 μm) can be obtained, with the gas flow rate used to control the size.

2.14. DESOLVATION

The capsule wall material (maltodextrin, sugars or gelatin) is first dissolved in a small quantity of solvent. The active agent is dispersed in this solution (carrier phase) and the dispersion extruded or sprayed into an excess of non-solvent (ethanol). This non-solvent desolvates the fluid carrier phase and causes the wall material to solidify. However, the decrease in droplet volume and viscosity during the process disturb the steady state droplet size equilibrium thereby leading to destabilization of the droplet suspension and coagulation of the droplets at the early stages of solvent removal. Particles with a size distribution in the range 0.1 to 500 μm may be produced in this way [20].

2.15. CENTRIFUGAL EXTRUSION

In this procedure the core and coating material must be immiscible. Both are pumped through a spinning two-fluid nozzle to produce droplets. The shell material is usually an aqueous polymer solution, which gels rapidly once the droplets fall into a hardening bath, due to, amongst other methods, solvent evaporation. This method produces capsules greater than 250 μm with a 15% size distribution, with the size dependent on nozzle diameter and the susceptibility of the encapsulant to the relatively high shear forces [1].

2.16. SPINNING DISC ATOMIZATION

The core material is dispersed in a liquid shell and fed onto a rotating disk that usually has a toothed edge. As a result core particles coated with the coat polymer fly from the rotating disk in the form of droplets [30]. The size of the droplets is determined by the rotation rate of the disc, flow rate of the dispersion and the surface tension [31].

2.17. JET CUTTING

In this method a fluid is pressed out of a nozzle to form a solid jet that is then cut into uniform segments by a rotating tool made of wires [32]. The size of the segments depends on the number of wires, rotation speed of the tool and the fluid flow rate through the nozzle. Due to the frictional forces acting on the surface of the segments as they fall through the air, beads are formed which drop into a hardening bath. This technique may be used to generate beads with diameters less than 300 μm, with a narrow size distribution, from highly viscous polymers at high production rates [33]. The main disadvantages are the inability to operate under sterile conditions and the relatively important loss of polymer at each cut, even though part may be recycled. To limit this loss to the cutter, the linear velocity of the jet (u_j) and of the cutting wire (u_w) should be equal. Under these conditions the fraction lost will be almost equal to $2d_w/\lambda$, where d_w is the diameter of the cutting wire and λ the length of the cut jet section.

2.18. ELECTROSTATIC DROPLET GENERATION

A polymer solution is extruded through a charged needle into a collecting solution that is earthed or has an opposite charge [34]. As the polymer solution passes through the needle it accumulates charge and the droplets formed at the tip are pulled off by electrostatic forces between it and the collecting bath [35]. If the electrodes are parallel plates, a uniform electric field is generated with respect to direction and strength, and thus a uniform force is applied to the droplets at the tip of the needle [36]. The droplet diameter is dependent on the needle diameter, the polymer flow rate, electrostatic pulse amplitude duration, the electrode spacing, wavelength and polymer concentration and viscosity. Since the needle gauge limits the size of the beads, the smallest bead diameter will be about 185 μm.

2.19. PRILLING

Prilling is a method by which a liquid is extruded through a nozzle subjected to mechanical vibration [37]. Liquid exiting a nozzle forms a jet, which naturally breaks-up to form droplets [38]. A range of flow regimes is obtained depending on the liquid velocity. Thus in the laminar regime drop formation is almost static, occurring at the nozzle tip, when the gravitational force is sufficient to overcome the surface tension adhering the liquid to the nozzle. As the flow rate is increased the frequency of droplet generation increases and the size dispersion is reduced. The liquid exiting the nozzle ultimately becomes a continuous jet. A turbulent domain is attained by increasing the liquid flow even more and results in the formation of a spray. By vibrating the nozzle a sinusoidal wave is generated in the extruded liquid and results in the break-up of the liquid jet, particularly in the laminar flow regime, to form droplets with a considerably reduced size distribution, with one droplet formed per hertz frequency applied. If a single nozzle is used then the droplets (such as of alginate) are allowed to fall into a hardening bath to form polymer beads [39]. These can subsequently be layered with a second polymer, such as poly-*L*-lysine, to form capsules [40]. Alternatively a dual nozzle device may be used in which the encapsulant is extruded through the central nozzle and the capsule wall polymer is extruded through the second nozzle directly onto the forming droplet's surface. In principle any polymer may be extruded onto the droplet surface, providing that the dynamic viscosity is less than about 300 mPas and that there is a difference in viscosity between the core material and wall polymer. Polymers with higher viscosity may be used by the addition of an electrostatic force to ionize the air through which the droplets form [41]. This combined prilling/electrostatic device is capable of forming beads or capsules of less than 100 μm in diameter with a size distribution of less than 3-5% under completely sterile operating conditions. The technique has been successfully applied to the immobilization of microbial cells [42] and the encapsulation of hydrophobic phases for the *in situ* extraction of aroma compounds formed in fermentation processes [43], as well as for the production of medical implants. The method is scalable by simply increasing the number of nozzles in the system.

2.20. HEAT DENATURATION

In this process capsules are formed by dispersion of the core material in a soluble protein slurry [12]. The protein is then denatured using heat, pH changes or coagulating enzymes, to obtain a solid protein layer around the core material [44].

2.21. CROSS-LINKED REVERSE-SOLUBILITY CELLULOSICS

This technique has been applied to the encapsulation of dyes or dye precursors. It is based on the use of certain cellulose derivatives that are soluble in water at low temperature but insoluble at temperatures above 40-44°C [20]. Droplets of the dye precursor are first emulsified in a solution of hydroxypropyl cellulose. The solution is heated such that the polymer forms a coating on the surface of the droplets in the emulsion. The system is then cooled to harden the polymer and produce capsules. In order to prevent solubilisation of the wall polymer at low temperatures, a cross-linking agent is added to the organic phase as well as the external aqueous phase.

2.22. LIPOSOMES

Liposomes are small (< 1 μm) microcapsules or nanocapsules, in which the core is aqueous and the wall is made of a phospholipid bilayer. Such capsules have very thin walls, and thus limited mechanical resistance, but may be used to encapsulate aqueous solutions, or lipophilic materials and are frequently used in the cosmetics industry [45].

2.23. SURFACTANT CROSS-LINKING

Volatile organic liquids containing reactants such as toluene diisocyanate are emulsified in aqueous solutions containing surface active agents that may be polymerized or cross-linked (PVA, cellulosics, starch) and the solvent evaporated to form small microcapsules or nanocapsules (diameter < 1 μm) [20].

2.24. UREA-FORMALDEHYDE POLYMERIZATION

This technique is mainly used for the coating of hydrophobic liquids [19]. It involves emulsification of the hydrophobic liquid in a solution of low molecular weight urea-formaldehyde pre-polymer. The pH of the solution is then lowered to around pH 2 in order to initiate polymerization. Capsules made in this way have high thermal stability.

2.25. LIQUID MEMBRANE CAPSULES

Such capsules are produced by passing droplets of one liquid, such as water, through an organic phase containing additives, into a second water phase. This results in the formation of a droplet surrounded by a thin coating of the organic liquid. Reversal of the procedure, by passing an organic phase through water into a second organic phase, may be used to produce capsules containing an organic phase surrounded by a hydrophilic liquid film. An example of this is the encapsulation of sodium hydroxide in an organic coating, followed by dispersal of the droplets in waste-water containing phenolic compounds. The phenolics diffuse through the capsule wall and react with the sodium hydroxide to form sodium phenolate that is trapped within the core [46].

A very wide variety of techniques exist for the formation of beads or capsules containing liquid cores. However no single technique appears to be suitable for producing such beads or capsules over a wide size range, with monodisperse size distribution, using a wide range of wall polymers. Furthermore, for biotechnological applications, in which the encapsulation procedure must involve a closed system, which can be operated under strictly sterile conditions, using polymers, and reactants, which are biocompatible, under physiological conditions of temperature and pH, few techniques are readily suitable. Of the ones described the prilling technique has proved to be the most versatile, being capable of producing monodisperse microcapsules containing cells, aqueous or organic phases. As a result the remainder of this paper will describe the technique in more detail, particularly for the production of lipophilic core capsules surrounded by hydrophilic capsule walls for use in *in situ* product recovery (ISPR).

3. Production of liquid core capsules for biotechnological applications

The immobilization and encapsulation of microbial, plant and animal cells for diverse applications ranging from waste water treatment to human implants, together with drug delivery and gene therapy applications, has been expanding rapidly over the last few years. In particular liquid-core capsules have been used in the biotechnology, food and cosmetic industries. Examples include alginate microcapsules containing culture medium and cells for shear protection and high cell density culture in bioreactors and organic core microcapsules with hydrophilic walls for the controlled release of flavours and fragrances and for the *in situ* extraction of lipophilic compounds from fermentation processes (capsular perstraction).

In a later chapter in this book (Heinzen *et al.*) laminar jet break-up technology [37] was shown to produce monodisperse beads and liquid core capsules as small as 100 μm in diameter. By applying a sinusoidal disturbance to the nozzle, through vibration of the nozzle or pulsation of the extruded fluid, the liquid jet breaks up into monodisperse droplets [38]. By combining jet break-up technology with a concentric nozzle, liquid core capsules can be produced in a single step and the membrane thickness varied by changing the production conditions.

Alternatively capsules may be produced by creating beads of a polymer, such as alginate, followed by coating of the beads with a second polymer, such as poly-*L*-lysine, and solubilisation of the alginate core using citrate [7]. However, while this method is very simple and may be achieved using a wide range of extrusion or atomization devices, it is more difficult to control the wall thickness and properties of the capsules.

3.1. ENCAPSULATION OF ANIMAL CELLS

Microencapsulation of cells offers several advantages over conventional suspension cultures, achieving higher cell densities, enhanced product recovery and protection of cells from hydrodynamic shear forces created by agitation and aeration. Many gel entrapment methods [47,48] have been applied to mammalian cell cultures including alginate, agarose, poly-*L*-lysine (PLL) [48] and acrylate copolymers [49]. The success of the alginate/poly-*L*-lysine gel encapsulation technique is mainly due to the very mild conditions under which encapsulation may be carried out, involving non-toxic components at a pH, osmolarity and temperature suitable for preserving animal cell viability. This procedure entraps viable cells within a semi-permeable polysaccharide-polycation membrane, which selectively allows small molecules such as nutrients and oxygen to diffuse through, while preventing the passage of large molecules and cells. If the permeability and membrane composition of these microcapsules have been extensively reported [50-52], the mechanical stability of such capsules, although a critical requirement, remains poorly characterised [53]. Indeed, capsule resistance is one of the critical factors when the capsules are intended for bioreactor cultures or as implants [54]. The bursting of capsules would result in the release of cells, cell components and genetic material, which could severely damage the whole process. Furthermore resistance is a key variable of microcapsules, being directly responsible for controlling the permeability. Therefore it is vital to quantitatively understand capsule stability as a function of time under different conditions such as cell culture media. Such results permit comparison of alginate/poly-*L*-lysine capsules with those produced using other polymer materials. Qualitative methods to measure the mechanical properties of alginate capsules have been described by a number of authors [54]. A common technique, the "explosion assay", [52] involves transferring capsules from a saline solution to water, the sudden influx of water causing the capsules to swell and break. However, this method is not applicable to all polymer systems such as chitosan-alginate [55], sulphoethylcellulose-PDADMAC and others, in which cases water actually increases the resistance of the capsules. Mechanical techniques measure the breakage of microcapsules in a turbine reactor [56], in a bubble column [57,58], in a shaking system [59] or in a cone-and-plate flow device [60]. These indirect methods give an indication of the resistance *versus* shear forces, however in the majority of cases the magnitude of these forces makes them difficult to quantify. Another approach consists in measuring the elastic properties of a single microcapsule using a micropipette aspiration device [61]. This method gives information on the membrane elasticity and the resistance itself. It is also possible to calculate the membrane thickness, from the wet membrane weight, and use this as an indicator of membrane strength [54].

An alternative, and highly sensitive technique, is to study the resistance or deformation of capsules under compression. These measurements enable determination of the burst force, the strain at bursting and the Young modulus while giving an indication of the capsule core structure (gelled, partially gelled or liquid). This technique has been successfully used by several authors [62-64,58] and provides results that may be used to compare different kinds of capsules. This method has the advantage of being fast, accurate, reproducible and requires only small quantities of capsules. The utilisation of a micro-manipulation technique [65] allows determination of burst force for capsules smaller than 1 μm in diameter. While the burst force is often used as a parameter to describe the global resistance, it provides no indication of the behaviour of capsules to abrasion and fatigue [58] when subjected to mechanical agitation in cell culture media.

Models for the compression of spherical particles, such as sea urchin eggs, rubber shells, red blood cells, have been reported [66,67]. However, it is difficult to describe the strength *versus* deformation by a mathematical model since hydrogel capsules have a complex structure with the core water being pressed out of the pores during compression.

Thus a major problem associated with the use of polyelectrolytes in the production of capsules is their relatively poor mechanical and long term stability [53,52], which has led to few successful industrial [68,69] or medical applications. A further, relatively unexplained feature, which is believed to play a major role in limiting cell viability in microcapsules, is the observation that cell growth and product recovery may be limited by the presence of significant amounts of the intracapsular polymer [50,51]. In the case of alginate, the intracapsular alginate content could be reduced by the use of multiple-membrane microcapsules resulting in enhanced intracapsular cell densities and product concentrations [51].

In a recent work by the authors [71], the problem of capsule instability was overcome by the cross-linking of the polyelectrolyte membrane complex. In this work, a mixture of alginate, propylene glycol alginate (PGA), bovine serum albumin (BSA) and polyethylene glycol (PEG) were extruded through an encapsulator system (Inotech Encapsulation model IEM) into a bath containing calcium chloride/PEG. After gelation and washing, the beads were incubated in a solution of poly-*L*-lysine (PLL) and the pH of the solution gradually increased. At neutral pH PLL is positively charged and reacts with the alginate (AG). Increasing the pH, results in a transacylation reaction between the amino groups of PLL and the ester groups of propylene glycol alginate (PGA), resulting in covalently bound amide. The BSA added to the core-polymer solution hinders the diffusion of NaOH through the bead. BSA is also neutralized by the NaOH and reacts with the ester groups to form amide bonds, resulting in a partially gelled core. The encapsulation system proposed consists of four principal steps, which are described schematically in Figure 1. The most important parameters influencing the membrane characteristics are the pH of the solution and the time of reaction [70].

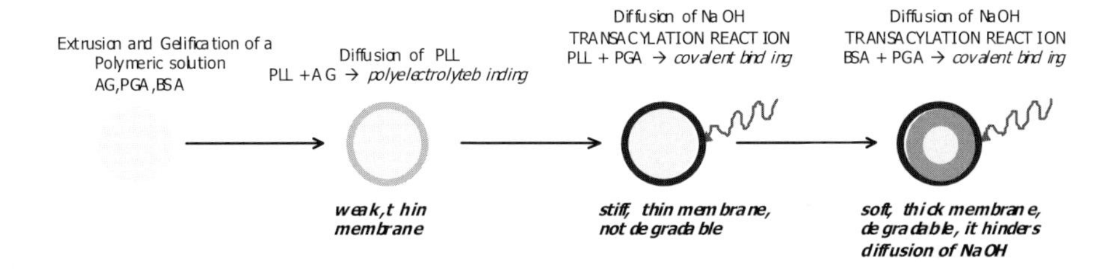

Figure 1. Schematic representation of an encapsulation system for the production of cross-linked alginate-propylene glycol alginate-PEG-BSA microcapsules.

The effect of transacylation time and pH of the alkaline solution on the capsule mechanical resistance are shown in Figure 2 and Figure 3. It can be observed that increasing the transacylation time had little effect on the resistance of the membranes. However this time was reduced to 10 seconds in order to limit the diffusion of NaOH through the core, and thus reduce the potential deleterious effects on cell viability. On the other hand at pH values between 9 and 10 there was a high increase in mechanical resistance, due to the transacylation reaction. The pH was thus fixed to 11, by addition of a small excess of NaOH, to be sure that most of the amino groups of PLL were deprotonated.

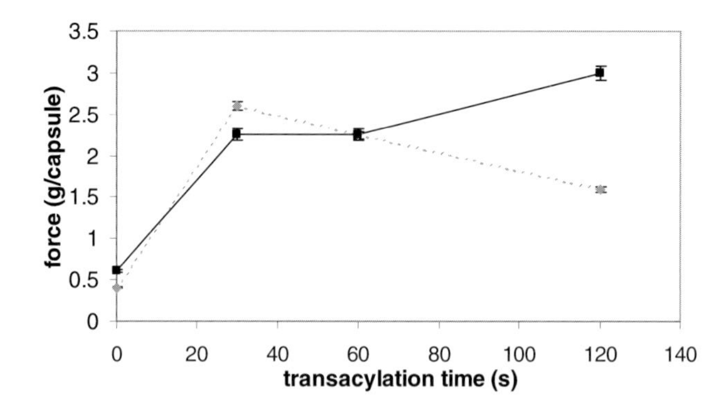

Figure 2. Mechanical resistance of capsules as a function of transacetylation time for a constant incubation time in PLL of 20 min. Average diameter of capsules, 1000 μm. Symbols: Capsules made from 1.2% alginate and 1.8% PGA (dark squares); capsules made from 1.2% alginate, 1.8% PGA, 4% BSA and 1% PEG (diamonds)

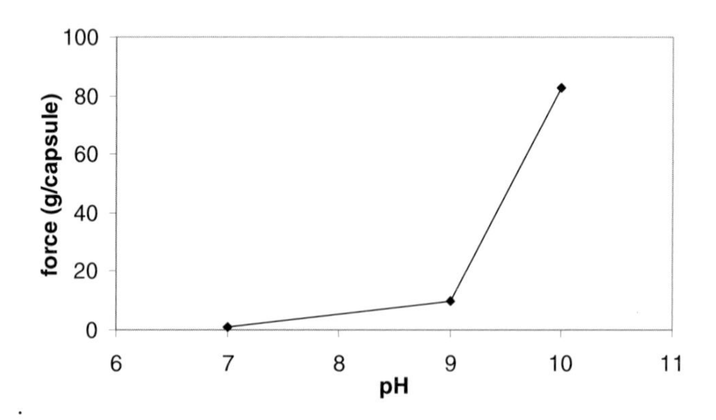

Figure 3. Mechanical resistance of capsules as a function of the pH of the transacetylation reaction. Average diameter of capsules, 1.66 mm. Symbols: Capsules made from 1.2% alginate, 1.8% PGA, 4% BSA and 1% PEG (diamonds).

Using this method, capsules with a diameter of 1.16 mm and mechanical resistance of 7.6 g/capsule were prepared containing recombinant Chinese hamster ovary (CHO) that secreted a medically important protein. These capsules, initially containing 2 x 10^5 cell/ml, were incubated in an agitated bioreactor with a suitable growth supporting medium and cultivated over a period of 30 days. During this period the cells completely colonized the core of the capsules (Figure 4) thus showing that they were neither affected by the transacylation reaction nor subject to any strong diffusion limitations. The mechanical resistance of the capsules decreased by only 11.3% after 30 days (720 hours) in the culture medium with mechanical agitation of 150 rpm. This indicates a relatively high long-term stability compared to capsules of alginate-PLL, which gave a decrease in mechanical resistance of 85% after 420 hours of batch culture [40].

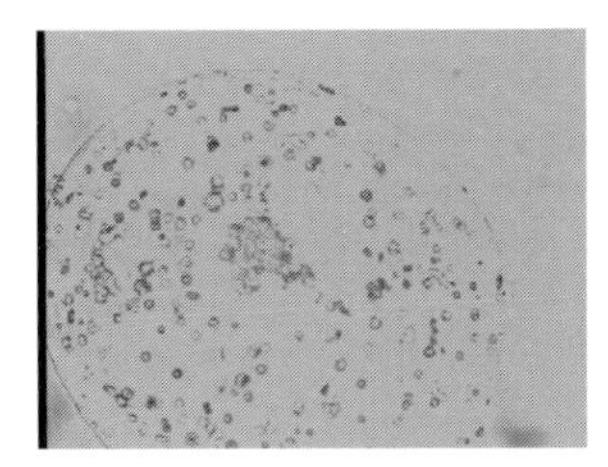 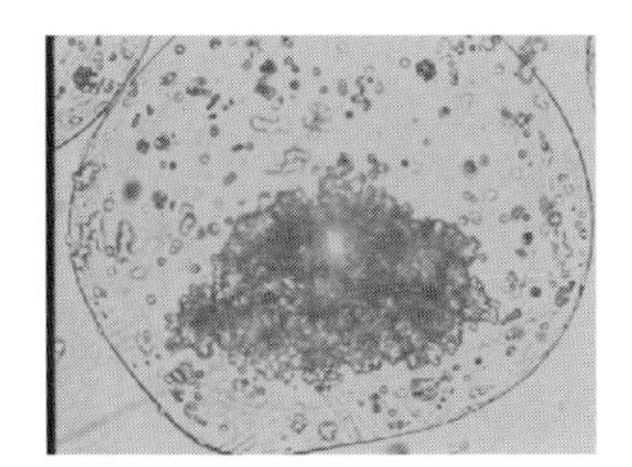 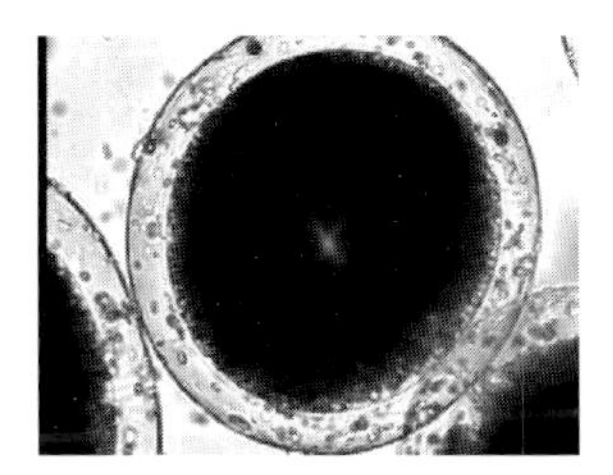

Figure 4. Photomicrographs of CHO cells encapsulated in the alginate-PGA-BSA-PEG system. Initial cell density $2x10^5$ cell/ml; final cell density 4×10^7 cell/ml. Capsule diameter 1.16 mm.

This work clearly shows that it is possible to create microcapsules having a liquid core and surrounded by cross-linked polymers, which can support the growth and viability of highly sensitive animal cells over extended periods with negligible loss of mechanical resistance or indeed physical properties such as porosity. As a result it envisaged that cell encapsulation for the production of recombinant proteins or for use as medical implants should become increasingly important.

4. Production of organic core liquid capsules for food and pharmaceutical applications

Using the jet break-up technique (Encapsulator, Inotech Encapsulation AG, Dottikon, Switzerland) a range of liquid core capsules could be produced in which the central core contained an oil while the wall material was a polyelectrolyte such as calcium alginate. These were produced by extrusion of a range of common oils through the central nozzle and supplying an alginate (1.5% sodium alginate, Inotech IE-1010) solution through the external nozzle directly onto the oil droplet surface. The capsules were dropped into a bath containing 0.1 M $CaCl_2$ solution (Inotech IE-1020) and after gelling the size and size distribution of the resulting capsules were determined using a Coulter counter (Coulter Electronics GmbH, Germany).

In order to avoid a large size distribution of the capsules, due to coalescence of the capsules during flight from the nozzle to the hardening bath and at the surface of the hardening solution, the use of the Encapsulator unit with an electrostatic generator is essential [41].

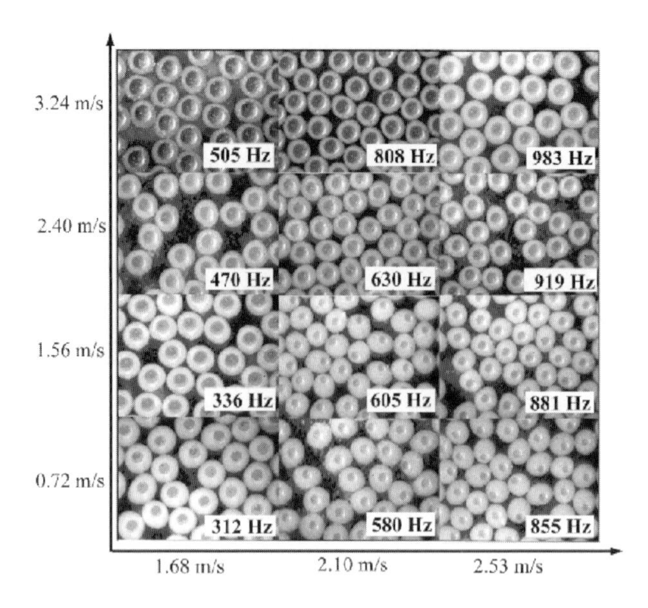

Figure 5. Liquid core capsules made of sunflower oil (core) and 1.5% alginate (shell).The x-axis and y-axis represent the flow rates of the polymer solution (alginate) and oil respectively. The values inset represent the frequency applied to the inner core fluid during extrusion through the Encapsulator device (Inotech Encapsulator model IEM).

In order to specify optimal production parameters for the liquid core capsules the velocities of the core and shell phases and vibration frequency applied were varied. The diameter of the central nozzle was 200 μm and 500 μm for the outer nozzle.
Monodisperse droplet formation and encapsulation of sunflower oil in alginate capsules can be varied over a large range of production parameters using the same nozzle. Depending on the velocities of the core and the shell liquid the vibration frequency required optimization for optimal and monodisperse droplet formation according to the theory of Rayleigh and Weber [39]. The borders of production limits are defined by fluid dynamics, with stable production of uniform capsules only obtained when the liquid jet shows laminar fluid behaviour.

Figure 5 shows sunflower oil-alginate capsules after gelation in the calcium chloride solution. The diameter of the core varied between 400 and 800 μm while the external capsule diameter varied between 550 and 1000 μm. The relative standard deviation of the capsule size varied for the different samples between 2.8 and 4.9%, indicating a very narrow size distribution compared with other techniques. Based on the results of the sunflower oil-alginate system a number of further systems were tested (Table 1).

Table 1. Production of liquid core capsules containing different encapsulants surrounded by walls made of alginate, gelatin or cellulose sulphate.

Shell material	**Core material**	**Core size (μm)**	**Shell size (μm)**	**Characterization**
Alginate 1.5%	Coffee extract	500	800	Monodisperse
Alginate 1.5%	Menthol	600	1200	<10%
Alginate 1.5%	Cells + culture medium	600	800	Monodisperse
Alginate 1.5%	Citrus fruit oil	400	600	<10%
Alginate 1.5%	Heptane	250	1200	Monodisperse
Gelatin	Sunflower oil	400	1000	Monodisperse
Cellulose sulphate	Cell culture medium	400	600	Monodisperse
Cellulose sulphate	Sunflower oil	400	600	Monodisperse

All systems generated monodisperse capsules except for menthol in alginate and heptane in alginate. The latter was due to the very small difference in viscosity between the alginate solution and encapsulant or partition coefficients for the two phases being too similar, resulting in multi-nucleate capsules or diffusion of the wall material into the capsule core.

In a recent work [43] capsules containing an organic solvent (dibutyl sebacate, DBS) surrounded by a membrane composed of calcium alginate have been prepared and applied for the *in situ* extraction of the natural aroma compound 2-phenylethanol, from a bioconversion reaction catalyzed by yeast. In this process the yeast converts *L*-phenylalanine to 2-phenylethanol, however the accumulation of the product above a value of 2.8 g/l results in the inhibition of the growth and PEA production by the yeast. Due to the higher solubility of PEA in organic solvents than cell culture media, liquid-liquid extraction would be generally used to extract the PEA continuously from the culture medium thereby preventing the inhibition. However, many organic solvents are toxic to the yeast, and can form stable emulsions, which makes the recovery of the organic phase and back-extraction of PEA difficult.

In order to overcome this Stark *et al.* [43] showed that liquid core capsules containing the organic solvent (DBS) could be sterilized and added directly to the bioreactor containing yeast (Figure 6). The PEA formed by the bioconversion diffused into the capsules and accumulated in the organic phase. Since there was no direct contact between the organic phase and the culture medium containing yeast, the toxic effects of the solvent and formation of stable emulsions could be avoided. In this way the productivity of the system could be increased significantly. Furthermore, the very high surface area to volume ratio of the capsules, means that there is a high mass transfer surface for diffusion of the PEA. As a result concentrations of more than 37 g/l could be achieved in a much lower volume of solvent (factor of 10) than that used in liquid-liquid extraction.

One problem in the use of such capsules for bioprocesses is associated with the requirement for sterile capsules. Figure 6 shows that while it is possible to heat sterilize capsules composed of alginate and use them for *in situ* extraction, the membrane surrounding the capsules can become very fragile and lead to release of the solvent. As a result it is imperative to develop polymeric membranes for liquid core capsules, which resist the sterilization procedure and the mechanical shear stresses associated with bioreactors over extended periods.

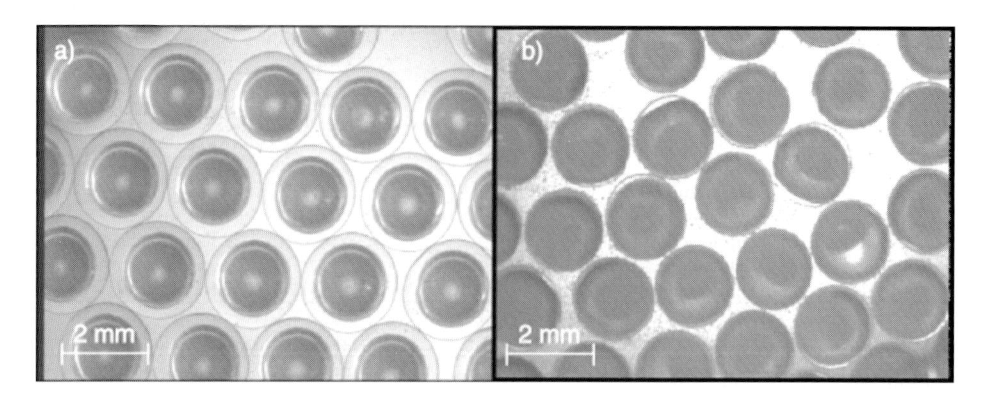

Figure 6. Photomicrographs of a) non- autoclaved and b) autoclaved liquid core capsules containing dibutyl sebacate. Membrane calcium alginate.

5. Conclusions

Many methods exist for the encapsulation of liquids within a range of polymers. However, the method used must be chosen as a function of the desired application, capsule properties and polymer(s). For biotechnological processes, the applications may be categorized as cell encapsulation or the encapsulation of organic solvents for use in extractive processes. In the first case the polymers must be biocompatible and non-cytotoxic, allowing the free diffusion of small molecules (nutrients and metabolic products) while selectively retaining the cells. In the second case the aim is to create a barrier, which allows for no direct contact between the cells and organic solvent, yet which selectively allows the diffusion of molecules based on their partition coefficients. In both cases the capsules must be chemically and physically resistant to the environment in which they are to be used over extended periods of time. Examples include the environment found in bodily fluids, such as the acidic pH of the stomach (for probiotics and nutraceuticals), the ionic composition of the peritoneal or cerebrospinal fluids. In these cases the long term resistance and ability to selectively exclude the endodiffusion of components of the immune system, while avoiding fibrosis are the key parameters. Conversely the encapsulation of cells or organic phases for use in bioprocesses requires that the capsules have long term mechanical resistance with respect to the sterilization process and shear stresses encountered in the bioreactor, together with stability and maintenance of physical characteristics in the highly ionic culture media, while avoiding release of cells or solvent into the culture. Transfer phenomena are important for both applications and thus the ability to make monodisperse capsules is a pre-requisite. Of the currently available techniques, the vibrating nozzle extrusion system appears to be the most appropriate, especially since it can be used with a wide variety of polymer systems.

Of the polymers routinely available applied to generate microcapsules, the most commonly used for biotechnological applications are those derived from natural sources due to their relatively non-toxic properties and mild conditions for polymerization. The major limit for their use involves the difficulty in obtaining such polymers in sufficiently pure form having defined molecular mass and viscosity, particularly between different batches. In addition many of them involve polyelectrolyte interactions to achieve the desired properties, however these are usually subject to important problems of instability in body fluids and culture media, as a result of the high ion content. This may be overcome by suitable chemical cross-linking of the polymers in the capsular membrane.

The technique, where cells are suspended in droplets of polymer matrix (such as alginate) and then capsules are obtained by the layering of further polymers onto the surface, followed by solubilisation of the core matrix, is unsuitable for many applications. This is primarily due to the inability to completely remove the core polymer from the capsule, with the result that the water activity and osmotic pressure within the capsule can be deleterious for cells. This may be overcome by the creation of liquid core capsules in which the core contains an aqueous phase in which there are no polymer materials. This may be achieved by the use of prilling techniques in which the hydrophilic core is extruded and the membrane polymer directly applied to the surface. Similarly such techniques enable the generation of a hydrophobic core, such as an organic solvent, surrounded by a hydrophilic polymeric membrane.

Due to the rapid developments in encapsulation technology and polymer chemistry it can therefore be envisaged that there will be a sharp increase of interest for the use of liquid core capsules for a wide range of bioprocess applications.

References

[1] Gibbs, B.F.; Kermasha, S.; Alli, I. and Mulligan, C.N. (1999) Encapsulation in the food industry: A review. Int. J. Food Sci. Nutr. 50: 213-224.

[2] King, A.H. (1995) Encapsulation of food ingredients: A review of available technology, focusing on hydrocolloids. In: Risch, S.J. and Reineccius, G.A. (Eds.) Encapsulation and Controlled Release of Food Ingredients. ACS Symposium Series 590; pp. 26-39.

[3] Chang, P.L. (1999) Encapsulation for somatic gene therapy. Ann. New York Acad. Sci. 875: 146-158.

[4] Chang, T.M. (1999) Artificial cells, encapsulation and immobilization. Ann. New York Acad. Sci. 875: 71-83.

[5] Blandino, A.; Macias, M. and Cantero, D. (2000) Glucose oxidase release from calcium alginate gel capsules. Enz. Microb. Technol. 27: 319-324.

[6] Kim, H.J.; Kim, J.H. and Shin, C.S. (1999) Conversion of *D*-sorbitol to *L*-sorbose by *Gluconobacter suboxydans* cells co-immobilized with oxygen- carriers in alginate beads. Proc. Biochem. 35: 243-248.

[7] Lim F. and Sun A.M. (1980) Microencapsulated islets as bioartificial pancreas. Science 210: 908-910.

[8] Uludag, H.; De Vos, P. and Tresco, P.A. (2000) Technology of mammalian cell encapsulation. Adv. Drug Del. Syst. 1-2: 29-64.

[9] Chia, S.M.; Leong, K.W.; Li, J.; Xu, X.; Zeng, K.; Er, P.N.; Gao, S. and Yu, H. (2000) Hepatocyte encapsulation for enhanced cellular functions. Tissue Eng. 5: 481-495.

[10] De Vos, P. and Marchetti, P. (2002) Encapsulation of pancreatic islets for transplantation in diabetes: the untouchable islets. Trends Mol. Med. 8: 363-366.

[11] Risch, S.J. (1995) Encapsulation: An overview of uses and techniques. In: Risch, S.J. and Reineccius, G.A. (Eds.) Encapsulation and controlled Release of Food Ingredients. ACS Symposium Series 590; pp. 2-25.
[12] Benita, S. (1996) Drugs and the pharmaceutical sciences. In: Benita, S. (Ed.) Microencapsulation, Methods and Industrial Applications. Marcel Dekker Inc., New York; Vol 73; pp. 587-632.
[13] Frangione-Beebe, M.; Rose, R.T.; Kaumaya, P.T. and Schwendeman, S.P. (2001) Microencapsulation of a synthetic peptide epitope for HTLV-1 in biodegradable poly (D, L- lactide-co- glycolide) microspheres using a novel encapsulation technique. J. Microencap. 18: 663-677.
[14] Lin, W.J. and Yu, C.C. (2001) Comparison of protein loaded poly (epsilon- caprolactone) microparticles prepared by the hot- melt technique. J. Microencap. 18: 585-592.
[15] Esquisabel, A.; Hernandez, R. and Iguartua, M. (1997) Production of BCG alginate-PLL microcapsules by emulsification/internal gelation. J. Microencap. 14: 627-638.
[16] Poncelet, D. (2001) Production of alginate beads by emulsification internal gelation. Ann. New York Acad. Sci. 944: 74-82.
[17] Lamprecht, A.; Schäfer, U.F. and Lehr, C.-M. (2000) Characterization of microcapsules by confocal laser scanning microscopy: structure, capsule wall composition and encapsulation rate. Europ. J. Pharma. Biopharma. 49: 1-9.
[18] Bachtsi, A.; Boutris, C. and Kiparissides, C. (1996) Production of oil- containing cross- linked poly(vinyl alcohol) microcapsules by phase separation: Effect of process parameters on the capsule size distribution. J. Appl. Polymer Sci. 60: 9-20.
[19] Kulkarni, A.R.; Soppimath, K.S.; Aminabhavi, T.M.; Dave, A.M. and Mehat, M.H. (2000) Glutaraldehyde crosslinked sodium alginate beads containing liquid pesticide for soil applications. J. Control. Rel. 63: 97-105.
[20] Levy, M.-C and Andry, M.-C. (1991) Mixed- walled mcirocapsules made of cros- linked proteins and polysaccharides: preparation and properties. J. Microencap. 8: 335-347.
[21] Jang, J. and Lee, K. (2002) Facile fabrication of hollow polystyrene nanocapsules by microemulsion polymerization. Chem. Commun. (Camb.) 10: 1089-1099.
[22] Heurtault, B.; Saulnier, P.; Pech, B.; Proust, J.-E. and Benoit, J.-P. (2002) A novel phase inversion- based process for the preparation of lipid nanocarriers. Pharma. Res. 19: 875-880.
[23] Kim, C.K.; Yoon, Y.S. and Kong, J.Y. (1995) Preparation and evaluation of flurbiprofen dry elixir as a novel dosage form using a spray- drying technique. Int. J. Pharma. 120: 21-31.
[24] Lee, S.-W.; Kim, M.-H. and Kim, C.-K. (1999) Encapsulation of ethanol by spray drying technique: effects of sodium lauryl sulfate. Int. J. Pharma. 187: 193-198.
[25] Ariga, O.; Itoh, K.; Sano, Y. and Nagura, M. (1994) Encapsulation of biocatalyst with PVA capsules. J. Ferment. Bioeng. 78: 74-78.
[26] Rogers, T.L.; Hu, J.; Yu, Z.; Johnston, K. and Williams, R.O. (2002) A novel particle engineering technology: spray- freezing into liquid. Int. J. Pharma. (in press).
[27] Dewettinck, K. and Huyghebaert, A. (1999) Fluidized bed coating in food technology. Trends Food Sci. Technol. 10: 163-168.
[28] Eshra, A.G.; Elkhodairy, K.A.; Mortada, S.A. and Nada, A.H. (1994) Preparation and evaluation of slow-release pan-coated indomethacin granules. J. Microencap. 11: 271-278.
[29] Abrahan, S.; Vieth, R. and Diane, J. (1996) Novel technology for the preparation of sterile alginate polylsyine microcapsules in a bioreactor. Pharma. Dev. Technol. 1: 63-68.
[30] Senuma, Y.; Lowe, C.; Zweiffel, Y.; Hilborn, J.G. and Marison, I. (2000) Alginate hydrogel microspheres and microcapsules prepared by spinning disk atomization. Biotechnol. Bioeng. 67: 616-622.
[31] Ogbonna, J.C.; Matsumura, M.; Yamagata, T.; Sakuma, H. and Kataoka, H. (1989) Production of micro-gel beads by a rotating disk atomizer. J. Ferment. Bioeng. 68: 40-48.
[32] Prusse, U.; Bruske, F.; Breford, J. and Vorlop, K.D. (1998) Improvement of the jet cutting method for the preparation of spherical particles from viscous polymer solutions. Chem. Eng. Technol. 21: 153-157.
[33] Prusse, U.; Fox, B.; Kirchhoff, M.; Bruske, F.; Breford, J and Vorlop, K.-D. (1998) New Process (jet cutting method) for the production of spherical beads from highly viscous polymer solutions. Chem. Eng. Technol. 21: 29-33.
[34] Bugarski, B.; Li, Q.L.; Goosen, M.F.A.; Poncelet, D.; Neufeld, R.J. and Vunjak, G. (1994) Electrostatic droplet generation- mechanism of polymer droplet formation. AICHE J. 40: 1026-1031.
[35] Halle, J.P.; Leblond, F.A.; Pariseau, J.F.; Jutras, P.; Brabant, M.J. and Lepage, Y. (1994) Studies on small (<300 μm) microcapsules: II - Parameters governing the production of algiante beads by high voltage electrostatic pulses. Cell Transplantation 3: 365-372.

[36] Poncelet, D.; Bugarski, B.; Amsden, B.; Zhu, J.; Neufeld, R.J. and Goosen, M.F.A. (1994) A parallel plate electrostatic droplet generator- parameters affecting microbead size. Appl. Microbiol. Biotechnol. 42: 251-255.
[37] Hulst, A.C.; Tramper, J.; Vantriet, K. and Westerbeek, J.M.M. (1985), A new technique for the Production of Immobilized Biocatalyst in Large Quantities, Biotechnol. Bioeng. 27: 870–876.
[38] Lord Rayleigh (1878) On the stability of jets. Proc. London Math. Soc. 10: 4-13.
[39] Serp, D.; Catana, E.; Heinzen, C.; von Stockar, U. and Marison, I.W. (2000) Characterization of an encapsulation device for the production of monodisperse alginate beads for cell immobilization. Biotechnol. Bioeng. 70: 41-53.
[40] Gugerli, R.; Catana, E.; Heinzen, C.; von Stockar, U. and Marison, I.W. (2002) Quantitative study of the production and properties of alginate/poly-*L*-lysine microcapsules. J. Microencap. 19: 571-590.
[41] Brandenberger, H. (1999) Monodisperse particle production: a method to prevent drop coalescence using electrostatic forces. Journal of electrostatics 45: 227-238.
[42] Serp, D.; von Stockar, U. and Marison, I.W. (2002) Immobilized bacterial spores for use as bioindicators in the validation of thermal sterilization processes. J. Food Protect. 65: 1134-1141.
[43] Stark, D.; Kornmann, H.; Münch, T.; Sonnleitner, B.; Marison, I.W. and von Stockar, U. (2002) A novel type of *in situ* extraction: The use of solvent containing microcapsules for the bioconversion of 2-phenylethanol from *L*-phenylalanine by *Saccharomyces cerevisiae*. Biotechnol. Bioeng. (in press).
[44] Ahmed, A.; Bonner, C. and Desai, T.A. (2002) Bioadhesive microdevices with multiple reservoirs: a new platform for oral drug delivery. J. Control. Rel. 81: 291-306.
[45] Walde, P. and Ichikawa, S. (2001) Enzymes inside lipid vesicles: preparation, reactivity and applications. Biomol. Eng. 18: 143-177.
[46] Arshady, R. (1989) Preparation of nano- and microspheres by polycondensation techniques. J. Microencap. 6: 1-12.
[47] Nigam, S.C.; Tsao, I.F.; Sakoda, A. and Wang, H.Y. (1988) Techniques for preparing hydrogel membrane capsules. Biotechnol. Tech. 2: 271-276.
[48] Jen, A.C.; Wake, M.C. and Mikos, A.G. (1996) Review: Hydrogels for cell immobilization. Biotechnol. Bioeng. 50: 357-364.
[49] Sefton, M.; Dawson, R. and Broughton, R. (1987) Microencapsulation of mammalian cells in a water-insoluble polyacrylate by coextrusion and interfacial precipitation. Biotechnol. Bioeng. 29: 1135-1143.
[50] Goosen, M.F.A.; O'Shea, G.M.; Gharapetian, H.M.; Shou, S. and Sun, A.M. (1985) Optimization of microencapsulation parameters: Semipermeable microcapsules as a bioartificial pancreas. Biotechnol. Bioeng. 27: 146-150.
[51] King, G.; Daugulis, A. and Faulkner P. (1987) Alginate-polylysine microcapsules of controlled membrane molecular weight cut-off for mammalian cell culture engineering. Biotechnol. Prog. 3: 231-241.
[52] Thu, B.; Bruheim, P.T.E.; Soon-Shiong, P.; Smidsrod, O. and Skjåk-Bræk, G. (1996) Alginate polycation microcapsules. I. Interaction between alginate and polycation. Biomaterials 17: 1031-1040.
[53] Lacik, I.; Brissova, M.; Anilkumar, A.; Powers, A. and Wang, T. (1996) New capsule with tailored properties for the encapsulation of living cells. J. Biomed. Mat. Res. 39: 52-60.
[54] Ma, X.J.; Vacek, I. and Sun, A. (1994) Generation of alginate-poly-*L*-lysine-alginate (APA) biomicrocapsules - The relationship between the membrane strength and the reaction conditions. Art. Cells Blood Substit. Immob. Biotechnol. 22: 43-69.
[55] Bartkowiak, A. and Hunkeler, D. (1999) Alginate-oligochitosan microcapsules: A mechanistic study relating membrane and capsule properties to reaction conditions. Chem. Mater. 11: 2486-2492.
[56] Poncelet, D. and Neufeld, R.T. (1989) Shear breakage of nylon membrane microcapsules in a turbine reactor. Biotechnol. Bioeng. 33: 95-103.
[57] Lu, G.Z.; Thompson, F.G. and Gray, M.R. (1992) Physical modelling of animal cell damage by hydrodynamic forces in suspension cultures. Biotechnol. Bioeng. 40: 1277-1281.
[58] Dos Santos, V.; Vasilevaka, T. and Kajuk, B. (1997) Production and characterisation of double layer beads for co-immobilization of microbial cells. Biotechnol. Ann. Rev. 3: 227-244.
[59] Chen, Z.; Bao, Y. and Gorczyca, W. (1995) Study of microencapsulation for pituitary transplantation: Capsule preparation and *in vitro* study. Artif. Cells Blood Substit. Immob. Biotechnol. 23: 597-604.
[60] Peirone, M.; Ross, C.; Hortelano, G.; Brash, J. and Chang, P. (1988) Encapsulation of various recombinant mammalian cell types in different alginate microcapsules. J. Biomed. Mat. Res. 42: 587-596.
[61] Jay, A.W.L. and Edwards, M.A. (1968) Mechanical properties of semi-permeable microcapsules. Can. J. Physiol. Pharm. 46: 731-737.
[62] Schoichet, M.; Li, R. and White, M. (1995) Stability of hydrogels used in cell encapsulation: an *in vitro* comparison of alginate and agarose. Biotechnol. Bioeng. 50: 374-381.

[63] Martinsen, A.; Skjåk-Bræk, G. and Smidsrod, O (1989) Alginate as immobilization material: I. Correlation between chemical and physical properties of alginate gel beads. Biotechnol. Bioeng. 33: 79-89.
[64] Chen, R. and Tsaih, M. (1997) Effect of preparation method and characteristics of chitosan on the mechanical and release properties of the prepared capsule. J. Appl. Polymer Sci. 66: 161-169.
[65] Zhang, Z.; Saunders, R. and Thomas, C.R. (1999) Mechanical strength of single microcapsules determined by a novel micromanipulation technique. J. Microencap. 16: 117-124.
[66] Andrei, D.; Briscoe, B. and Williams, D. (1996) The deformation of microscopic gel particles. J. Chim. Phys. 93: 960-976.
[67] Liu, K.K.; Williams, D.R. and Briscoe, B.J. (1996) Compressive deformation of a single microcapsule. Phys. Rev. 54: 6673-6680.
[68] Posillico, E.G. (1986) Microencapsulation technology for large-scale antibody production. Biotechnology 4: 114-117.
[69] Duffy, S.J.B. and Murray, W.D. (1996) Bioconversion of forest products industry waste cellulosics to fuel ethanol: A review. Biores. Technol. 55: 1-33.
[70] Lévy, M.-C. and Edwards-Lévy, F. (1999). Serum albumin-alginate coated beads: mechanical properties and stability. Biomaterials 20: 2069-2084.
[71] Gugerli, R. (2003) Polyelectrolyte- complex and covalent- complex microcapsules for encapsulation of mammalian cells: potential and limitations. PhD Thesis, Swiss Federal Institute of Technology (EPFL), Lausanne, Switzerland.

DIRECT COMPRESSION – NOVEL METHOD FOR ENCAPSULATION OF PROBIOTIC CELLS

ENG SENG CHAN AND ZHIBING ZHANG
Department of Chemical Engineering, University of Birmingham, Edgbaston, B15 2TT, United Kingdom – Fax: 44 121 414 5334 – Email: Z.Zhang@bham.ac.uk

1. Introduction

Probiotics may be defined as a mono or mixed culture of live microorganisms that applied to man or animal affects beneficially the host by improving the properties of the indigenous microflora [1]. They confer many healthy benefits upon consumption such as suppressing the growth of pathogens, reducing the risk of cancer formation and reducing the serum cholesterol levels of the host [2,3]. Other reported therapeutic or nutritional values of probiotic bacteria include improvement of lactose digestion and immune system as well as ability to synthesize various vitamins and enhancement of bioavailability of many minerals to the host.

The understanding of beneficial effects of probiotic bacteria has led to the increasing use of them as dietary adjunct. Many probiotic products are available in the market and they exist in different forms such as fermented drink, freeze-dried powder or capsule. However, most of the probiotic products have a short shelf life even when they are stored at low temperature [4]. Several studies have shown that the number of viable bacteria in some of the commercial products was actually below the desired level [1, 3]. It has also been reported that a significant number of cells were killed when they were exposed to a low pH medium [5,6,7]. This remains as a concern to either manufacturers or consumers as the probiotic effects of the bacteria can only be exerted if a sufficient number of viable bacteria survive through the stomach and are delivered to the site of action.

Among many attempts to stabilise the cells during storage and/or when exposed to an acidic medium, encapsulation has seemed to be a promising technological approach. Several studies have shown successful encapsulation of probiotic bacteria using various materials and methods. One of the popular methods is by filling the probiotic cells in powder forms or liquid suspension in hard capsules. For example, it was demonstrated that freeze-dried cell powders placed in gelatin capsules could retain their activity better than those without being encapsulated during storage [8]. However, it is still necessary for the capsules to be coated with an enteric-film in order to enhance the resistance of the cells against gastric acid [9], which can be very expensive.

V. Nedović and R. Willaert (eds.), Fundamentals of Cell Immobilisation Biotechnology, 205-227.

In addition, Kim *et al.* [10] developed a cell encapsulation method via film coating using a Wurster coater. It shows improvement in cell survival after storage and during exposure to an acidic medium. Yokota *et al.* [11] employed the same coating method for tablets containing freeze-dried lactic acid bacteria and the results also showed improved cell stability during storage and when exposed to an acidic medium. Furthermore, Appelgren and Eskilson [12] developed a process named 'Continuous Multi-Purpose Melt Technology (CMT)' to encapsulate cells. Freeze-dried cells are mixed with a solid or viscous material, which is heated until it is melted and an acceptable viscosity is reached. The fluid is then disintegrated into a fine mist in order to envelope the cells. In-vitro tests of acid tolerance of the hot-melt coated freeze-dried *L. acidophilus* revealed that more than 10 % of the cells survived after one-hour exposure to a simulated gastric juice of pH 2.

Cell entrapment in matrix was also tested for stabilizing probiotic bacteria. Champagne *et al.* [13] shows that the storage stability of freeze-dried *Lactococcus lactis* encapsulated in calcium alginate matrix was significantly higher than that of the freeze-dried free cells. Sultana *et al.* [7] also demonstrates enhancement of storage stability of cells in yoghurt when they were in an encapsulated state. However, the cells encapsulated by this method did not show improvement to acid tolerance [6,7].

The advantages and disadvantages of using each encapsulation method is summarised in Table 1. Despite of their successes, there is still a need to develop new techniques, which are more efficient or cost effective. This work describes the use of compression coating as a cell encapsulation method in order to enhance the stability of cells after storage and during exposure to an acidic medium.

Table 1. Advantages and disadvantages of each encapsulation method.

Method	Advantages	Disadvantages	Ref.
Hard capsule ***e.g.* gelatin capsule**	**Maintain cell quality** Maintain the activity of freeze-dried cells. **Simple process** Process is easy, cheap and gentle to cells.	**Not resistant to acid** The gelatin capsule is permeable to acidic media. Hence, the capsule has to be coated with an enteric coating material. **Problems relating to coating of capsules** Many problems are related to the coating of hard gelatin capsules. During coating, the gelatin shell might be softened and becomes sticky due to solubilisation. The gelatin shell might also become brittle due to water evaporation and drying. Therefore, a precoating is necessary to avoid such problems. In addition, the capsule might be separated into halves due to the movement in the coater.	[8], [9], [14]

Hot-melt coating	**Cell stabilisation** Offer cell protection in an acidic medium. Believed to have improved storage stability of cells. **Environmental friendly** No dust generated if compared to film coating. **Less process variables** Less process complication if compared to film coating.	**High contact temperature** The temperature employed to keep the coating material melted was reported to be more than 100°C. This might create a problem for mesophile microorganisms. The survival of cells could depend on the length of contact time at deleterious temperature. **Lack of literature** The data on the hot-melt coating of cells is rare such as the range of coating materials suitable for this application and the methodology. **Required specialised equipment** The success of employing hot-melt coating to encapsulate the cells was believed to be partly due to the use of specially designed equipment, which is not widely available.	[9], [12], [18]
Cell entrapment in matrix ***e.g.* calcium alginate**	**Prolong shelf-life** Could improve storage stability of cells whether it is freeze-dried or in yoghurt. **Simple process** Simple process where large-scale production is possible with existing technology such as jet cutting. **Cheap** The material used is cheap, readily available. Doesn't require highly skilled workers or specialised equipment.	**Acid intolerance** Did not improve acid tolerance due to porosity of the matrix. Further coating is required to achieve gastro-resistant property.	[6], [7], [13]

2. Brief history of compression coating

P.J. Noyes acquired the patent of compression coating in 1896. However, early attempts were defeated by the mechanical problems involved due to lack of engineering skills. Major advances were made in early 1950s as machines for compressing a coating around a tablet core appeared on the market. They were accepted enthusiastically thorough 1960s. However, up to date, only non-biological ingredients have been coated in this manner.

3. Recent applications of compression coating

Over recent years, there is a renewed interest in using the technique to coat drugs with a gel-forming polymer for controlled-release purposes. Several studies [19,20,21] demonstrated the potential use of pectin formulations as compression-coating materials for delivery of drugs to the colon. The results show a marked and significantly rapid release of drugs in the presence of pectinolytic enzymes, which exist in the colon. In addition, the coating was also reported to prevent diffusion of the dissolution liquid towards the cores due to formation of a gel barrier at the beginning of dissolution. Krishnaiah *et al.* [22,23] also developed colon-specific delivery systems by compression-coating the drugs with guar gum formulations. In vitro studies showed that a significant amount of drugs was released in the presence of rat caecal contents, which suggests the susceptibility of guar gum to the action of colonic bacterial enzymes. The finding was also confirmed by studies on human volunteers. In another work, Kaneko *et al.* [24] evaluated the use of sodium alginate as a compression-coating material to coat a soluble drug, theophylline. A sustained release profile of the drug was observed due to the formation of a water-soluble hydrogel and diffusion of the drug across the barrier to a dissolution liquid. It was also found that there was a lag time of around 2 hours before any drug was released. This suggests that the formulation of the coating material could retard the penetration of the dissolution medium towards the core.

4. Compression coating as potential cell encapsulation method

Compression coating may also be applicable to cell encapsulation because of its advantages over reported coating methods and also possible operation on an industrial scale. When this method is used in conjunction with a gel-forming polymer, it may be possible to protect the cells from an acidic medium due to the delayed permeation of the liquid as suggested by past workers, as well as to stabilise the cells during storage. Hence, this forms a basis for employing the method and relevant materials for cell encapsulation, which has been implemented by the authors. The principle of this method is first to compress cell-containing powers into a pellet, which is then encapsulated by further compression, as shown in Figure 1, and the details are described as follows.

5. Compression of cell containing powder into a pellet

Lactobacillus acidophilus ATTC 4356 from human origin was used as model probiotic bacteria in this study. The cells were suspended in a protective solution as described by Couture *et al* [25] prior to freeze drying. The solution consisted of 28% w/v skimmed milk (Fluka, UK), 4% w/v sucrose (Sigma, UK) and 0.3% w/v ascorbic acid (Sigma, UK). The freeze-dried cells were then ground into fine powders. Their water activity was measured by a water activity meter (Rotronic, A1H, Switzerland). For compression of the cell powders, they were poured into a die with a diameter of 6mm. Compression of the powders was performed using a flat-faced punch, which was attached to a Lloyd Material Testing Machine (6000R, UK). A 30kN load cell was used. The speed of compaction was controlled by setting the displacement of the punch to 2mm per minute. The compression process was terminated as soon as the required load was reached. The compression pressure was calculated based on the compression load over the contact area of the punch.

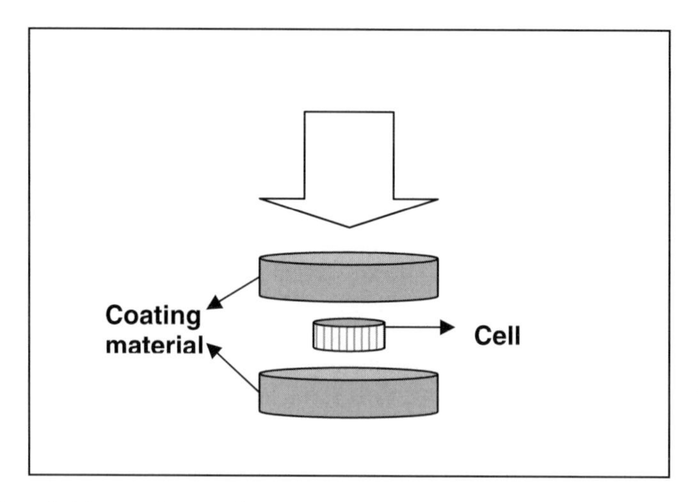

Figure 1. Schematic diagram showing compression coating of cells

The viability and sub-lethal injuries of the bacteria were determined by a well-established method [26]. The cell pellet was homogenised with 100ml of phosphate buffer, pH 6.8 in a stomacher (Seward, UK) for 4 minutes. The cell suspension was then serially diluted in 0.1% peptone solution (Sigma, UK) and spread onto pre-dried MRS agar (MRSA) (Oxoid, UK), MRS agar plus sodium chloride (MRSANa) (Sigma, UK) and MRS agar plus oxgall (MRSAO) (Difco, UK). NaCl or Oxgall was added to the MRSA prior to sterilisation. The colonies formed on the plates were counted with the aid of a colony counter after incubation of at least 40 hours at 37 °C. The survival of the cells in MRSANa or MRSAO is expressed in

Survival (%) = (CFUs formed on MRSANa or MRSAO / CFUs formed on MRSA) × 100 (CFU Colony Forming Units)

The concentration of the salts used in the MRS agar was pre-determined by spreading the untreated (healthy) cells onto MRS agar with different concentrations of respective salts. Hence, the minimum inhibition concentration of the salts could be determined from the survival-concentration curves and used as the salt concentrations for the detection of sub-lethally injured cells.

6. Selection of compression pressure for coating

Figure 2 shows the cell viability versus compression pressure when the cell containing powders were compacted. As can be seen, there was no significant loss of cell viability when the powders were compressed up to a pressure of 30 MPa. When this pressure was exceeded, the survival of the cells gradually decreased to about 85% when a pressure of 90 MPa was reached. The drop in the cell viability was almost linear above 90 MPa where only about 33% of the cells survived after they were compressed at 180 MPa.

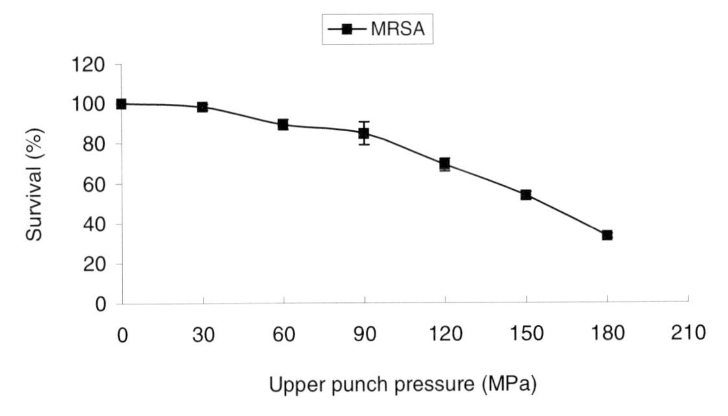

Figure 2. Effect of compression pressure on the viability of freeze-dried L. acidophilus. *Error bars in all figures represent the standard error of the mean.*

The resistance of *L. acidophilus* to NaCl and Oxgall after they were compressed at different pressures is presented in Figure 3. The bacteria did not develop any significant sensitivity to NaCl, i.e. there was no significant damage to their membrane after they were compressed up to a pressure of 90 MPa. Beyond that pressure, the resistance of the cells to the salt decreased gradually and about 80% of the cells showed growth on the MRSANA after being compressed at 180MPa. In contrast, the cells developed sensitivity to oxgall even before the compression, and only about 75% of the freeze-dried cells grew on the MRSAO. This indicates that the wall of cells was more susceptible to freeze-drying and compression than the membrane for the given formulation of the cell containing powders and processing conditions investigated. After a pressure was applied to cells, their resistance to oxgall decreased gradually. Only 30% of the cells survived in the MRSAO after being compressed at 180 MPa.

Figure 4 shows the effect of compression pressure on the displacement of the top surface of the powder bed, and the relationship between the two parameters is fitted by Kawakita equation [27]:

$$L = \frac{0.15P}{1 + 0.064P} \tag{1}$$

and the regression coefficient is 0.998. As can be seen, the displacement increases with the compression pressure monotonically, but the slope of the curve decreases, which indicates the powder bed became more and more compact as the compression pressure was raised

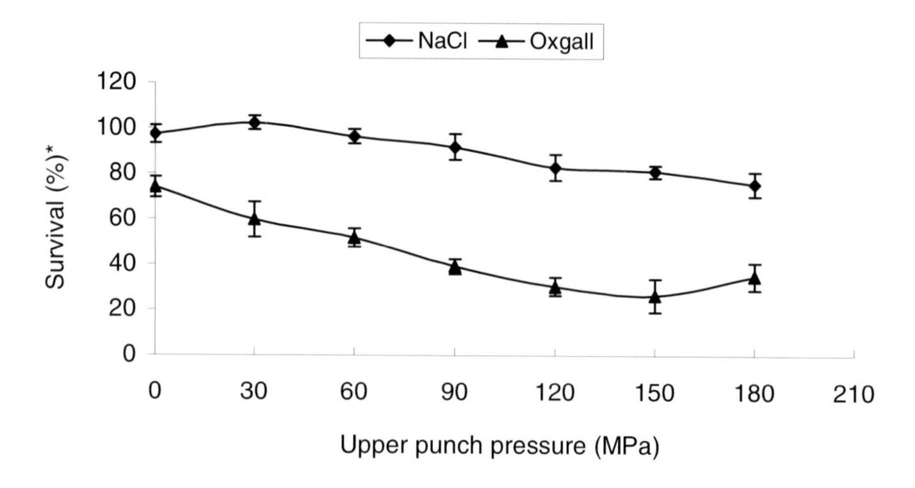

Figure 3. Effect of compression pressure on the viability of L. acidophilus on salted agar.

**Survival (%) was calculated based on the CFUs formed on salted agar over CFUs formed on MRS agar after the cells were compressed.*

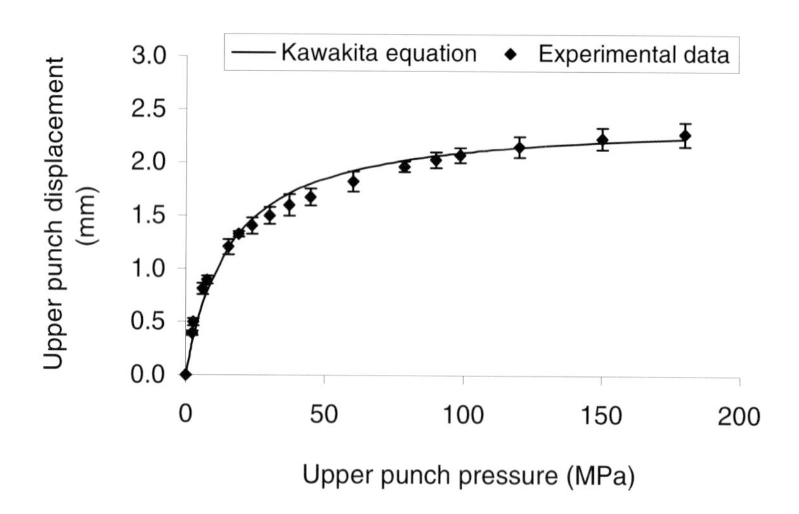

Figure 4. The displacement of the top surface of the powder bed versus compression pressure.

7. Compression coating of cell pellet

7.1. COATING MATERIAL

Sodium alginate (BDH, UK) powder was used as a model coating material. However, it was found that the powder alone could not form a rigid compact as it crumbles easily. The same phenomenon was observed when pectin powder alone was compressed [28]. This is because polymeric materials are generally elastic and poorly compressible [29]. Therefore, a binder was added to sodium alginate powder in order to obtain a more rigid compact. In this work, hydroxypropyl cellulose (HPC) (Hercules, UK) was used as a binder and the tablets show better rigidity than if no binder was used. HPC is widely used as a binder in tablet formulation. The sodium alginate and HPC were used in the weight ratio of 9:1. The coating materials were mixed on a roller for 10 minutes to obtain a homogeneous mixture. The sterility of the powders was checked by dissolving some samples with sterilised distilled water. The solution was then plated on MRS agar to examine contaminants. No contaminants were found in the powders during the course of experiments.

7.2. COATING PROCESS

The die and punch were assembled to provide a platform to which half of the total amount of the coating material was poured onto. The amount of coating used ranged from 250 mg to 450 mg. The compressed cell pellet was then carefully positioned on the centre of the die before the rest of the coating material was poured on top of it, as illustrated in Figure 1. The coating process of the tablet was performed according to the compression procedure described in Section 1.5.

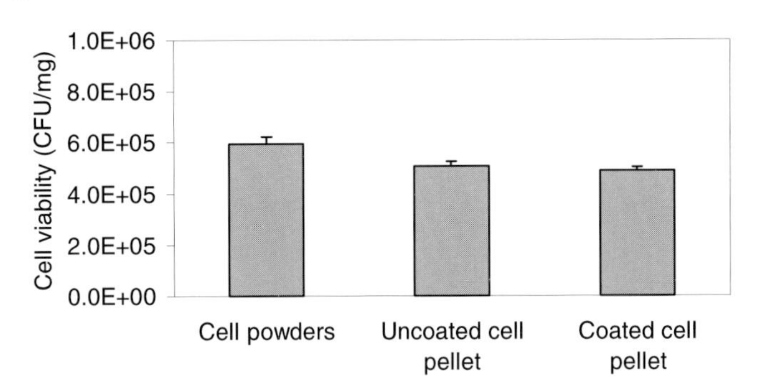

Figure 5. Relative viability of L. acidophilus *after being compressed at 60 MPa.*

Figures 2 and 3 show that the compression pressure could have harmful effects on the cells during compaction. Therefore it is necessary to employ a compression pressure as low as possible in order to minimise the loss of cell viability or sub-lethal injuries. On the other hand, the tablets produced should be rigid enough for further handling. It was suggested by Yanagita *et al.* [30] that conventional tabletting process employs a pressure of about 30 to 50 MPa. As a result, pressures up to 60 MPa were chosen to produce cell pellets and compression coating in this work.

The relative viability of the cell containing powders (as a control), of the uncoated cell pellets and of the coated cells resulting from compression at 60 MPa is shown in Figure 5. The compression of cell containing powders at 60 MPa to form pellets caused a loss in cell viability by about 11%. However, the compression coating did not result in any further reduction in cell viability.

7.3. EFFECT OF COMPRESSION COATING ON CELL STABILITY

Figure 6 shows the influence of compression coating on the stability of the bacteria during storage. Clearly, the stability of the coated bacteria was significantly improved, with approximately 10 times higher than that of the uncoated cell pellets and freeze-dried bacteria in powders at the end of 30 days. The bacteria coated at the 30 MPa seemed to be more stable than those coated at 60 MPa. This was also applicable to the uncoated cell pellets that resulted from the compression at the two different pressures.

Furthermore, the results show that compression pressure could affect the stability of the cells, as the survival of uncoated cells was less than the cell powders. The storage results confirm that the isolation of cells from the surrounding environment could result in greater stability. This might be because the coating materials provide a barrier to retard the movement of moisture into or from the cells, to keep the cells away from light and oxygen. However, the results also showed that pressure is another factor, which affected the stability of the bacteria apart from the environment factors. This could be due to that the bacteria sustained some sub-lethal injuries during the compression process, which made them more vulnerable during storage.

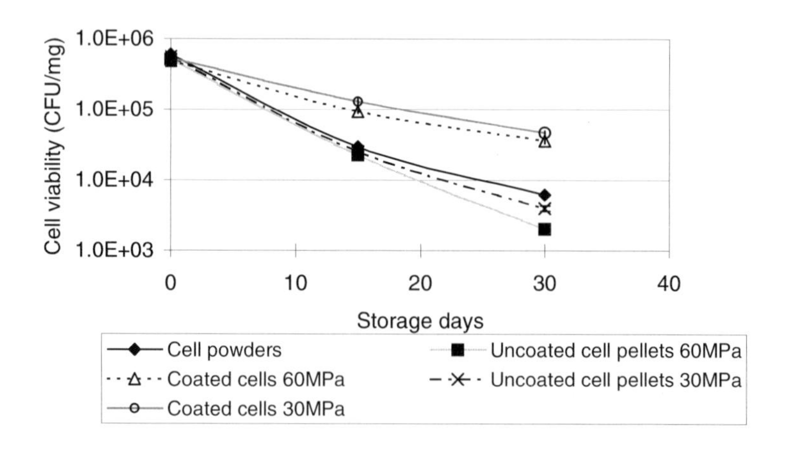

Figure 6. Effect of compression coating on the stability of L. acidophilus after storage.

The effect of the amount of coating materials on the cell stability during storage has also been investigated, and the results are shown in Figure 7. As can be seen, there is no

Table 1. Advantages and disadvantages of each encapsulation method. (continued)

Method	Advantages	Disadvantages	Ref.
Film-coating	**Stabilisation of cells** Proven to be able to enhance storage stability and offer protection to cells when exposed to an acidic medium. **Extensively used** Widely used in pharmaceutical industry and well reported. **Versatile** Equipment is readily available and versatile (used for other purposes *e.g.* drying, granulation).	**Wet granulation of dried cells** Prior to film-coating, wet granulation of cell powders with other excipient might be required. This is to obtain a homogenous powder mixture and to enable the use of a spheronisation-extrusion method for formation of spherical particles. It was shown that this method could result in 3 log loss of cell survival. **Many process variables** There are many process variables to be considered during the coating process such as: • Spray rate (nozzle design, pumping system, liquid/air flow-rate). • Drying conditions (air flow, temperature, humidity). • Coating liquid (formulations, viscosity, surface tension). • Particle properties (porosity, surface roughness). In order to have a successful coating, all operating variables have to be optimised. If this is not achieved, problems such as blocking of nozzles, improper film forming, attrition of solid particles, agglomeration of particles could easily occur. **Batch to batch variation** Due to its process complication, batch-to-batch variation might easily occur. This could also be due to the skill of the operators or variation in size distribution of the solid particles, which could up-set the parameters set-up. **Presence of dust in exhaust gases** No coating process is 100% efficient in terms of the amount of solids incorporated into the coating that actually deposits on the subject. Hence, some of coating material will generate dust and escape with the drying air in the exhaust system. This posses an environmental problem and health hazard to the operators. Therefore, treatment of exhaust gas is necessary in film-coating process.	[10], [11], [12], [15], [16], [17], [18]

significance difference between survival of the cells when they were coated with 350 mg and 250 mg of coating material as both show approximately 10-fold increase in survival if compared to cell powders after 30 days of storage at 25°C.

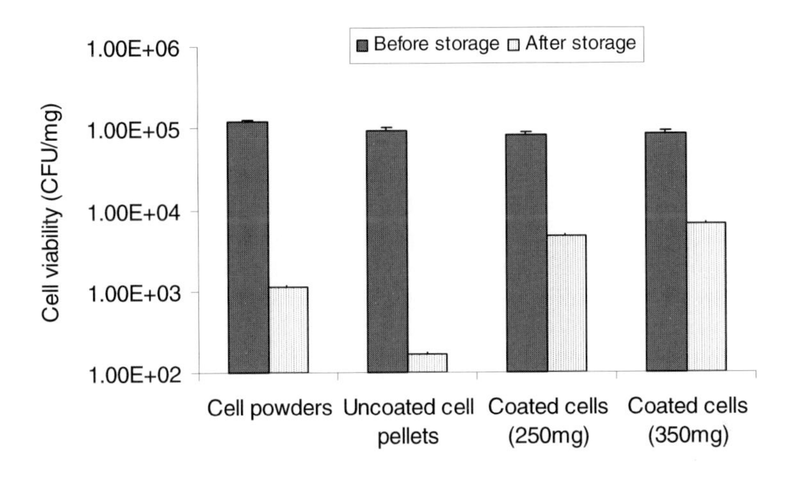

Figure 7. Effect of the amount of coating on cell stability after storage at 25°C for 30 days under atmospheric conditions.

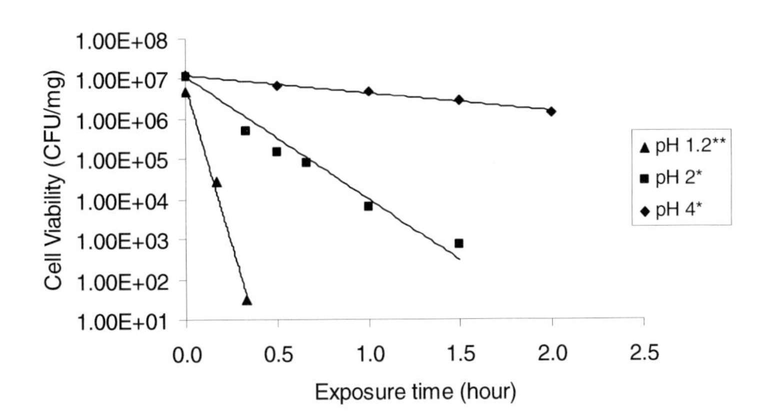

Figure 8. Survival of uncoated cell powders in SGF.

** Uncoated cells were exposed to 600 ml of SGF.*
*** Uncoated cells were exposed to 20 ml of SGF.*

8. *In-vitro* studies of cells

8.1. SURVIVAL OF UNCOATED CELLS DURING EXPOSURE TO A SIMULATED GASTRIC FLUID (SGF)

Figure 8 shows the survival profile of uncoated cell powders over time when they were exposed to a SGF of pH 1.2, 2 and 4, which was made of phosphate buffer solution (PBS) with pH adjusted by addition of HCl. Generally, there was a linear reduction in log of number of viable cells for all cases. The results indicate that the cells were very vulnerable when exposed to SGF of pH 2 and below. The number of cell survival decreased dramatically from 10^6 CFU/mg to 10^1 CFU/mg in less than 30 minutes after they were exposed to pH 1.2. The cells survived better at pH 2 where the cell viability deceased from 10^7 CFU/mg to 10^3 CFU/mg after 90 minutes of exposure. The death of cells was less intense at pH 4 where they lost around 1 log cycle of their initial viability when they were exposed to the SGF for 2 hours.

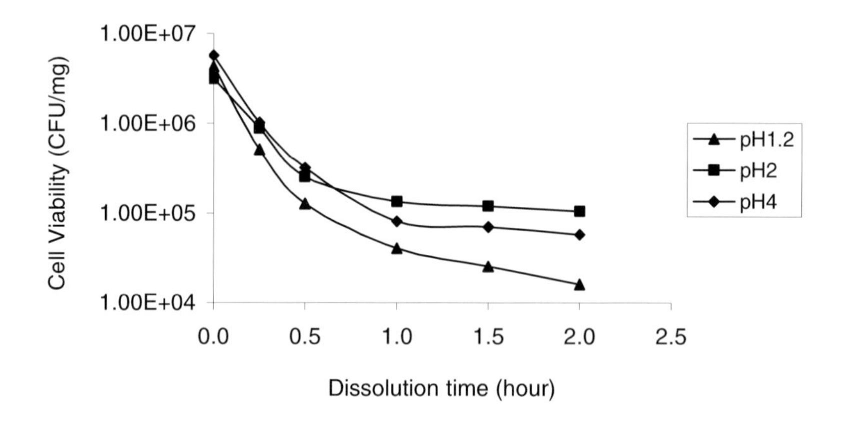

Figure 9. Effect of the SGF pH on survival profile of cells coated with 250 mg of coating.

8.2. SURVIVAL OF COATED CELLS DURING EXPOSURE TO THE SGF

Figure 9 shows the effect of pH of the SGF on the survival of coated cells after 2 hours exposure to the medium. Generally, it shows that compression-coated cells could survive better than the uncoated cells when exposed to pH 1.2 and pH 2 (also see Figure 8). The survival of coated cells at pH 2 was relatively higher than that at pH 1.2. This suggests that pH could be a factor in killing the cells. Interestingly, the exposure of coated cells to less detrimental acidic fluid (pH 4) has resulted in more cell death if compared to the coated cells exposed to pH 2. It can also be seen that the cell death became less intense after some time of dissolution in all cases.

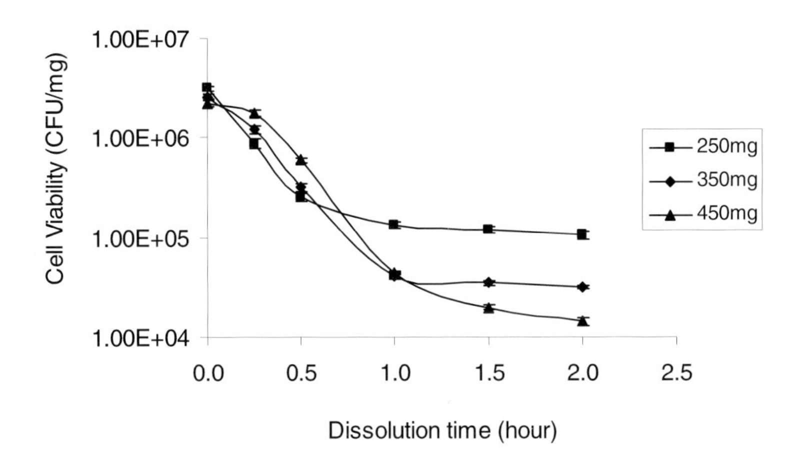

Figure 10. Effect of amount of coating on survival of cells during exposure to the SGF of pH 2.

The effect of amount of the coating materials on the viability of cells when exposed to the SGF of pH 2 was also investigated, and the result is shown in Figure 10. All curves except those coated with 450mg of coating materials show an exponential decrease in cell viability at the beginning of exposure to the SGF. The cells became more stable and the death of the cells was less intense thereafter. The cells coated with 450mg of coating materials showed a distinct phase at the beginning of exposure to the SGF. The death of the cells was delayed before an exponential loss of cell viability could be observed. Figure 10 also shows that the duration of exponential death of cells increases with increasing amount of the coating materials. At the end of 2hr, the cells, which were coated with the least amount of coating materials (250 mg), suffered the minimum loss of their viability. In all cases, the reduction in cell viability was only 1.5 to 2.5 log.

The formation of hydro-gel around the cell pellet was thought to be the basis to offer a protection to cells, as the acidic fluid needs to permeate through the layer before reaching the cells. However, the results are not what originally expected as exposure to a less detrimental acidic medium (pH 4) or increasing amount of coating did not improve cell survival, but resulted in more cell death. This indicates that the formation of hydro-gel could have a protective effect as well as killing effect on cells during exposure to the acidic medium. It is hypothesized that the rehydration rate of cells in the SGF might depend on the value of its pH or the amount of coating materials, which can affect the cell survival. It has been shown by Monk *et al.* [31] that a critical water content exists for cells to survive when rewetted from the dry state. In addition, Leach and Scott [32] lends support to Monk *et al.* [31] by showing that extreme slowness of rehydration caused a significant reduction in viable cells. Therefore, in our case, it is thought that the cell pellet has to achieve a certain moisture content reasonably quickly in order to prevent significant cell death. However, the formation of hydro-gel on contact with the aqueous medium might reduce significantly the rate of penetration of the fluid across the gel barrier towards the cell pellet.

8.3. MOISTURE CONTENT OF CELL PELLETS DURING EXPOSURE TO THE SGF

In order to verify this hypothesis, the moisture content of cell pellets during dissolution was monitored. Figure 11 shows the effect of pH of the SGF on the change of moisture content of cell pellets with a coating of 250 mg. Generally, the moisture content profiles of cell pellets, which were exposed to pH 1 and pH 2, are relatively similar. Both show a higher liquid uptake if compared to the moisture content profile of cell pellets which was exposed to pH 4. Cotrell [33] and King [33] reported that pH of the medium could influence the rheology of alginate gel and hydration, due to the interconversion between the carboxylate anions (sodium alginate) and free carboxyl groups (alginic acid). At neutral pH, sodium alginate is soluble, but below pH 3, it forms water swellable but insoluble alginic acid. In fact, it was observed that sodium alginate compact formed an insoluble but relatively porous, less retarding gel layer when hydrated at pH 1.2, relative to those of soluble and continuous gel layer formed at pH 7.5 [33]. In our case, the gel layer at pH 4 could be less porous than those formed at pH 2, thus offered more resistance to the diffusion of the medium to the cell pellet. This could reduce the rate of rehydration, prolong the time taken for the cells to achieve their critical moisture level, and consequently result in higher cell death. This argument is further supported by the results in Figure 12, which shows the effect of amount of coating on the moisture content of cell pellets during exposure to the SGF of pH 2. As expected, the fluid ingress towards the cell pellets was quicker when a less amount of coating material was applied to the cell pellets. Therefore, the increasing amount of coating prolonged the time taken for the cells to gain necessary amount of water, thus resulted in higher cell death.

8.4. QUANTITATIVE ANALYSIS OF THE REHYDRATION FEATURES OF COATED CELLS

Since it is hypothesized that the survival of dried cells could be dependent on the rate of rehydration of cells and also level of moisture content, the re-hydration features of cells shown in Figures 9 to Figure 12 are quantitatively analysed and summarised in Table 2. The finding is in good agreement with Monk *et al.* [31] regarding the existence of a critical water content during rehydration of dry cells. It was observed that the reduction in cell death was substantial when the moisture content of cell pellets were below the critical level (20%-30%) but became stable (stabilisation phase) when the moisture content of the cell pellet was above that level. Therefore, this confirms the hypothesis that the time needed for the cell pellet to be re-hydrated above this critical level would dictate the magnitude of cell death, as indicated by the prolonged exponential phase as exposure to the SGF of pH 4 or the amount of coating increases. This is because it took longer for the fluid to reach the cell pellet, which resulted in the decrease of rehydration rate.

Leach and Scott [32] reported that the optimum re-hydration rate for cell recovery lies between 10^{-2} to 10^{-1} mg/sec/mg dry matter. They also found that extreme slowness of rehydration, less than 10^{-2} mg/sec/mg dry matter, could be very detrimental to cell survival. In this case, the calculated rates of re-hydration of the cell pellets are far below the suggested detrimental rate, in the region of 10^{-5} mg/sec/mg dry matter. That may explain why the rate of cell death in the exponential phase seems to be independent of the rate of re-hydration of cells, as the cell death rate constants remain relatively consistent.

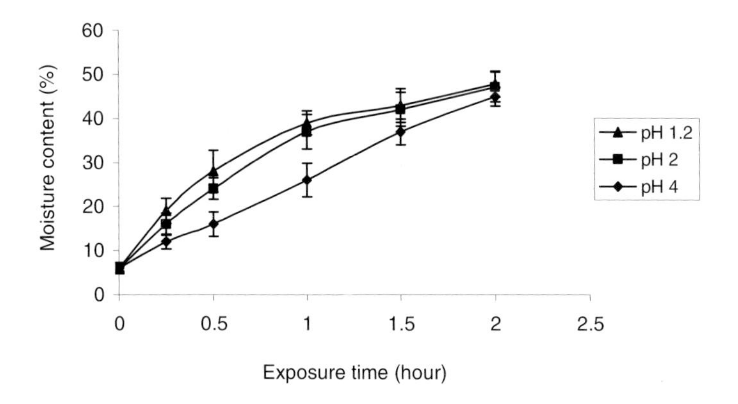

Figure 11. Effect of SGF pH on the moisture content of cell pellets coated with 250 mg of coating.

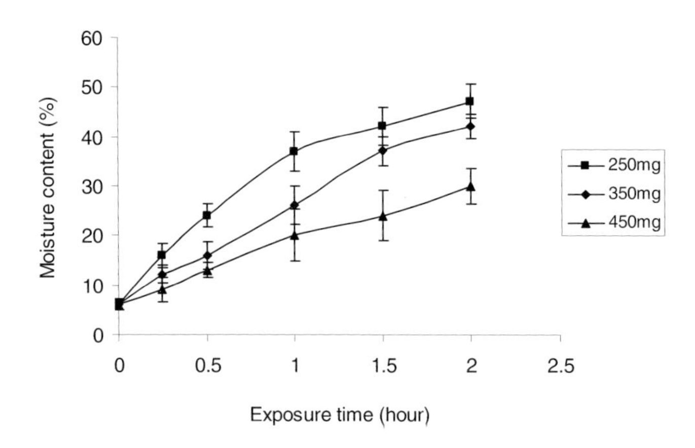

Figure 12. Effect of amount of coating on the moisture content of cell pellets during exposure to SGF of pH 2.

Furthermore, it was an interesting phenomenon that the cell death was less intense in the stabilisation phase even though the cells were constantly re-hydrated with the SGF of pH 2 (Figures 9 and 10). The stabilisation of the cells could be attributed to them having achieved the critical moisture level. In addition, it is speculated that stabilisation of the cells could also be resulted from the buffering capacity of the cell pellets. The dried cells were incorporated with a large amount of skimmed milk powders, which when dissolved, have a pH of around 6.5. Due to the slow re-hydration rate of cell pellets, the number of hydrogen ions that could reach the cell pellets is limited, thus could not significantly alter the pH of the cell pellets. To prove the hypothesis, a tablet was exposed to a medium of pH 2 for 2 hours and subsequently the cell pellet was removed from the hydrated coating layer. The re-hydrated cell pellet was cut into 2 halves and the cross-section was then gently rubbed on a pH paper. The result supported the hypothesis as it reveals a pH near neutrality (picture not shown). On the other hand, the results also indicate that the cells might be killed by acid, as the reduction in cell viability was more substantial when they were exposed to pH 1.2 than to pH 2 or pH 4. This could be mainly due to the increased concentration of hydrogen ions. Hence the cell pellet might not be able to cope with the excessive amount of hydrogen ions and it might eventually lose some of its buffering capacity.

Table 2. Rehydration features of cells during exposure to the SGF.

Amount of coating (mg)	SGF pH	Duration (hour), t			Cell death rate constant k_d* (log hr^{-1})	Critical moisture content (%)**	Rehydration rate k_r (10^{-5}mg/sec/ mg dry cell pellet)***
		Lag phase	Exponential phase	Stabilisation phase			
250	1.2	Not observed	≤ 0.5	≥ 1.5	$\approx 3.2 \pm 0.02$	≈ 28	≈ 12.3
250	2.0	Not observed	≈ 0.5	≥ 1.5	$\approx 2.2 \pm 0.04$	≈ 24	≈ 8.5
250	4.0	Not observed	≈ 1.0	≈ 1.0	$\approx 2.0 \pm 0.04$	≈ 26	≈ 5.4
350	2.0	< 0.25	≥ 0.75	≈ 1.0	$\approx 1.9 \pm 0.01$	≈ 26	≈ 5.4
450	2.0	≤ 0.25	$0.75 \leq t \leq 1.25$	≤ 0.75	$\approx 2.1 \pm 0.03$	≈ 22	≈ 3.3

* k_d was calculated based on the cell death during the exponential phase in Figures 9 and 10
** Moisture content of cell pellet (%) at the beginning of stabilisation phase
*** k_r was calculated based on duration of exponential phase

8.5. POSTULATION OF MECHANISMS OF CELL DEATH DURING EXPOSURE TO THE SGF

Several studies have shown the influence of rehydration conditions on cell viability during resuscitation of dried cells. It is generally agreed that rapid re-hydration of dried cells could lead to a loss in viability, possibly due to osmotic shock [32,34-37]. Leach and Scott [32] speculates that fast rates could disrupt cellular components as a result of differential expansion or swelling. Therefore, slow rehydration is preferred in recovery of dried cells and it is shown to have resulted in higher cell viability [32,34,36]. However, Leach and Scott [32] found that extreme slowness in rehydrating dried cells could be very detrimental. The deleterious effect is attributed to the presence of a limited amount of water. However, the exact mechanism to cause cell death has not been successfully identified. Monk and McCafferey [38] speculated that a drastic physical distortion in cells could occur at a low water content due to the non-uniform swelling if some parts of the dried cells are more hygroscopic than others. They also suggested that the death of cells at a low moisture content might be due to unbalanced metabolism of cells as not all components of the metabolic system are hydrated to their activity, thus the reaction in damp cells would result in depletion of a necessary substance or excess production of a toxic material. Furthermore, it is possible that a hygroscopic substance could dissolve at low water contents and it could act as a toxin because of its high concentration. These hypotheses might be responsible for the death of cells in the exponential phase as seen in Figures 9 and 10.

However, as the water content increases, the toxin could be diluted and it is more likely that the cells could achieve nearly their normal metabolic balance. This could indeed happen where stabilisation of the cells commenced when the cell pellets attained a moisture content in the range of 20% to 30% as shown in Figures 9 and 10. The moisture level should correspond to a water activity (Aw) of 0.8-0.9 as shown by the sorption isotherm curve (Figure 13). It can be noticed from the curve that when the Aw is around 0.85 and greater, the increases in moisture content are relatively larger than the changes in Aw. Harris attributed this range of water activities to a more metabolically active state of cells [36]. Therefore, stabilisation of the cells might have occurred after they were hydrated to their normal metabolic state, which might be required for maintenance of cell functions.

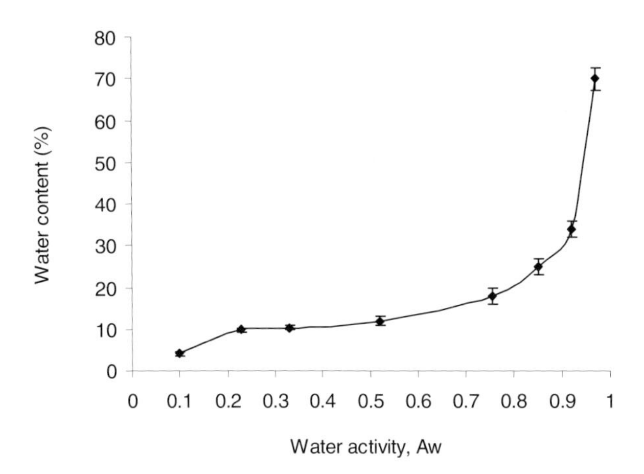

Figure 13. Sorption isotherm curve for freeze-dried cells.

On the other hand, Poirier *et al* [34] shows that the water activity range of 0.117-0.455 must be crossed with caution during re-hydration in order to maintain the viability of

Saccharomyces cerevisiae. It is suggested that the range of Aw corresponds to membrane phospholipid phase transition where the cell membrane is unstable and sensitive to the water flow into the cell. Rapid crossing of this water activity range could cause the membrane to leak whereas slow rehydration could allow slow water uptake through an unstable membrane, hence preserving cell viability. However, this observation supports the existence of a critical moisture level during rehydration of dried cells, although further explanation of cell death in relation to cell membrane cannot be offered at this point. Kosanke *et al.* [36] reinforced the idea of the existence of a critical moisture range by showing that a gradual increase of Aw (from typically Aw of 0.6 to 0.99) during rehydration of gram-negative bacteria in clay formulations could be responsible for the increase in cell viability. Therefore, it must be clarified that the requirements for optimal rehydration (rehydration rate, media, moisture content) can differ between strains and species [35]. This could prevent generalisations such as the range of water activity or water content, which are critical to cell viability during rehydration. In addition, the different methods and conditions used in preparing, drying and rehydrating cells as well as the various additives used could add to the variations between observations among different workers.

8.6. EFFECT OF SIMULATED INTESTINAL FLUID ON CELL SURVIVAL

Figure 14 shows the influence of amount of coating on survival of cells after exposure to the SGF at pH 2 for 2 hours and then to a simulated small intestinal fluid made of phosphate buffer solution at pH 6.8 until they were fully released.

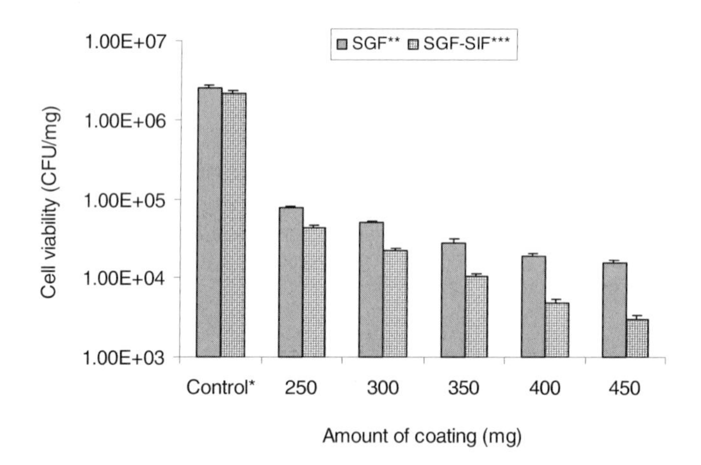

Figure 14. Effect of amount of coating on the cell survival.

* *Viability of coated cells without treatment*
** *Cell viability was recorded after exposure to SGF of pH 2 for 2 hours*
*** *Cell viability was recorded after exposure to SGF of pH 2 for 2 hours and subsequent exposure to SIF until coating was totally dissolved*

The results confirm earlier observations that the cell survival was inversely related to the amount of coating where the lowest amount of coating (250mg) registered the highest cell survival. The relationship is also true when the cells were first exposed to the SGF and subsequently released to the SIF. It could also be observed by comparing Figures 10 and 14 that the cell survival after being released to the SIF was lower than that being exposed to SGF only. The difference in cell survival between them seems to increase with increasing amount of coating.

The loss of viable cells upon release might not be resulted from the effect of coating. The total dissolution time of tablets (or total cell release time) was found to increase with the amount of coating materials (more details are given in Section 1.9.7). The release time of cells coated with 250 mg to 450 mg of coating materials ranged from 5 to 9 hours. The amount of time of each tablet spent in the SGF before being transferred to the SIF was 2 hours. By subtracting the number of viable cells in the SIF from the number of viable cells in the SGF for each amount of coating, the reduction in cell viability due to the effect of the SIF was determined, which is plotted versus the exposure time to the SIF in Figure 15. The exposure time in the SIF for coated cells means the time taken for their compete dissolution to occur. For free cells, it just means the time spent in the SIF. Figure 15 clearly shows that both coated cells and uncoated cells gave proximity in the magnitude of cell death during exposure to the SIF for the same period of exposure time. This suggests that the coating did not cause further reduction in cell viability in subsequent exposure to the SIF after exposure to the SGF for 2 hours.

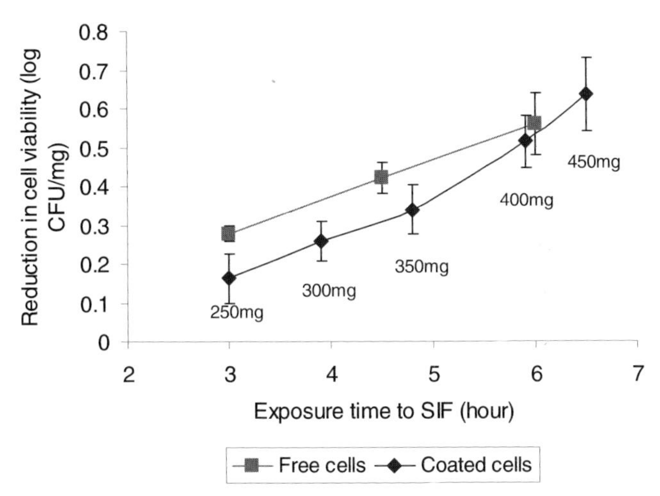

Figure 15. Effect of amount of coating on the survival of cells when exposed to the SIF.*

** All coated cells were pre-exposed to SGF for 2 hours before being transferred to SIF. The exposure time of all coated cells to SIF means the total time spent in the media before total dissolution of coating material occurred.*

8.7. RELEASE TIME OF COATED CELLS

Table 3 shows the vertical thickness of coating and the corresponding release time of cells to the SIF. The horizontal thickness of the tablet is approximately 2 mm and does not change with the amount of coating because it is determined by the diameter of the punches for compressing the cell pellet (6mm) and coating material (10mm) (Figure 16). Therefore, the increasing amount of coating has resulted in increasing vertical thickness. A tablet with 250mg of coating material is approximately 10mm and 3.5mm in diameter and thickness respectively whereas a tablet with 450mg of coating material is 10mm and 6mm in diameter and thickness respectively.

Table 3. Release times of coated cells.

Amount of coating (mg)	SGF pH	Vertical thickness (mm) (± 0.05)	Total release time* (hour) (± 0.25)
250	2	0.79	5.1
300	2	1.10	5.9
350	2	1.39	6.8
400	2	1.67	7.9
450	2	1.95	8.5
250	1.2	0.79	5.1
250	4	0.79	4.6

** Recorded based on 2 hours exposure to the SGF and subsequent amount of time spent for total dissolution in the SIF pH 6.8.*

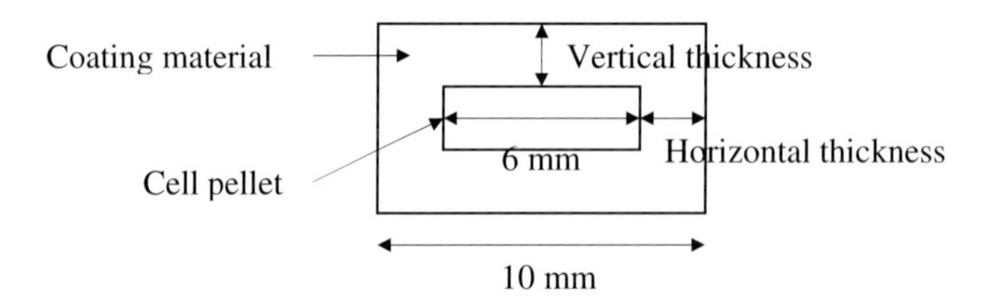

Figure 16. Schematic diagram of coated cells.

The amount of coating and pH of the SGF have been shown to have significant influences on the release time of the cells (Table 3). The release time increases with the increasing amount of coating. The release mechanism of cells to SIF is primarily due to erosion of the alginate gel layer. The cells coated with 250mg and 450mg of coating were completely released approximately 5 hours and 8 hours of dissolution respectively. When the coated cells were exposed to the SGF pH 4, the release time of cells was approximately 30 minutes shorter than that of the coated cells exposed to pH 1.2 or pH 2. The release time of the cells could be used as an indicator of the site of cell release in human gastro-intestinal tract. It is assumed that the mean residence time of dosage form in stomach is 2 hours. Gruber [39] and Phillips [39] reported that the mean transit time of dosage form in the small intestine of a healthy subject is about 3 to 4 hours. Therefore, it is deduced that the cells would probably be released in the region between the near end of small intestine and the beginning of colon. This region can be a good site of delivery of probiotic bacteria because it is indigenous to many microflora in human. In fact, Conway *et al.* [5] tested 3 Lactobacillus strains and it was found that they showed better adherence to caecum and colon epithelial cells than to the ileum epithelial cells.

9. Conclusion

A novel encapsulation method based on direct compression has been developed. The method and formulation has shown to improve survival of cells during storage as well as during exposure to an acidic medium if compared to control. Although the experiments were carried out on a laboratory scale, it is believed that the method is applicable to an industrial scale since the tabletting process has been well established. It is believed that the method and formulation can also be applied to pharmaceutical, food, and chemical industries. Future work includes new formulations to form rigid tablets at lower pressures and to increase the rehydration rate of cell pellets without compromising their buffering capacity.

References

[1] Shortt, C. (1998) Living it up for dinner. Chemistry & Industry: 300-303.

[2] Gomes, A.M.P. and Xavier Malcata, F. (1999) *Bifidobacterium spp.* and *Lactobacillus acidophilus*: biological, biochemical, technological and therapeutical properties relevant for use as probiotics. Trends in Food Sci. and Technol. 10: 139-157.

[3] Kailasapathy, K. and Chin, J. (2000) Survival and therapeutic potential of probiotic organisms with reference to *Lactobacillus acidophilus* and *Bifidobacterium spp.*. Immunology and Cell Biology 78: 80-88.

[4] Lee, Y.K. and Salminen, S. (1995) The coming of age of probiotics. Trends in Food Sci. and Technol. 6: 241-245.

[5] Conway, P.L.; Gorbach, S.L. and Goldin B.R. (1987) Survival of lactic acid bacteria in the human stomach and adhesion to intestinal cells. J. Diary Sci. 70(1): 1-12.

[6] Favaro Trindade, C.S. and Grosso, C.R.F. (2000) The effect of the immobilisation of *Lactobacillus acidphilus* and *Bifidobacterium lactis* in alginate on their tolerance to gastrointestinal secretions. Milchwissenschaft 55(9): 496-499.

[7] Sultana, K.; Godward, G.; Reynolds, N.; Arumugaswamy, R.; Peiris, P. and Kailasapathy, K. (2000) Encapsulation of probiotic bacteria with alginate-starch and evaluation of survival in simulated gastrointestinal conditions and in yoghurt. Int. J. Food Microbiol. 62: 47-55.

[8] Lal, M.; Tiwari, M.P.; Sinha, R.N. and Ranganathan, B. (1978) Activity of the freeze-dried *Streptococcus lactis* sub sp *Diacetilactic* DRC in gelatin capsules and strip packed tablets during storage. Egyptian J. Diary Sci. 6: 33-37.

[9] Lauland, S. (1994) Commercial aspects of formulation, production and marketing of probiotic products. In: Gibson, S.A.W. (Ed.) Human Health: The contribution of microorganisms. Springer-Verlag, London; pp. 159-173.

[10] Kim, H.S.; Kamara, B.J.; Good, I.C. and Enders, G.L. Jr. (1998) Method for the preparation of stabile microencapsulated lactic acid bacteria. J. Ind. Microbiol. 3: 253-257.

[11] Yokota, T.; Sato, T.; Uemitsu, N. and Kitabatake, K. (1990) Development of enteric coated tablets containing lactic acid bacteria. In: Proceedings of the International Bifidobacterium Conference, Tokyo (Japan), September 12-13 1990; Japan Bifidus Foundation, Tokyo; p. 53

[12] Appelgren, C. and Eskilson, C. (1990) A novel method for the granulation and coating of pharmacologically active substances. Drug Development and Industrial Pharmacy 16(15): 2345-2351.

[13] Champagne, C.P.; Morin, N.; Couture, R.; Gagnon, C.; Jelen, P. and Lacroix, C. (1992) The potential of immobilized cell technology to produce freeze-dried, phage-protected cultures of *Lactococcus lactis*. Food Research International 25: 419-427.
[14] Thoma, K. and Bechtold, K. (1992) Enteric coated hard gelatin capsules. Capsugel Library: 1-16.
[15] Cole, G.C. (1995) Environmental considerations: treatment of exhaust gases from film-coating processes. In: Cole, G.C. (Ed.) Pharmaceutical coating. Taylor & Francis Ltd.; pp. 240-247.
[16] Wesdyk, R.; Joshi, Y.M.; De Vincentis, J.; Newman, A.W. and Jain, N.B. (1993) Factors affecting differences in film thickness of beads coated in fluidized bed units. Int. J. Pharmaceutics 93: 101-109.
[17] Porter, S.C. and Bruno, C.H. (1990) Coating of Pharmaceutical solid-dosage form. In: Lieberman H.A. *et al.* (Eds.) Pharmaceutical dosage forms: Tablets. Volume 3. Marcel Dekker, US; pp. 77-158.
[18] Barthelemy, P.; Laforet, J.P.; Farah, N. and Joachim, J. (1999) Compritol® 888 ATO: an innovative hot-melt coating agent for prolonged-release drug formulations. Eur. J. Pharmaceutics and Biopharmaceutics 47: 87-90.
[19] Ashford, M.; Fell, J.; Attwood, D.; Sharma, H. and Woodhead, P. (1993) An evaluation of pectin as a carrier for drug targeting to the colon. J. Controlled Release 26: 213-220.
[20] Ashford, M.; Fell, J.; Attwood, D.; Sharma, H. and Woodhead, P. (1994) Studies on pectin formulations for colonic drug delivery. J. Controlled Release 30: 225-232.
[21] Semde, R.; Amighi, K.; Devleeschouwer, M.J. and Moes, A.J. (1999) *In vitro* evaluation of pectin HM/ethylcellulose compression-coated formulations intended for colonic drug delivery. S.T.P. Pharma Sciences 9(6): 561-565.
[22] Krishnaiah, Y.S.R.; Satyanarayana, S.; Rama Prasad, Y.V. and Narashimha, R.S. (1998) Evaluation of guar gum as a compression coat for drug targeting to colon. Int. J. Pharmaceutics 171: 137-146.
[23] Krishnaiah, Y.S.; Satyanarayana, S. and Rama Prasad, Y.V. (1999) Studies of guar gum compression-coated 5-aminosalicylic acid tablets for colon-specific drug delivery. Drug Develop. Ind. Pharmacy 25(5): 651-657.
[24] Kaneko, K.; Kanada K.; Miyagi, M.; Saito, N.; Ozeki, T.; Yuasa, H. and Kanaya, Y. (1998) Formation of water-insoluble gel in dry-coated tablets for the controlled release of theophylline. Chem. and Pharmaceutical Bulletin 46(4): 728-729.
[25] Couture, R.; Gagne D. and Champagne, C.P. (1991) Effet de divers additifs sur la survie a la lyophilisation de *Lactococcus lactis*. Canadian Institute of Food Sci. Technol. 24(5): 224-227.
[26] Johnson, M.; Ray, B. and Speck, M.L. (1984) Freeze-injury in cell wall and its repair in *Lactobacillus acidophilus*. Cryo-letters 5: 171-176.
[27] Paronen, P. and Iikka, J. (1996) Porosity-pressure functions. In: Alderborn, G. and Nystrom, C. (Eds.) Pharmaceutical Powder Compaction Technology. Marcel Dekker, New York; pp. 55-97.
[28] Kim, H.; Venkatesh, G. and Fassihi, R. (1998) Compactibility characterization of granular pectin for tableting operation using a compaction simulator. Int. J. Pharmaceutics 161: 149-159.
[29] Takeuchi, H.; Yasuji, T.; Hino, T.; Yamamoto, H. and Kawashima, Y. (1998) Spray-dried composite particles of lactose and sodium alginate for direct tabletting and controlled releasing. Int. J. Pharmaceutics 174: 91-100.
[30] Yanagita, T.; Miki, T.; Sakai, T. and Horikoshi, I. (1978) Microbiological studies on drugs and their raw materials. I. Experiments on the reduction of microbial contaminants in tablets during processing. Chem. and Pharmaceutical Bulletin 26: 185-190.
[31] Monk, G.W.; Elbert, M.L.; Stevens, C.L. and McCafferey, P.A. (1956) The effect of water on the death rate of *Serratia marcescens*. J. Bacteriology 72: 368-372.
[32] Leach, R.H. and Scott, W.J. (1959) The influence of rehydration on the viability of dried micro-organisms. J. General Microbiol. 21: 295-307.
[33] Hodsdon, A.C.; Mitchell, J.R.; Davies, M.C. and Melia, C.D. (1995) Structure and behaviour in hydrophilic matrix sustained release dosage forms: 3. The influence of pH on the sustained-release performance and internal gel structure of sodium alginate matrices. J. Controlled Release 33: 143-152.
[34] Poirier, I.; Marechal, P.A.; Richard, S. and Gervais, P. (1999) *Saccharomyces serevisiae* viability is strongly dependant on rehydration kinetics and the temperature of dried cells. J. Appl. Microbiol. 86: 87-92.
[35] De Valdez, G.F.; Giori, G.S.; Ruis Holgado, A.P. and Oliver, G. (1985) Effect of drying medium on residual moisture content and viability of freeze-dried lactic acid bacteria. Appl. and Environmental Microbiol. 49(2): 413-415.
[36] Kosanke, J.W.; Osburn, R.M.; Shuppe, G.I. and Smith, R.S. (1992) Slow rehydration improves the recovery of dried bacterial populations. Canadian J. Microbiol. 38: 520-525.

[37] Kearney, L.; Upton, M. and McLoughlin, A. (1990) Enhancing the viability of *Lactobacillus plantarum* inoculum by immobilizing the cells in calcium-alginate beads incorporating cryoprotectants. Appl. Environmental Microbiol. 56: 3112-3116.
[38] Monk, G.W. and McCafferey, P.A. (1957) The effect of sorbed water on the death rate of washed *Serratia marcescens*. J. Bacteriology 73: 85-88.
[39] Reddy, S.M.; Sinha, V.R. and Reddy, D.S. (1999). Novel oral colon-specific drug delivery systems for pharmacotherapy of peptide and nonpeptide drugs. Drugs of Today 35 (7): 537-580.

CELL IMMOBILISATION IN PRE-FORMED POROUS MATRICES

GINO V. BARON[1] AND RONNIE G. WILLAERT[2]
[1]*Department of Chemical Engineering, Vrije Universiteit Brussel, Pleinlaan 2, B-1050 Brussel, Belgium – Fax: 32-2-6293248 – Email: gvbaron@vub.ac.be*
[2]*Department of Ultrastructure, Flanders Interuniversity Institute for Biotechnology, Vrije Universiteit Brussel, Pleinlaan 2, B-1050 Brussel, Belgium – Fax 32-2-6291963 – Email: Ronnie.Willaert@vub.ac.be*

1. Introduction

Immobilised cell systems are usually subdivided into 4 major categories according to the physical mechanism of cell location, i.e. surface attachment, entrapment within "porous matrices", "containment behind a barrier" and "self-aggregation of cells" [1]. The category "entrapment within porous matrices" can be further subdivided – based on the type of the used immobilisation support material – into "gel entrapment" and " pre-formed porous supports". In this chapter, the category "cell immobilisation in pre-formed porous matrices" will be reviewed. Firstly, a few examples of the use and application of porous immobilisation matrices are given. Next, the adhesion of cells to a support material is reviewed. This chapter ends with an in depth discussion of the cell immobilisation process, i.e. immobilisation kinetics and mass transport.

The cells are entrapped in a matrix, which protects them from the shear field outside the matrix. This is of particular importance for fragile cells such as mammalian cells. Unlike gel entrapment systems, porous supports can be inoculated directly from the bulk medium. As with the adsorption method, cells are not completely separated from the effluent in these systems, which can be a disadvantage for some applications. Mass transport of substrates and products can be achieved by molecular diffusion as well as convection, by proper matrix design and organisation of external flow. Consequently, mass transport limitations can become less severe under optimal conditions. When ideally, the colonised porous matrix retains some free space for flow, immobilisation occurs partly by attachment to the internal surface, self-aggregation and retention in dead-end pockets within the material. This is only possible when cell adhesion is not very strong and the application of high external flow rates reversibly removes cells from the matrix. When high cell densities are obtained, convection is no longer possible and the cell system behaves as dense cell agglomerates with strong diffusion limitations. Cell immobilisation methods are simple and a high degree of cell viability is retained

V. Nedović and R. Willaert (eds.), Fundamentals of Cell Immobilisation Biotechnology, 229-244.

upon entrapment. The pre-formed matrix is chemically inert, resistant to microbial attack and incompressible. Often steam sterilisation is possible and the matrix can be reused. Usually, the matrix takes up a significant volume fraction resulting in a lower immobilised cell density compared to other immobilisation methods.

2. Porous immobilisation matrices

Various porous matrices have been described to be used for living cell immobilisation. Table 1 (a and b) gives some illustrative examples. The choice will usually depend on the used cell type and the kind of application. For example, the immobilisation of microbial cells for the production of biochemicals in large packed bioreactors requires a matrix with excellent mechanical characteristics to withstand the high-pressure drop in the reactor; or tissue engineering porous matrices need to be an excellent scaffold for cell attachment and growth, and must be characterised by excellent biocompatibility characteristics. Usually, a choice have to be made from different suited matrices. This choice will be mostly based on an economic basis: What is the cheapest matrix material? Is the matrix for the specific application patented? Is it reusable?

Table 1a. Examples of cell systems with immobilisation by entrapment within porous pre-formed supports: synthetic organic polymers

Material	Void volume (%)	Cells	Product/objective	Reference
Polyethylene terephthalate	85-90	Human trophoblast	Tissue engineering	[5]
Poly (*D,L*-lactic-*co*-glycolic acid)	89	Enterocytes	Tissue engineering	[6]
Poly-*L*-lactic acid	91	Chondrocytes	Neocartilage formation	[2]
Polyglycolic acid	92-97	Chondrocytes	Neocartilage formation	[2-4]
Polystyrene	> 90	Vero and hybridoma cells	Cell growth	[7]
Polyvinyl foam		Cyanobacteria	Photosynthetic e-transport	[8]
Polyurethane foam		Active sludge	Wastewater treatment	[9]
		Anabaena variabilis	Orthophosphate removal	[10]
		Aspergillus niger	Citric acid	[11]
		Yarrowia lipolytica	Oil degradation	[12]
		Bacillus sp.	Dimethylphtalate degradation	[13]
		Methanogen species	Methane	[14]
		Porphyridium cruentum	Polysaccharides	[15]
		Pseudomonas sp.	Naphthalene degradation	[16]
		Ralstonia sp. and *Pseudomonas putida*	Aromatic compounds in olive mill wastewaters	[17]
		Capsicum frutescens	Capsaicinoids	[18-20]
		Hybridoma cells	Monoclonal antibodies	[21]
Silicone rubber	32-65	Various (animal) cells	Improved O_2-permeation	[22]
		Saccharomyces cerevisiae	Kinetic activity	[23]

In tissue engineering, some applications require porous biodegradable scaffolds. The repair of large cartilage defects requires the use of a three-dimensional scaffold to provide a structure for cell proliferation and control the shape of the regenerated tissue. For example, cartilage implants based on chondrocytes and 3D fibrous polyglycolic acid

3. Adhesion mechanisms

The immobilisation mechanism of living cells in a porous matrix is initially mainly by adsorption on the surfaces of the internal pores and external surfaces. Later on, also self-aggregation and entrapment in dead-end pores occurs. The adsorption of cells to an organic or inorganic support material is rendered by Van der Waals forces and ionic interactions. In the special case when microbial exopolymers are involved, covalent interactions are also involved [56]. In this case, the adsorption process can be divided in several successive steps [57-60]: (1) adsorption of extracellular organic macromolecules on the surface (not observed in all cases); (2) transport of cells from the bulk phase to the surface; (3) reversible attachment of cells; and (4) biosynthesis of polymers by the cell which leads to an irreversible attachment of the microorganism to the surface caused by covalent and ionic bonds.

The adhesion behaviour of viable cells is influenced by the following factors [56]: (a) the physical and chemical parameters of the adsorption matrix; (b) the micro-organisms which should be immobilised (especially the biochemical characteristics of the outer surface of their cell wall is important); (c) the composition, the chemical and physical parameters and the fluid conditions of the surrounding liquid phase. The adhesion mechanisms of different bacteria or the effects of different environmental and/or physiological conditions have been intensively studied [61].

The physical chemistry of interfaces provides two theoretical approaches for describing adsorption of a particle to a surface [62]. The first is the DLVO (Derjaguin, Landau, Verwey, Overbeek) theory, which takes the interactions between two solids approaching each other into consideration: an attractive term is due to dispersion (London) van der Waals forces and a second term is due to electrostatic interactions (attractive or repulsive), resulting from the overlap of electrical double layers. This approach which considers only long-range forces, neglects polar forces that may be important at close separation and let the solvent play a direct role: non-dispersive van der Waals forces (dipole-dipole: Keesom; dipole-induced dipole: Debeye), ion-dipole forces and hydrogen bonding. The second approach is provided by thermodynamic considerations based on the Gibbs free energy of adhesion per unit area interface:

$$\Delta\overline{G} = \gamma_{CS} - \gamma_{CL} - \gamma_{SL} \qquad (1)$$

where γ is the interfacial tension at the cell-support (CS), cell-liquid (CL), and support-liquid (SL) interfaces. Adhesion should take place when $\Delta\overline{G} < 0$. The variation of the amount of adsorbed cells as a function of the surface energy of the support or of the liquid has been found to follow the same trend as $\Delta\overline{G}$ ([63-65]). Various methods allow deducing the interfacial tensions from the surface tension of the liquid and contact angle measurements. In all methods, the evaluation of γ_{CS} – a solid-solid interfacial tension – relies on experimental determinations concerning solid-liquid interfaces. Therefore, the thermodynamic approach does not allow to account for the existence of electrostatic interactions between surfaces, either cell-support or cell-cell interactions.

The adhesion of a single, dense layer of microbial cells can be induced by reducing the electrostatic repulsion between the cells and the support material or by stimulating the electronic attraction. Several methods have been optimised to adhere microbial cells with retention of their metabolic activity to different supports (like glass, polycarbonate, polystyrene): pH control, starvation of yeast cells in pure water, adsorption of metallic ions on the cell surface or on the support, or coating the support by a layer of positively charged colloidal particles [66-70]. The influence of cations on the adsorption of two negatively charged strains of *Pseudomonas fluorescens* on negatively charged glass has been investigated by interference reflection microscopy (IRM) [71]. The addition of cations caused changes in the IRM image that went along with a decrease in separation distance by neutralising negative charges on bacterial surface polymers. Ion-exchange supports have also been used to immobilise microbial cells ([72-75]).

To obtain some guidelines for the immobilisation by adsorption of microorganisms, Mozes and co-workers [62] studied the adhesion of *S. cerevisiae*, *Acetobacter aceti* and *Moniliella pollinis* to different support materials (glass, metals, and plastics), where some were treated by a Fe (III) solution. The surface properties of the cells were characterised by the zeta potential and an index of hydrophobicity. The supports were characterised by surface chemical analysis (XPS) and contact angle measurements. Cell suspensions in pure water at a given pH were left to settle on plates. The plates were rinsed and examined microscopically. *S. cerevisiae* and *A. aceti* adhered to metals under certain pH conditions, but did not adhere to any of the other materials tested unless it was previously treated by ferric ions. Adhesion of these hydrophilic cells is essentially controlled by electrostatic interactions. *M. pollinis* adhered spontaneously to glass and polymeric materials, but its attachment is also influenced by cell-cell or cell-support electrostatic repulsions. Near the cell isoelectric point, cell flocculation is competing with adhesion to a support.

The influence of the hydrophobicity/hydrophilicity as well as the ionic properties of the support material on the adsorption of fungal *Trichoderma reesei* cells have been studied by coating a fibrous carrier with hydrophilic and hydrophobic copolymers generated by irradiation [76-80]. Relatively mild hydrophilic polymers – such as the hydroxypropyl methacrylate polymer – gave the best adsorption results. The effect of the ionic properties of the polymers was investigated by changing the monomer composition in a trimethylpropane triacylate – acrylic acid or methacrylic acid diethylaminoethyl ester system. Cationic polymers were better than anionic polymers in the adhesion of cells; and more positive charge or less negative charge in the polymers led to an increase in the growth of the cells immobilised on their surface.

The process of yeast cell immobilisation onto fired bricks has been known to occur by adsorption [30]. Yeast cells have been found to possess a negative charge on the cell wall, possibly due to the presence of carboxyl and phosphate groups [68]. The charge on fired brick is not exactly known; however, clay – which forms a large proportion of the brick – is positively charged in an acid medium. It has also been reported that fired brick contains reactive hydroxyl groups [81]. In an acid medium, the excess H^+ ions present neutralise the reactive hydroxyl groups and leave a net positive charge on the surface of brick particles due to excess H^+ ions in solution. Spontaneous adsorption has therefore been demonstrated [30].

Most animal cells from solid tissue grow as adherent monolayers, unless they have transformed and become anchorage independent. Following tissue disaggregation or subculture, they will need to attach and spread out on the substrate before they will start to proliferate. Anchorage dependent cells are often diploid and exhibit contact inhibition, wherein cell division is gradually inhibited by cell-cell contact as the culture reaches confluence. The growth of these cells on a flat and/or spherical surface can be quantitatively analysed using an appropriate mathematical model ([82-86]).

Cell adhesion is mediated by specific cell surface receptors for molecules in the extracellular matrix [87]. It seems likely that the secretion of extracellular matrix proteins and proteoglycans by the cells may precede spreading. Cell substrate interactions are mediated primarily by integrins, receptors for matrix molecules such as fibronectin, entactin, laminin and collagen, which bind them via a specific motif usually containing the arginine-glycine-aspartic acid (RDG) sequence [88]. Each integrin comprises one α and one β subunit, both of which are highly polymorphic, thus generating considerable diversity among integrins. Transmembrane proteoglycans also interact with matrix constituents such as other proteoglycans or collagen, but not via the RGD motif. Some transmembrane and soluble proteoglycans also act as low-affinity growth factor receptors [89,90] and may stabilise, activate and/or translocate the growth factor to the high-affinity receptor, participating in its dimerisation [91].

The first step of the cell-surface interaction of anchorage dependent cells is attachment, in which the cells retain the round shape they possessed in suspension. Usually, attachment of cells to a surface is then followed by a type of conformational change – known as spreading – in which the cells increase their area in contact with the surface. The spreading is essential for their correct attachment and early start of proliferation [92-95]. The kinetics of attachment and spreading have been studied by allowing the animal cells to attach to the surface of a planar waveguide [96]. Measuring the effective refractive index of the waveguide allows the number of cells per unit area and a parameter uniquely characterising their shape, such as the area in contact with the surface, to be determined.

4. The cell immobilisation process in porous carriers

4.1. IMMOBILISATION KINETICS

The immobilisation kinetics of *Saccharomyces cerevisiae* cells in open-pore sintered porous glass beads have been studied in detail [97-100]. Different techniques can be used to immobilise cells on or in a porous support material. The simplest way is to shake a cell suspension with the carrier, and remove the excess cells afterwards by decantation [30,101-103]. After the immobilisation step, the particles have to be introduced aseptically in the bioreactor, which can be problematic. A more straightforward approach is to immobilise the cells *in situ* in the bioreactor [29,104]: a cell suspension is fed to the bioreactor – containing the sterilised carrier – and after the immobilisation, the free cells can eventually be removed by passage of a washing solution. The feeding of the cells to the bioreactor can be accomplished either in a recirculation or a flow through mode.

The entrapment of hybridoma cells producing IgG in packed beds has been investigated by Murdin and co-workers [105]. Various immobilisation materials were tested: polyester foam, sintered glass, nylon fibre, nylon wool, polyether foam and stainless steel sponge. Cells introduced into the medium reservoir of the reactor were recirculated through the packed bed, and counts of cells remaining in suspension were made over several hours. Polyester foam trapped cells most efficiently, whereas stainless steel wool and sintered glass trapped cells more slowly but had a high capacity. After 24 hours, few cells were observed in suspension in reactors using these materials.

4.1.1. Influence of the flow rate on the immobilisation process

The external fluid flow rate is an important process parameter during the immobilisation process. The physical interpretation and explanation of this influence is not straightforward because the fluid flow rate may affect different physical phenomena at the same time, sometimes in an opposite way. An increase in fluid flow rate is accompanied by an increase in intraparticle velocities, leading to an improved transport of cells from the extraparticle to the intraparticle zone. Moreover, more cells are fed to the column per unit of time, and for a concentration dependent immobilisation rate, a higher biomass loading should be obtained during a certain time interval.

The dependency of the biomass loading on the flow rate for the immobilisation of *S. cerevisiae* in porous glass beads has been studied for the flow through mode [98,99]. In this mode, a well-mixed yeast cell suspension was fed to a packed bed column. At lower fluid flow rates, cells do not easily enter the interior of porous glass beads in a packed bed reactor, and as a consequence mainly the outer part of the matrix is colonised by the cells. If the flow rate through the reactor is too low (i.e. < 1 ml/ml for *S. cerevisiae* cells), column plugging is visually observed since cells are deposited in the interstitial spaces of the packed bed. Higher flow rates lead to enhanced intraparticle fluid velocities, so that the cells can penetrate further into the matrix. More cells are fed to the packed bed per unit of time at higher flow rates, and more cells are immobilised during the same time interval. However, a smaller fraction of quiescent intraparticle zones will be present at high fluid velocities, and adsorbed cells will detach more easily. Eventually, the latter effect gets the upper hand resulting in a decreased biomass loading. A major drawback of the flow through mode is the low immobilisation efficiency: less than 1% of the cells fed at the reactor inlet are immobilised.

The influence of the flow rate on the immobilisation process has also been investigated for the recirculation mode [98,99]. In this case, a yeast cell suspension was well-mixed in a stirred tank reactor and fed to a packed bed column. The outlet cell suspension was recirculated to the stirred tank reactor. Upon immobilisation, the free-cell recirculation concentration decreases exponentially with time. The decrease is steeper when the flow rate is increased. A mathematical model has been developed to predict the immobilisation time and to characterise quantitatively the immobilisation process. This model is based on the following assumptions: (i) plug flow conditions in the packed bed reactor, (ii) the stirred reactor can be regarded as an ideal continuous stirred tank reactor (CSTR), (iii) the cell immobilisation process follows a first order kinetics. The mass balance for biomass over the packed bed column and the stirred tank were written as:

$$\frac{\partial C_X}{\partial t} + u\frac{\partial C_X}{\partial z} = -\frac{1-\varepsilon_{ext}}{\varepsilon_{ext}}\frac{\partial C_{X'}}{\partial t} \qquad (2)$$

$$\frac{dC_{X_{CSTR}}}{dt} = D\left(C_{X_{out}} - C_{X_{CSTR}}\right) \tag{3}$$

where C_X is the biomass concentration, u is the interstitial fluid velocity, z the axial coordinate, ε_{ext} the extraparticle porosity of the packed bed, and D is the dilution rate. The increase of the immobilised cell concentration $C_{X'}$ during the immobilisation process is proportional to the cell concentration: $\partial C_{X'} / \partial t = kC_X$, where k is the first order rate constant for cell immobilisation. The initial and boundary conditions are:

$$\text{packed bed}: \; t = 0, \; z = 0: \; C_X = C_{X_{in}} \; \text{ and } \; z > 0: \; C_X = 0 \tag{4}$$

$$\text{stirred tank}: \; t = 0: \; C_{X_{CSTR}} = C_{X_{CSTR,0}} \tag{5}$$

where $C_{X_{in}}$ and $C_{X_{out}}$ are the biomass concentrations at the column inlet and outlet respectively, $C_{X_{CSTR}}$ is the biomass concentration in the stirred tank and $C_{X_{CSTR,0}}$ is the initial cell concentration in the stirred tank.

The entrapment mechanism of hybridomas in a packed bed using polyester foam cubes as matrix material has been investigated by Murdin *et al.* [105]. Culture medium containing various cell concentrations was pumped downwards at various flow rates through the packed bed in an open (rather than recirculating) system. The packed bed acted as a filter in the way in which it removed cells from the medium. Following an initial period of coating, the bed stabilised and thereafter trapped a constant proportion of the input cells. The proportion of the cells passing through the bed was independent of the concentration of cells applied to the bed, but varied depending on the flow rate through the bed. Immobilisation is more efficient at slower flow rates, in terms of the overall efficiency, allowing the use of an immobilisation protocol to reduce shear effects experienced by hybridomas. The capture of the cells by the packed bed material could be described by using an equation, which described the gravitational capture behaviour of fixed bed filters.

4.1.2. Influence of the initial cell concentration and biomass loading

When a higher initial cell concentration is used in an immobilisation experiment, the cell retention rate is decreased and the immobilisation time is increased [98,99]. As a porous glass particle has only a finite number of immobilisation sites, the ultimate maximum being the void space of the particles; a decreasing number of these sites become available for the free cells. Moreover, the increasing number of immobilised cells impedes the transport of free cells from the interstitial spaces of a packed bed into the porous particles. This means that the immobilisation kinetics is characterised with a variable rate constant k. The influence of the amount of immobilised cells on the parameter k was expressed through a reduction of the intraparticle porosity ε_{int} which expresses the hindered cell transport and the reduction of the available immobilisation sites:

$$\varepsilon_{\text{int}} = \frac{V_{voids} - v_{cell} B_{particle}}{V_{particle}} \tag{6}$$

where V_{voids} is the volume of the intraparticle voids, v_{cell} the specific cell volume, $V_{particle}$ the particle volume and $B_{particle}$ the biomass loading per particle. An empirical correlation for k as a function of ε_{int} was found:

$$k = \frac{A\varepsilon_{\text{int}} - B}{\varepsilon_{\text{int}} - C} \tag{7}$$

where A, B and C are constants.

The low intraparticle fluid velocities, characteristic for porous particles of low permeability, leads to dead intraparticle zones. Consequently, the mechanism of hydrodynamic deposition may not be ignored. Therefore, the immobilisation of yeast cells in a column of porous particles is probably a combined effect of cell attachment by adsorption and cell entrapment in the quiescent zones of the packed bed.

At high external flow velocities, less dead zones for entrapment are available, and the absorbed cells are more prone to detachment due to the higher internal flow rate. It has been shown that at high external flow rates, the internal fluid velocity reaches a value $1\ 10^{-5} - 1\ 10^{-4}$ m/s, which corresponds with the detachment velocity of the yeast cells [98]. At the highest flow rates, the latter effect dominates and results in a low biomass loading.

Based on these immobilisation experiments, the mechanism of yeast cell immobilisation in porous glass beads can be described. Initially, the matrix is free of cells. The initial rate constant for cell immobilisation has the same order of magnitude as the inverse of the time constant for intraparticle convection. As immobilisation proceeds, the intraparticle porosity decreases as a consequence of cell immobilisation. The cells are retained preferentially in pores of small diameters where the intraparticle velocities are small enough to allow attachment or retention. The strong reduction of the rate constant k in the first phase of cell immobilisation reflects the decrease in the number of available "preferential" sites for immobilisation. In the second phase, cell attachment occurs mainly in the larger pores where the velocities are higher and the probability of immobilisation is lower. This is reflected in a lower value of k. At this point, the parameter k is hardly influenced by the number of remaining available sites, although a small decrease may be expected as more cells are immobilised. Additional immobilised cells do not significantly alter the flow path in the larger pores.

Opara and Mann [30] studied the influence of the porosity (56 to 72%) of ultraporous fired-bricks and particle size (0.805 mm to 0.075 mm) on the cell-loading capacity and cell growth inside the support. A batch immobilisation method, where a yeast cell suspension was shaken in the presence of the brick support, was used. It was found that cell saturation of the surface area available within the support matrix was completed within four hours of contact between the cell and the adsorbing surface, especially for the most porous samples. Cell growth was therefore not observed in such cases. However, the less porous samples supported some cell growth upon incubation. Cell hold-up was also observed to increase exponentially when either the porosity was increased or the particle size was decreased. The influence of particle size became insignificant at very high porosities.

4.2. MASS TRANSPORT IN POROUS GLASS BEADS

4.2.1. Apparent effective diffusivity

Dependent on the external flow conditions of a porous particle and the degree of porosity, intraparticle forced convection can have a significant influence on the mass transport in the particle. This influence has been investigated in "diffusivity" experiments where the mass transport is characterised by an apparent effective diffusion coefficient. In these experiments, the diffusion equation could still adequately predict the evolution of the solute profiles when the value of the apparent effective diffusivity was adjusted appropriately.

The effective diffusivity of glucose in porous glass beads has been determined using a transient method [106]. The value of the diffusivity was expected to be lower than the value of the corresponding diffusion coefficient in water, but the opposite was observed. This effect resulted from intraparticle fluid flow, leading to high values of the "apparent" effective glucose diffusivity. To rule out the intraparticle forced convection effect, the pores of the glass beads were filled with Ca-alginate gel (1%). Since glucose can diffuse as freely into the gel as in water, a reduction in effective diffusivity of glucose in the beads as compared to the corresponding diffusivity in water can be entirely attributed to the reduction of the available volume to a fraction ε of the total bead volume and an increase in path length for diffusion by a tortuosity factor τ. The value of the effective diffusivity of glucose was 68% reduced compared to the value for diffusion in pure water. A tortuosity factor of 1.7 was estimated.

Membrane mass spectrometry with reduced sample withdrawal has been used to determine the value of the apparent effective diffusion coefficient of oxygen in a porous glass disc under well-defined experimental conditions [107]. It was shown that in this case the mass transfer of oxygen in porous glass was reduced by 29% compared to the oxygen diffusivity in pure water. The tortuosity factor was estimated and a low value of 1.2 was found which reflects the limited extent of internal convection under the used experimental conditions.

4.2.2. Influence of intraparticle convection on the mass transfer process

Useful information about fluid flow and mass transfer phenomena in porous particles can be obtained directly from the residence time distribution (RTD) by performing stimulus-response experiments: a pulse or step in tracer concentration is applied at the inlet of a packed bed of porous beads and the tracer concentration is measured as a function of time in at least one location of the reactor (usually at the reactor outlet).

The flow pattern in the interstitial spaces can be represented as axially dispersed plug flow [108]. As the fluid velocities inside the porous particles are small compared to the interstitial fluid velocities, the fluid phase inside the porous particles can be considered as a stagnant volume. The solute (tracer or substrate) is exchanged between the extraparticle fluid phase and the stagnant volume by diffusion, convection or a combination of both. But, the mass transfer process can be described by a diffusion equation characterised by an apparent effective diffusivity.

The mass balance of the tracer component in the external fluid phase (with tracer concentration C_1) and in a porous glass bead (with concentration C_2) can be written as [97-99]:

$$\varepsilon_{ext}\frac{\partial C_1}{\partial t}=\varepsilon_{ext}D_{ax}\frac{\partial^2 C_1}{\partial z^2}-\varepsilon_{ext}v\frac{\partial C_1}{\partial z}-(1-\varepsilon_{ext})\varepsilon_{int}\frac{\partial C_{2,av}}{\partial t} \quad (8)$$

$$\varepsilon_{int}\frac{\partial C_2}{\partial t}=D_{e,app}\left(\frac{\partial^2 C_2}{\partial r^2}+\frac{2}{r}\frac{\partial C_2}{\partial r}\right) \quad (9)$$

where ε_{int} is the internal bead porosity, D_{ax} the axial dispersion coefficient in the extraparticle fluid phase, $D_{e,app}$ the apparent effective diffusion coefficient and $C_{2,av}$ the value of C_2 averaged over the porous spherical particle:

$$C_{2,av}=\frac{3}{R^3}\int_0^R C_2 r^2 dr \quad (10)$$

with R the bead radius. Dankwerts boundary conditions were used:

$$z=0:\ \frac{\partial C_1}{\partial z}=\frac{v}{D_{ax}}(C_0-C_1) \quad (11)$$

$$z=L:\ \frac{\partial C_1}{\partial z}=0 \quad (12)$$

where C_0 is the tracer concentration at the column inlet and L the column length. The initial conditions for a step input type experiment are:

$$t=0:\ C_1=C_2=0 \quad (13)$$

Residence time distribution experiments were performed at different liquid flow rates. The axial dispersion coefficient (D_{ax}) and the interstitial velocity (v) were found to be related as:

$$D_{ax}=\frac{vd}{Pe} \quad (14)$$

with d the diameter of the spherical particle.

scaffolds closely resembled normal cartilage histologically, as well as with respect to cell density and tissue composition [2-4].

Table 1b. Examples of cell systems with immobilisation by entrapment within porous pre-formed supports: natural organic polymers, anorganic materials, metallics

Material	Void volume (%)	Cells	Product/objective	Reference
Natural organic polymers				
Collagen (Microsphere™)	75	Animal cells	Biologics	[24-25]
Collagen-glucos-amino-glycan (Informatrix™)	99.5	Animal cells	Biologics	[24-25]
Gelatin (Cultipher™)	50	Animal cells	Biologics	[24-26]
Hyaluronic acid gelatin composites	–	Human cells for tissue engineering	Wound dressing, scaffold	[27]
Sponge		*Saccharomyces cerevisiae*	Ethanol	[28]
Anorganic materials				
Bentonite		Active sludge	Wastewater treatment	[9]
Brick		*Saccharomyces uvarum*	Metabolic activity study	[29]
	56-72	Yeast cells	Cell growth	[30]
		Clostridium beijerinckii	Butanol	[31]
Ceramics (P2-131/R3-130)	36-46	*Saccharomyces cerevisiae*	Ethanol	[32]
Clay		Nitrifying bacteria	Ammonia removal	[33]
Cordierite			Oxygen transfer study	[34]
Diatomaceous earth (Celite, kiezelguhr)		*Acinetobacter calcoaceticus*	Extracellular polysaccharides	[35]
		Burkholderia cepacia	Degradation of 4-nitro- and 4-aminobenzoate	[36]
		Fusarium flocciferum	Phenol degradation	[37]
		Gordona sp. + *Nocardia* sp.	Desulphurization of light gas oil	[38]
	70	*Penicillium chrysogenum*	Penicillin	[39]
		Streptomyces kasugaensis	Kasugamycin	[40]
		Rhodococcus erythropolis	Phenol biodegradation	[41]
		Streptomyces cattleya	Thienamycin	[42]
		Tolypocladium inflatum	Cyclosporin A	[43-46]
		Xanthomonas campestris	Xanthan gum	[47]
Glass (Siran™)	60	Animal cells	Biologics	[25,48,49]
		Hybridoma cells	IgG mAb's	[50]
Saponite		Active sludge	Wastewater treatment	[9]
Sepiolite		Active sludge	Wastewater treatment	[9]
		Saccharomyces cerevisiae	Pesticide determination	[51]
Silica		*Saccharomyces uvarum*	Metabolic activity study	[29]
Zeolite		Active sludge	Wastewater treatment	[9]
Metallics				
Alumina pellets		*Zymomonas mobilis*	Ethanol	[52]
Stainless steel mesh	80	*Saccharomyces cerevisiae*	Ethanol	[53,54]
particles		*Trichoderma reesei*	Cellulase	[55]

The value of the apparent effective diffusivity was obtained by fitting the experimental data to model predictions [98]. A higher fluid flow rate, and as a consequence a higher intraparticle fluid velocity, leads to an increase in mass transfer rate between the extraparticle and intraparticle fluid phase due to intraparticle convection.

The RTD profiles obtained from experiments performed in the presence of immobilised *S. cerevisiae* cells were hardly influenced by the presence of these cells [97]. Only for a high biomass loading at high liquid flow rates was a noticeable change in RTD profiles observed. Computer simulations revealed that due to the presence of immobilised cells, the reduction in mass transfer rate between extraparticle and intraparticle fluid phase is compensated partially by a decrease in intraparticle porosity.

References

[1] Karel, S.F.; Libicki, S.B. and Robertson, C.R. (1985) The immobilization of whole cells: engineering principles. Chem. Eng. Sci. 40: 1321-1354.

[2] Freed, L.E.; Marquis, J.C.; Nohria, A.; Emmanual J.; Mikos, A.G. and Langer, R. (1993) Neocartilage formation in vitro and in vivo using cells cultured on synthetic biodegradable polymers. J. Biomed. Mat. Res. 27: 11-23.

[3] Freed, L.E.; Marquis, J.C.; Vunjak-Novakovic, G.; Emmanual, J. and Langer, R. (1994) Composite of cell-polymer cartilage implants. Biotechnol. Bioeng. 43: 605-614.

[4] Freed, L.E.; Vunjak-Novakovic, G.; Marquis, J.C. and Langer, R. (1994) Kinetics of chondrocyte growth in cell-polymer implants. Biotechnol. Bioeng. 43: 597-604.

[5] Ma, T.; Li, Y.; Yang, S.-T. and Kniss, D.A. (2000) Effects of pore size in 3-D fibrous matrix on human trophoblast tissue development. Biotechnol. Bioeng. 70: 606-618.

[6] Mooney, D.J.; Organ, G.; Vacanti, J.P. and Langer R. (1994) Design and fabrication of biodegradable polymer devices to engineer tubular tissues. Cell Transplant. 3: 203-210.

[7] Lee, D.W.; Piret, J.M.; Gregory, D.; Haddow, D.J. and Kilburn, D.G. (1992) Polystyrene macroporous bead support for mammalian cell culture. Ann. NY Acad. Sci. 665: 137-145.

[8] Affolter, D. and Hall, D.O. (1986) Long-term stability of photosynthetic electron transport in polyvinyl foam immobilised bacteria. Photobiochem. Photobiophys. 12: 193-201.

[9] Borja, R. and Banks, C.J. (1994) Kinetic study of anaerobic digestion of fruit-processing wastewater in immobilized-cell bioreactors. Biotechnol. Appl. Biochem. 20: 79-92.

[10] Gaffney, A.M.; Markov, S.A. (2001) Utilization of cyanobacteria in photobioreactors for orthophosphate removal from water. Appl. Biochem. Biotechnol. 91-93: 185-193.

[11] Lee, Y.H.; Lee, C.W. and Chang, H.N. (1989) Citric acid production by *Aspergillus niger* immobilised on polyurethane foam. Appl. Microbiol. Biotechnol. 30: 141-143.

[12] Oh, Y.S.; Maeng, J. and Kim, S.J. (2000) Use of microorganism-immobilized polyurethane foams to absorb and degrade oil on water surface. Appl. Microbiol. Biotechnol. 54: 428-423.

[13] Niazi, J.H. and Karegoudar, T.B. (2001) Degradation of dimethylphtalate by cells of *Bacillus* sp. immobilized in calcium alginate and polyurethane foam. J. Environ. Sci. Health Part A Tox. Hazard Subst. Environ. Eng. 36: 1135-1144.

[14] Fynn, G.H. and Whitmore, T.N. (1982) Colonisation of polyurethane reticulated foam biomass support particles by methanogen species. Biotechnol. Lett. 4: 577.

[15] Thepenier, C.; Gudin, C. and Thomas, D. (1985) Immobilisation of *Porphyridium cruentum* in polyurethane foams for the production of polysaccharides. Biomass 7: 225-240.

[16] Manohar, S.; Kim, C.K. and Karegoudar, T.B. (2001) Enhanced degradation of naphthalene by immobilization of *Pseudomonas* sp. strain NGK1 in polyurethane foam. Appl. Microbiol. Biotechnol. 55: 311-316.

[17] Bertin, L.; Majone, M.; Di Gioia, D. and Fava, F. (2001) An anaerobic fixed-phase biofilm reactor system for the degradation of low-molecular weight aromatic compounds occurring in the effluents of anaerobic digestors treating olive mill wastewaters. J. Biotechnol. 87: 161-177.

[18] Lindsey, K.; Yeoman, M.M.; Black, G.M. and Mavituna, F. (1983) A novel method for the immobilisation of plant cells. FEBS Lett. 155: 143-149.

[19] Mavituna, F.; Park, J.M.; Wilkinson, A.K. and Williams, P.D. (1987) Characteristics of immobilised plant-cell reactors. In: Webb, C. and Mavituna, F. (Eds.) Plant and animal cells, process possibilities. Ellis Harwood, Chichester, UK; pp. 92-115.

[20] Holden, M.A. and Yeoman, M.M. (1987) Optimisation of product yield in immobilised plant cell cultures. In: Moody, G.W. and Baker, P.B. (Eds.) Bioreactors and biotransformations. Elsevier Applied Science Publishers, London, UK; pp. 1-11.

[21] Lazar, A.; Silberstein, L.; Mizrahi, A. and Reuveny, S. (1988) An immobilized hybridoma culture perfusion system for the production of monoclonal antibodies. Cytotechnol. 1: 331-337.

[22] Muscat, A.; Bettin, A. and Vorlop, K.-D. (1996) Herstellung und Charakterisierung hochporöser und elestischer Silicon-Trägermaterialien mit einer verbesserten Sauerstoffversorgung für immobilisierte Zellen. Chem. Ing. Techn. 68: 584-586.

[23] Knights, A.J. (1993) Porous silicone rubber: a novel material for a matrix to immobilize microbial, mammalian and plant cells. Proceedings I. Chem. E. Event.

[24] Looby, D. and Griffiths, B. (1990) Immobilization of animal cells in porous carrier culture. TIBTECH 8: 204-209.

[25] Cahn, F. (1990) Biomaterials aspects of porous microcarriers for animal cell culture. TIBTECH 8: 131-136.

[26] Ohlson, S.; Branscomb, J. and Nilsson, K. (1994) Bead-to-bead transfer of Chinese hamster ovary cells using macroporous microcarriers. Cytotechnol. 14: 67-80.

[27] Chio, Y.S.; Hong, S.R.; Lee, Y.M.; Song, K.W.; Park, M.H. and Nam, Y.S. (1999) Studies on gelatin-containing artificial skin: II. Preparation and characterization of cross-linked gelatin-hyaluranate sponge. J. Biomed. Mater. Res. 48: 631-639.

[28] Ogbonna, J.C.; Mashima, H. and Tanaka, H. (2001) Scale up of fuel ethanol production from sugar beet juice using loofa sponge immobilized bioreactor. Bioresour. Technol. 76: 1-8.

[29] Monsan, P.; Durand, G. and Navarro, J.-M. (1987) Immobilisation of microbial by adsorption to solid supports. Methods Enzymol. 135: 307-318.

[30] Opara, C.C. and Mann, J. (1987) Development of ultraporous fired bricks as support for yeast cell immobilization. Biotechnol. Bioeng. 31: 470-475.

[31] Lienhardt, J.; Schripsema, J.; Qureshi, N. and Blaschek, H.P. (2002) Butanol production by *Clostridium beijerinckii* BA101 in an immobilized cell biofilm reactor: increase in sugar utilization. Appl. Biochem. Biotechnol. 98-100: 591-598.

[32] Demuyakor, B. and Ohta, Y. (1992) Promotive action of ceramics on yeast ethanol production, a,d its relationship to pH, glycerol and ethanol dehydrogenase activity. Appl. Microbiol. Biotechnol. 36: 717-721.

[33] Shan, H. and Obbard, J.P. (2001) Ammonia removal from prawn aquaculture water using immobilized nitrifying bacteria. Appl. Microbiol. Biotechnol. 57: 791-798.

[34] Kornfield, J; Stephanopoulos, G. and Voecks, G.E. (1986) Oxygen transfer in membrane-ceramic composite materials for immobilized-cell monolithic reactors. Biotechnol. Prog. 2: 98-104.

[35] Wang, S.-D. and Wang, D.I.C. (1990) Mechanisms for biopolymer accumulation in immobilized *Acinetobacter calcoaceticus* system. Biotechnol. Bioeng. 36: 402-410.

[36] Peres, C.M.; Van Aken, B.; Naveau, H. and Agathos, S.N. (1999) Continuous degradation of mixtures of 4-nitrobenzoate and 4-aminobenzoate by immobilized cells of *Burkholderia cepacia* strain PB4. Appl. Microbiol. Biotechnol. 52: 440-445.

[37] Anselmo, A.M.; Cabral, J.M.S. and Novais, J.M. (1989) The adsorption of *Fusarium flocciferum* spores on Celite particles and their use in the degradation of phenol. Appl. Microbiol. Biotechnol. 31: 200-203.

[38] Chang, J.H.; Chang, Y.K.; Ryu, H.W. and Chang, H.N. (2000) Desulfurization of light gas oil in immobilized-cell systems of *Gordona* sp. CYKS1 and *Nocardia* sp. CYKS2. FEMS Microbiol. Lett. 182: 309-312.

[39] Gbewonyo, K. Meier, J. and Wang, D.I.C. (1987) Immobilization of mycelial cells on celite. Methods. Enzymol. 135: 318-333.

[40] Kim, C.J.; Chang, Y.K.; Chun, G.T.; Jeong, Y.H. and Lee, S.J. (2001) Continuous culture of immobilized *Streptomyces* cells for kasugamycin production. Biotechnol. Prog. 17: 453-461.

[41] Prieto, M.B.; Hidalgo, A.; Rodriguez-Fernandez, C. and Serra, J.L. (2002) Appl. Microbiol. Biotechnol. 58: 853-859.

[42] Baker, E.E.; Prevoznak, R.J.; Drew, S.W. and Buckland, B.C. (1983) Thienamycin production by *Streptomyces cattleya* cells immobilized in Celite beads. Dev. Ind. Microbiol. 24: 805-815.

[43] Chun, G.-T. and Agathos, S.N. (1989) Immobilization of *Tolypocladium inflatum* spores into porous celite beads for cyclosporin production. J. Biotechnol. 9: 237-254.

[44] Chun, G.-T. and Agathos, S.N. (1991) Comparative studies of physiological and environmental effects on the production of cyclosporine A in suspended and immobilized cells of *Tolypocladium inflatum*. Biotechnol. Bioeng. 37: 256-265.
[45] Chun, G.-T. and Agathos, S.N. (1993) Dynamic response of immobilized cells to pulse addition of L-valine in cyclosporine A biosynthesis. J. Biotechnol. 27: 283-295.
[46] Lee, T.H.; Chun, G.T. and Chang, Y.K. (1997) Development of sporulation/immobilization method and its application for the continuous production of cyclosporin A by *Tolypocladium inflatum*. Biotechnol. Prog. 13: 546-550.
[47] Nilsson, K.; Buzsaky, F. and Mosbach, K. (1986) Growth of anchorage-dependent cells on macroporous microcarriers. Bio/Technol. 4: 989-990.
[48] Robinson, D.K. (1987) Ph.D. thesis. Massachusetts Institute of Technology, Cambridge, USA.
[49] Looby, D. and Griffiths, B. (1988) Fixed bed porous glass sphere (porosphere) bioreactors for animal cells. Cytotechnol. 1: 339-346.
[50] Reiter, M.; Blüml G.; Zach, N.; Gaida, T.; Kral, G.; Assadian, A.; Schmatz, C.; Strutzenberger, K.; Hinger, S. and Katinger, H. (1992) Monoclonal antibody production using porous glass bead immobilization technique. Ann. NY Acad. Sci. 665:146-151.
[51] Tunceli, A.; Bag, H. and Turker, A.R. (2001) Spectrophotometric determination of some pesticides in water samples after preconcentration with *Saccharomyces cerevisiae* immobilized on sepiolite. Fresenius J. Anal. Chem. 371: 1134-1138.
[52] Koutinas, A.A.; Kanellaki, M.; Lykourghiotis, A.; Typas, M.A. and Drainas, C. (1988) Ethanol production by *Zymomonas mobilis* entrapped in alumina pellets. Appl. Microbiol. Biotechnol. 28: 235-239.
[53] Atkinson, B.; Black, G.M.; Lewis, P.J.S. and Pinches, A. (1979) Biological particles of given size, shape and density for use in biological reactors. Biotechnol. Bioeng. 21: 193-200.
[54] Black, G.M.; Webb, C.; Matthews, T.M. and Atkinson, B. (1984) Practical reactor systems for yeast cell immobilization using biomass support particles. Biotechnol. Bioeng. 26: 134-141.
[55] Webb, C.; Fukuda, H. and Atkinson, B. (1986) The production of cellulase in a spouted bed fermenter using cells immobilised in biomass support particles. Biotechnol. Bioeng. 28: 41-50.
[56] Klein, J. and Ziehr, H. (1990) Immobilization of microbial cells by adsorption. J. Biotechnol. 16: 1-16.
[57] Marshall, K.C.; Stout, R. and Mitchell, R. (1971) Mechanisms of initial events in the sorption of marine bacteria to surfaces. J. Gen. Microbiol. 68: 337-338.
[58] Characklis, W.G. (1981) Fouling biofilm development: a process analysis. Biotechnol. Bioeng. 23: 1923-1960.
[59] Duddridge, J.E. and Pritchard, A.M. (1982) Factors affecting the adhesion of bacteria to surfaces. Microb. Cor. Proc. Conf., Met. Soc. London, 28-25.
[60] Ellwood, D.C.; Keevil, C.W.; Marsh, P.D.; Brown, C.M. and Wardell, J.N. (1982) Surfaces associated growth. Phil. Trans. R. Soc. London. B, 512-532.
[61] Marshall, K.C. (1985) Mechanisms of bacterial adhesion at solid-water interfaces. In: Savage, D.L. and Fletcher, M. (Eds.) Bacterial Adhesion. Plenum Publishing Corp., New York, N.Y., USA; pp. 133-161.
[62] Mozes, N.; Marchal, F.; Hermesse, M.P.; Van Haecht, J.L.; Reuliaux, L.; Leonard, A.J. and Rouxhet, P.G. (1987) Immobilization of microorganisms by adhesion: interplay of electrostatic and nonelectrostatic interactions. Biotechnol. Bioeng. 30: 439-450.
[63] Fletcher, M. and Loeb, G.I. (1979) Influence of substratum characteristics on the attachment of a marine pseudomonad to solid surfaces. Appl. Environ. Microbiol. 37: 67-72.
[64] Büsscher, H.J.; Weerkamp, A.H.; van der Mei, H.C.; Van Pelt, A.W.J.; DeJong, H.P. and Arends, J. (1984) Measurements of the surface free energy of bacterial cell surfaces and its relevance for adhesion. Appl. Environ. Microbiol. 48: 980-983.
[65] Absolom, D.R.; Lamberti, F.V.; Policova, Z.; Zingg, W.; van Oss, C.J. and Neumann, A.W. (1983) Appl. Environ. Microbiol. 46: 90.
[66] Thonard, P.; Custinne, M. and Paquot, M. (1982) Zeta potential of yeast cells: application in cell immobilization. Enz. Microb. Technol. 4: 191-194.
[67] Mozes, N. and Rouxhet, P.G. (1984) Dehydrogenation of cortisol by *Arthrobacter simplex* immobilized as supported monolayer. Enzyme Microb. Technol. 6: 497-502.
[68] Van Haecht, J.L.; Bolipombo, M. and Rouxhet, P.G. (1985) Immobilization of *Saccharomyces cerevisiae* by adhesion: treatment of cells by Al ions. Biotechnol. Bioeng. 27: 217-224.
[69] Mozes, N. and Rouxhet, P.G. (1985) Metabolic activity of yeast immobilized as supported monolayer. Appl. Microbiol. Biotechnol. 22: 92-97.

[70] Champluvier, B.; Kamp, B. and Rouxhet, P.G. (1988) Immobilization of β-galactosidase retained in yeast: adhesion of the cells on a support. Appl. Microbiol. Biotechnol. 27: 464-469.

[71] Fletcher, M. (1988) Attachment of *Pseudomonas fluorescence* to glass and influence of electrolytes on bacterium-substratum separation distance. J. Bacteriol. 170: 2027-2030.

[72] Hattori, R. and Hattori, T. (1985) Adsorptive phenomena involving bacterial cells and an anion exchange resin. J. Gen. Appl. Microbiol. 31: 147-165.

[73] Bar, R.; Gainer, J.L. and Kirwan, D.J. (1986) Immobilization of *Acetobacter acetii* on cellulose ion exchangers: adsorption isotherms. Biotechnol. Bioeng. 28: 1166-1171.

[74] Yoshiota, T. Shimamura, M. (1986) Studies of polystyrene-based ion-exchange fibres. V. Immobilization of microorganisms by adsorption on a novel fibre-form anion-exchanger. Bull. Chem. Soc. Japan 59: 77-82.

[75] Glassner, D.A.; Grulke, E.A. and Oriel, P.J. (1989) Characterization of an immobilized biocatalyst system for production of thermostable amylase. Biotechnol. Prog. 5: 31-39.

[76] Kumakura, M.; Tamada, M.; Kasai, N. and Kaestu, I. (1989) Enhancement of cellulase production by immobilization of *Trichoderma reesei* cells. Biotechnol. Bioeng. 33: 1358-1362.

[77] Zhao Xin, L. and Kumakura, M. (1992) Cellulase activity of *Trichoderma reesei* immobilized on gauze covered with hydrophilic and hydrophobic copolymers. J. Chem. Tech. Biotechnol. 54: 129-133.

[78] Zhao Xin, L. and Kumakura, M. (1993) Immobilization of *Trichoderma reesei* cells on paper covered by hydrophilic and hydrophobic copolymers generated by irradiation. Enzyme Microb. Technol. 15: 300-303.

[79] Zhao Xin, L. and Kumakura, M. (1994) Effect of carrier ionic properties on cellulase productivity by immobilized filamentous *Trichoderma reesei*. J. Chem. Tech. Biotechnol. 60: 183-187.

[80] Zhao Xin, L. and Kumakura, M. (1994) Characterisation of filamentous cells immobilized with ionic-hydrophobic polymers prepared by a radiation polymerization method. Process Biochem. 29: 651-656.

[81] Drucker, D.B. (1981) Microbiological applications of gas chromatography, Cambridge University Press, Cambridge, UK.

[82] Frame, K.K. and Hu, W.-S. (1988) A model for density-dependent growth of anchorage-dependent mammalian cells. Biotechnol. Bioeng. 32: 1061-1066.

[83] Cherry, R.S. and Papoutsakis, E.T. (1989) Modeling of contact-inhibited animal cell growth on flat surfaces and spheres. Biotechnol. Bioeng. 33: 300-305.

[84] Lim, J.H.F. and Davies, G.A. (1990) A stochastic model to simulate the growth of anchorage dependent cells on flat surfaces. Biotechnol. Bioeng. 36: 547-562.

[85] Forestell, S.P.; Milne, B.J.; Kalogerakis, N. and Behie, L.A. (1992) A cellular automaton model for the growth of anchorage-dependent mammalian cells used in vaccine production. Chem. Eng. Sci. 47: 2381-2386.

[86] Hawboldt, K.A.; Kalogerakis, N. and Behie, L.A. (1994) A cellular automaton model for microcarrier cultures. Biotechnol. Bioeng. 43: 90-100.

[87] Freshney, R.I. (2000) Biology of cultured cells. In: Culture of animal cells – A manual of basic techniques. Wiley-Liss, New York, USA; pp. 9-18.

[88] Yamada, K.M. and Geiger, B. (1997) Molecular interactions in cell adhesion complexes. Curr. Opin. Cell Biol. 9: 76-85.

[89] Subramanian, S.V.; Fitzgerald, M.L. and Bernfield, M. (1997) Regulated shedding of syndecan-1 and –4 ectodomains by thrombin and growth factor receptor activation. J. Biol. Chem. 272: 14713-14720.

[90] Yevdokimova, N. and Fresney, R.I. (1997) Activation of pacrine growth factors by heparan sulphate induced by glucocorticoid in A549 lung carcinoma cells. Brit. J. Cancer 76: 261-289.

[91] Schlessinger, J.; Lax, I. and Lemmon, M. (1995) Regulation of growth factor activation by proteoglycans: What is the role of the low affinity receptors? Cell 83: 357-360.

[92] Maroudas, N.G. (1973) Chemical and mechanical requirements for fibroblast adhesion. Nature 244: 253-254.

[93] Grinnell, F. (1978) Cellular adhesiveness and extracellular substrata. Int. Rev. Cytol. 53: 65-144

[94] Curtis, A.S.G. and Pitts, J.D. (1980) Cell adhesion and cell motility. Cambridge University Press, Cambridge, UK.

[95] Griffiths, J.B. and Riley, P.A. (1985) Cell biology: basic concenpts. In: Spier, R.E. and Griffiths, J.B. (Eds.) Animal Cell Biotechnology. Vol. 1, Academic Press, New York, USA; pp. 17-48.

[96] Ramsden, J.J.; Li, S.-Y.; Prenosil, J.E. and Heinzle, E. (1994) Kinetics of adhesion and spreading of animal cells. Biotechnol. Bioeng. 43: 939-945.

[97] De Backer, L. (1994) Porous glass as a cell immobilisation matrix for packed bed bioreactors. PhD dissertation, Vrije Universiteit Brussel, Brussels, Belgium.

[99] De Backer, L. (1996) Immobilisation of cells in porous carriers. In: Willaert, R.G.; Baron, G.V. and De Backer, L. (Eds.) Immobilised living cell systems: modelling and experimental methods. John Wiley & Sons, Chichester, UK; pp. 237-254.

[100] Willaert, R.; De Backer, L. and Baron G.V. (1996) Modelling immobilised bioprocesses. In: Wijffels, R.H.; Buitelaar, R.M.; Bucke, C. and Tramper, J. (Eds) Immobilized cells: basics and applications. Elsevier Science, Amsterdam, The Netherlands; pp. 154-161.

[101] Sitton, O.C. and Gaddy, J.L. (1980) Ethanol production in immobilized cell reactor. Biotechnol. Bioeng. 22: 1735-1748.

[102] Krekeler, C.; Ziehr, H. and Klein, J. (1991) Influence of physicochemical bacterial surface properties on adsorption to inorganic porous supports. Appl. Microbiol. Biotechnol. 35: 484-490.

[103] Tyagi, R.D.; Gupta, S.K. and Chand, S. (1992) Process engineering studies on continuous ethanol production by immobilized *S. cerevisiae*. Process Biochem. 27: 23-32.

[104] Bisping, B. and Rehm, H.J. (1986) Glycerol production by cells of *Saccharomyces cerevisiae* immobilized in sintered glass. Appl. Microbiol. Biotechnol. 23: 174-179.

[105] Murdin, A.D.; Thorpe, J.S.; Kirkby, N.; Groves, D.J. and Spier, R.E. (1987) Immobilisation and growth of hybridomas in packed beds. In: Woody, G.W. and Baker, P.B. (Eds.) Bioreactors and Biotransformations. Elsevier, Amsterdam, The Netherlands; pp. 99-110.

[106] De Backer, L. and Baron, G.V. (1993) Effective diffusivity and tortuosity in a porous glass immobilization matrix. Appl. Microbiol. Biotechnol. 39: 281-284.

[106] De Backer, L. and Baron, G.V. (1994) Residence time distribution in a packed bed bioreactor containing porous glass particles: influence of the presence of immobilized cells. J. Chem. Tech. Biotechnol. 59: 297-302.

[107] Willaert, R.G. and Baron, G.V. (1994) The dynamic behaviour of yeast cells immobilised in porous glass studied by membrane mass spectrometry. Appl. Microbiol. Biotechnol. 42: 664-670.

[108] Levenspiel, O. (1972) Chemical reaction engineering. 2nd edition, John Wiley & Sons, New York, USA.

WHOLE CELL IMMOBILIZATION IN CHOPPED HOLLOW FIBRES

KAMALESH SIRKAR[1] AND WHANKOO KANG[2]
[1]*Otto H. York Department of Chemical Engineering, New Jersey Institute of Technology, Newark, NJ 07102, USA – Fax: 973 642 4854 – Email: sirkar@adm.njit.edu*
[2]*Department of Chemical Engineering, Hannan University, Ojung Dong 133, Daejon, Korea*

1. Introduction

Whole cell-based bioreactors are being increasingly employed in fermentation, waste treatment as well as biomedical processes. For extended periods of whole cell activity at an acceptable level, whole cell immobilization is preferred. There are a number of ways that have been studied/pursued/employed for whole cell immobilization. These include surface attachment, entrapment in porous matrices, self-aggregation and containment behind a barrier [1]. There are advantages and disadvantages of each technique *vis-à-vis* the particular application.

Such bioreactors frequently require separation of products or by-products of the bioreaction processes. If product inhibition exists, local removal of the product is highly beneficial. On occasions, gases have to be supplied and product gases have to be withdrawn. Besides, substrate has to be continuously supplied. Hollow fibre-based bioreactors have been used often in large or small scale to these ends. In addition, hollow fibres acting as barriers have successfully contained whole cells either in the fibre lumen or in the extracapillary space. On rare occasions, the spongy substrate of a hollow fibre has also been used to immobilize whole cells. Further, hollow fibres are being used to exchange O_2 and CO_2 in bioartificial devices like blood oxygenators [2] and bioartificial liver devices [3].

All such functions are carried out with continuous strands of hollow fibres that are potted into tube-sheets at the ends of the bioreactors. Sometimes such diverse functions of the hollow fibres prove demanding. For example, hollow fibre wall rupture has been known to occur due to uncontrolled growth of yeast cells [4]; in addition there are diffusional limitations, gas supply and removal problems, *etc*. To bypass such problems in hollow fibre bioreactors carrying out multiple functions, it is beneficial to decouple appropriate hollow-fibre-based functions in continuous-hollow-fibre-strand-based bioreactors. We developed a novel microporous hollow fibre-based cell immobilization

V. Nedović and R. Willaert (eds.), Fundamentals of Cell Immobilisation Biotechnology, 245-256.

technique that possesses the advantages of matrix entrapment and yet could be efficiently used in a continuous-hollow-fibre-strand-based bioreactor [5-7].

2. Chopped hollow fibre-based immobilization

The technique is briefly as follows. Individual chopped microporous/porous hydrophilic or hydrophobic hollow fibres or bundles of such chopped fibres having a particular length are usually located in the extracapillary space of a hollow fibre bioreactor. The chopped fibre length may vary over a wide range *e.g.* 0.12 – 5 cm; the fibre internal diameter and the fibre wall pore dimensions may also vary greatly. Under appropriate conditions, whole cells are spontaneously immobilized in the lumen and on the outside surface of such chopped hollow fibres after seeding. The fermentation medium is allowed to perfuse through the extracapillary space around such chopped hollow fibres. The medium enters through the device shell side at one end and leaves through the shell side at the other end. Continuous strands of hollow fibres along the length of the bioreactor can carry on a number of different functions: gas supply, gas removal, and product removal by nondispersive solvent extraction, pervaporation, permeation, *etc.* Such continuous strands of hollow fibres can be either porous or nonporous whereas the chopped hollow fibres are invariably porous/microporous. Figure 1 schematically illustrates the general concept of a bioreactor in such a context.

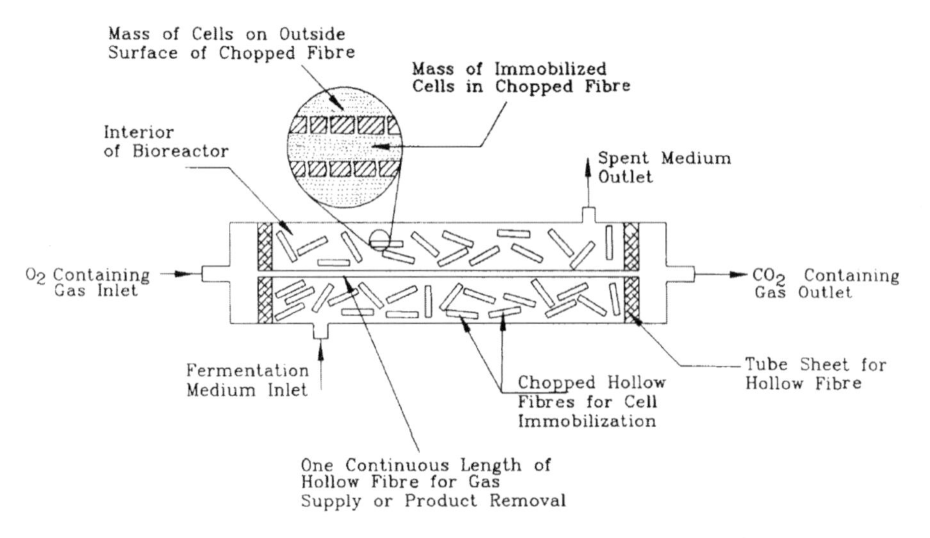

Figure 1. Schematic of bioreactor containing whole cells immobilized in chopped hollow fibres.

Such a system possesses a number of beneficial characteristics. First, the observed cell density that can be achieved in a bioreactor can be quite high. For example, Kang *et al.* had obtained respectively a cell density of 4.5×10^9 cells per unit of chopped fibre volume [6] and 140 g/l of chopped fibre volume [8] in the case of yeast cells and alcohol production. Second, these fermentation processes were carried out for extended periods of time, 130 h in [5], 170 h in [6] and three weeks in [8] with no observed deterioration of the immobilization process. Third, the chopped hollow fibres with immobilized cells are neutrally buoyant when hollow fibres based on polypropylene are used. Fourth, these immobilization structures do not suffer from CO_2-induced bloating nor need medium pH control as in hydrocolloidal gel-based processes [9]. Fifth, the chopped hollow fibre support volume is limited due to the low thicknesses of fibre walls employed traditionally in membrane separations. This is unlike that in conventional porous matrix entrapment techniques. Sixth, these biocatalyst supports are easily biodegradable and do not create a waste disposal problem as diatomaceous earth particles do. Seventh, in the case of polypropylene-based fibres, they have been reused after steam sterilization [5].

Biomass support particles (BSPs) consisting of an interconnecting void within an open network of matrix support material [10] have some similarities to our proposed chopped hollow fibre based systems. Two types of BSPs used, namely spheres of stainless steel knitted mesh and cubes of reticulated polyester foam, are not likely to be compatible with continuous hollow fibre strand-based bioreactors. The rough edges of the BSPs will most likely damage continuous strands of hollow fibres, which are often thin and fragile. The large dimensions of the individual BSP, namely, 6 mm diameter or 6 mm side of a cube [10], will preclude higher density of continuous strands of hollow fibres in the bioreactor. A higher density of continuous strands of hollow fibres in the bioreactor facilitates efficient withdrawal/addition of reactants/gases/products throughout the reactor volume and leads to higher volumetric production efficiency. In bioreactors without any continuous strands of hollow fibres, chopped hollow fibres may be used singly or in bundles [5] as effectively as other BSPs.

Efficient and proper utilization of the chopped hollow fibres for whole cell immobilization requires an understanding of various features/characteristics of the chopped hollow fibres *vis-à-vis* cell growth, cell immobilization, substrate supply/product removal etc. We will briefly describe first some observed characteristics [5-6, 8] in this cell immobilization technology. We will then describe the results of a few studies to provide a fundamental basis for the observed characteristics. The following discussion is based primarily on yeast cells although it has been successfully employed with mammalian cells also [11].

3. Observed characteristics of cell immobilization in chopped hollow fibres

Chopped polymeric hollow fibres having porous/microporous walls can be made of hydrophobic polymers or hydrophilic polymers or may have one side hydrophobic and the other side hydrophilic. Typical examples amongst many of hydrophobic hollow fibres are Celgard® fibres of polypropylene or polysulphone hollow fibres. Cuprophan® is a particular example of a hydrophilic hollow fibre. In shake-flask studies, Shukla *et al.* [5] observed that there was no growth or immobilization of cells of *Saccharomyces cerevisiae* (NRRL Y-132) on the outside surface or in the lumen of chopped Celgard® X-20 polypropylene fibres when these were not wetted with water (conventionally polypropylene is not wetted by water; therefore special wetting techniques have to be adopted to wet the pores in the fibre wall as well as the fibre lumen with water). On the other hand, wetted chopped X-20 fibres showed considerable growth of cells in the fibre lumen as well as on the fibre outside surface. Hydrophobic Celgard® X-20 fibres whose outer surface has been hydrophilized showed cell growth but the hydrophobic lumen did not show any. Further, no cell growth into the micropores of the fibre wall was observed.

Hydrophilic Cuprophan® hollow fibres of regenerated cellulose showed considerable growth of yeast cells on both outside surface and lumen of the fibres [5]. In shake-flask studies, the extent of cell growth calculated as percentage of dry weight of cells in the fibres to the total weight of cells and fibres was found to be significantly less for hydrophilic Cuprophan® fibres compared to the wetted polypropylene Celgard® fibres X-20 and X-10. Cuprophan® fibres have a density much higher than that of the medium unlike the Celgard® fibres; therefore they settle to the bottom reducing the availability of a part of the surface for cell growth. More importantly, the pore size of Cuprophan® fibres (~ 40-50 Å) is at least an order of magnitude smaller than those in Celgard® fibres; thus the resistance to pore transport of substrate and nutrients through Cuprophan® fibres are likely to be much higher leading to reduced cell growth [5].

The dependences of cell growth on the chopped fibre length and fibre diameter have also been studied [5,6]. As time progresses, cells grow from the end of the fibres to the inside of the fibre lumen [6]. Cell growth decreases as the fibre middle section is approached. Length of the chopped fibres was varied from 0.159 cm to 5.08 cm in different studies. The smaller the chopped fibre length, the higher is the extent of cell growth in the fibre. For example, 0.635 cm long fibres of Celgard® X-10 had dry cell weight of 63.30% of the total weight of cells and fibres compared to 31.25% for 1.27 cm long fibres of the same type. Obviously, the diffusional resistance to substrate/nutrient transport along the fibre length from the fermentation medium at the two ends of the chopped fibre is crucial here; the shorter the chopped fibre length, the lower is the transport resistance and the higher is the cell growth [5,6].

As the inside diameter of the chopped fibre decreases, the cell concentration based on unit total fibre volume increases [6]. Chopped fibres of smaller internal diameter are likely to facilitate higher cell growth by providing a higher surface area of cell attachment per unit volume of the chopped fibre. Bundled chopped fibres of Celgard® X-20 of length 2.54 cm with 100 fibres in each bundle were almost as effective in developing cell growth and achieving cell density as loose Celgard® X-20 chopped fibres. Each fibre bundle was tied at the middle. Such fibre bundles after cell growth were washed and employed in ethanol fermentation as the source for immobilized yeast cells. As demonstrated in Shukla *et al.* [5], ethanol fermentation progressed well and when arbitrarily stopped at 90 hours, had an ethanol yield of 0.425 g/g.

As mentioned earlier, the immobilized cell densities achieved in tubular bioreactors in ethanol fermentation were high. Shukla *et al.* [5] achieved a density of 9.36×10^9 cells per unit of fibre lumen volume using wetted 0.635 cm long hydrophobic chopped Celgard® X-20 fibres. Kang *et al.* [6] employed hydrophilic Cuprophan® 0.318 cm long hollow fibres and obtained a similar cell density (~ 9.2×10^9). Kang *et al.* [8] employed similar hydrophilic chopped hollow fibres in a tubular bioreactor having 420 continuous hydrophobic microporous hollow fibres for gas supply and ethanol removal by nondispersive solvent extraction; the immobilized cell density achieved was as much as 300 g cells/l of fibre lumen volume at the reactor entrance decreasing to around 100 at the exit of the 38 cm long reactor. This last reactor, a locally integrated extractive bioreactor, performed successfully with a 300 g/l glucose substrate feed; the immobilization requirements and product separation requirements were locally decoupled everywhere along the reactor. The success with this tubular bioreactor prompted the following question. What dimensions of the chopped fibre diameter and length are beneficial for rapid growth of cell density in the tubular reactor?

4. Modelling immobilization of whole cells in chopped hollow fibre systems

Prediction of the ultimate density of immobilized *Saccharomyces cerevisiae* cells, achieved in chopped microporous hollow fibres located in a bioreactor is a demanding goal. A first step in that direction was taken by Kang [12]. The focus was on predicting the cell density during the early stages of the formation of immobilized cell layers in the fibre lumen and on the surface of the chopped hollow fibres as observed in shake-flask studies. The two regions in the chopped fibre, the outside surface and the fibre lumen, are, however, subject to different dynamics. Cells on the total outside surface of the fibres can be exchanged with those in the bulk solution; cells in the fibre lumen can only be exchanged at the fibre ends with those in the bulk. Prediction of the density of cells attached and growing on the outside surface is going to be different from that trapped in the chopped fibre lumen. Further, the behaviour of cell attachment-detachment *vis-à-vis* a hollow fibre surface is likely to be influenced by growth-nongrowth conditions.

A limited initial effort by Kang [12] considered the development of cell attachment-detachment kinetics on the membrane surface under nongrowth condition. Such a condition is present in a medium without any glucose or yeast extract but having only suspended yeast cells and salts to the extent present in the real fermentation medium. The phenomenon of cell attachment and growth in a monolayer on the outside surface involves a number of steps: attachment, detachment and growth. Adoption of a nongrowth medium allowed potentially the creation of conditions for only attachment-detachment. Assuming Langmuir-type adsorption isotherm, the following adsorption rate equation was used for less than monolayer coverage of the membrane surface (Figure 2a):

$$r_a = k_a c_{xb}\left(1 - \frac{c_{xs}}{c_t}\right) - k_d c_{xs} \quad (1)$$

Here r_a, k_a, c_{xb}, c_{xs}, c_t and k_d are respectively the rate of cell adsorption (g/cm^2/h), rate constant of cell adsorption, bulk cell concentration (g/cm^3), concentration of cell attached to surface (g/cm^2), maximum possible cell concentration on surface for monolayer coverage (g/cm^2) and rate constant for cell desorption. It was experimentally observed *via* scanning electron micrographs (SEMs) of membrane surface after cell adsorption that very few double-layered cells were present. A direct experimental conclusion therefore is that the immobilized cells density is less than that covering the surface as a monolayer.

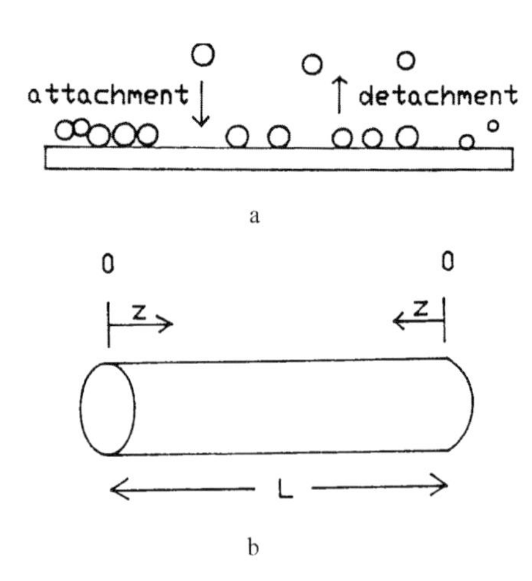

Figure 2. a) Schematic diagram of cell attachment-detachment behaviour; b) coordinate system for a chopped hollow fibre.

To simulate the microporous hollow fibre membrane surface, flat microporous hydrophobic Celgard® 2400 membranes having essentially the same characteristics as the hollow fibres were chosen to experimentally study the density of the cells attached to the fibre outside surface (without any growth). Flat Celgard® 2400 membrane samples 1 cm × 1 cm in dimensions (25 μm thick) were wetted and then transferred into flasks containing a solution without any glucose and yeast extract but containing suspended cells and salts in concentrations present in a regular fermentation medium [12]. For a system where the cells were attached and growing on the surface, the procedure was similar, except a regular fermentation medium containing glucose and yeast extract was used. Detailed experimental procedures for measuring the cell density on the membrane surfaces and cell density in the solution are available in Kang [12].

The behaviour of cell attachment-detachment to wetted flat microporous Celgard® 2400 membrane surface for nongrowth conditions is illustrated in Figures 3 and 4. Figure 3 shows the density of cells attached on the surface as a function of time. For the data of Figure 3, the medium did not have glucose and yeast extract and had only suspended cells and salts; thus cell attachment-detachment phenomenon was isolated from growth. During the experiment, bulk cell concentration was kept at 3 g/l. Around 50 h, the cell density reaches a plateau. The parameters, adsorption constant (k_a) and desorption constant (k_d) of equation 1, may be obtained from the data in Figure 3 by plotting r_a *vs.* c_{xs} (g/cm^2). Figure 4 shows adsorption rate *vs.* the concentration of cells attached on the surface. From Figure 4, one can estimate that k_a is 6.17×10^{-4} (cm h^{-1}) and k_d is 9.495×10^{-4} (h^{-1}).

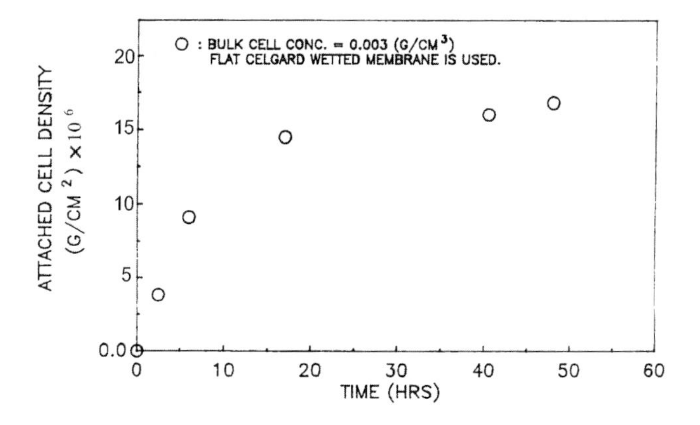

Figure 3. Density of cells attached to Celgard 2400® flat membrane surface as a function of time under nongrowth conditions.

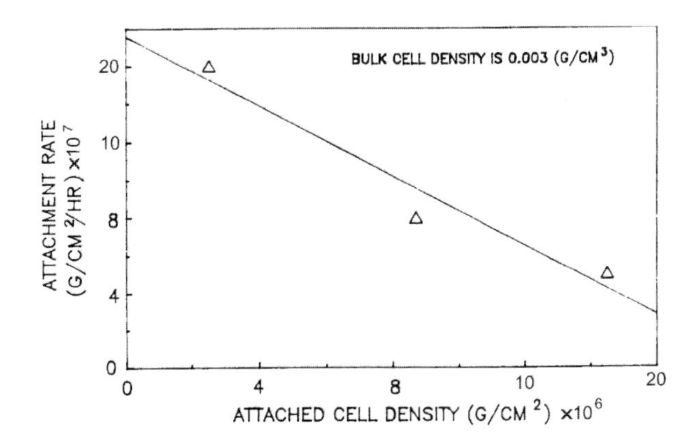

Figure 4. Cell concentration rate vs. the concentration of cells attached to flat membrane surface under nongrowth conditions.

Once these rate constants k_a and k_d are experimentally obtained, this cell attachment and detachment kinetics can be combined with the cell growth kinetic equations for describing the density of cells attached and growing on the outside surface. For the yeast cell growth on fibre surfaces, the Michaelis-Menten kinetic equation with an ethanol inhibition term has been used. Some constants needed in this Michaelis-Menten equation have been obtained by batch studies [12] and are provided at the end of this article in Table 1. The overall governing equations with cells attaching, detaching and growing on the surface are the following:

$$\frac{dc_{xb}}{dt} = \mu_m \left(1 - \frac{c_{pb}}{p_m}\right)\left(\frac{c_{sb}}{K_s + c_{sb}}\right)\left(c_{xb} + c_{xs} f \frac{a}{v}\right) - \left(r_a \frac{a}{v}\right) \quad (2)$$

$$\frac{dc_{sb}}{dt} = \frac{1}{Y_{x/s}} \mu_m \left(1 - \frac{c_{pb}}{p_m}\right)\left(\frac{c_{sb}}{K_s + c_{sb}}\right)\left(c_{xb} + c_{xs}\frac{a}{v}\right) \tag{3}$$

$$\frac{dc_{pb}}{dt} = v_m \left(1 - \frac{c_{pb}}{p'_m}\right)\left(\frac{c_{sb}}{K'_s + c_{sb}}\right)\left(c_{xb} + c_{xs}\frac{a}{v}\right) \tag{4}$$

$$\frac{dc_{xs}}{dt} = \mu_m \left(1 - \frac{c_{pb}}{p_m}\right)\left(\frac{c_{sb}}{K_s + c_{sb}}\right) c_{xs}(1 - f) + r_a \tag{5}$$

where c_{xb}, c_{sb}, c_{pb}, c_{xs}, a, v, and f respectively mean the bulk cell concentration, bulk substrate concentration, bulk ethanol concentration, surface attached cell concentration, membrane surface area, volume of fermentation medium, and the fraction of daughter cells coming back to bulk medium out of total daughter cells generated on the membrane surface. Further $Y_{x/s}$, p_m, K_s, p'_m, K'_s, μ_m, and v_m mean respectively the yield of biomass, ethanol concentration above which cells do not grow, saturation constant for growth, ethanol concentration above which cells do not produce ethanol, saturation constant for product formation, maximum specific growth rate for immobilized cells, and maximum specific ethanol productivity for immobilized cells. The following assumptions have been made in deriving these equations: (1) f is constant during fermentation; (2) bulk concentration of cell, substrate and ethanol are uniform in batch fermentation; (3) for simplicity, the values of maximum specific growth rate and maximum specific ethanol productivity obtained in shake-flask study [12] can be used.

Flat microporous wetted Celgard® 2400 membranes with characteristics similar to hollow fibres were also used experimentally to simulate the density of cells attached and growing on the outside surface of the fibres. By varying the value of f, the value of f can be determined that matches the predictions with experimental data. Once such an f value is obtained, the equations which describe the density of cells attached and growing on the outside surface can be applied to the kinetics of whole cell immobilization in the chopped hollow fibre, which includes fibre the outside surface as well as fibre lumen. Ideally, f will be valid only for the environment from which its value was determined.

It is useful now to turn to observations in solutions containing glucose and yeast extract. From the SEM photographs of cells attached to the flat membrane surface, it appears probable that cells are attached in an evenly distributed manner [12]. Further, the whole membrane surface is not covered over the duration of experiments discounting the possibility of multilayer coverage, except due to growth of a cell. In addition, one is likely to infer from observations of dented scars that these result from the escape of the produced daughter cells to bulk medium. This provides some justification for the use of the parameter f, the fraction of daughter cells coming back to the bulk medium out of the total daughter cells generated on the membrane surface.

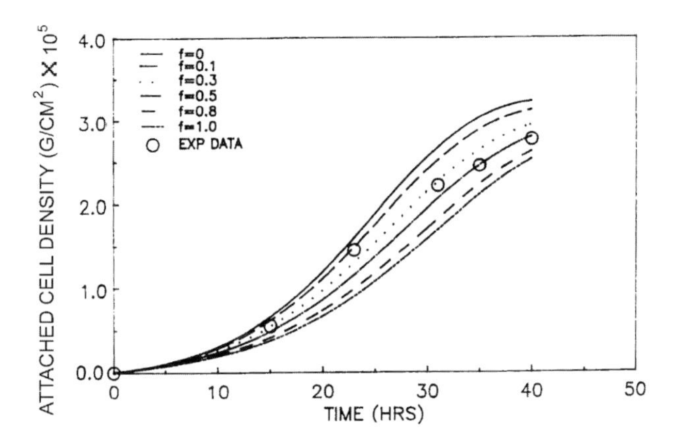

Figure 5. Cell concentration adsorbed on flat Celgard 2400® membrane as a function of time under growth conditions.

The observed variation in cell concentration adsorbed on the flat Celgard® membrane as a function of time is shown in Figure 5 by the circles. The medium containing 100 g/l glucose was inoculated with yeast. The inoculum level was about 5% of total volume. The different lines shown in Figure 5 represent predicted values obtained by solving equations 1-5 with different f values. It appears that an average value of f = 0.3 described the data without great error; thus f = 0.3 may be used in modelling, provided similar conditions are maintained in all experiments.

One can now using appropriate assumptions proceed towards developing a model to determine the overall cell density in the chopped hollow fibre located in a fermentation medium. The configuration of a single chopped hollow fibre of total length L is shown in Figure 2b. A number of simplifying assumptions were employed:

- There is no convective flow through the lumen in spite of outside flow fluctuations.
- There is no radial concentration gradient of substrate, product and cells inside the fibre lumen at any z-coordinate location. Essentially, we are dealing with a radially averaged concentration.
- Neglect transmembrane transport of solute. The presence of a cell layer on the outside surface would consume most of the substrate that would have otherwise diffused into the lumen.
- Cells diffuse axially only *via* the cell concentration gradient.
- Each fibre has its own surrounding compartment for mass balance and diffusion. There is no interaction between the compartments of neighbouring fibres.
- Neglect the cohesiveness of the cells for the time being: no clumping together of the cells exists.

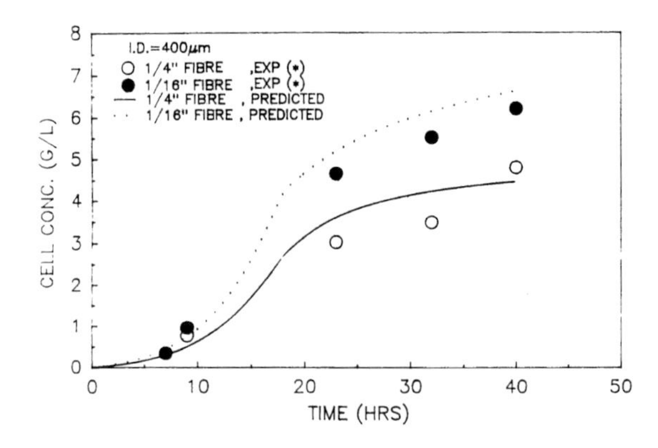

Figure 6. Length dependence of yeast cell growth in chopped hollow fibres with times.
[*] Reference [6].

Some of these assumptions need to be evaluated. However, these allow the development of appropriate differential equations, for example, for the time-dependant behaviour of bulk substrate concentration, substrate concentration in the lumen side, bulk cell concentration and cell concentration in the lumen side. Couple these with the rate equation (1) for surface adsorption-desorption of cells as well as axial diffusion equations for c_s, c_p, and c_x inside the chopped fibre lumen. Kang [12] has summarized these equations and provided a solution of these equations.

Briefly illustrated here is one of the model results, namely, the variation of the total cell concentration in a chopped hollow fibre with time and its dependence on the fibre length. Figure 6 illustrates the data from 0.159 cm and 0.635 cm long chopped hollow fibres in batch culture [6] using wetted hydrophobic Celgard® fibres. It appears that the model predictions of the cell concentration based on the fibre lumen volume are quite close to the experimentally observed values. The variation of cell growth with length is obvious. As the length of the chopped hollow fibre is decreased, a large fraction of the lumen space of the fibre is being used by yeast cells.

The diffusional resistances along the fibre length to the supply of substrate and cells from the bulk medium at two ends of a chopped fibre are very likely to be critical to the cell growth in the fibre lumen. SEM photographs [12] taken at different distances from the fibre end also show that the cell population decreases along the length in the fibre lumen of a 2.54 cm long fibre. Predictions from the model equations of the cell concentration at any location inside the fibre show an increase with time; however for fibres of length 2.54 cm, the middle part of the fibre lumen is not at all used. On the other hand, the cell concentration on the outside of the fibre surface is same irrespective of the fibre length. The parameters employed in the model simulations are provided in Table 1. Note: D_s, D_p and D_x are respectively the diffusion coefficients for the substrate, product and the cell.

Table 1: The values of parameters used in the model.

Parameter	Value
μ_m	0.22 (h^{-1})
v_m	1.3 (h^{-1})
K_s	0.476 (g/l)
K'_s	0.666 (g/l)
p_m	87 (g/l)
p'_m	114 (g/l)
$Y_{x/s}$	0.065 (g cell/g substrate)
c_t	1.333×10^{-4} (g/cm^2)
k_a	6.17×10^{-4} (cm/h)
k_d	9.495×10^{-2} (h^{-1})
D_s	6.9×10^{-6} (cm^2/s)
D_p	1.28×10^{-5} (cm^2/s)
D_x	9.5×10^{-9} (cm^2/s)

5. Concluding remarks

Chopped porous/microporous hollow fibres singly or in bundles provide a convenient structure for spontaneous whole cell immobilization in the hollow fibre lumen and on the outside surface of the hollow fibre. The shorter the fibre length, the higher is the immobilized cell density in the fibre lumen. Smaller fibre diameters facilitate the achievement of a higher cell density due presumably to a higher surface area per unit fibre lumen volume. Porous hydrophobic chopped fibres can be used for cell immobilization only when wetted by water. In tubular bioreactors, chopped porous hydrophilic as well as hydrophobic fibres are convenient vehicles for achieving a high immobilized-cell density. Experimentally observed adsorption-desorption characteristics of yeast cells on the surface of microporous membranes have been described by a Langmuir-type adsorption for nongrowth conditions. For growth conditions, an estimate of an empirical factor *f*, which is the fraction of the daughter cells coming back to the bulk medium out of total daughter cells generated on the support membrane's surface, has been obtained to describe the observed cell growth behaviour on a microporous membrane. This factor and the adsorption rate equation are essential to developing a model to describe the initial cell growth with time. The model predictions of the dependence of cell growth with chopped fibre length describe the major observed characteristics well. Short length and small diameter of chopped fibres yield higher immobilized cell density. Chopped fibres are likely to be quite useful for cell immobilization in bioreactors with or without long continuous strands of hollow fibres.

References

[2] Strain, A.J. and Neuberger, J.M. (2002) A bioartificial liver – State of the art. Science 295(5557): 1005-1009.

[3] Jasmund, I.; Langsch, A.; Simmoteit, R. and Bader, A. (2002) Cultivation of primary porcine hepatocytes in an Oxy-HFB for use as a bioartificial liver device. Biotechnol. Prog. 18(4): 839-846.

[4] Inloes, D.S.; Taylor, D.P.; Cohen, S.N.; Michaels, A.S. and Robertson, C.R. (1983) Ethanol production by *Saccharomyces cerevisiae* immobilized in hollow-fibre membrane bioreactors. Appl. Environ. Microbiol. 46: 264-278.
[5] Shukla, R.; Kang, W.K. and Sirkar, K.K., (1989) Novel hollow fibre immobilization techniques for whole cells and advanced bioreactors. Appl. Biochem. Biotechnol. 20/21: 571-586.
[6] Kang, W.K.; Shukla, R. and Sirkar, K.K., (1990) Novel membrane-based immobilization technique for bioreactors. Ann. N. Y. Acad. Sci. 589: 192-202.
[7] Sirkar, K.K. and Shukla, R.K. (1996) Improved hollow fibre immobilization. U.S. Patent 5,510,257.
[8] Kang, W.K.: Shukla, R. and Sirkar, K.K. (1990) Ethanol production in a microporous hollow-based extractive fermentor with immobilized yeast. Biotechnol. Bioeng. 36: 826-833.
[9] Scott, C.D. (1987) Techniques for producing monodispersed biocatalyst beads for use in columnar bioreactors. Ann. N.Y. Acad. Sci. 501: 487-493.
[10] Black, G.M.; Webb, C.; Matthews, T.M. and Atkinson, B. (1984) Practical reactor systems for yeast cell immobilization using biomass support particles. Biotechnol. Bioeng. 26: 134-141.
[11] Shukla, R.K. (1993) Personal communication; Immunomedics Inc.
[12] Kang, W.K. (1990) Hollow fibre membrane-based extractive bioreactors and a whole cell immobilization technique. PhD Dissertation. Chem. and Chem. Eng. Stevens Institute of Technology.

USE OF VIBRATION TECHNOLOGY FOR JET BREAK-UP FOR ENCAPSULATION OF CELLS AND LIQUIDS IN MONODISPERSE MICROCAPSULES

CHRISTOPH HEINZEN[1], ANDREAS BERGER[1] AND IAN MARISON[2]
[1]Inotech Encapsulation AG, Kirchstrasse 1, CH-5605 Dottikon, Switzerland – Fax: +41 56 624 29 88 – Email: heinzen@inotech.ch
[2]Laboratory of Chemical and Biochemical Engineering, Swiss Federal Institute of Technology (EPFL), CH-1015 Lausanne, Switzerland – Fax: +41 21 693 31 94 – Email: ian.marison@epfl.ch

Summary

Applying a vibration on a laminar jet for controlled break-up into monodisperse microcapsules is one among different extrusion technologies for encapsulation of animal and plant cells, microbes, enzymes and liquids [1].

The vibration technology is based on the principle that a laminar liquid jet breaks up into equally sized droplets by a superimposed vibration. In the late 19th century, Lord Rayleigh theoretically analysed the instability of liquid jets [2]. He showed that the frequency for maximum instability is related to the velocity of the jet and the nozzle diameter.

The optimal vibration parameters are easily and quickly determined in the light of a stroboscope. Once determined, the parameters can be reset in the future, making the process highly reproducible.

Optimal production parameters for beads and capsules including monodispersity parameter for narrow size distributions and examples of encapsulation of cells and liquids in microcapsules will be described.

1. Introduction

1.1. OVERVIEW OF DIFFERENT IMMOBILISATION TECHNOLOGIES

Microcapsules are defined as particles, spherical or irregular, in the size range of about 1 μm to 1 mm, and composed of an excipient polymer matrix (shell or wall) and an incipient core substance [3]. Depending on material, manufacturing technique and application, microcapsules are divided into different classes: solid particles are

V. Nedović and R. Willaert (eds.), Fundamentals of Cell Immobilisation Biotechnology, 257-275.

dispersed in a matrix, while droplets are encapsulated either in mono- or polynuclear form. A second wall layer can be coated around the capsules to modify its permeability and stability.

Based on different criteria immobilisation techniques are divided in two major groups (see also Figure 1) [4]:

- Immobilisation based on binding at a carrier
- Immobilisation based on encapsulation in matrices or membranes

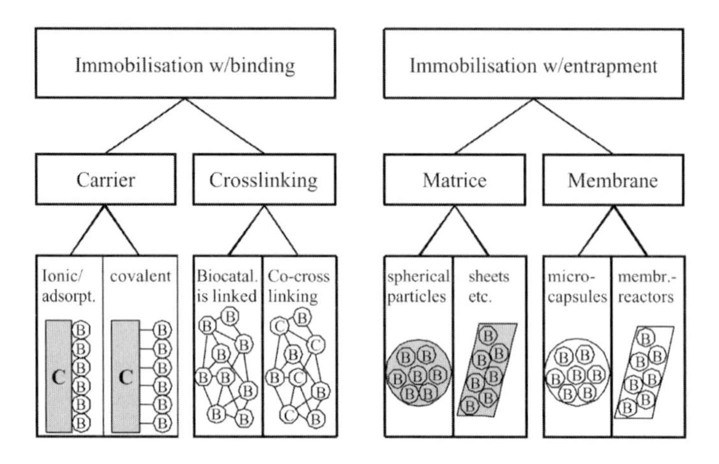

Figure 1. Overview of different immobilisation techniques [4].

Microcapsules have been prepared by a variety of different methods that are often combined or overlap to a certain extent. For the sake of simplicity, the main microencapsulation techniques are categorised as mechanical (see below), physico-chemical (simple and complex coacervation, phase separation) or chemical operations (*in-situ* polymerization, interfacial polycondensation).

Mechanical microencapsulation operations result rather from mechanical procedures than from a well defined physical or chemical phenomenon. Various coating and spray drying methods are routinely used in industry. Extrusion of polymer solutions through nozzles to produce either beads or capsules is mainly used on laboratory scale, where often, simple devices such as syringes are applied. If the droplet formation occurs in a well-controlled way (contrary to spraying) the technique is known as prilling. This is preferably done by pulsation of the jet or vibration of the nozzle [5,6]. The use of coaxial air flow [7] or an electrostatic field [8] is other common techniques to form droplets. Mass production of beads can either be achieved by multi-nozzle systems [9], rotating disc atomiser [10] or by the recently developed jet-cutting technique [11]. Centrifugal systems using either a multi-nozzle system or a rotating disk have also been developed for the mass production of microcapsules [12].

This review focuses on the sterile cell encapsulation production technology based on vibrating nozzles, whereby the cells or the enzyme are mixed with a hydrogel, as *e.g.* alginate, and this suspension is dropped in a gelation solution, *e.g.* $CaCl_2$ bath. Of course, the presented technology can also be used for other polymer systems, such as gelatine, organic polymers *etc*. In such cases, the polymer must be thermostated for bead formation.

1.2. NOZZLE EXTRUSION TECHNOLOGIES FOR PRODUCTION OF BEADS AND CAPSULES

For the formation of beads or capsules and the encapsulation of cells the following production technologies based on the extrusion of a liquid through a nozzle or an orifice are used:

a) Simple dripping (Figure 2A)
Dripping is the simplest technology for production of beads: a polymer – cell suspension is dropped from a capillary or an orifice (O) to a gelation solution (G) for the formation of beads. Due to its simplicity this technology is widely used. The important disadvantage of this technology is the fact, that beads or capsules are formed with a very large diameter of 2 – 5 mm, which is too large for many biotechnological or medical applications.

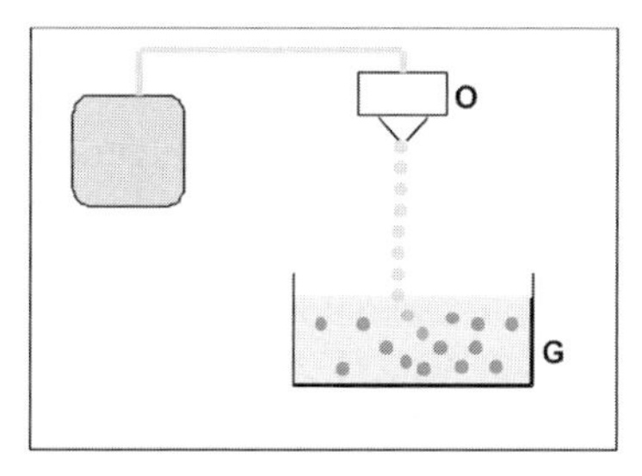

A: Simple dripping

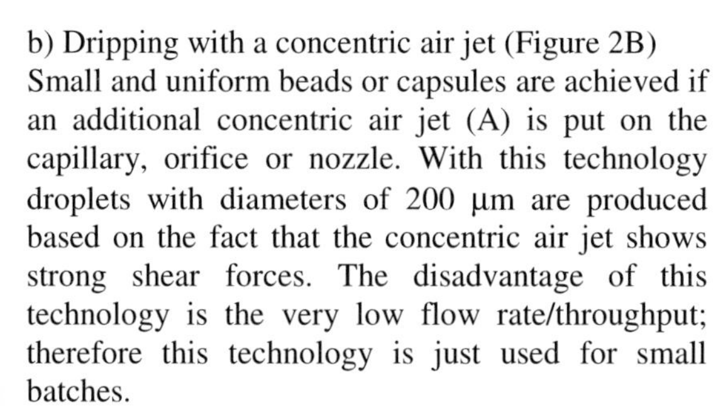

b) Dripping with a concentric air jet (Figure 2B)
Small and uniform beads or capsules are achieved if an additional concentric air jet (A) is put on the capillary, orifice or nozzle. With this technology droplets with diameters of 200 µm are produced based on the fact that the concentric air jet shows strong shear forces. The disadvantage of this technology is the very low flow rate/throughput; therefore this technology is just used for small batches.

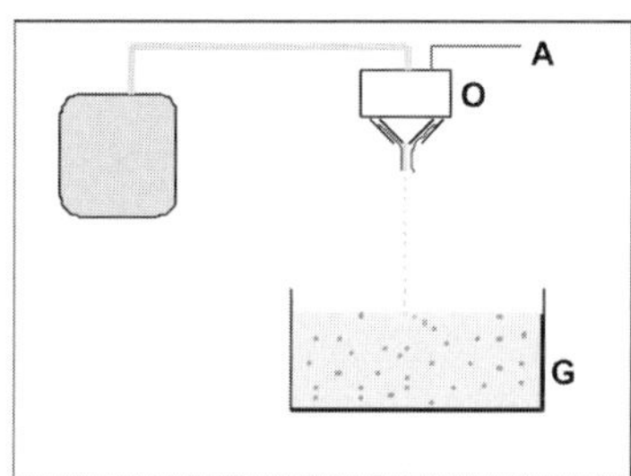

B: Concentric air jet

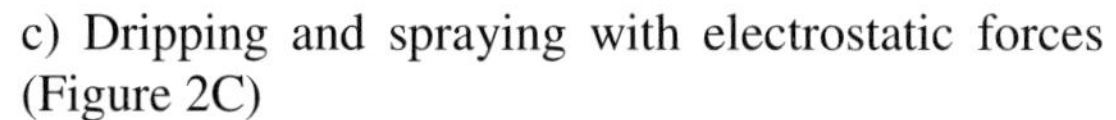

c) Dripping and spraying with electrostatic forces (Figure 2C)
Another technology to obtain small beads or capsules is the application of electrostatic forces. The capillary, orifice or nozzle and the gelation bath are put in opposite charges, thus this force forms the beads. Similar to technology b) the flow rate here is very low and this technique is just used for small batches. The potentials used are between 5 and 25 kV. The distance between the orifice and the surface of the gelation bath has to be constant in order to get constant production conditions.

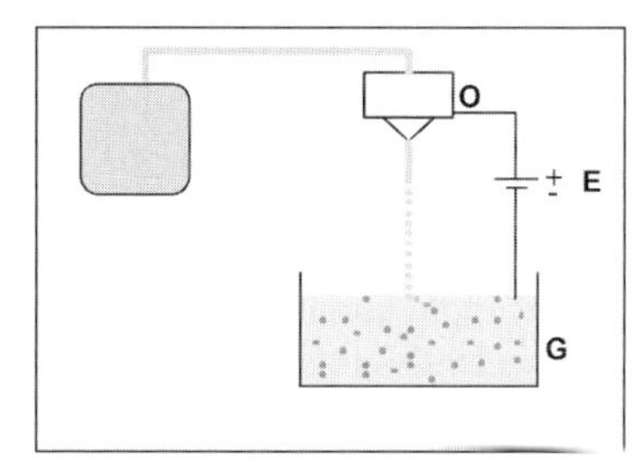

C: Electrostatic forces

d) Rotating disk and jet cutter technologies (Figure 2D)
For the production of larger batches of beads and capsules this technology is widely used. Droplets are formed either with a rotating nozzle (RN) or with a rotating disk based on centrifugal forces. Another possibility is that a stationary jet is cut with a rotating device (JC), described by Prüsse [11].

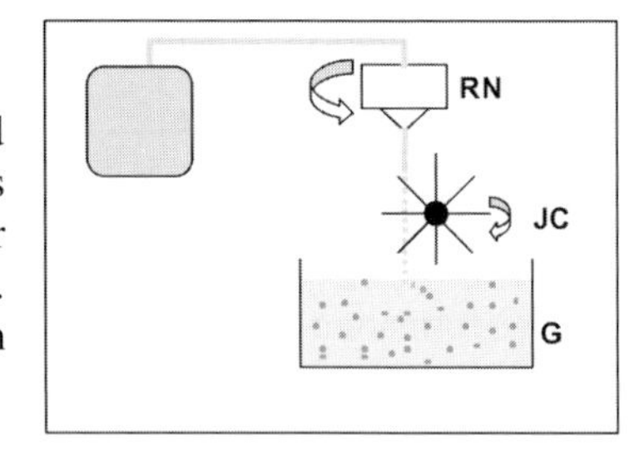

D: Rotating atomizers

e) Vibration technology (Figure 2E)
If one need uniform, monodisperse and small beads or capsules and an up-scalable technology, the vibration technology is the answer. Based on a vibration that is applied to the nozzle (VN) the laminar jet breaks down into droplets of equal sizes. The advantage of this process is the reproducible control of bead formation, the application under sterile conditions and the possibility of easy scale-up. This technology is described in details in the following sections.

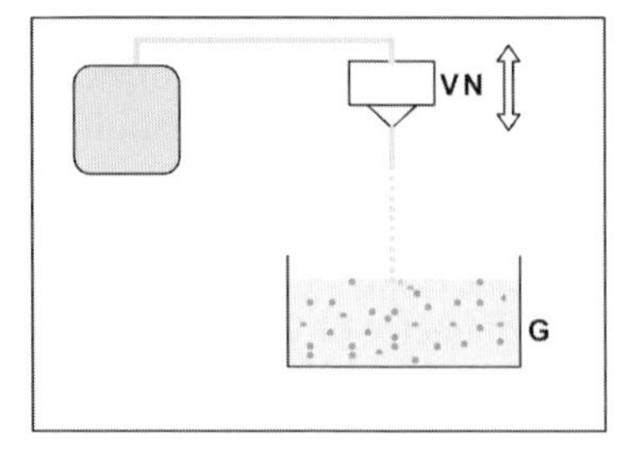

E: Vibration

Figure 2. Extrusion techniques for bead/capsule production.

1.3. OVERVIEW ON DIFFERENT APPLICATIONS OF ENCAPSULATION

Why are cells and other ingredients encapsulated? The encapsulation of human, animal and plant cells, microbes, enzymes and drugs or other ingredients into microbeads may be performed for different reasons: the encapsulated item has to be protected from environmental influences, *e.g.* immune system of a patient, or it has to resist shear forces in a bioreactor or has to be protected against oxidation. Following these demands, the encapsulation technology has its use in many different applications. A selection of known applications produced with the vibration technology is presented in Table 1.

Especially interesting are medical applications of encapsulation utilised in cell therapies. Many diseases are caused by the inability of the body to produce the necessary amount of a specific molecule, such as a hormone, growth factor, or enzyme. Cell therapy offers an enormous potential for the treatment of such diseases. Encapsulated cell systems consisting of living cells immobilised and protected inside micro- or macrocapsules, are implanted into a patient, where the cells produce the required therapeutic substances. A selection of the most important diseases that could be treated by encapsulated cells [13] is:

Alzheimer's, ALS, affective disorders, Huntington's, hypoparathyroidism, haemophilia, anaemia, enzymatic defects, liver failure, syringomyelia, infertility, atherosclerosis, muscular dystrophy, wound healing, AIDS, cancer, diabetes, kidney failure, spinal cord injuries, chronic pain, strokes, dwarfism, epilepsy, Parkinson's.

Table 1. Applications of vibration technology

Type of encapsulation	Selection of known applications	Advantages
Encapsulation of "living" material (Figure 2A)	Cell transplantation for treatment of cancer, diabetes or liver diseases Production of drugs with animal cells or microbes Starter cultures within food production *In-vitro* test systems HTS	Protection against immune system, shear forces *etc.* Confined genetic drift High cell density Higher productivity Cell retention system Bead = mini-fermentor Bead offers 3D structure
Encapsulation of liquids (Figure 3B)	Extraction of organic compounds in fermentation and in cosmetic processes Stereoselective reactions Release of organic compounds for cosmetic and food applications	Increase of product conc. Solvent extraction without stable emulsions and direct contact of solvent Use of less organic solvent Low cost extraction process Easy handling
Capsules with special features	Taste masking Binding of compounds Specific transition of compounds	Controlled release Sudden release Induced release Optimised storage

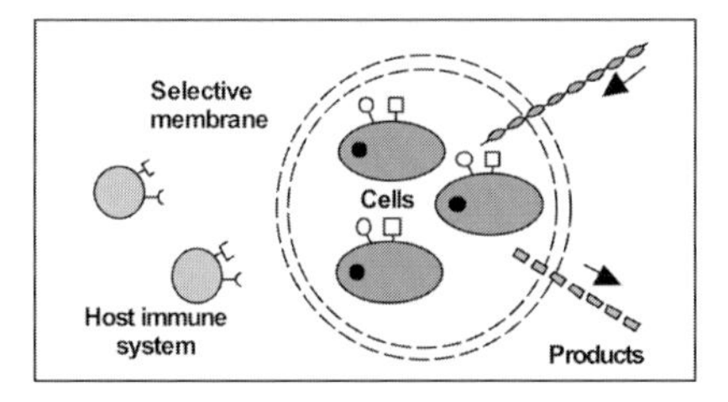

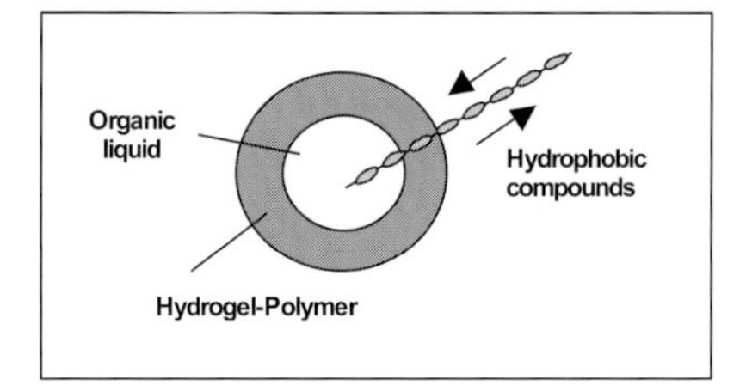

Figure 3. Encapsulation of: A) cells B) liquids.

1.4. IMPORTANT FEATURES OF CELL ENCAPSULATION

In most cases beads and capsules with immobilised cells must satisfy several requirements:

- Small diameter, preferentially less than 0.7 mm in order to prevent diffusion limitations and necrotic areas in the centre of the beads. Cells must be provided with nutrients otherwise cells in the centre of such beads die and they send necrotic messages to living cells [14].

- Narrow size distribution of the diameters: by using a suboptimal production technique large droplets are produced and they create diffusion limitations and send necrotic messages to living cells (see above)
- Short production time: if the production time - formation of droplets and their solidification - exceeds a certain time, the viability of the cells decreases dramatically.

All these demands are fulfilled by the vibration technology for production of encapsulated cells in beads or capsules.

The vibration technology has the following potentials and advantages:

Versatile technology	Flavours, fragrances, enzymes, cells, vitamins *etc.* as solid or liquid compound to be encapsulated in different polymer matrices (*e.g.* gelatine, alginate, PVA, chitosan, cellulose) resulting in capsules or beads stored as dry particles or in a storage liquid produced with the same technology
Uniform microbeads	Spherical shape, narrow size distribution resulting in good granular flow
Selectable bead size	Bead diameters from 0.1 mm to 2 mm
Sterile production	Technology designed for sterile production
High efficiency	No loss of product material and full viability of encapsulated living material
Easy scalable	From 10 g up to several 100 kg/batch

2. Technical background

2.1. MECHANISMS OF DROPLET FORMATION

As mentioned above, this chapter focuses on the production of beads and capsules by using the vibration technology.

Five different mechanisms of droplet formation occur at the nozzle outlet as a function of the jet outflow velocity (Figure 4). They arise through the interaction of gravity, impulse, surface tension and friction forces [15]. At a low outflow velocity ($v < v_a$), single droplets are directly formed at the orifice outlet (mechanisms 0 and I). The droplet diameter d_D is calculated in equation (1) from the equilibrium of the main forces present, gravity and surface tension, whereas d_N is the diameter of the nozzle (m), $\Delta\rho$ is the density difference (kg/m^3) and σ is the surface tension (N/m):

$$d_D = \sqrt[3]{\frac{6 d_N \sigma}{g \Delta\rho}} \tag{1}$$

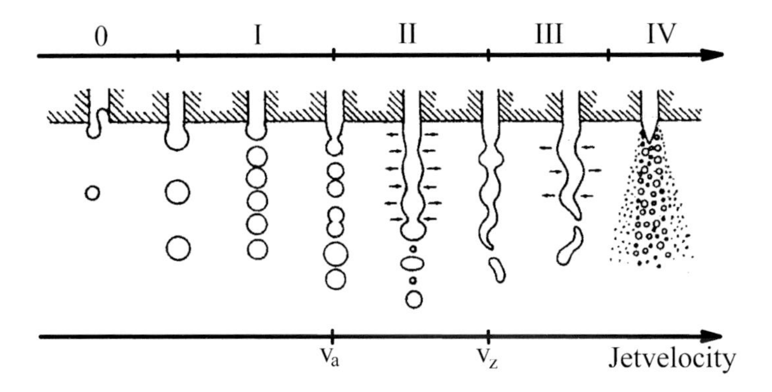

Figure 4. Different mechanisms of droplet formation [15].

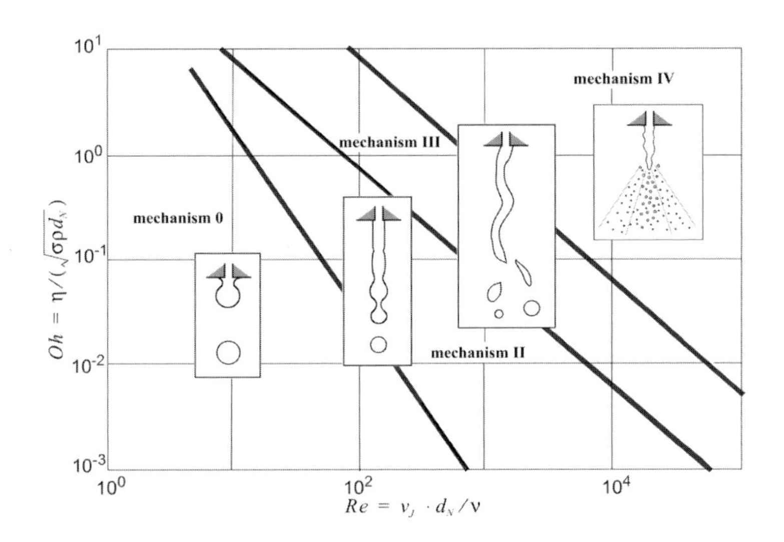

Figure 5. Ohnesorge number versus *Reynolds number (Oh / Re) graph for optimal working conditions [25].*

Increasing the kinetic force causes the uninterrupted outflow of the jet, which brakes up afterwards by axial symmetrical vibrations and the surface tension (mechanism II). A further increase of the jet velocity ($v > v_z$) leads to statistical distribution of the droplet size, that is caused either by spiral symmetrical vibrations (mechanisms III) or by the high friction forces that are present, when the jet is sprayed (mechanism IV). The mechanisms of droplet formation as function of operating conditions presented by Reynolds and Ohnesorge numbers are shown in Figure 5.

2.2. DROPLET FORMATION WITH VIBRATION

The controlled formation of monodisperse droplets is only possible *via* mechanisms I and II. However, only the latter is of industrial importance. By introducing an additional sinusoidal force on the droplet formation mechanism II, the laminar jet breaks up in droplets, which size can be freely chosen in a certain range depending on the applied frequency. This is in contrast to the droplet formation without pulsation described in equation 1, where the droplet diameter of a certain liquid depends only on the nozzle diameter. The sinusoidal force can be applied to the prilling system by the following methods:

- Vibrating of the nozzle
- Periodic changes of the nozzle / orifice diameter
- Pulsation of the polymer-cell suspension

The choice of the optimal method is dependent on the used system. For liquid-liquid systems, it could be shown that the pulsation of the liquid is the best one [5,6,16]. For bead formation in a gas phase – as it will be discussed in this chapter – all methods can be applied.

The technology is based on the principle that a laminar liquid jet is broken into equally sized droplets by a superimposed vibration. In the late 19th century, Lord Rayleigh theoretically analysed the instability of liquid jets. He showed that the frequency for maximum instability is related to the velocity of the jet and the nozzle diameter (Figure 6) whereas d_D is the droplet diameter (m); d_N is the nozzle diameter (m); v_J is the jet velocity (m/s); f is the frequency (Hz); λ_{opt} is the optimal wavelength (m); η is the dynamic viscosity (kg/m s); ρ is the density (kg/m^3) and σ is the surface tension (N/m). The optimal vibration parameters can be determined in the light of the incorporated stroboscope. Once determined, the parameters can be reset in the future, making the process highly reproducible. The bead diameter can be set between 0.1 - 1.5 mm.

$$\lambda = \frac{v_J}{f} \quad (2)$$

$$\lambda_{opt} = \pi\sqrt{2}d_N\sqrt{1+\frac{3\eta}{\sqrt{\rho\sigma d_N}}} \quad (3)$$

$$d_D = \sqrt[3]{1.5 d_N^2 \lambda_{opt}} \quad (4)$$

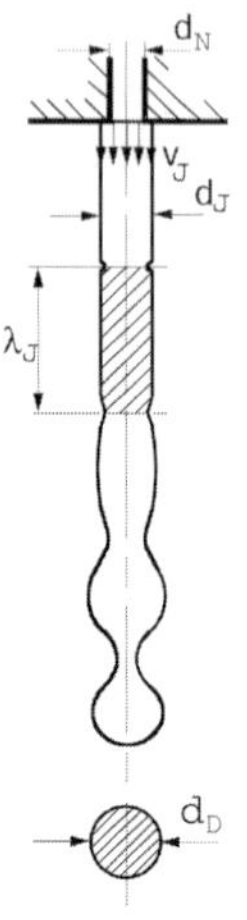

d_N v_J d_J λ_J d_D

Figure 6. Droplet formation parameters [17].

Based on the equations (2) to (4) and referring to the optimal production conditions for bead formation under the assumption of a given nozzle diameter, there are two parameters to be determined: the frequency, f; and the jet velocity, v_J. These two parameters trigger the production conditions and they have to be optimised in a certain range. Figure 7 shows optimal production conditions for bead formation by the vibrating process [18]. The optimal range of stable bead formation (c) is limited by the following factors:

a) Too small jet velocity, no jet formation
b) Too small vibration frequencies resulting in satellite droplet formation
d) Too large vibration frequencies resulting in an unstable droplet formation
e) Too large jet velocity, resulting in a wavy jet

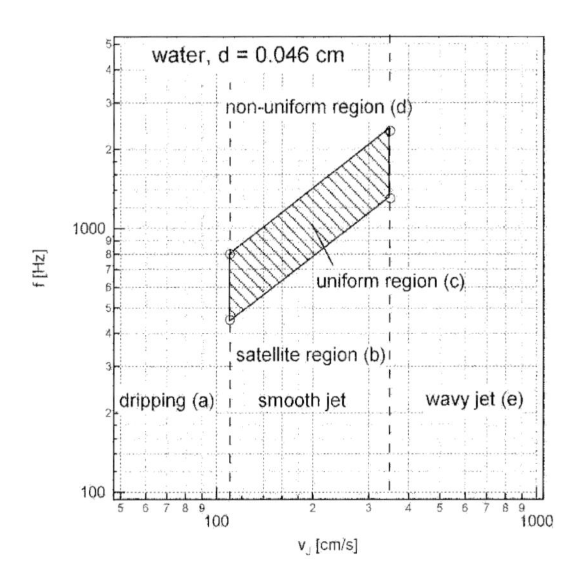

Figure 7. Mechanism of droplet formation as functions of vibration frequency (f) and jet velocity (v_J) for water and d = 0.046 cm (d = nozzle diameter) [18].

Figure 8 shows the bead diameter as a function of the vibration frequency for different flow rates calculated by equation 2 and 4. Lower flow rates corresponding to lower pumping rates produce smaller beads. Higher vibration frequencies produce smaller beads too.

2.3. DISPERSION OF DROPLETS

The production of uniform beads or capsules is a key factor in encapsulation processes. To avoid large size distributions due to coalescence effects during the flight and the hitting phase at the surface of the hardening solution the use of the dispersion unit with the electrostatic dispersion unit is essential (Figure 9). Electrostatic fields are used on the one hand for droplet formation (see section 2.2c) and on the other hand for separation of droplets, *e.g.* used in a FACS device (fluorescence activated cell sorting [19] or the bubble jet printer [20]).

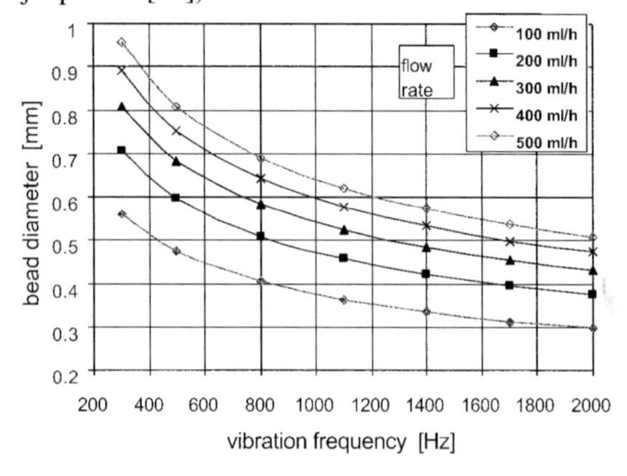

Figure 8. Influence of the vibration frequency and the flow rate on the bead diameter [Data Inotech].

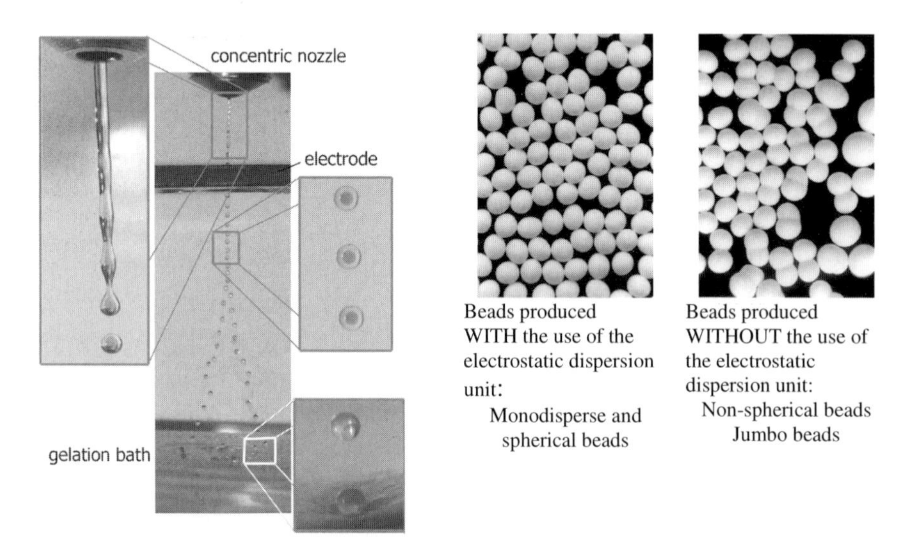

Figure 9. Preventing of coalescence by using an electrostatic dispersion unit [Data Inotech].

If droplets are formed by the break-up of a jet based on vibration technology they fly through an electrostatic field and are charged. As a result these droplets don't hit each other during the flight and are spread over a larger surface of the gelation bath thus resulting in monodisperse beads. Droplet formation without an applied potential of the electrostatic generator shows several particles of double or triple size caused by coalescence [21].

3. Materials and methods

3.1. ENCAPSULATION DEVICES

3.1.1. Lab-scale devices

All the lab scale experiments were performed with a lab scale device, described in detail in Heinzen [6]. With such a device it is possible:

- To encapsulate cells, microorganisms, enzymes and drugs (liquid or solid) without any significant loss of cell viability.
- To produce beads or capsules with equal diameters between 100 up to 2000 μm.
- To set diameter of these beads reproducibly by an external production parameter.
- To achieve beads or capsules with a very narrow size distribution (below 4%), due to the highly precise syringe pump and the Inotech bead dispersion system.
- To work under sterile working conditions (reaction vessel can be autoclaved).
- To get production rates from 500 up to 3000 beads per second depending on the bead diameter.
- To process very little amounts of polymer-ingredient solutions (encapsulation devices show low dead volume of approximately 300 μl).
- To work with different encapsulation polymers, *e.g.* alginate, cellulose sulphate, gelatine, organic polymers.

The lab devices are designed to work either with a single or a concentric nozzle. With the single nozzle "homogenous" beads with equally distributed cells or ingredients over the whole cross-section could be obtained, *e.g.* there are cells in the centre and at the surface of the bead.

With the concentric nozzle it is possible to produce capsules with a defined core region (having the cells or ingredient) and a shell region with "pure" polymer. The resulting capsules show therefore a liquid core.

The main parts of the Encapsulation lab scale device with a concentric nozzle are shown in Figure 10. All parts of the instrument, which are in direct contact with the capsules, can be sterilised by autoclaving. The product to be encapsulated (hydrophobic or hydrophilic liquid) is put into the syringe (1) or the product delivery bottle (2). The shell material (polymer) is put into the syringe (3) or the product delivery bottle (4). Both liquids are forced to the concentric nozzle by either a syringe pump (S) or by air pressure (P). The liquids then pass through a precisely drilled concentric-sapphire-nozzle (7) and separate into concentric, equal sized droplets on exiting the nozzle. These droplets pass an electrical field between the concentric nozzle (7) and the electrode (8) resulting in a surface charge. Electrostatic repulsion forces disperse the capsules as they drop to the hardening solution.

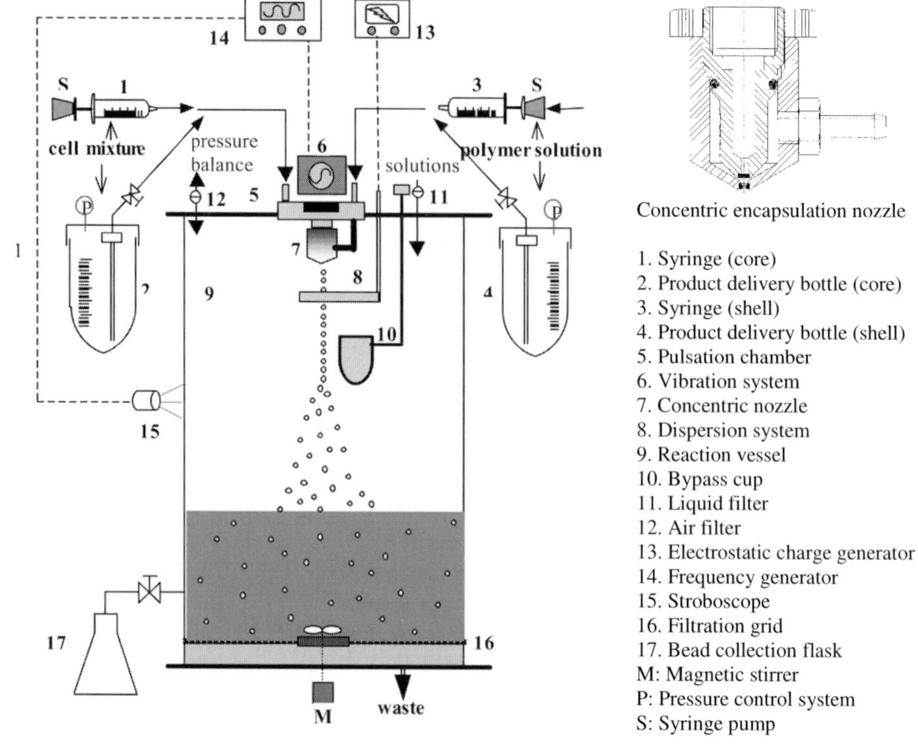

Figure 10. Schematic representation of the concentric nozzle lab scale encapsulation system [Data Inotech].

Capsule size is controlled by several parameters including the vibration frequency, nozzle size, flow rate, and physical properties of the polymer-product mixture. Optimal parameters for capsule formation are indicated by visualisation of real-time droplet formation in the light of a stroboscope lamp (15). When optimal parameters are reached, a standing chain of droplets is clearly visible.

Once established, the optimal parameters can be pre-set for subsequent capsule production runs with the same encapsulating polymer-product mixture. Poorly formed capsules, which occur at the beginning and end of production runs, are intercepted by the bypass-collection-cup (10). Depending on several variables, 50 – 4000 capsules per second are generated and collected in a hardening solution within the reaction vessel (9). Solutions in the reaction vessel are continuously mixed by a magnetic stirrer bar (M) to prevent capsule clumping. At the conclusion of the production run, the hardening solution is drained off (waste port), while the capsules are retained by a filtration grid (16). Washing solutions, or other reaction solutions, are added aseptically through a sterile membrane filter (11). The capsules can be further processed into capsules with additional membranes, or transferred to the capsule collection flask (17). A commercially available lab scale encapsulation device for monodisperse production of beads and capsules under sterile and reproducible production conditions is presented in Figure 11.

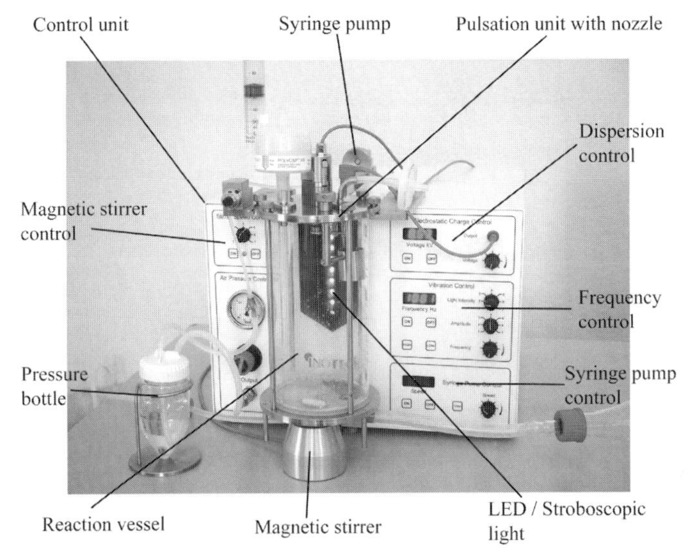

Figure 11. Major parts of a commercially available lab scale encapsulation device [Data Inotech].

3.1.2. Pilot – and industrial scale devices

Scale up of the vibration technology is "simple": due to the fact that the maximum flow rate per nozzle is limited, a scale up is the multiplication of the single nozzle configuration, *e.g.* if the flow rate per nozzle is 0.5 l/h, a multinozzle plant with 20 nozzles gives a through put of 10 l/h. The only challenge is that each nozzle of the multinozzle plant must show similar production conditions, meaning equal flow rate, frequency and amplitude. Comparing to other granulation and bead formation technologies the vibration technology has an important advantage: scale-up directly from lab to industrial scale is possible, *e.g.* from one nozzle to 200 nozzles without doing experiments in pilot scale, *e.g.* 20 nozzles.

The main parts of the Encapsulation Multinozzle plant are shown in Figure 12. The encapsulation procedure of producing microbeads and microcapsules with the Encapsulation Multinozzle plant is similar to the process with one nozzle of a lab scale Encapsulator (see section 4.1.1). Beads and capsules produced with the Encapsulation Multinozzle device show the same features as beads and capsules produced with the lab scale device.

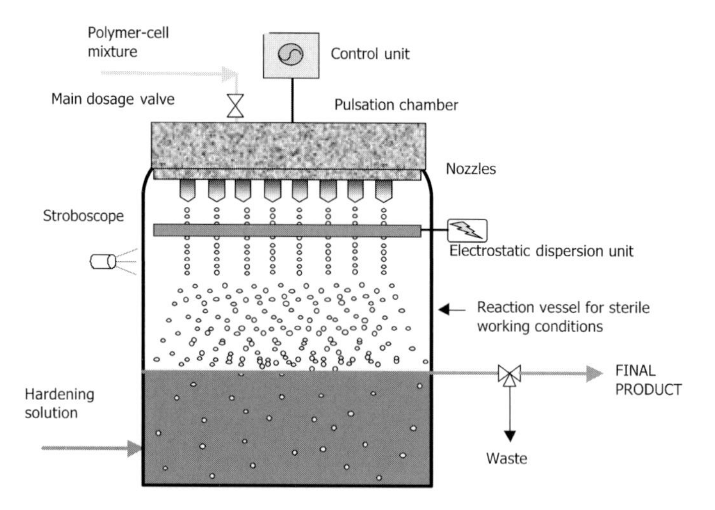

Figure 12. Major parts of Encapsulation Multinozzle device [Data Inotech].

3.3. TEST SOLUTIONS AND ANALYTICS

In general, all experiments were performed with 1.5% Na-alginate (IE-1010, Inotech Encapsulation AG, Switzerland). The hardening bath was a 0.1 M solution of $CaCl_2$ (IE-1020, Inotech Encapsulation AG, Switzerland). The determination of bead and capsules size, size distribution of the capsules, mechanical resistance and pore size are described in [22]. Detailed encapsulation protocols are described elsewhere for CHO cells [23], and for liquid core capsules [24].

4. Results

4.1. OPTIMAL WORKING CONDITIONS (SINGLE NOZZLE)

The parameters, which influence the bead and capsules formation process, are shown in Table 2. For further information and mathematical considerations please consult the work of Brandenberger [25] showing all the above-mentioned input parameters and their impact on the bead and capsule formation process including some mathematical models.

Of course, there are some additional input parameters like *e.g.* the distance between the nozzle and a gelation bath or the speed of a stirrer. These parameters are strongly depending on the encapsulation polymer used and no general rule can be determined.

Figure 13 shows the working limits of a 1.5% alginate solution (100 mPa s viscosity) for different nozzle sizes. Within the lines optimal bead formation is possible and depending on the applied frequency and flow rate, a defined bead size diameter is generated.

Table 2. Major production parameters influencing bead formation process.

Input parameter	**Impact on bead / capsule formation process**
Nozzle diameter	Defines bead diameter range: Bead diameter is approx. twice as big as the nozzle diameter.
Flow rate or speed of polymer-ingredient mixture	Too little flow rate results in wetting of the nozzle Too large flow rate results in spraying, secondary droplets and uncontrolled bead formation
Material parameters as viscosity, density and surface tension	Density and surface tension show minimal impact on bead formation. Viscosity is a major player: Due to the fact that the vibration technology is theoretically based on liquids with Newton fluid dynamics, liquids with a dynamic viscosity larger than 300 mPa s won't be able to be separated with a 0.5 mm vibrating nozzle into droplets. Of course this rule depends on the nozzle size and working temperature.
Frequency	For a given nozzle diameter, material parameters and flow rate, there is a range of +/- 20% to 40% of the calculated optimal frequency based on equation 2 and 3. If frequencies used are not within this range, no influence on bead formation will be seen.
Amplitude	Based on the Rayleigh's theory the amplitude has no impact on bead formation. We know from our experience that liquids with a larger viscosity need a larger amplitude compared to *e.g.* water.

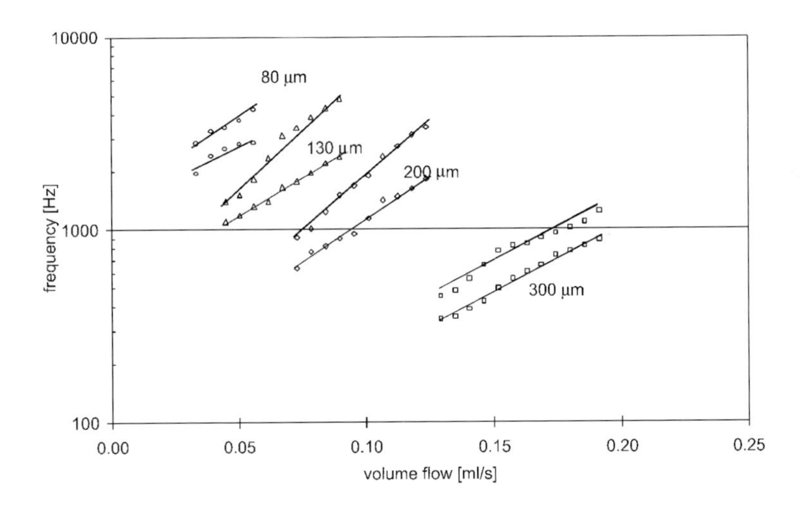

Figure 13. Optimal bead formation working limits for different nozzle diameters [25].

Figure 14 shows alginate beads after gelation if a too high flow rate was applied [25]: a lot of distorted beads, twin beads, bead-chains and "spaghettis". Figure 15 (left) shows alginate beads produced with the lab scale encapsulation unit. They show a very narrow size distribution (2.9%), due to the use of the dispersion unit. Figure 15 (right) displays the multinozzle encapsulation plant. In the stroboscopic light uniform bead formation can be observed.

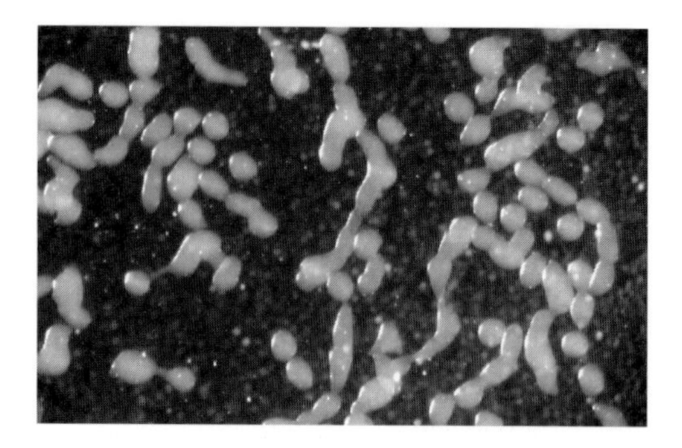

Figure 14. Bead produced at too high flow rates (speed of jet: 6 m/s; nozzle diameter: 0.2 mm) [25].

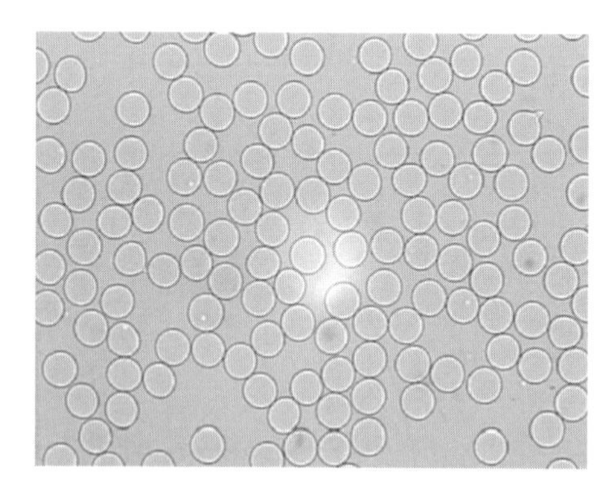

Figure 15. Alginate beads produced with the vibration technology (left). Optimal working conditions with a multinozzle encapsulation unit (2 x 17 nozzles). Device shows very uniform bead formation conditions (right) [Data Inotech].

4.2. OPTIMAL WORKING CONDITIONS (CONCENTRIC NOZZLE)

Referring to the table given in section 5.1 (input parameters and their impact on bead formation), there are following additional input parameters for the use of concentric nozzles:

- Material parameters and flow rate of the second liquid.
- Geometry of the nozzle, nozzle diameter ranges.

Figure 16 shows the working limits for different core (sun flower oil) and shell (4% wt alginate solution) flow rates and applied frequencies. Within the dashed lines, stable capsule formation could be measured. As easily can be seen from Figure 16, the flow rate of the shell liquid has the major impact on the working limits. Please find in-depth details in Berger [26].

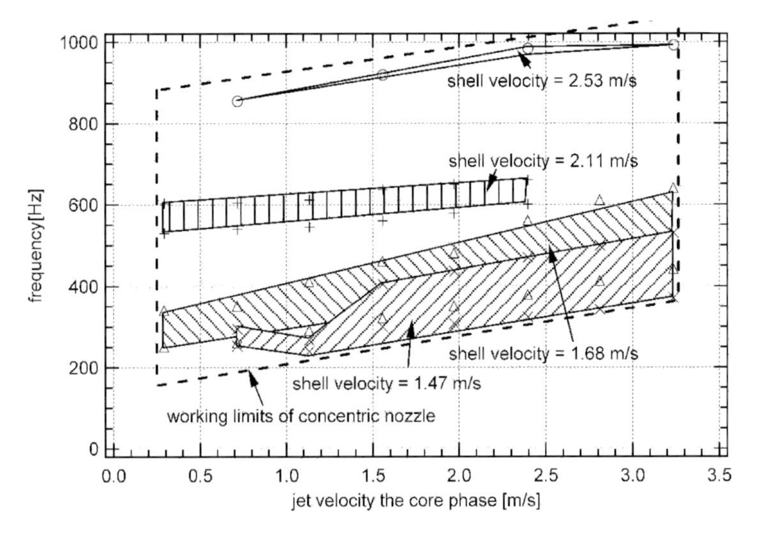

Figure 16. Working limits for a concentric nozzle [26].

Figure 17 displays one example of the use of a vibrating concentric nozzle: gelatine capsules with a liquid core of neutral oil.

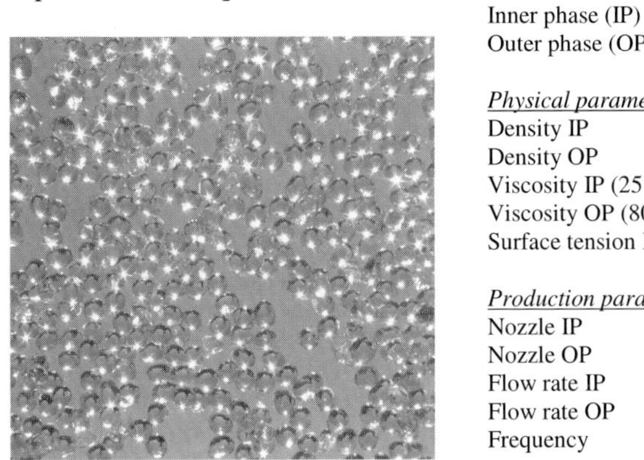

Figure 17. Liquid core capsules made of gelatine with a liquid core of neutral oil [Data Inotech].

Inner phase (IP)	=	neutral oil
Outer phase (OP)	=	10 % gelatine solution
<u>Physical parameters</u>		
Density IP	=	917 kg / m^3
Density OP	=	1020 kg / m^3
Viscosity IP (25°C)	=	27 m Pa s
Viscosity OP (80°C)	=	90 m Pa s
Surface tension IP/OP	=	30.0 dyne / cm
<u>Production parameters</u>		
Nozzle IP	=	0.6 mm
Nozzle OP	=	1.0 mm
Flow rate IP	=	308 ml/h
Flow rate OP	=	4160 ml/h
Frequency	=	305 Hz
<u>Bead parameter (dry capsules)</u>		
Diameter inner bead	=	0.93 mm
Diameter outer bead	=	0.99 mm
Standard deviation	=	9.5%

More examples of encapsulation results made with the vibration technology are listed in this book within the chapter of Marison *et al.* "liquid core capsules for applications in biotechnology" [27]. Different polymer-ingredient systems and applications were already studied and performed resulting in dry microcapsules made of:

- Shell matrix polymers like alginate, gelatine, pectin, agarose, chitosan, cellulose derivatives, starch and their derivatives, paraffin and other organic polymers, acrylic amides, waxes based on modified vegetable polymers, PVA [Data Inotech].
- Core compounds like coffee oil, sunflower oil, citrus fruit oil, vitamins, enzymes, cells (yeast, microbes, animal cells), drugs, flavour and fragrances [Data Inotech].

5. Conclusions

With the vibration technology it is possible to reproducibly produce beads and capsules in spherical shape with diameters set in range from 2 mm down to 0.1 mm with narrow size distribution. The technology provides work under sterile conditions and easy scale-up to flow rate of 200 l/h. In this manner it is possible to encapsulate islets, human T-cells, fibroplasts, hepatocytes, CHO and many other cells without viability loss as well as to encapsulate liquids and other ingredients in different polymer matrices.

The most important advantage of the vibration technology as compared to other immobilization techniques is the possibility of an exact determination of droplet (and thus bead or capsule) diameter and their narrow size distribution. A limitation of the vibration technology is a limited use of liquids showing a too large viscosity.

References

[1] Poncelet, D.; Bugarski, B.; Amsden, B.G.; Zhu, J.; Neufeld, R. and Goosen, M.F.A. (1994) A parallel-plate electrostatic droplet generator - parameters affecting microbead size. Appl. Microbiol. Biotechnol. 42: 251-255.

[2] Lord Rayleigh (1878) On the stability of jets. Proc. London Math. Soc. 10: 4-13.

[3] Gutcho, M.H. (1979) Microcapsules and other capsules: advances since 1975. Noyes Data Corporation, New Jersey.

[4] Hartmeier, W. (1986) Immobilisierte Biokatalysatoren. Springer-Verlag, Berlin.

[5] Kegel, B.H.R. (1988) Mikroverkapselung durch Hydroprillierung. Thesis, Swiss Federal Institute of Technology, Zurich.

[6] Heinzen, Ch. (1995) Herstellung von monodispersen Mikrokugeln durch Hydroprillen. Thesis, Swiss Federal Institute of Technology, Zurich.

[7] Lee, G.M. and Palsson, B.O. (1993) Stability of antibody productivity is improved when hybridoma cells are entrapped in calcium alginate beads. Biotechnol. Bioeng. 42: 1131-1135.

[8] Poncelet, D. (1997) Fundamentals of dispersion in encapsulation technology. In: Godia, F. and Poncelet, D. (Eds.) Proceedings of the Int. Workshop Bioencapsulation VI, Barcelona, Spain, August 30-September 1, 1997; UAB Barcelona; T1.1, pp. 1-4.

[9] Brandenberger, H. and Widmer, F. (1998) A new multinozzle encapsulation / immobilization system to produce uniform beads of alginate. J. Biotechnol. 63: 73-80.

[10] Goodwin, J.T. and Somerville, G.R. (1974) Microencapsulation by physical methods. CHEMTECH 4: 623-626.

[11] Prüsse, U.; Bruske, F.; Breford, J. and Vorlop, K. D. (1998) Improvements to the jet cutting process for manufacturing spherical-particles from viscous polymer solutions. Chem. Ing. Tech. 70: 556-560.

[12] Somerville, G.R. and Goodwin, J.T. (1980) Microencapsulation using physical methods. In: Kydonieus, A.F. (Ed.) Controlled release technologies: methods, theory and applications. CRC press, Boca Raton; pp. 155-164.

[13] Kühtreiber, W.M.; Lanza, R.P. and Chick, W.L. (Eds.) (1999) Cell encapsulation technology and therapeutics. Birkhäuser, Boston.

[14] Vorlop, K.-D. and Klein, J. (1982) New developments in the field of cell immobilization formation of biocatalysts by inotoropic gelation. In: Enzyme technology. Springer, Berlin; pp. 219-235.

[15] Müller, C. (1985) Bildung einheitlicher Feststoffpartikel aus Schmelzen in einer inerten Kühlflüssigkeit (Hydroprilling). Thesis, Swiss Federal Institute of Technology, Zurich.

[16] Buettiker, R. (1978) Erzeugung von gleich grossen Tropfen mit suspendiertem und gelöstem Feststoff und deren Trocknung im freien Fall. Thesis, Swiss Federal Institute of Technology, Zurich.

[17] Weber, C. (1931) Zum Zerfall eines Flüssigkeitsstrahls. Zeit für angewandte Mathematik und Mechanik, 11: 136.

[18] Sakai, T.; Sdakata, M.; Saito, M.; Hoshino, M. and Satoshi, S. (1985) Uniform size droplets by longitudinal vibration of newtonian and non-newtonian fluids. ICLASS-82: 37-45.

[19] Bonner, W.A.; Hulett, H.R.; Sweet, R.G. and Herzenberg, L.A. (1972) Fluoroescence activated cell sorting. Rev. Sci. Instr. 43: 404-409.

[20] Sweet, R.G. (1964) High frequency recording with electrostatically deflected ink jets. Rev. Sci. Instr. 36: 131-136.

[21] Brandenberger, H.; Nüssli, D.; Piëch, V. and Widmer, F. (1999) Monodisperse particle production: a method to prevent drop coalescence using electrostatic forces. J. Electrostatics 45: 227-238.

[22] Serp, D.; Catana, E.; Heinzen, C.; von Stockar, U. and Marison, I.W. (2000) Characterization of an encapsulation device for the production of mono-disperse alginate beads for cell encapsulation. Biotechnol. Bioeng. 70: 41-53.

[23] Pernetti, M.; Gugerli, R.; von Stockar, U.; Heinzen, C.; Marison, I.W. and Annesini, M.C. (2001). Animal cell encapsulation within polyelectrolyte and covalent membranes. Proceedings of the IX Int. BRG Workshop, Warsaw, May 11-13 2001; S. VI-3, pp. 1-4.

IMMOBILIZATION OF CELLS AND ENZYMES USING ELECTROSTATIC DROPLET GENERATION

BRANKO M. BUGARSKI[1], BOJANA OBRADOVIC[1], VIKTOR A. NEDOVIC[2] AND DENIS PONCELET[3]
[1]Department of Chemical Engineering, Faculty of Technology and Metallurgy, University of Belgrade, Karnegijeva 4, 11000 Belgrade, Yugoslavia – Fax: + 381113370472 – Email: branko@elab.tmf.bg.ac.yu
[2]Department of Food Technology and Biochemistry, Faculty of Agriculture, University of Belgrade, Nemanjina 6, 11081 Belgrade-Zemun, Yugoslavia – Fax: +38111193659 – Email: vnedovic@eunet.yu
[3]ENITIAA, Rue de la Géraudiere, BP 82 225, Nantes, 44322, France – Fax: +33 2 51 78 54 67 – Email: poncelet@enitiaa-nantes.fr

Abstract

Selection of support material and method of immobilization is made by weighing the various characteristics and required features of the cell/enzyme application against the properties and limitations of the combined immobilization and support. A number of practical aspects should be considered before embarking on experimental work to ensure that the final immobilized cell or enzyme preparation is fit for the planned purpose or application to operate at optimum effectiveness. The mechanism of alginate droplet formation as well as experimental parameters for producing small hydro gel beads using an electrostatic droplet generator was investigated. It was found that microbead size was a function of needle diameter, charge arrangement (*i.e.* electrode geometry and spacing) and strength of the electric field. The process of alginate droplet formation under the influence of electrostatic forces was assessed with an image analysis/video system and revealed distinct stages; after a voltage was applied the liquid meniscus at the needle tip was distorted from a spherical shape into an inverted cone-like shape. Alginate solution flowed into this cone at an increasing rate causing formation of a neck-like filament. When this filament broke away, producing small droplets, the meniscus relaxed back to a spherical shape until flow of the alginate caused the process to start again. Various cells suspensions and enzymes were subjected to a high voltage immobilization process in order to assess the effects of electric fields on animal cell viability and enzyme activity. There was no detectable loss in cell viability or enzyme activity after the voltage was applied.

V. Nedović and R. Willaert (eds.), Fundamentals of Cell Immobilisation Biotechnology, 277-294.

2.4. EXTRUSION OF AN ANIMAL CELL SUSPENSION USING ELECTROSTATIC DROPLET GENERATOR

Animal cell sensitivity directly exposed to a high potential (6-8 kV) was examined by extruding an insect cell (SF-9) suspension using charged needle set-up with a 26 g. needle. The cell viability was assessed by staining the cells with trypan blue dye. The cells were cultured in shake flasks in IPL41 medium prior to extrusion.

To assess cell growth and functional activity as well as the enzyme activity, several types of cells (hybridoma [1,27], insect [1,28], Langerhans islet [1,29,30], yeast [32-34]), and enzymes (lipases [35]) were subjected to immobilization in defined electrostatic field.

3. Results and discussion

3.1. INVESTIGATION OF PARAMETERS AFFECTING MICROBEAD SIZE

In the case of the positively charged needle set-up, the effect of electrode spacing on alginate bead size, produced with a 22 gauge needle, is shown in Figure 3A. The electrode spacing was not found to be significant over the range investigated. For example, at an applied voltage of 6 kV the mean bead size decreased from 530 to 600 μm with a standard deviation of approximately 100 μm, as the electrode spacing decreased from 4.8 to 2.5 cm, respectively. When the applied voltage to the alginate solution was increased to 12 kV, at a distance between needle tip and collecting solution of 4.8 cm, the average bead diameter decreased to 340 μm. Keeping the applied potential constant at 12 kV, but reducing the electrode distance to 2.5 cm, resulted in only a slightly smaller bead size (300 μm).

The relationship between the applied voltage, alginate concentration and droplet diameter was also investigated (Figure 3B,C). When the alginate concentration was decreased from 1.5% to 0.8% the average bead diameter decreased for the most 20%. At the lower polymer concentration the standard deviation decreased, due to a more uniform bead size distribution. For example, at an applied voltage of 5 kV the mean bead diameter decreased from 440 ± 200 μm to 380 ± 80 μm, for the beads prepared with alginate concentrations of 1.5% and 0.8%, respectively.

While the alginate concentration was not found to be a significant parameter, bead diameter could be readily controlled by the needle size and applied voltage (Figure 3B,C). For example at an applied voltage of 4 kV the mean bead diameter was reduced by a factor of 3 from approximately 1600 μm to 500 μm when the needle size was reduced from 22 gauge to 26 gauge. In the case of the 26-gauge needle as the voltage increased above 6 kV, natural harmonic oscillation of the needle was observed. This resulted in bimodal distribution of microbead sizes with a large fraction of 50 μm in diameter.

A study of a similar electrostatic method of alginate bead formation [37] showed that with a 23 gauge needle the minimum bead size that could be achieved varied between 600 μm and 1000 μm using a sodium alginate solution when 5 kV was applied to the capillary tip. In our studies at the same voltage and a slightly larger needle

IMMOBILIZATION OF CELLS AND ENZYMES USING ELECTROSTATIC DROPLET GENERATION

BRANKO M. BUGARSKI[1], BOJANA OBRADOVIC[1], VIKTOR A. NEDOVIC[2] AND DENIS PONCELET[3]

[1]*Department of Chemical Engineering, Faculty of Technology and Metallurgy, University of Belgrade, Karnegijeva 4, 11000 Belgrade, Yugoslavia – Fax: + 381113370472 – Email: branko@elab.tmf.bg.ac.yu*

[2]*Department of Food Technology and Biochemistry, Faculty of Agriculture, University of Belgrade, Nemanjina 6, 11081 Belgrade-Zemun, Yugoslavia – Fax: +38111193659 – Email: vnedovic@eunet.yu*

[3]*ENITIAA, Rue de la Géraudiere, BP 82 225, Nantes, 44322, France – Fax: +33 2 51 78 54 67 – Email: poncelet@enitiaa-nantes.fr*

Abstract

Selection of support material and method of immobilization is made by weighing the various characteristics and required features of the cell/enzyme application against the properties and limitations of the combined immobilization and support. A number of practical aspects should be considered before embarking on experimental work to ensure that the final immobilized cell or enzyme preparation is fit for the planned purpose or application to operate at optimum effectiveness. The mechanism of alginate droplet formation as well as experimental parameters for producing small hydro gel beads using an electrostatic droplet generator was investigated. It was found that microbead size was a function of needle diameter, charge arrangement (*i.e.* electrode geometry and spacing) and strength of the electric field. The process of alginate droplet formation under the influence of electrostatic forces was assessed with an image analysis/video system and revealed distinct stages; after a voltage was applied the liquid meniscus at the needle tip was distorted from a spherical shape into an inverted cone-like shape. Alginate solution flowed into this cone at an increasing rate causing formation of a neck-like filament. When this filament broke away, producing small droplets, the meniscus relaxed back to a spherical shape until flow of the alginate caused the process to start again. Various cells suspensions and enzymes were subjected to a high voltage immobilization process in order to assess the effects of electric fields on animal cell viability and enzyme activity. There was no detectable loss in cell viability or enzyme activity after the voltage was applied.

V. Nedović and R. Willaert (eds.), Fundamentals of Cell Immobilisation Biotechnology, 277-294.

accomplished by establishing an electric field between a positively charged needle and a grounded plate containing 1.5% $CaCl_2$ hardening solution (Figure 2B).
Finally, for scale-up purposes a 1.5 litre cylindrical reservoir with twenty needles for the continuous production of polymer beads was designed (Figure 2C). The liquid flow rate was kept constant at 0.7 l/h (36 ml/h per needle) by adjusting the air pressure head above the polymer solution. A ground collecting plate of $CaCl_2$ solution was placed 2.5 cm below the needles. Twenty stainless steel needles (22 gauge) 1.2 cm radially apart were connected to the cylindrical reservoir containing 1 litre of polymer solution and then attached to the high potential unit.

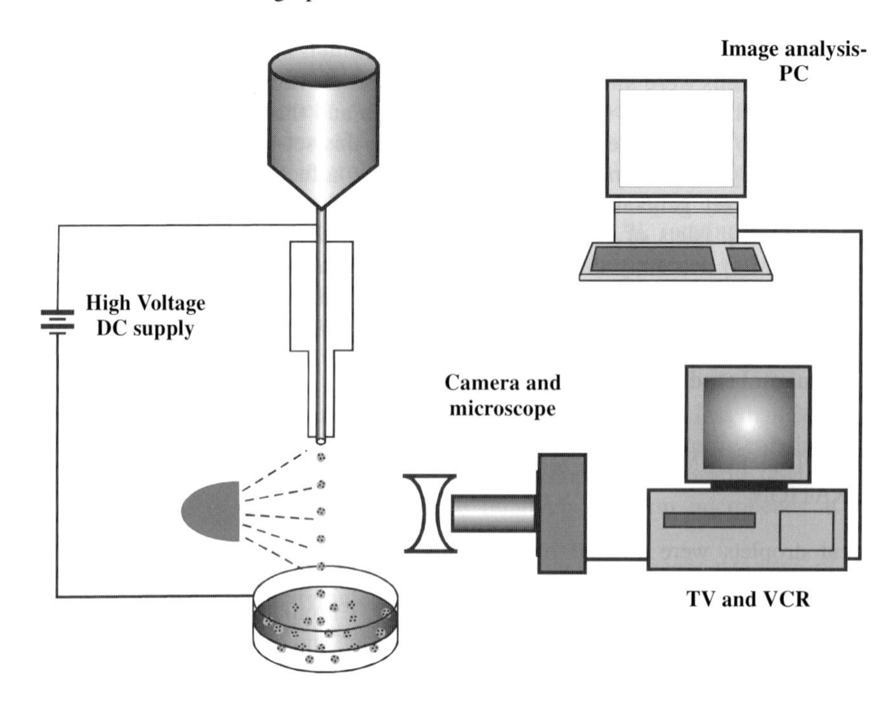

Figure 1. Schematic diagram of the experimental set-up.

2.2. ANALYSIS OF THE DROPLET FORMATION USING IMAGE ANALYSIS

The image analysis system consisted of a video camera (Panasonic Digital 51000m) video Adapter (Sony Trinitrom PVM 1342Q) and VHS recorder (Panasonic NV8950), and the results were analyzed with java version 1.3 software for image analysis (Jandel scientific, CA). For close-up studies of droplet formation, the video was connected to a microscope lens (Olympus SYH, Japan). Droplet images were frozen under a strobe light (Stobatac, GRC, MA) at defined frequencies between 50 and 400 Hz.

2.3. DETERMINATION OF MICROBEAD SIZE DISTRIBUTION

Volumetric (volume of micro spheres in each diameter class) and volumetric cumulative bead size distributions were determined by laser light scattering using a 2602-LC particle analyzer (Malvern Instruments) and HR 850 (Cilas-Alcatel) according to a log-normal distribution model. The mean diameter d_{50} was evaluated at 50% of the cumulative volume fraction. Resolution of the size has been evaluated to less than 10% by replications of the measurement.

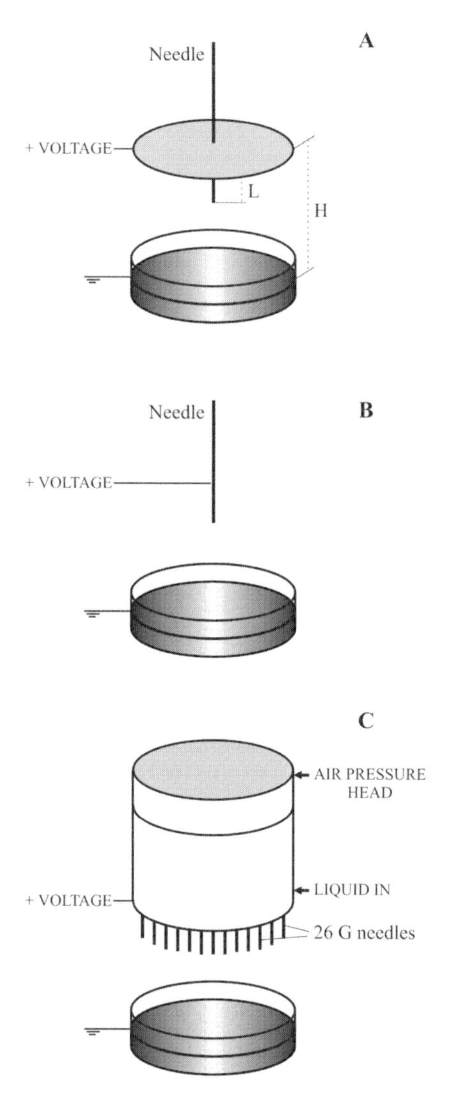

Figure 2. Electrode arrangements: A) Parallel plate set-up with a positively charged plate, B) Positively charged needle, C) Multineedle parallel plate device.

2.4. EXTRUSION OF AN ANIMAL CELL SUSPENSION USING ELECTROSTATIC DROPLET GENERATOR

Animal cell sensitivity directly exposed to a high potential (6-8 kV) was examined by extruding an insect cell (SF-9) suspension using charged needle set-up with a 26 g. needle. The cell viability was assessed by staining the cells with trypan blue dye. The cells were cultured in shake flasks in IPL41 medium prior to extrusion.

To assess cell growth and functional activity as well as the enzyme activity, several types of cells (hybridoma [1,27], insect [1,28], Langerhans islet [1,29,30], yeast [32-34]), and enzymes (lipases [35]) were subjected to immobilization in defined electrostatic field.

3. Results and discussion

3.1. INVESTIGATION OF PARAMETERS AFFECTING MICROBEAD SIZE

In the case of the positively charged needle set-up, the effect of electrode spacing on alginate bead size, produced with a 22 gauge needle, is shown in Figure 3A. The electrode spacing was not found to be significant over the range investigated. For example, at an applied voltage of 6 kV the mean bead size decreased from 530 to 600 μm with a standard deviation of approximately 100 μm, as the electrode spacing decreased from 4.8 to 2.5 cm, respectively. When the applied voltage to the alginate solution was increased to 12 kV, at a distance between needle tip and collecting solution of 4.8 cm, the average bead diameter decreased to 340 μm. Keeping the applied potential constant at 12 kV, but reducing the electrode distance to 2.5 cm, resulted in only a slightly smaller bead size (300 μm).

The relationship between the applied voltage, alginate concentration and droplet diameter was also investigated (Figure 3B,C). When the alginate concentration was decreased from 1.5% to 0.8% the average bead diameter decreased for the most 20%. At the lower polymer concentration the standard deviation decreased, due to a more uniform bead size distribution. For example, at an applied voltage of 5 kV the mean bead diameter decreased from 440 ± 200 μm to 380 ± 80 μm, for the beads prepared with alginate concentrations of 1.5% and 0.8%, respectively.

While the alginate concentration was not found to be a significant parameter, bead diameter could be readily controlled by the needle size and applied voltage (Figure 3B,C). For example at an applied voltage of 4 kV the mean bead diameter was reduced by a factor of 3 from approximately 1600 μm to 500 μm when the needle size was reduced from 22 gauge to 26 gauge. In the case of the 26-gauge needle as the voltage increased above 6 kV, natural harmonic oscillation of the needle was observed. This resulted in bimodal distribution of microbead sizes with a large fraction of 50 μm in diameter.

A study of a similar electrostatic method of alginate bead formation [37] showed that with a 23 gauge needle the minimum bead size that could be achieved varied between 600 μm and 1000 μm using a sodium alginate solution when 5 kV was applied to the capillary tip. In our studies at the same voltage and a slightly larger needle

(22 gauge) we have obtained a bead diameter 50 % smaller than that reported by Keshavarz *et al.* [37]. This difference may be partly explained by different geometry and lower viscosity of alginate solution used in our studies. With a 26 gauge needle under the same conditions the bead diameter, in our studies, decreased further, to about 150 ± 100 μm, suggesting a strong influence of needle size on bead diameter.

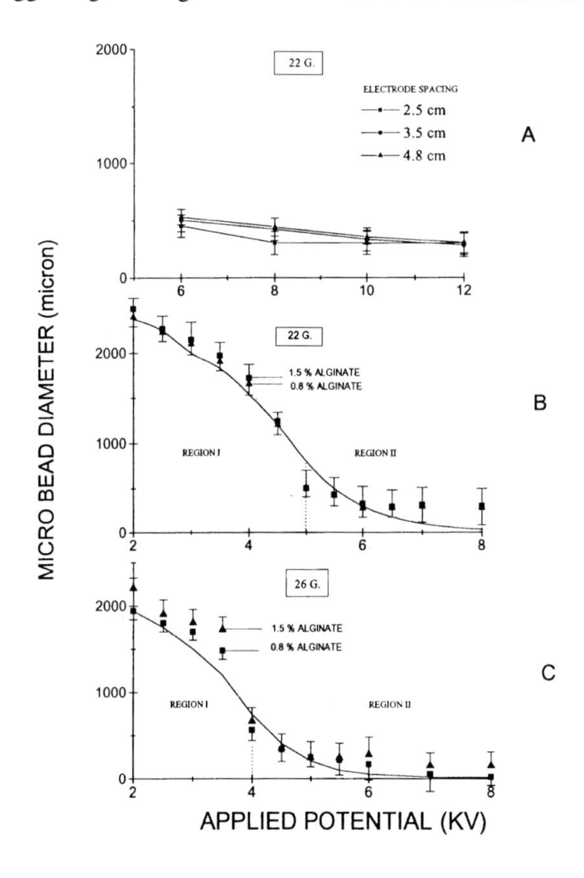

Figure 3. Effects on microbead size of A) Applied potential and electrode spacing; B) Alginate concentration and C) Needle size with the positively charged needle set-up. A 22 gauge needle was used in A) and B) and a 26 gauge needle in C).

3.2. EFFECT OF ELECTRODE GEOMETRY ON MICROBEAD SIZE

In the case of the parallel plate set-up, the effect of electrode spacing and charge arrangement (*i.e.* different electric field and surface charge intensity) on polymer bead size is shown in Figure 4A. As the potential between the electrodes in the parallel plate set-up for example, increased from 6 to 12 kV at 4.8 cm, the average bead diameter decreased from 2300 μm to 700 μm using a 22-gauge needle. Reducing the electrode distance resulted in even smaller bead sizes suggesting a strong influence of distance and voltage on microbead size with this charge arrangement. For example, at 10 kV, reducing the distance from 4.8 cm to 2.5 cm resulted in a decrease in polymer bead diameter from 1500 μm to 350 μm. However increasing the applied potential above 12 kV and decreasing the electrode spacing did not result in further decrease in bead diameter. This was probably due to a discharge between the plates accompanied by sparking as a result of air ionization in the space between the electrodes.

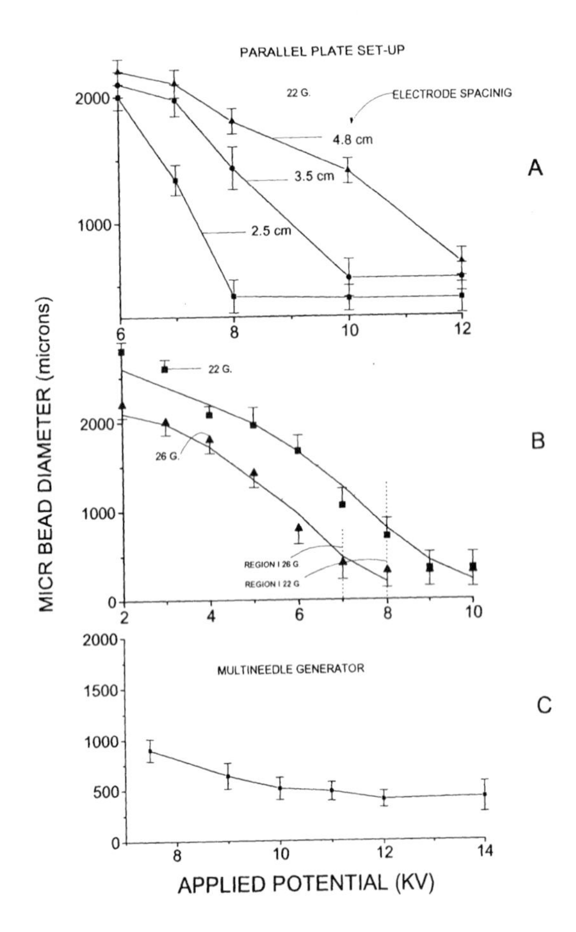

Figure 4. Effects on microbead diameter of A) Applied potential and electrode spacing, B) Needle size, and C) Scale–up with a parallel plate set-up.

For a fixed electrode distance (2.5 cm), needle length (1.3 cm) and alginate concentration (1.5%) the bead diameter could be reduced by decreasing the needle size from 22 to 26 gauge (Figure 4B). A decrease by a factor of two in bead diameter was observed for a wide range of applied potentials (2-8 kV).

The multiple needle device was essentially a scaled-up version of the parallel plate set-up (Figure 2C). Results where similar to that found with the single needle setup (Figure 2A). At an electrode distance of 2.5 cm increasing the potential from 7 to 12 kV resulted in a decrease in bead diameter from 950 μm ± 100 μm, to 400 ± 150 μm (Figure 4C). The device for the continuous beads production had a processing capacity of 0.7 L/h.

3.3. INVESTIGATION OF MECHANISM OF DROPLET FORMATION WITH IMAGE ANALYSIS/VIDEO SYSTEM

In the absence of an electrostatic field with gravitational force acting alone, the mean bead diameter was 2400 μm ± 200 μm at a constant alginate flow rate of 36 ml/h and using a 22 gauge needle. In this case a droplet was produced every one to two seconds. Each drop grew at the tip of the needle until its weight overcame the net vertical component of the surface tension force (Figure 5A).

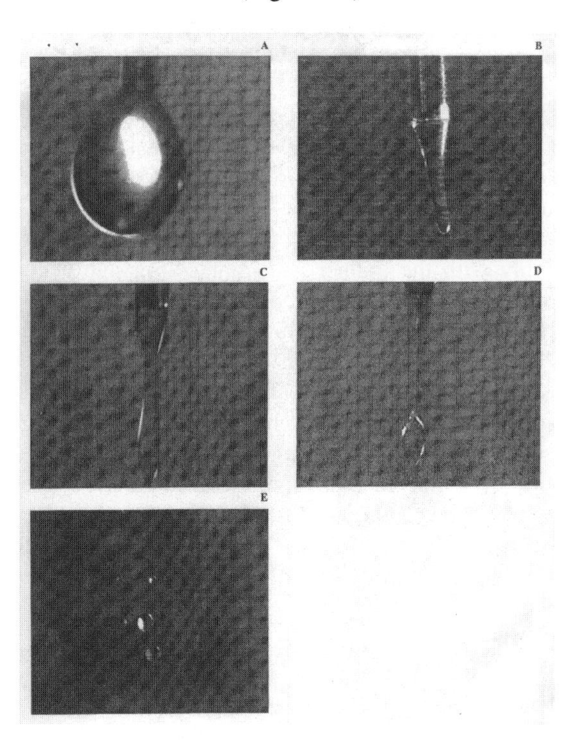

Figure 5. Electrostatic droplet formation at the needle tip with 1.5% sodium alginate without applied potential (A) and at applied potentials between 4 and 5 kV: (B) Meniscus formation; (C) Neck formation; (D) Stretching of the liquid filament; (E) Brake-up of the liquid filament.

Examination of the formation of droplets under the influence of electrostatic forces revealed that an elongated cone formed as the droplet meniscus advanced (Figure 5B). The forming droplet was drawn out into a long slender filament (voltage of 4-5 kV,

2.5 cm electrode distance, 1.5 % sodium alginate concentration, flow rate of 36 ml/h, 22 gauge needle). A high charge density at the tip of the inverted cone reduces the surface tension of the alginate solution [38] resulting in neck formation (Figure 5C). For the more concentrated alginate (*i.e.* 1.5%) we observed that the neck elongated up to one 1 mm before detachment (Figure 5D). While the main part of the liquid neck, quickly coalesced into a new drop, the long linking filament broke up into a large number of smaller drops (Figure 5E). It was also observed that small (satellite) droplet formation usually accompanied higher voltages (above 6 kV), because the elongation of the liquid neck prior to rupture was much more pronounced. The largest of these drops was one half of the main drop diameter while the smallest was less than 20 μm.

When the concentration of the alginate was decreased from 1.5% to 0.8%, a difference was observed in the formation of the long thin neck or a filament linking the new droplet and the meniscus at the tip of the needle. For the low viscosity alginate, neck elongation was not as pronounced, resulting in a more uniform bead size (Figure 6A-C).

Briefly, at the early stage, the shape of liquid meniscus is almost spherical. After the voltage is applied the meniscus is distorted into a conical shape as shown in Figure 6. Consequently the alginate solution flows through this weak area at an increasing rate causing the formation of a neck. When the filament breaks away the meniscus of the liquid on the needle is suddenly decreased for a short period until flow of the liquid causes the process to start again.

When the voltage was increased above 6 kV, harmonic natural needle oscillation was observed, but only with the thinner and lighter 26-gauge needle. A high surface charge and an electric field on the surface of capillary tip, gave rise to a mechanical force causing needle vibration and resulting in an oscillating thread-like filament (Figure 7A).

The periodic oscillation of the electrically stressed meniscus at the capillary tip caused a sinusoidal shaped filament to detach from the meniscus at the needle tip. The long tapered filaments are formed as a result of the surface energy component arising from the presence of the surface charge. From a consideration of the minimum surface energy, molecular forces tend to decrease the surface to volume ratio, whereas increasing this ratio minimizes the energy component due to electrostatic charges. Figure 7B indicates the manner in which the stream disintegrates by a vigorous whipping action. Fragmentation of the filament results in a bimodal size distribution with peaks at 50 μm and 190 μm in diameter. This phenomenon was not observed with a 22-gauge needle.

Let us examine the different mechanisms of droplet formation in regions I and II (Figure 3) for the charged needle set-up (*i.e.* Figure 2B). In the region I, the primary factor regulating droplet size is probably the intensity of the electric field in the vicinity of the forming droplet. Since the electric field increases with decreasing diameter, the intensity of the field is the highest at the meniscus tip. Formation of a continuous thin liquid filament and its break-up into small droplets is probably induced by an interaction between the surface charge on the liquid meniscus and the external electric field.

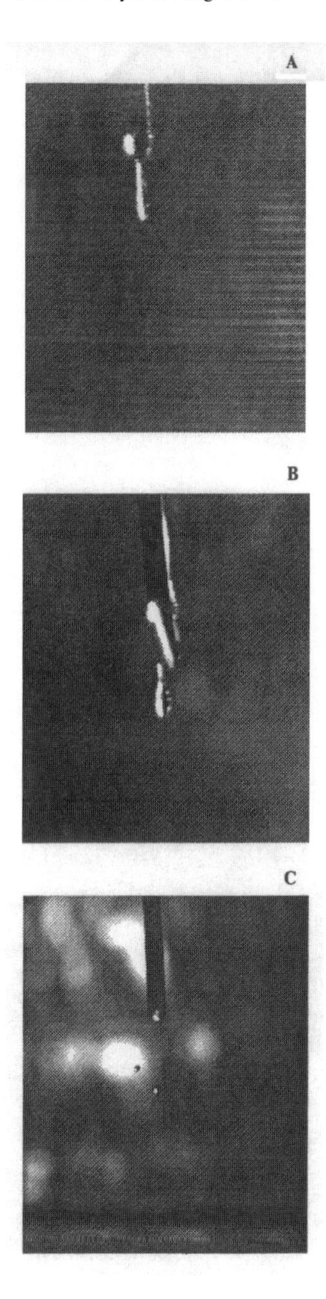

Figure 6. Droplet formation using low alginate concentration (0.8%). A) Meniscus formation; B) Filament detachment from the tip of the needle; C) Dispersion of the liquid filament.

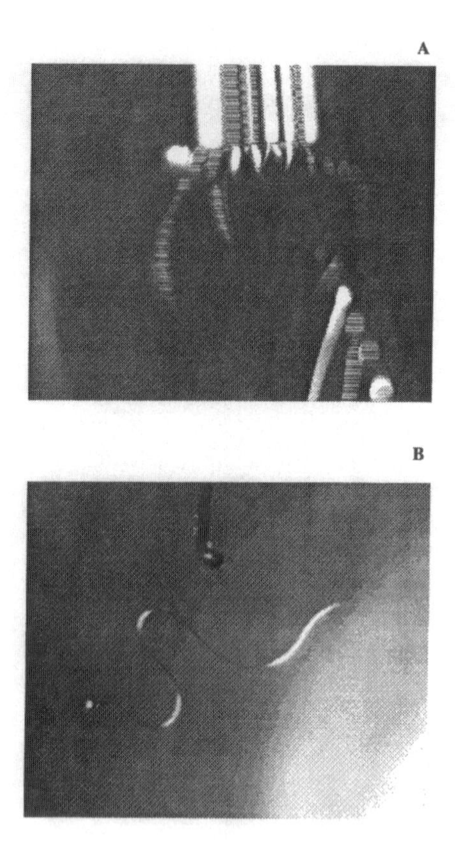

Figure 7. Needle oscillation (26 gauge, 7 kV). A) Needle vibration due to electrostatic forces, B) Detachment of liquid filament.

The electric charge on the surface of droplets gives rise to the mechanical force that is directed normally outward from the surface (*i.e.* in direct opposition to the inward acting surface forces). Also, gravity acts in the same direction as the electric field. Consequently, the terminal velocity of the charged droplet will increase. In addition, liquid surface tension may change in electric field as first proposed by Sample and Bollini [26]. They showed that the dynamic surface tension was generally higher than the static surface tension for the same liquid. For example, the surface tension of water reached a dynamic value of 0.110 N/m, as compared to a static value of only 0.074 N/m. On the other hand, distribution of the charge can take place only on the surface of the liquid so that mutual repulsion forces may cause a reduction of the surface tension. The result of these two effects (*i.e.* – increased velocity and reduced surface tension) is the increase of the Reynolds and Weber numbers of the charged droplet and the observed effect of a decrease in the droplet diameter with the increase in the electric field in region I (Figure 3).

Further increase in the voltage in region II (above 8 kV in our set-up) resulted in a slight increase in the mean bead diameter before discharge and air ionization occurred. A possible increase in surface tension in region II at higher applied potentials (> 8 kV) did not result in any further decreases in droplet diameter. Based on video/image analysis, the properties of surface charge can be used to explain the levelling off of the droplet diameter with increasing potential in region II. When a sufficiently large charge is added to the droplet, the latter may not have enough time for break-up due to the high velocity induced by the intense electric field.

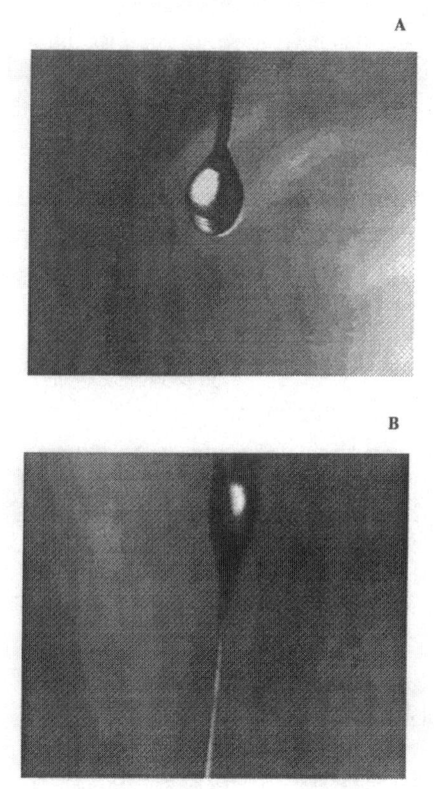

Figure 8. Comparative analysis of parallel plate and positively charged needle at the same potential (6 kV) and electrode distance (2.5 cm). A) Parallel plate set-up, B) Positively charged needle. Note the formation of the jet spray in Figure B.

A comparative analysis of the two charge set-ups (Figure 2A,B) at the same applied potential and needle size was carried out in order to give insight into the droplet formation mechanism. Looking at the formation of the droplet at the needle tip using the image analysis/video system, it was observed that each charge setup produced a different mode of droplet formation. At the same relatively high potential difference (6 kV), needle size (26 gauge), electrode spacing (2.5 cm) and flow rate (36 ml/h) there was a noticeable difference in bead diameters. With the parallel plate charge set-up, for example, we observed that at the given potential difference, the frequency of droplets leaving the tip of the needle was below that required to initiate spraying. The average mean diameter was found to be 1100 µm ± 200 µm (Figure 8A). In contrast, with a positively charged needle a Taylor cone like meniscus [36] was observed with a well developed jet (80 µm diameter) ejecting droplets of 170 ± 70 µm in diameter (Figure 8B), a decrease by factor seven.

3.4. MICRO BEAD SIZE DISTRIBUTION

The mean bead size distribution curves obtained by plotting relative frequencies *versus* bead diameters typically resulted in a continuous function symmetrical about the mean value (Figure 9A). At 4.5 kV and 6 kV, the mean bead size distribution was found to vary about the mean of 225 µm (Figure 9A) and 170 µm (Figure 9B) respectively. However, in case of the naturally vibrating needle (7 kV), a bimodal bead size distribution was observed, one peak at 50 µm and a second peak at 190 µm (Figure 9C).

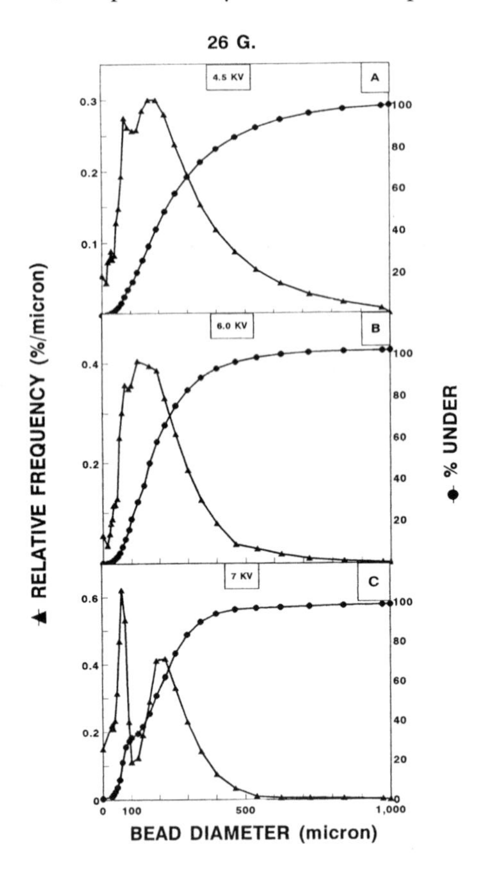

Figure 9. Droplet size distribution produced with positively charged 26 gauge needle, 2.5 cm electrode distance, 0.8% alginate concentration and electric potentials of: A) 4.5 kV, B) 6 kV, C) 7 kV.

3.5. IMMOBILIZATION OF CELLS AND ENZYMES USING ELECTROSTATIC DROPLET GENERATOR

To assess the effect of an electrostatic field on animal cells viability, an insect cell suspension was directly exposed by extrusion to the electrostatic field. No detectable change in insect cell viability was observed after extrusion. The initial cell density, 4×10^5 cell/ml, remained essentially unchanged at 3.85×10^5 cell/ml and 3.8×10^5 cell/ml immediately after passing trough the generator with an applied potential difference of 6 kV and 8 kV, respectively. Prolonged cultivation of these cells did not show any loss of cell density or viability [15].

Results of a short time cultivation study have indicated an optimal diameter range of 500 to 600 μm for alginate microbeads loaded with brewing yeast cells. When parameters of electrostatic field such as applied potential, needle size and electrode distance were adjusted a uniform beads (500 ± 50 μm) with immobilized yeast were obtained which reached a maximum cell concentration of 2.3×10^9 cell/ml bead after one week of batch cultivation (Figure 10A) [33].

Electrostatically obtained microbeads less than 150 μm in diameter, coated with poly-l-lysine were found to be effective as microcarriers for culturing surface attached insect cells, which reached a density of 1.5×10^7 cells/ml beads (Figure 10B) [28].

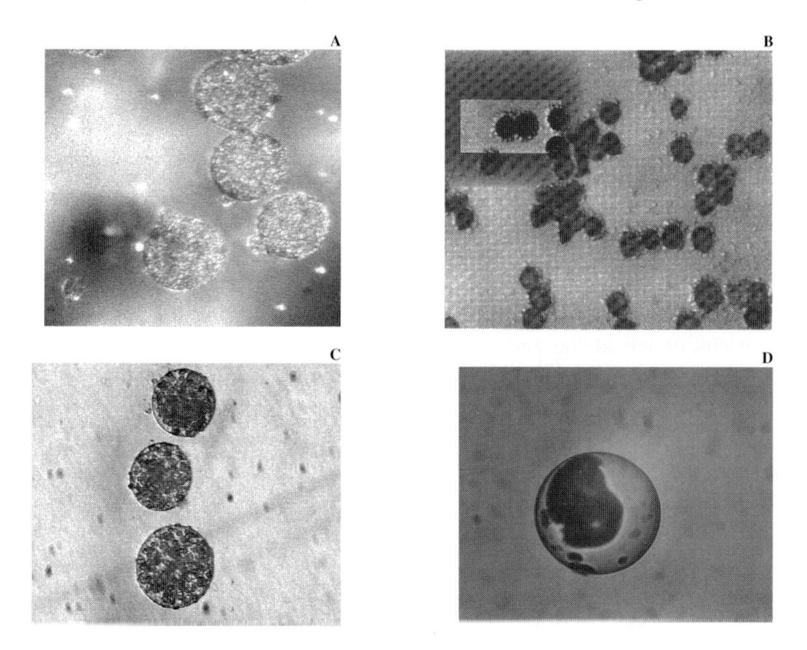

Figure 10. Different cell types immobilized using electrostatic droplet generation: A) Yeast cells, B) Insect cells/surface attachment, C) Hybridoma cells, D) Langerhans islet/single islet encapsulation.

Biological outputs such as cell density and productivity of IgG, were used to test suitability of the immobilization method and a bioreactor system for monoclonal antibody production. In this system utilising alginate-poly-l-ornithine (PLO) microcapsules (300-400 μm in diameter) cell density reached 1.5 x 10^8 cells/microcapsule, with IgG product concentration at a level of 700 μg/ml (Figure 10C) [4,27].

An artificial cell system of encapsulated islets of neonatal rat pancreas was tested for functional activities, after utilizing spraying mode to produce alginate-PLO microcapsule. Spherical microcapsules of approximately 200 μm each containing 1 islet were obtained using electrostatic field. Islets maintained functional insulin response to glucose stimulation and preserved integrity of the cell surface antigens (Figure 10D) [29,30].

Lipase from *Candida rugosa* was immobilized in alginate beads in the electrostatic field for the application in a non-aqueous reaction system. Electrostatic droplet immobilization provided monodispersed immobilized lipase alginate beads (600 μm) with negligible loss of enzyme and was proven to be more efficient in comparison to a free lipase system [39]. The beads retained high lipase activity in a reaction system of palm oil hydrolysis with the immobilization efficiency of about 99% [35].

4. Conclusions

The formation of droplets from a charged needle and a parallel plate arrangement, in a defined electric field, was examined. In the parallel plate set-up, reducing the electrode distance and increasing the applied voltage resulted in smaller beads suggesting a strong influence of distance and applied potential on bead diameter. Further reduction in bead size, with both set-ups, was achieved by decreasing the needle size. Electrode spacing was not important with the charged needle arrangement. The greatest decrease in microbead size was observed when a natural needle oscillation, caused by surface charge and a high electric field in the vicinity of the needle, resulted in whip-like liquid filaments breaking off at the end of the needle. This phenomenon produced a bimodal size distribution with a large fraction of droplets below 50 μm in diameter. Modification of the parallel plate set-up system in the form of a multi-needle device showed that it is possible to continuously produce uniform microbeads at a high processing capacity. Finally, there was no detectable loss in viability after passing animal cells through the electrostatic droplet generator. This is a promising result as it proves the technique amenable for cell immobilization. The size dispersion remains low in spraying mode [15,22,23,40]. Further experiments are required to more fully understand and control droplet formation. By combining jet formation under electrostatic potential with jet breakage by vibration, it is expected that small microcapsules with a narrow size distribution may be prepared. For laboratory applications, the present design is suitable, but it is necessary to design a pilot plant system with higher flow rates to obtain basic parameters for implementation at the industrial scale.

Acknowledgements

This work was funded by the Ministry of Science, Technologies and Development of the Republic of Serbia.

References

[1] Goosen, M.F.A.; Mahmud, E.S.C.; M-Ghafi, A.S.; M-Hajri, H.A.; Al-Sinani, Y.S. and Bugarski, M.B. (1997) Immobilization of cells using electrostatic droplet generation. In: Bickerstaff, G.F. (Ed.) Immobilization of Enzymes and Cells. Humana Press. Totowa, NJ, Chapter 20; pp. 167-174.
[2] Balchhandran, W. and Bailey, A.G. (1984) The dispersion of liquids using centrifugal and electrostatic forces. IEEE Trans. Ind. App. IA20: 682-686.
[3] Fillimore, G.L. and Lokeren, D.C. (1982) Multinozzle drop generator which produces uniform break-up of continuous jets. Institute of Electrical and Electronics Engineers, Annual Meeting of the Industrial Application Society: 991-998.
[4] Bugarski, B.; Jovanović, G. and Vunjak G. (1993) Bioreactor systems based on microencapsulated animal cell cultures. In: Goosen, M.F.A. (Ed.) Fundamentals of Animal Cell Encapsulation and Immobilization. CRC Press Inc., Boca Raton, Florida, Chapter 12; pp. 267-296.
[5] Romo, S. and Perezmartinez, C (1997) The use of immobilization in alginate beads for long-term storage of Pseudanabaena-Galeata (Cyanobacte-ria) in the laboratory. J. Phycol. 33: 1073–1076.
[6] Walsh, P.K.; Isdell, F.V.; Noone, S.M.; Odonovan. M.G. and Malone, D.M. (1996) Growth patterns of *Saccharomyces cerevisiae* microcolonies in alginate and carrageenan gel particles: effect of physical and chemical properties of gels. Enzyme Microb. Technol. 18: 366–372.
[7] Green, K.D.; Gill, I.S.; Khan, J.A. and Vulfson, E.N. (1996) Microencapsulation of yeast cells and their use as a biocatalyst in organic solvents. Biotechnol. Bioeng. 49: 535–543.
[8] Poncelet, D.; Bugarski, B.; Amsden, B.G.; Zhu, J.; Neufeld, R and Goosen, M.F.A. (1994) A parallel-plate electrostatic droplet generator: parameters affecting microbead size. Appl. Microbiol. Biotechnol. 42: 251–255.
[9] Poncelet, D.; Desmet, B.P.; Beaulieu, C.; Huguet, M.L.; Fournier, A. and Neufeld, R.J. (1995) Production of alginate beads by emulsification internal gelation. Appl. Microbiol. Biotechnol. 43: 644–650.
[10] Begin, F.; Castaigne, F. and Goulet, J. (1991) Production of alginate beads by a rotative atomizer. Biotechnol. Tech. 5: 459–464.
[11] Ogbonna, J.C.; Matsumura, M. and Kataoka, H. (1991) Effective oxygenation of immoblized cells through reduciton in bead diameter: a review. Process Biochem. 26: 109–121.
[12] Klein, J.; Stock, J. and Vorlop, K.D. (1983) Pore size and properties of spherical Ca-alginate biocatalysts. Eur. J. Appl. Microbiol. Biotechnol. 18: 86–91.
[13] Levee, M.G.; Lee, G.M.; Paek. S.H. and Palsson, B.O. (1994) Microencapsulated human bone-marrow cultures: a potential culture system for the clonal outgrowth of hematopoietic progenitor cells. Biotechnol. Bioeng. 43: 734–739.
[14] Kwok, K.K.; Groves, M.J. and Burgess, D.J. (1991) Produciton of 5–15 mm diameter alginate polylysine microcapsules by air-atomization technique. Pharm. Res. 8: 341–344.
[15] Bugarski, B.; Li, Q.L.; Goosen, M.F.A.; Poncelet, D.; Neufeld, R.J. and Vunjak, G. (1994) Electrostatic droplet generation: mechanism of polymer droplet formation. AICHE J. 40: 1026–1031.
[16] Halle, J.P.; Leblond, F.A.; Pariseau, J.F.; Jutras, P.; Brabant, M.J. and Lepage, Y. (1994) Studies on small (less than 300 mm) microcapsules. II. Parameters governing the production of alginate beads by high-voltage electro-static pulses. Cell Transplant. 3: 365–372.
[17] Prusse, U.; Fox, B.; Kirchhof, M.; Brueke, F.; Breford, J. and Vorlop, K.D. (1998) New process (jet cutting method) for the production of spherical beads from highly viscous polymer solutions. Chem. Eng. Technol. 21: 29–33.
[18] Brandenberger, H. and Widmer, F. (1997) Monodisperse particle production: a new method to prevent drop coalescence using electrostatic forces. J. Electrostat. 45: 227–238.
[19] Ghosal, S.K.; Talukdar, P. and Pal, T.K. (1993) Standardization of a newly designed vibrating capillary apparatus for the preparation of microcapsulses. Chem. Eng. Technol. 16: 395–398.
[20] Seifert, D.B. and Phillips, J.A. (1997) Production of small, monodispersed alginate beads for cell immobilization. Biotechnol. Prog. 13: 562–568.

[21] Serp, D.; Cantana, E.; Heinzen, C.; von Stockar, U. and Marison, I.W. (2000) Characterization of encapsulation device for the production of monodisperse alginate beads for cell immobilization. Biotechnol. Bioeng. 70(1): 41-53.

[22] Bugarski, B.; Amsden, B.; Neufeld, R.; Poncelet, D. and Goosen, M.F.A. (1994) Effect of electrode geometry and charge on the production of polymer micobeads by electrostatics. Can. J. Chem. Eng. 72: 517-522.

[23] Poncelet, D.; Babak, V.G; Neufeld, R.J.; Goosen, M. and Bugarski, B. (1999) Theory of elecrostatic dispersion of polymer solution in the production of microgel beds containing biocatalyst. Adv. Colloid Interface Sci. 79(2-3): 213-228.

[24] Rayleigh, Lord (1882) On the equilibrium of liquid condusting masses charged with electricity. Phil. Mag. 14: 184-186.

[25] Nawab, M.A. and Mason, S.G. (1958) The preparation of uniform emulsions by electrical dispersion. J. Colloid Sci. 13: 179-187.

[26] Sample, S.B. and Bollini, R. (1972) Production of liquid aerosols by harmonic electrical spraying. J. Colloid. Sci. 41: 185-193.

[27] Bugarski, B.; Vunjak, G. and Goosen, M.F.A. (1999) Principles of bioreactor design for encapsulated cells. In: Kuhtreiber, W.M.; Lanza, R.P. and Chick, W.L. (Eds.) Cell Encapsulation Technology and Therapeutics. Birkhauser, Boston; Chapter 30; pp. 395- 416.

[28] Bugarski, M.B.; Smith, J.; Wu, J. and Goosen, M.F.A. (1993) Methods for animal cell immobilization using electrostatic droplet generation. Biotechnol. Tech. 7(9): 677-682.

[29] Bugarski, M.B.; Sajc, L.; Plavšić, M.; Goosen, M.F.A. and Jovanović, G. (1997) Semipermeable alginate-PLO microcapsule as a bioartificial pancreas. In: Funatsu, K.; Shirai, Y. and Matsushita, T. (Eds.) Animal Cell Technology, Basic and Applied Aspects. Volume 8, Kluwer Academic Publishers, London, Boston, Dordrecht; pp. 479-486.

[30] Rosinski, S.; Lewinska, D.; Migaj, M.; Wozniewicz, B. and Werynski, A. (2002) Electrostatic microencapsulation of parathyroid cells as a tool for the investigation of cell's activity after transplantation. Landbauforschung Völkenrode SH 241: 47-50.

[31] Pjanović, R.; Goosen, M.F.A.; Nedović, V. and Bugarski, M.B. (2000) Immobilization/encapsulation of cells using electrostatic droplet generation. Minerva Biotechnology 12: 241-248.

[32] Nedović, A.V.; Obradović, B.; Leskošek, I.; Pešić, R. and Bugarski, B. (2001) Electrostatic generation of alginate microbeads loaded with brewing yeast. Process Biochem. 37: 17-22.

[33] Nedović, V.A.; Obradović, B.; Poncelet, D.; Goosen, M.F.A.; Leskošek-Čukalović, I. and Bugarski, B. (2002) Cell immobilisation by electrostatic droplet generation. Landbauforschung Volkenrode SH 241: 11-18.

[34] Knežević, Z.; Bobić, S.; Milutinović, A.; Obradović, B.; Mojović, Lj. and Bugarski, B. (2002) Alginate immobilized lipase by electrostatic extrusion. Process Biochem. 38: 313-318.

[35] Bugarski, B. and Goosen, M.F.A. (1996) Methods for animal cell immobilization using electrostatic extrusion. In: Kobayashi, T.; Kitagawa, Y. and Okumura, K. (Eds.) Animal Cell Technology, Basic and Applied Aspects. Volume 6, Kluwer Academic Publishers, London, Boston, Dordrecht; pp. 157-160.

[36] Taylor, G.I. and Van Dyke M.D. (1969) Electrically driven jets. Proc. R. Soc. A. 313: 453-475.

[37] Keshavarz, T.; Ramsden, G.; Phillips, P.; Mussenden, P. and Bucke, C. (1992) Application of electric field for production of immobilized biocatalysts. Biotech. Tech. 6: 445-450.

[38] Hendricks, C.D.Jr. (1962) Charged droplet experiments. J. Colloid Sci. 17: 249-259.

[39] Mojović, Lj.; Šiler-Marinković, S.; Kukić, G.; Bugarski, M.B. and Vunjak-Novaković, G.V. (1994) *Rhizpus arrhizus* lipase-catalyzed interesterification of palm oil midfraction in a gas-lift reactor. Enzyme Microb. Technol. 16:159-162.

[40] Poncelet, D.; Neufield, R.J.; Goosen, M.F.A.; Bugarski, M.B. and Babak, V. (1999) Formation of microgel beads by electrostatic dispersion of polymer solutions. AIChE J. 45: 2018-2023.

THE JETCUTTER TECHNOLOGY

ULF PRUESSE AND KLAUS-DIETER VORLOP
Institute of Technology and Biosystems Engineering, Federal Agricultural Research Centre, Bundesallee 50, 38116 Braunschweig, Germany – Fax: +49 531 596 4199 – Email: klaus.vorlop@fal.de

1. Introduction

Solid particles (pellets, beads) in the size range between μm and mm play an important role in various industries like biotechnology, agriculture, chemical, pharmaceutical and food industry. Thus, plenty of particle production technologies exist and their further development is of major interest both from the economic and scientific point of view.

Generally, single and discrete solid particles may be produced by three different approaches:

- From larger solid entities by grinding,
- from smaller solid entities by agglomeration, granulation, pressing or tabletting – small fluid entities may also be used if in-situ drying is applied –, or
- from fluid entities in the same size range in addition with an immediate physical or chemical solidification step.

Despite the necessity to be produced in industrial amounts, in recent years, solid particles are more and more required to have an ideal spherical shape as such beads are much easier to dose, pose less danger to humans and equipment during manufacturing (less respirable dust resulting from abrasion and lower explosion risk) and last but not least, look much nicer from an aesthetic point of view, which is very important if the beads are part of a final product.

From the three different approaches named above, the third one, in principle, is best suited for the production of ideal spherical beads, since only by this approach the solid bead has been a more or less equally sized liquid droplet – which is perfectly round due to the surface tension – directly prior to its solidification. Correspondingly, numerous different techniques exist which use the principle of generating a droplet which immediately afterwards is solidified to a spherical bead by physical means, *e.g.* cooling or heating, or chemical means, *e.g.* gelation, precipitation or polymerisation. These techniques include emulsion techniques, simple dropping, electrostatic-enhanced dropping, jet break-up (vibration) or rotating disc and rotating nozzle processes. Some of them, *e.g.* dropping and vibrational techniques are especially useful for lab-scale applications, whereas others, *e.g.* rotating disc and nozzle techniques and to some extent also the vibration technique, may also be used for large-scale applications. Anyway, all

V. Nedović and R. Willaert (eds.), Fundamentals of Cell Immobilisation Biotechnology, 295-309.

techniques have in common that the fluids, which are processable, have to be low in viscosity. Further, these techniques either allow monodisperse beads to be produced in small quantities or beads with an undesired broad size range to be produced in large quantities, but none of these processes enables the production of monodisperse beads in large quantities.

A new and simple technology for bead production that meets the requirement of producing monodisperse beads with a high production rate is the JetCutter technology. This technique is especially capable of processing medium and highly viscous fluids up to viscosities of several thousands mPas. Monodisperse beads originating from solutions, melts or dispersions in the size range from approx. 200 μm up to several millimetres are accessible. Therefore, the JetCutter is not only a valuable completion to the palette of bead production technologies but might even have a great potential to replace existing technologies.

2. Principle of function

For bead production by the JetCutter the fluid is pressed with a high velocity out of a nozzle as a solid jet. Directly underneath the nozzle the jet is cut into cylindrical segments by a rotating cutting tool made of small wires fixed in a holder. Driven by the surface tension the cut cylindrical segments form spherical beads while falling further down, where they finally can be gathered (Figure 1).

Bead generation by JetCutting is based on a mechanical impact of the cutting wire on the liquid jet. This impact leads to the cut together with a cutting loss, which, in a first approach, can be regarded as a cylindrical segment with the height of the diameter of the cutting wire. This segment is pushed out of the jet and slung aside where it can be gathered and recycled. The losses will be described in detail later.

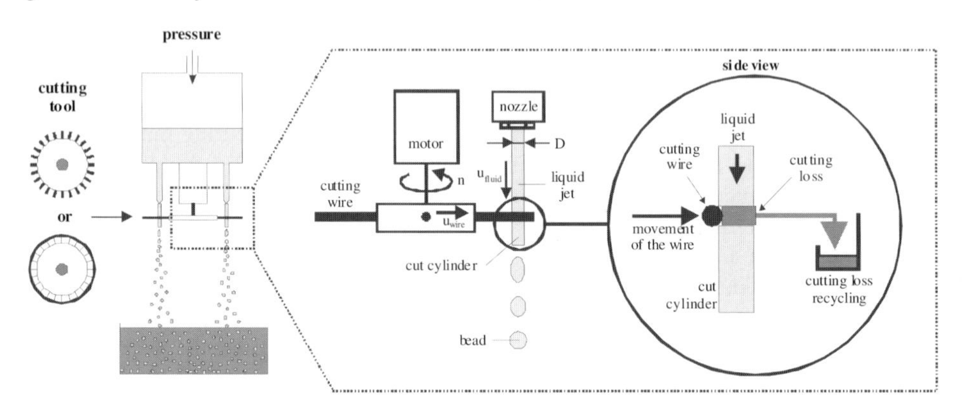

Figure 1. Scheme of the cutting process, simplified model.

As only a mechanical cut and the subsequent bead shaping driven by the surface tension are responsible for bead generation, the viscosity of the fluid has no direct influence on the bead formation itself. Thus, the JetCutter technology is capable to process fluids with viscosities up to several thousands mPas.

The size of the beads can be adjusted within a range of between approx. 200 μm up to several millimetres. The main parameters are the nozzle diameter, the flow rate through the nozzle, the number of cutting wires and the rotation speed of the cutting tool. In order to get narrowly distributed beads one has to take care of a steady flow through the nozzle and an uniform rotation speed of the cutting tool.

3. Description of the JetCutter device

3.1. GENERAL EQUIPMENT

A JetCutter device consists of five central elements:

- Pump or pressure vessel,
- solid jet nozzle,
- cutting tool,
- motor,
- spraying shield.

The first element is related to fluid feeding. This may be achieved either by using a pressurised vessel or a simple storage tank connected to a pulsation-free pump, *e.g.* a screw pump, and the necessary tubes ending up inside the nozzle. Special attention has to be paid to ensure a steady flow through the nozzle.

A major requirement for the JetCutter is that a solid jet is formed. Therefore, special solid jet nozzles have to be applied which are commercially available. Not the whole particle size range (200 μm up to several millimetres) is accessible with only one single nozzle diameter. Thus, nozzle diameters ranging from some tens of microns up to a few millimetres have to be applied to cover the whole size range. Even if solid jet nozzles are used, the liquid jet disintegrates after a certain length. This length depends on the viscosity and velocity of the fluid. In order to ensure a perfectly shaped jet at the point where it is cut by the wires, the cutting tool should not be too far away from the nozzle outlet, *e.g.* only a few millimetres (Figure 2, left).

The cutting tool itself is the major part of the whole JetCutter device. It is very important that the cutting wires are equally distributed around the tool in order to produce beads of the same size. This is best achieved by a circular stabilisation of the wires on the outer perimeter (Figure 2, right). This circular stabilisation is even essential if the wire diameter is reduced in order to decrease the cutting losses. Therewith, the diameter of the cutting wire may be decreased down to 30 μm. Usually, stainless steel wires are used but the application of polymer fibres is also practicable.

A circular stabilisation of the wires is not necessary for cutting wire diameters larger than 300 μm. Such large wire diameters seem to be detrimental as far as the losses are concerned but, nevertheless, are necessary for the production of larger beads. As a rule of thumb it can be said that small wires are to be used for the production of small beads, as they cause only small losses and do not transfer such a large impulse to the liquid jet than thicker wires would do, which would lead to a serious deflection of the beads. Thick wires have to be used for the production of larger beads, as smaller ones are not able to cut a thick liquid jet into separate segments, since, in this case, the liquid jet flows together again after the wire has passed it.

Figure 2. Arrangement of nozzle and cutting tool (left); cutting tool with 48 stainless steel wires (Ø = 50 μm) and circular stabilisation (right); Ø = diameter.

Another central part of the JetCutter device is the motor, to which the cutting tool is mounted. No special care has to be taken concerning the motor unless it is capable to maintain a constant rotation speed. Usually between 3000 and 12000 rpm (rotations per minute) are used for bead production. The motor with the cutting tool may be inclined up to 70° with regard to the nozzle in order to reduce the losses (see section 4).

The spraying shield surrounds the cutting tool. Its task is to collect the losses, which were slung aside by the cutting tool. If necessary, the losses may be recycled through a drain inside the spraying shield's channel. The shield also covers the rotating cutting tool, which is important for safety reasons.

3.2. SPECIAL EQUIPMENT

The general equipment described before is sufficient for the production of beads from a broad variety of fluids, like polymer solutions, sols and dispersions. Nevertheless, for some applications special equipment is needed, *e.g.*:

- A spraying tunnel for larger beads,
- a heating device for melt processing,
- a 2-fluid nozzle for simultaneous coating.

The high fluid and therefore bead velocity is one of the advantages of the JetCutter, as high throughputs are easily realised. Nevertheless, this high droplet velocity is a problem regarding the collection of beads with a spherical shape, especially for larger beads. If the droplets were collected in a collection bath, *e.g.* a $CaCl_2$ bath for alginate beads, the droplets may be deformed at the liquid surface when entering the bath. For small droplets the problems are minor even at speeds of up to 30 m/s. However, larger droplets, which have such high speeds, will be deformed at the collection bath surface due to their higher weight.

In order to overcome this problem, the droplets have to be pre-gelled prior entering the collection bath. This pre-gelation is achieved by letting the droplets fall through a tunnel (5 m length) equipped with several spraying nozzles (Figure 3, left and middle). The hardening solution from the collection bath is permanently pumped through the spraying nozzles, which generate a fine mist (aerosol) of the hardening solution inside the tunnel. During falling, the spherical droplets are covered with the mist and, thus, are pre-gelled maintaining this spherical shape. The pre-gelation hardens the droplets – in fact they are no droplets anymore but capsules – so that they maintain their spherical shape when they enter the collection bath.

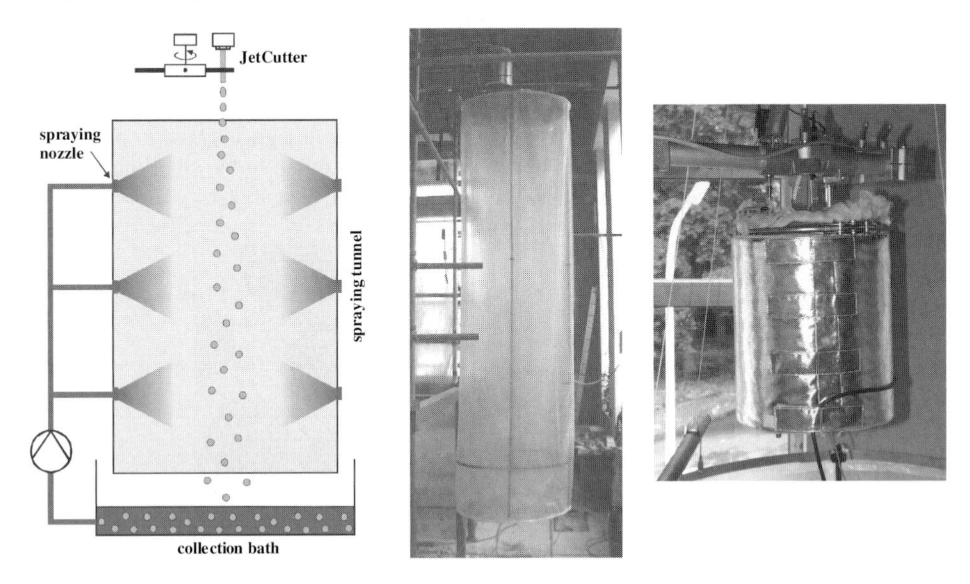

Figure 3. Scheme of the 5 m spraying tunnel (left); photo of the combined spraying/cooling tunnel (middle); photo of the heating device surrounding the JetCutter, the heating device sits on top of the 5 m tunnel.

Another special equipment is needed for the processing of all kinds of hot material, which might be melts, *e.g.* waxes, or hot solutions, *e.g.* gelatin solutions. In order to process such materials not only thermostated tanks, vessels, pumps and tubes have to be used but also the nozzle and the cutting tool have to be heated to avoid clogging. Therefore, the JetCutter has to be surrounded by a heating chamber (Figure 3, right).

The heated JetCutter sits on top of the 5 m tunnel, which in this case acts as cooling line. This cooling line is sufficient to harden small beads, which have high velocities. In order to harden also larger beads, the top of the tunnel might be additionally equipped with a device to distribute cold gas inside the tunnel.

For some applications a one-step coating process is desired, in which one substance has to be coated by another and the spherical shape is maintained. Such coated particles can be applied for protection of sensitive substances, for controlled release systems, for taste masking or for the encapsulation of liquids.

Common pharmaceutical coating technologies do not cover all applications, *e.g.*, they are not well suited for the encapsulation of liquids. In these cases dropping or vibration techniques with a 2-fluid-nozzle are often applied. These nozzles generally consist of a central cannula which is assigned to the core liquid and that is surrounded regularly by a second annular nozzle, which is assigned to the coating liquid. The Jet Cutter technology can also be used for simultaneous coating procedures by using a 2-fluid-nozzle (Figure 4, left).

So far, only preliminary experiments have been carried out concerning this topic. As model system the encapsulation of common vegetable oil inside calcium alginate beads has been investigated. For a better visualisation the oil has been coloured with charcoal. The coloured oil has been used as core liquid whereas a 2% sodium alginate solution has been the coating liquid. The beads have been gathered in a 2% calcium chloride solution, where ionotropic gelation of the alginate with the calcium ions has occurred finally leading to calcium alginate coated oil beads (Figure 4, right). Thus, in principle, the JetCutter can be used for one-step coating processes, but due to the small number of experiments carried out so far, no general conclusion can be drawn, yet.

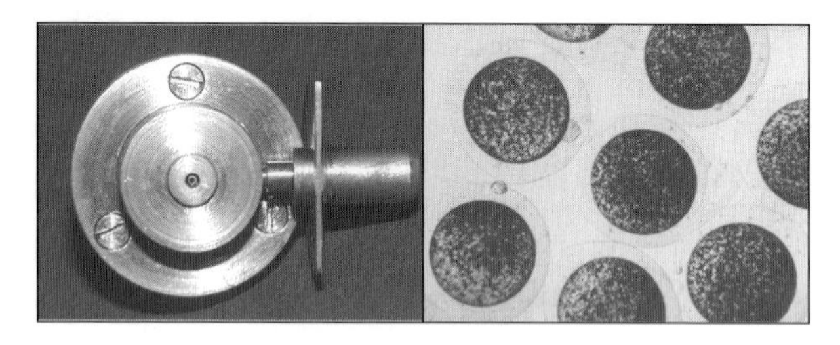

Figure 4. Photo of the 2-fluid nozzle used with the JetCutter (left); Ca-alginate coated vegetable oil beads produced with the JetCutter (right), the oil has been coloured with charcoal.

4. Model of the cutting process

Bead generation by JetCutting is achieved by the cutting wires, which cut the liquid jet coming out of a nozzle. Each cut leads to a cylinder, which afterwards becomes a bead, and a cutting loss (Figure 1). The cutting losses, although minor in their extent as it will be shown later, serve very well as a parameter to discuss the whole cutting process.

Usually, beads of a definite diameter are required. Such beads may be produced with plenty of different sets of parameters, which are the nozzle diameter, the cutting wire diameter, the number of cutting wires in the cutting tool, its rotation speed and the flow rate through the nozzle. In a first simple geometrical approach the influence of the nozzle and the cutting wire diameter shall be discussed.

Beads of a definite diameter can be produced by JetCutting with different nozzle diameters. This means that either short and thick or long and thin cylinders are cut from the jet. By adjusting the right parameters both cylinders have the same volume and, thus, the resulting beads have the same diameter. Obviously, these different ways of cutting, i.e. the nozzle diameter, will have a significant influence on the resulting cutting losses as it is displayed in Figure 5.

Figure 5 clearly demonstrates that the losses dramatically decrease with decreasing nozzle diameter, i.e. if longer and thinner cylinders are cut. Depending of the rheological behaviour of the fluid it is possible to get beads from cut cylinders having a length that is up to 30 times their diameter. The diameter to length ratios shown in Figure 5 range from 1:0.7 (left) to 1:10 (right).

It is even more obvious that the diameter of the cutting wire has an influence on the cutting loss, which is shown in Figure 6. Unfortunately, it is not possible just to always use very small cutting wires. As already mentioned, very small wires are not able to cut thick liquid jets, because the liquid flows together again after the wire has passed it. As a rule of thumb the diameter of the cutting wire should have at least 1/10 of the liquid jet's diameter.

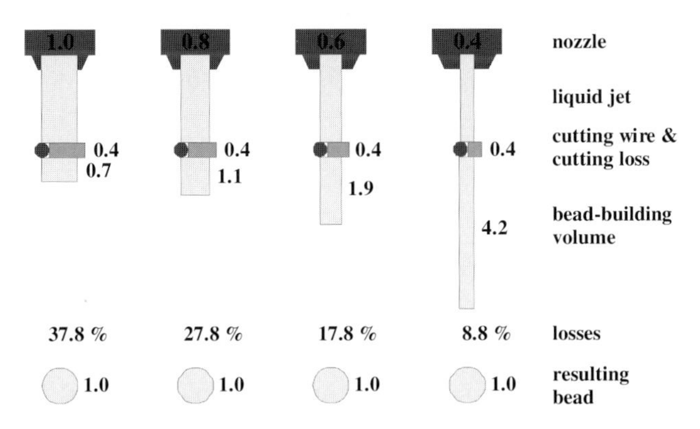

Figure 5. Influence of the nozzle diameter on the cutting losses, unless otherwise written the unit for all numbers is mm, cylinder height is displayed for the bead-building volume.

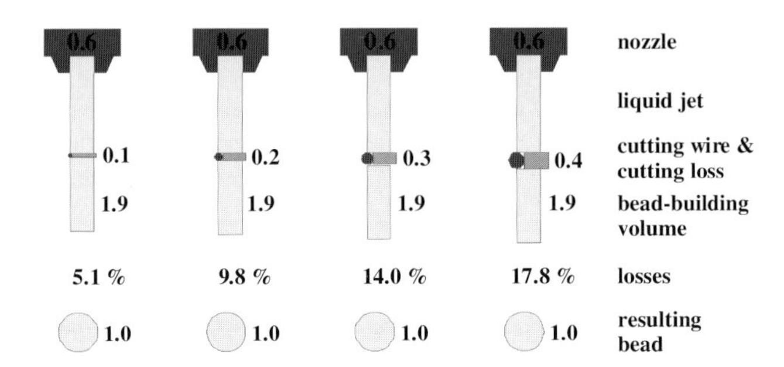

Figure 6. Influence of the diameter of the cutting wire on the cutting losses, unless otherwise written the unit for all numbers is mm, cylinder height is displayed for the bead-building volume.

At first view the cutting process according to this geometrical model is very simple. Nevertheless, at second view the cutting process is more complicated. As discussed so far the cutting process is idealised and only valid if the velocity of the cutting wire is much higher than the velocity of the liquid jet. Actually, these two velocities are in the same range so that the progressive movement of the liquid jet has to be taken into account for a proper description of the cutting process (Figure 7).

Figure 7 shows both a locally and temporally resolved scheme of the cutting process. On the very left hand side, the liquid jet and the positions of the cutting wires at the different times t_1 to t_5 (cutting plane) in relation to the liquid jet are shown. Further to the right, the progressive movement of the jet is shown at each time ($t_{1\text{-}5}$) as well as the actual position of the cutting wire at that time (black circle). Blank circles indicate prior and subsequent positions of the wire. The cutting loss that is pushed out of the jet is also displayed.

In case of a vertical nozzle and a horizontal cutting plane (Figure 7, top) it can be seen that the progressive movement of the liquid jet during the cutting process leads to a diagonal cut through the liquid jet. Accordingly, a proper inclination of the cutting tool should lead to a straight cut through the jet (Figure 7, bottom).

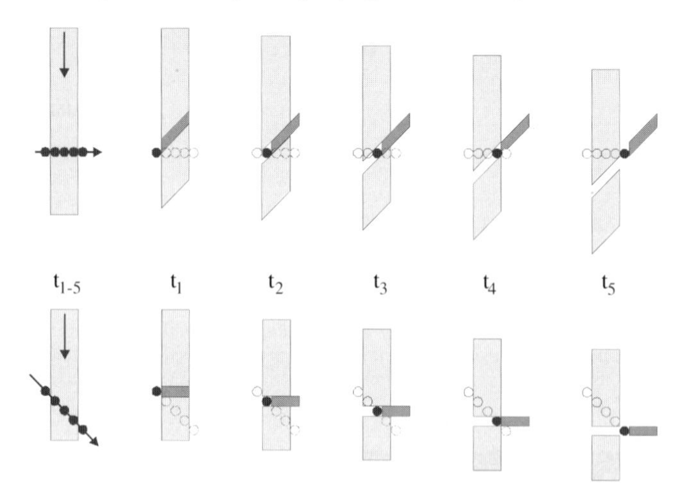

Figure 7. Locally and temporally resolved schematic representation of the cutting process for a horizontal cutting plane (top) and an inclined cutting plane (bottom), actual position of the cutting wire: black circles, former and future positions of the cutting wires: blank circles, cutting loss: dark grey, liquid jet: bright grey, t = time.

Figure 7 displayed that the progressive movement of the jet during cutting leads to a diagonal cut through the liquid jet. Thus, the cutting loss has a somewhat ellipsoidal shape – and is therefore larger than it is in the idealised model in Figure 5 and 6 – and the cut cylinders are distorted (Figure 8). Further, it is imaginable that the ends of the distorted cylinders might be torn off from the rest of the cylinder and form the so-called additional spraying losses. In that case the overall losses generated by the mechanical cut through the liquid jet would be quite high. Nevertheless, it is also shown in Figure 8 that a proper inclination either of the cutting tool or the nozzle leads to a straight cut through the jet with the 'normal' cutting loss and no additional spraying losses.

On the basis of this more sophisticated geometrical model a set of equations displayed in Table 1 has been derived which is capable of describing the cutting process both for a perpendicular arrangement of nozzle and cutting plane (horizontal cutting plane) as well as for an inclined arrangement (inclined cutting plane).

One of the most important parameters for the JetCutter is the ratio of the velocities of the fluid (u_{fluid}) and the cutting wire (u_{wire}). It determines the cutting angle β as well as the proper inclination angle α. It can be taken from Table 1 that for a horizontal cutting plane, the cutting angle not only influences the losses but also the bead diameter, as the bead-building volume is decreased by the additional spraying losses (see also Figure 8). For an inclined cutting plane, the overall losses are the same as the cutting losses, because no additional spraying losses are formed.

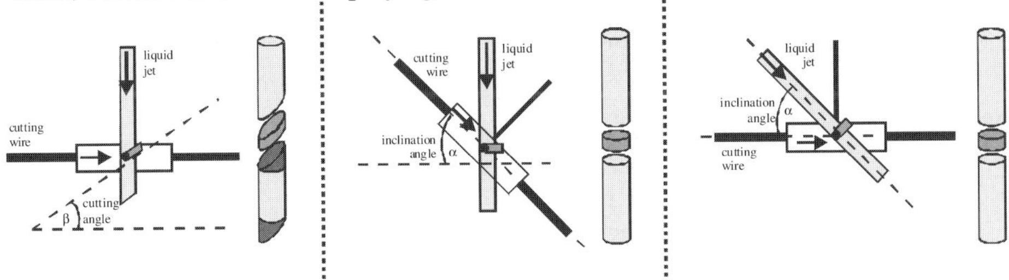

Figure 8. Possible arrangements of nozzle and cutting tool, from left to right: perpendicular arrangement, inclined cutting tool, inclined nozzle.

Table 1. Mathematical model for the cutting process (for details see [1-2]); V_{loss} = volume of the cutting loss, V^*_{loss} = volume of the overall loss, d_{wire} = cutting wire diameter, D = nozzle diameter, n = number of rotations, z = number of cutting wires, d_{bead} = bead diameter; for other abbreviations see text.

Parameter	Horizontal cutting plane	Inclined cutting plane
Angle	$\beta = \arctan\left(\frac{u_{fluid}}{u_{wire}}\right)$	$\alpha = \arcsin\left(\frac{u_{fluid}}{u_{wire}}\right)$
Cutting loss	$V_{loss} = \frac{\pi \cdot D^2}{4} \cdot \frac{d_{wire}}{\cos\beta}$	$V_{loss} = \frac{\pi \cdot D^2}{4} \cdot d_{wire}$
Overall loss	$V^*_{loss} = \frac{\pi \cdot D^2}{4} \cdot \left[\frac{u_{fluid}}{n \cdot z} - \frac{(d_{wire} + D \cdot \sin\beta)}{\cos\beta}\right]$	$V^*_{loss} = \frac{\pi \cdot D^2}{4} \cdot d_{wire}$
Bead diameter	$d_{bead} = \sqrt[3]{\frac{3}{2} \cdot D^2 \cdot \left[\frac{u_{fluid}}{n \cdot z} - \frac{(d_{wire} + D \cdot \sin\beta)}{\cos\beta}\right]}$	$d_{bead} = \sqrt[3]{\frac{3}{2} \cdot D^2 \cdot \left[\frac{u_{fluid}}{n \cdot z} - d_{wire}\right]}$

Table 2 shows the check-up of the model. Here, the experimental values of the overall losses during the production of PVA beads in dependence of the diameter of the cutting wires used are displayed. Two sets of experiments are shown, one with a horizontal cutting plane (cutting losses and additional spraying losses) and one with a properly inclined cutting plane (only cutting losses). For comparison, the theoretical values according to the equations in Table 1 are also given.

Table 2. Overall losses for horizontal and inclined cutting plane in dependence on the diameter of the cutting wire, Exp. = experimental, Calc. = calculation.

Wire diameter, mm	Overall losses, %			
	Horizontal cutting plane		Inclined cutting plane	
	Exp. losses	Calc. losses	Exp. losses	Calc. losses
0.1	8.8	7.9	2.4	2.0
0.2	10.4	10.1	4.0	3.9
0.3	10.2	12.5	6.6	5.8

Table 2 offers three important bits of information:

- The losses decrease if a proper inclination is applied.
- Experimental and theoretical values are in good agreement, so that the model is suited to describe the cutting process.
- By using small cutting wires and an inclined cutting plane, the losses can be decreased down to less than 3%, which means that more than 97% of the initial liquid is transformed into monodisperse beads. Such low losses are tolerable; no loss recycling has to be applied.

Anyway, although the model is able to describe the process in terms of losses and also the resulting bead diameter, it is still a model. Thus, it is not really surprising that reality still looks different (Figure 9).

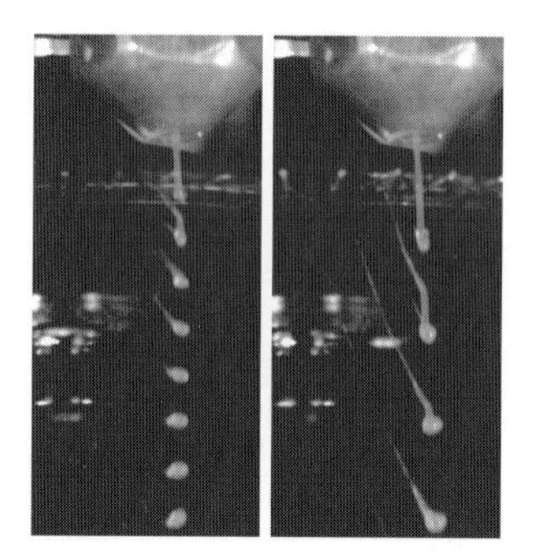

Figure 9. High-speed images of the cutting process during the production of alginate beads, left: high wire velocity and low fluid velocity = small beads, right: low wire velocity and high fluid velocity = large beads.

In Figure 9, two photos of the cutting process taken with a high-speed camera are shown. In both cases a horizontal cutting plane has been applied. It is obvious that the cut segments do not really have a cylindrical shape. Nevertheless, these segments are able to form spherical beads after a short way (Figure 9, left). Regarding the losses, there is a bend end of the cut segment at its top, but it is almost still in contact with the cutting wire, which pulls it from the cut segment like if it was molten cheese (Figure 9, right). But still in this case beads are formed after a while. Anyway, these photos clearly indicate that there is still a lot of work to do in order to fully understand the process of bead formation by JetCutting.

5. Throughput and scale-up

A necessary requirement for bead production in JetCutting is that really a solid jet is pressed out of the nozzle and that this solid jet is maintained until it is cut by the wires. This can be achieved with a combination of special solid jet nozzles and a high fluid velocity (up to 30 m/s), the latter with corresponding high flow rates. Due to the solid jet requirement, the flow rate per nozzle is considerably higher for the JetCutter than for any other bead production technology [3].

The throughput of the JetCutter is best described by the cutting frequency. It determines how often the jet is cut in a definite time period and, thus, how many beads are generated in that time. The cutting frequency is the product of the number of cutting wires in the cutting tool and its rotation speed. Usually, the JetCutter is used with cutting frequencies between 5000 and 10000 Hertz (Hz) (current maximum 14400 Hz), what means that 5000 up to 10000 beads per second are generated. As a rule of thumb it can be said that the higher the viscosity of the fluid is and the smaller the beads size shall be, the higher the cutting frequency might be.

The production rates of a single nozzle JetCutter device for common cutting frequencies are shown in Table 3. The rates are given in terms of L/(h·nozzle). Table 3 indicates that, depending on the desired particle size, even with a single nozzle JetCutter device and common cutting frequencies, the production rate per day can range between a kilogram and several tons of beads.

Table 3. Theoretical throughput of the JetCutter in L/(h·nozzle) for different bead diameters and cutting frequencies (5000, 7500 and 10000 Hz).

Bead diameter, mm	Throughput, L/(h·nozzle)		
	5000 Hz	7500 Hz	10000 Hz
0.2	0.08	0.11	0.15
0.4	0.60	0.90	1.2
0.6	2.1	3.1	4.1
0.8	4.8	7.2	9.7
1.0	9.4	14.1	18.8
1.5	31.8	47.7	63.6
2.0	75.4	113	151
2.5	147	221	295
3.0	254	382	509

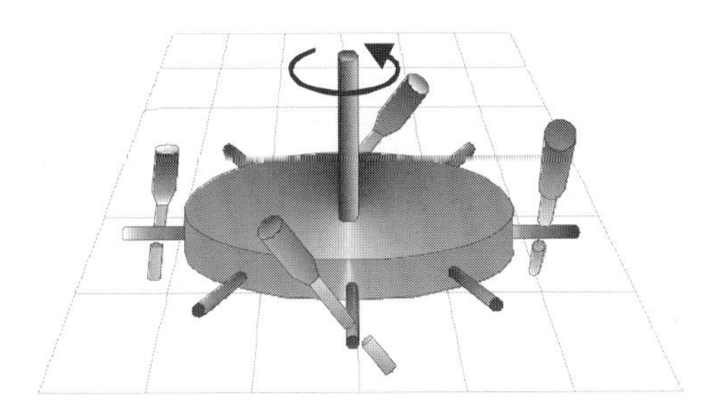

Figure 10. Scheme of a multi nozzle JetCutter operating without additional spraying losses.

Anyway, sometimes these throughputs might not be sufficient, so that a further scale-up is necessary. Generally, for the JetCutter technology two ways of scaling-up are possible. First, a multi-nozzle JetCutting device can be applied, in which the nozzles are staggered near the perimeter of the cutting tool [3]. In this case, special attention has to be paid to avoid additional spraying losses. If a horizontal cutting tool is used with vertically arranged nozzles considerable amounts of additional spraying losses will be obtained, which, of course, is undesirable. The application of an inclined cutting tool, although perfectly suited for a single-nozzle system, is not much better since the additional spraying losses can only be avoided at one single site on the circuit, whereas on the opposite side of the circuit these losses would be even higher than usual. The problem can be solved only if properly inclined nozzles are used together with a horizontal cutting tool. With this arrangement the additional spraying losses can be avoided at any site on the cutting tool's circuit (Figure 10).

The second way for a JetCutter scale-up is the increase of the cutting frequency. A further enhancement of the cutting frequency will be achieved when a motor drive with a higher rotation speed and a cutting tool with more wires are used. Cutting frequencies of up to 25000 Hz are within range. This approach needs not only higher rotations speeds but also a higher throughput per nozzle, which means a higher velocity of the jet and the beads. As already mentioned, the high speed of the beads might cause problems, as they might be deformed when they enter a collection bath. Thus, this approach might only be successful for some special applications.

6. Applications

Since spherical beads are intermediates or products in different industrial sections, *e.g.* pharmaceutical, chemical and food industry, biotechnology, agriculture, many applications exist for the JetCutter technology. Generally, each application field has its own requirements and restrictions concerning the materials to be encapsulated, the type and viscosity of the fluid, the desired particle size or the medium in which the beads should be gathered. In this connection it is advantageous that the JetCutter is capable to process all kind of liquid material covering

- Solutions (alginate, pectinate, chitosan, cellulose derivatives, polyvinyl alcohol (PVA), gelatin, carrageenan),
- melts (waxes, polymers, sugars, sugar alcohols) and
- dispersions (emulsions, suspensions, inorganic sols).

Usually, something is encapsulated inside the beads in order to get a formulation of an active agent, a controlled-release system, a protection against impacts from the environment, a taste-masking system or an immobilised catalyst. Such substances might be:

- Catalysts (enzymes, bacteria, fungi, chemical catalysts),
- active agents (pharmaceuticals, pesticides),
- ingredients (vitamins, amino acids, probiotics),
- aromas and fragrances,
- pigments and dyes,
- particles (titanium dioxide, zirconium dioxide, magnetite).

Some examples of beads of pure polymer solutions as well as some encapsulated substances produced with the JetCutter are displayed in Figure 11.

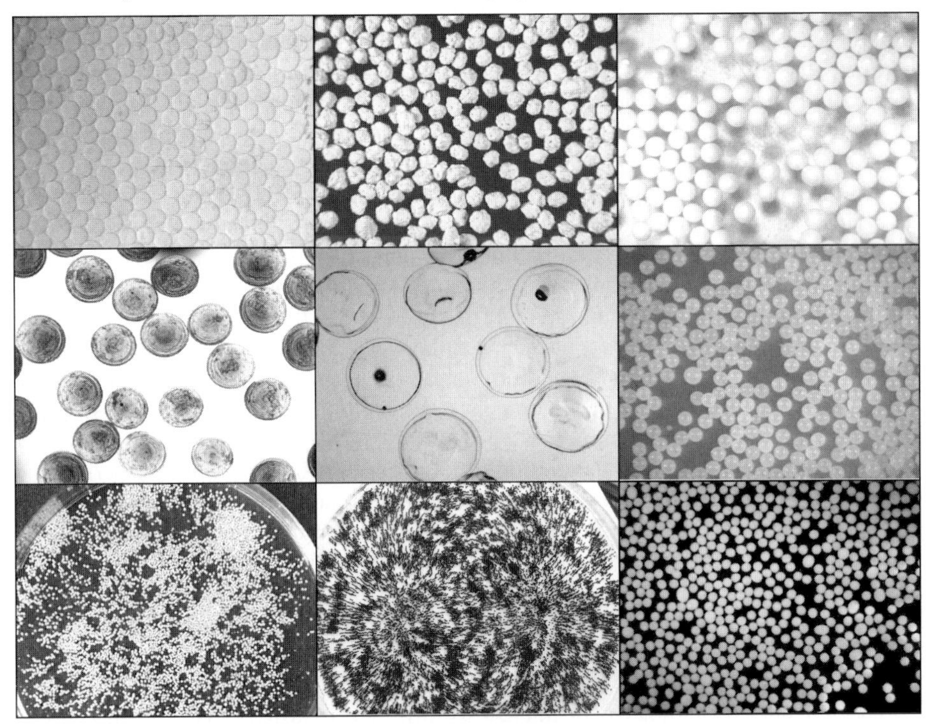

Figure 11. Photos of different types of beads prepared by Jet Cutting (not in true scale). From top left to bottom right: Ca-alginate (Ø = 0.6 mm), bacteria in Ca-alginate, freeze-dried (Ø = 0.3 mm), vitamin in Ca-pectinate (Ø = 0.5 mm), chitosan (Ø = 0.5 mm), gelatin (Ø = 0.8 mm), wax (Ø = 0.7 mm), PVA (Ø = 0.5 mm), 20 % magnetite in chitosan (Ø = 0.8 mm), 40 % TiO_2 in alginate (Ø = 1.3 mm).

As already mentioned, one of the major advantages of the JetCutter is that the viscosity of the fluid does not limit bead generation. That means that not only the formulation recipes used at the moment can be applied to bead production by JetCutting but also those whose transformation into products failed due to a too high viscosity. For the same reason biological matter can be treated at lower temperatures, i.e. more carefully, with the JetCutter since heating for viscosity reduction is not needed. Further on, encapsulation matrices originating from solutions with a high polymer content and corresponding high viscosities generally form mechanically superior beads. This is especially advantageous for the production of immobilised biocatalysts [4-8]. For this purpose encapsulation matrices based on polyvinyl alcohol (PVA) are particularly suited, as PVA hydrogels are very stable, do not show any abrasion and are not biodegradable. Such beads were also successfully used for the encapsulation of metal catalysts [9].

7. Summary and prospect

Solid particles (pellets, beads) in the size range between μm and mm play an important role in various industries like biotechnology, agriculture, chemical, pharmaceutical and food industry. Thus, plenty of particle production technologies exist. Some of them, like simple dropping, electrostatic-enhanced dropping or vibration are especially suited for lab-scale applications, others, like rotating disc and nozzle technologies and to some extent also the vibration technique, are also suited for technical applications.

With all these technologies either monodisperse beads in small amounts or broadly distributed beads in large amounts may be produced and all these technologies are limited concerning the viscosity of the fluids, which can be processed.

The only technology so far, which is able to transform not only low viscosity but also highly viscous solutions into monodisperse beads, is the JetCutter. The bead size, which is accessible by the JetCutter, range between approx. 200 μm up to several millimetres. The throughput is suited both for lab-scale and technical scale applications. The principle of function and the equipment is rather simple, which makes the JetCutter quite attractive for industrial purposes.

A mathematical model of the cutting process exists, which is able to describe very well the bead formation in dependence of the main parameters. *Vice versa*, suitable parameters can be estimated from this model in order to produce the desired beads.

Bead generation by JetCutting is not possible without the production of losses. Quite a lot has been written about the losses. It is necessary to point out that the reason for that is that the losses are very well suited to describe the whole cutting process and not that the losses are so important. In fact, the losses can be entirely regarded as negligible – less than 2% – if the JetCutter is run in the right way.

The JetCutter technology is of interest for different industries, like pharmaceutical, chemical and food industry, biotechnology or agriculture. As a broad variety of solutions, melts and dispersions are processable, the JetCutter is useful for manifold applications such as the formulation of active agents, the preparation of controlled-release and taste-masking systems, the encapsulation of ingredients or the immobilisation of (bio-) catalysts [10].

None of the other technologies for bead production shows this sum of advantageous characteristics as the JetCutter does. It is, at least, an alternative and completion to other techniques. Whether it is even more will be seen during the next years. Although the JetCutter is a quite novel technology, the first industrial scale bead production process based on the JetCutter technology has already started in summer 2002.

References

[1] Pruesse, U.; Fox, B.; Kirchhoff, M.; Bruske, F.; Breford, J. and Vorlop, K.-D. (1998) New process (JetCutting method) for the production of spherical beads from highly viscous polymer solutions. Chem. Eng. Technol. 21: 29-33.

[2] Pruesse, U.; Bruske, F.; Breford, J. and Vorlop, K.-D. (1998) Improvement of the JetCutting method for the preparation of spherical particles from viscous polymer solutions. Chem. Eng. Technol. 21: 153-157.

[3] Pruesse, U.; Dalluhn, J.; Breford, J. and Vorlop, K.-D. (2000) Production of spherical beads by JetCutting, Chem. Eng. Technol. 23: 1105-1110.

[4] Pruesse, U.; Fox, B.; Kirchhoff, M.; Bruske, F.; Breford, J. and Vorlop, K.-D. (1998) The JetCutting method as new immobilisation technique. Biotechnol. Tech. 12: 105-108.

[5] Muscat, A.; Pruesse, U. and Vorlop, K.-D. (1996) Stable support materials for the immobilisation of viable cells. In: Wijffels, R.H.; Buitelaar, R.M.; Bucke, C. and Tramper, J. (Eds.) Immobilized cells: Basics and applications. Elsevier Science, Amsterdam (The Netherlands); pp. 55-61.

[6] Leidig, E.; Pruesse, U.; Vorlop, K.-D. and Winter, J. (1999) Biotransformation of poly R-478 by continuous cultures of PVAL-encapsulated *Trametes versicolor* under non-sterile conditions. Bioprocess Eng. 21: 5-12.

[7] Reimann, C.; Pruesse, U.; Welter, K.; Willke, Th. and Vorlop, K.-D. (1997) Stoffkonversion nachwachsender Rohstoffe durch einschlussimmobilisierte Biokatalysatoren (in German). VDI-Berichte 1356: 211-214.

[8] Jahnz, U.; Wittlich, P.; Pruesse, U. and Vorlop, K.-D. (2001) New matrices and bioencapsulation processes. In: Hofman, M. and Thonart, P. (Eds.) Engineering and manufacturing for biotechnology, Volume 4; Hofman, M. and Anné, J. (Ser. Eds.) Focus on biotechnology. Kluwer Academic Publishers BV, Dordrecht (The Netherlands); pp. 293-307.

[9] Pruesse, U.; Morawsky, V.; Dierich, A.; Vaccaro, A. and Vorlop, K.-D. (1998) Encapsulation of microscopic catalysts in polyvinyl alcohol hydrogel beads. In: Delmon, B.; Jacobs, P.A.; Maggi, R.; Martens, J.A.; Grange, P. and Poncelet, G. (Eds.) Studies in Surface Science and Catalysis, Volume 118, Preparation of Catalysts VII. Elsevier Science, Amsterdam (The Netherlands); pp. 137-146.

[10] Pruesse, U.; Jahnz, U.; Wittlich, P.; Breford, J. and Vorlop, K.-D. (2002) Bead production with JetCutting and rotating disc/nozzle technologies. Landbauforsch. Völkenrode SH 241: 1-10.

INDUSTRIAL SCALE ENCAPSULATION OF CELLS USING EMULSIFICATION/DISPERSION TECHNOLOGIES

RONALD J. NEUFELD[1] AND DENIS PONCELET[2]
[1]*Department of Chemical Engineering, Queen's University, Kingston, Ontario, Canada, K7L 3N6 – Fax: +1 613 533 6637 – Email: neufeld@chee.queensu.ca*
[2]*ENITIAA, rue de la Géraudière, BP 82225, 44322 Nantes cedex 3, France – Fax: +33 2 51 78 54 67 – Email: poncelet@enitiaa-nantes.fr*

1. Introduction

Industrial scale immobilization of cells *via* entrapment within ionic or thermal gelling polymer systems, or *via* microencapsulation within semi-permeable membranes, necessitates the use of emulsification/dispersion steps to generate small and often micron-sized droplets, on a large scale. Individual droplets within the emulsion contain the biocatalyst, and lead to the final gelled microsphere or membrane-coated microcapsule. Widely used laboratory techniques involve the formulation of droplets and thus beads *via* individual droplet extrusion technologies, which are well suited to the small scale formulation of monodisperse beads and capsules, often in a millimetre size range. These techniques are poorly suited to very large scale formulation, particularly when droplets or microbeads are preferred, with diameters often extending to well under 500 μm [1]. Smaller diameter preparations are often desired to improve mass transfer characteristics, but droplet extrusion techniques become increasingly limited in productivity with decreasing bead size, particularly when using viscous polymer solutions.

Emulsion/dispersion systems overcome the limitations of single droplet extrusion methodologies, as they are generally not limited by scale, and control of the resulting microbead diameter is possible through selection of appropriate dispersion devices (vessels, baffles, impellers) and operating conditions (mixing rates, phase ratios, surfactants), leading to a wide range of possible bead diameters, extending from a few microns, up to within the millimetre size range. Scale is limitless, as vessel size becomes the design criterion.

There are five challenges associated with emulsion/dispersion systems:

- Challenge 1 involves the need for a non-toxic, biocompatible, non-aqueous dispersion phase, into which the gel sol is dispersed, droplets formed and

V. Nedović and R. Willaert (eds.), Fundamentals of Cell Immobilisation Biotechnology, 311-325.

gelation or polymerization initiated. Vegetable, mineral or silicone oils are commonly used.

- Challenge 2 requires that gelled beads, microspheres or microcapsules be separated easily and cleanly from the oil phase. The smaller the bead, the greater are the separation problems. As gelation polymers are generally hydrophilic, simple phase partitioning may be an easy and quick solution; however depending on the polymer, there may be some residual oil coating the microbeads or microcapsules, which is normally removed through washing with mild surfactant.
- Challenge 3 involves a better control over the microsphere size distributions, associated with emulsion/dispersion systems. While the operator has considerable flexibility over the control of mean bead diameter covering a considerable size range, thus far it has not been possible to produce monodisperse preparations of microspheres using emulsion technology. For some applications, this can be problematic, as a classification of the microspheres can lead to a serious reduction in yield, depending on the extent of the size dispersion.
- Challenge 4 is associated with the economic and environmental need to reuse, or recycle the oil phase used in the process. Disposal of large quantities of oil is costly, difficult and environmentally unacceptable.
- Challenge 5 relates to the ability to operate biocatalyst encapsulation operations as a continuous process. In fact, dispersion systems are readily adaptable to continuous operations, thus for the same level of productivity in comparison to a batch process, smaller scale of process equipment is required, and better uniformity in product quality possible.

The examples provided in this chapter, serve as case studies to illustrate several emulsion/dispersion systems used for cell encapsulation. The case studies are presented in the following order:

- simple thermal gelation
- thermal/ionotropic gelation
- continuous processing involving thermal/ionotropic gelation
- ionotropic gelation involving physicochemical reaction with membrane coating
- microencapsulation *via* interfacial cross-linking
- microencapsulation *via* interfacial polymerization reaction

All examples involve emulsion/dispersion systems. Detailed protocols are available in the references cited.

2. Thermal gelation of agarose microspheres containing recombinant *Saccharomyces cerevisiae* in bioconversion of fumarate to L-malate

Temperature setting gels are important for many applications in the food industry, because of their viscosifying and gel setting properties. Gel forming polymers are often heated to facilitate dissolution, then form a gel upon cooling. Most are familiar with the use of semi-solid agar medium, in the preparation of petri dishes for cell culture. Gel beads can be formed using a similar principle, however because of the need to form small diameter spheres, it is necessary to either extrude the gel forming solution dropwise into a chilled bath, or alternatively, form an emulsion/dispersion of the gel forming sol, into warm vegetable oil, then initiate gelation by dropping the temperature of the emulsion. The faster the temperature drop to below the gel point, the more likely that the resulting gel beads are discrete and spherical. In addition, through careful control of the conditions under which the emulsion is prepared, it is possible to design microspheres with fixed mean diameter, and narrow size distribution. The biocatalyst is added to the gel forming solution prior to emulsification, and it is obvious that the cells would need to tolerate the elevated gel temperature until such time as the gel is set.

One particular encapsulation application [2] involves the emulsion/dispersion of agarose sol containing cells, into corn oil at a ratio of 1:4. As agarose forms a gel below 40°C, the emulsion was held at 40°C in a temperature regulated, water-jacketed mixing vessel, and gelation initiated by rapid cooling of the emulsion. After 15 min, gelled microspheres were partitioned into water, separated by filtration, washed to remove traces of oil and assayed. Agarose microspheres readily partitioned in the aqueous phase with essentially 100% recovery. There was no residual oil remaining with the microspheres, confirmed microscopically, and through a complete absence of oil film during subsequent repeated assay of the encapsulated cell activity. It was estimated that the microspheres contained 10% by weight cells. The diameter of the microspheres was controlled by varying the mixing rate during emulsification. Wide size distributions resulted, with standard deviations representing approximately 50% of the microsphere mean diameter. Recycle of the corn oil was not tested, but would be readily achieved as the oil was not modified during the encapsulation process.

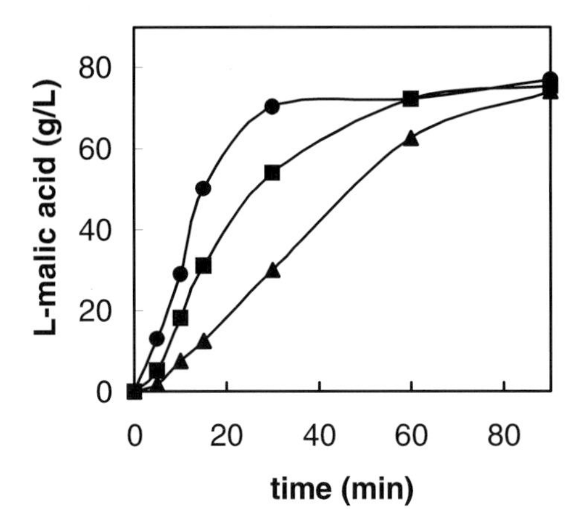

Figure 1. Bioconversion of fumaric to L-malic acid with an agarose-immobilized transformant yeast in 193 μm (●), 871 μm (■) and 2.4 mm (▲) diameter beads. Redrawn from Figure 5 in Reference [2].

The bioconversion of fumaric to L-malic acid was demonstrated with a recombinant *Saccharomyces cerevisiae* encapsulated in the agarose microspheres. Three bead mean diameters were tested, 193 μm, 871 μm and 2400 μm. The rate of bioconversion increased with decreasing bead diameter as shown in Figure 1, explained by reduced mass transfer limitations. Similar bioconversion rates were observed between cells encapsulated within the smallest diameter microspheres (193 μm), to that of free cells. Based on regression analysis, it was shown that activities similar to that of free cells could be obtained with microsphere diameters less than approximately 300 μm. Stable activities over a 48 h period were observed with the encapsulated cells.

3. Thermal/ionotropic gelation in formulation of gellan gum microbeads containing gasoline-degrading microorganisms

Gellan gum is an interesting polymer for encapsulation, even though it has not received the attention of more widely available polysaccharides such as alginate and carrageenan. It is known to have superior rheological properties [3] leading to excellent mechanical and thermal stability [4], yet is probably more expensive due to it being produced *via* a fermentation process. Also in contrast to other ion-sensitive gelling polysaccharides (e.g. alginate), gellan-ion interactions are non-specific and gels can be formed with various cations [5]. The formation of gellan gum microspheres requires emulsion/dispersion, but in contrast to the case of agarose described above, gellan gum is a thermally gelling polysaccharide, but in addition requires cationic interactions. The following procedure is outlined to illustrate the principle [6].

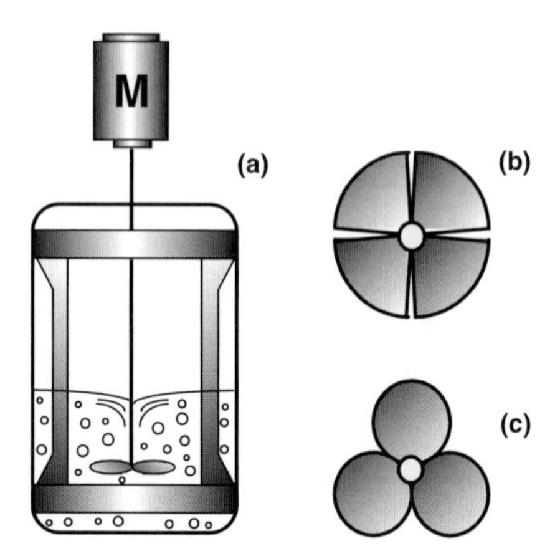

Figure 2. Design of baffled vessel and impellers used to form gellan gum microspheres by emulsification/internal gelation. Vessel is 10 cm wide, and quarter-circular paddle, and three-blade paddles are 5 cm diameter. (Figure courtesy of Dr. Peyman Moslemy)

The configuration of the baffled mixing vessel used to generate a gellan gum-canola oil emulsion under sterile and aseptic conditions is illustrated in Figure 2. The pre-gel solution consisted of 0.75% gellan gum in a solution of $CaCl_2$. A mixed bacterial culture was added to the sol at 45°C, and emulsified in warm canola oil facilitated with Span 80, an oil soluble surfactant. The temperature of the dispersion was quickly dropped to 15°C to initiate gelation, and the microbeads then partitioned into water, and washed with sterile 0.1% Tween 80. Cell mass loading ranged up to 20 g_{cells} L^{-1} sol, and mean diameter and size distribution were determined as a function of emulsion mixing rate, gellan sol volume fraction, emulsifier concentration and emulsification time.

Microbead mean diameter was predetermined, within the range of 12 μm to 135 μm to enable hydraulic distribution through a granular matrix. Control of mean diameter was exercised by varying the mixing rate, emulsifier concentration and volume fraction of gellan gum in oil dispersion. Microbeads were spherical, and dispersions were unimodal and followed a log-normal distribution. One particular batch (0.75% gellan, 0.06% $CaCl_2$, 8 g_{cells} L^{-1} sol, 0.1% emulsifier, gellan gum/oil volume fraction of 0.143, emulsification time 10 min, and stirring rate of 4500 rpm) showed a mean diameter of 21 μm and a distribution size range from 16 to 34 μm (span of 0.2). Mean diameter decreased with increasing mixing rate and emulsifier concentration and as an ascending function of gellan gum volume fraction in the emulsion.

A pre-acclimated, mixed bacterial culture (MBC) was encapsulated for the bioaugmentation of a gasoline-contaminated aquifer. The carrier microbeads were to be sufficiently small to be transported through a granular soil matrix, using a formulation methodology that would be readily scaled to that required for environmental application. The objective of immobilization was to protect cells from biotic and abiotic stresses, including the inhibitory effects of the petroleum substrate. It was also desired to improve on the distribution of microorganisms through the subsurface environment, as free microorganisms tend to adhere to soil grains [7]

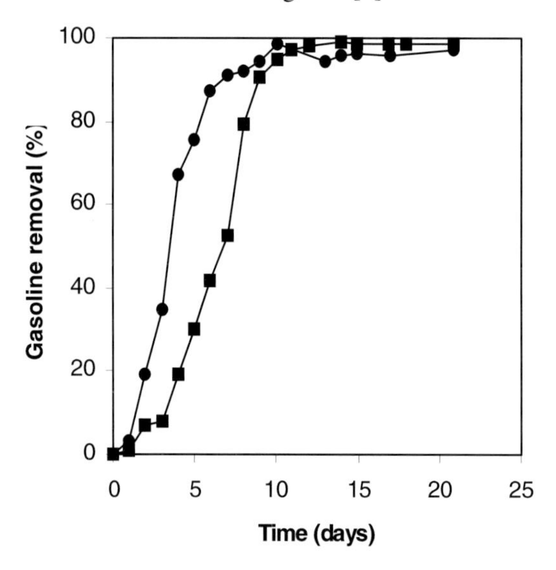

Figure 3. Biodegradation of gasoline by gellan gum encapsulated cells (●), and by free (■) mixed bacterial culture. The mean diameter of the microbeads was 23 μm. Redrawn from Figure 7 in Reference [6].

Encapsulated cells removed over 98% of gasoline hydrocarbons (Figure 3), and a reduced lag in comparison to free cells was explained by enhanced protection of the cells by the encapsulating matrix. In this application, the oil phase following emulsion gelation could potentially be recycled in the process, as it had not been modified in any way. The fate of the surfactant added to facilitate emulsification is uncertain, but since it is oil soluble, it was assumed to have remained with the oil phase, thus additional surfactant would not need to be added for the second immobilization cycle.

4. Continuous process for the thermal gelation of κ-carrageenan microspheres containing brewer's yeast, using static mixer technology

κ-Carrageenan is a thermal setting gel, with a gel point temperature, dependent on the concentration of both the polymer solution, and that of the countervailing cation - KCl in this case. Lower gelation temperature carrageenans are increasingly available, and through appropriate control of the polymer and KCl concentration, are better suited to biocatalyst encapsulation. This particular example is being provided to illustrate a large-scale encapsulation process with interesting potential for continuous operation. The process was also conducted aseptically, demonstrating its potential in pure culture encapsulation.

The unit operation of interest in this process involves the use of static mixers. The mixers were used to carry out two functions, firstly to disperse the yeast paste into the viscous carrageenan sol, and secondly, to achieve the emulsion/dispersion of the yeast-carrageenan sol into vegetable oil, at a temperature above the gel-point of the polymer. The emulsion was subsequently quick-cooled, causing the carrageenan droplets to gel. Static mixers are constructed with a series of elements, fixed transversely within a pipe as shown in Figure 4. The elements repeatedly cut, rotate and recombine the flowing fluid, where the mixing energy generated is sufficient to disperse, or to emulsify viscous mixtures such as carrageenan and vegetable oil. It is desired that the more uniform shear, in contrast to a turbine impeller in a baffled reactor for example, promote an increasingly homogeneous dispersion [8] at flow rates typical of those encountered at the industrial scale (several to hundreds L h^{-1}). Control of microsphere properties is thus possible through a wide range of operating properties, including number of mixer elements, fluid velocity, mixer diameter, and carrageenan volumetric fraction in the dispersion.

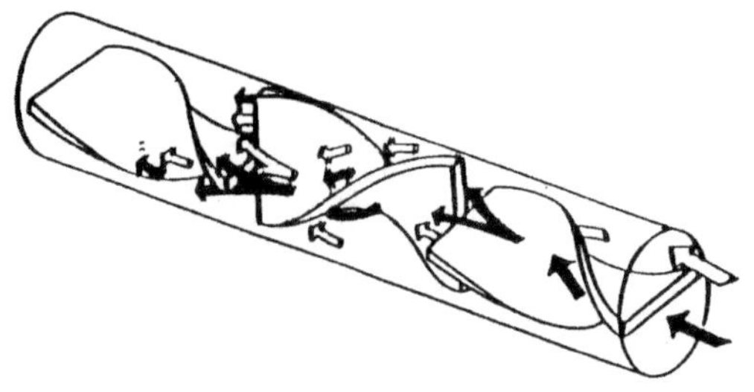

Figure 4. Design and operation of static mixer, used to disperse/emulsify two immiscible fluids. Note the static elements, which sequentially divide, rotate and recombine the flowing fluids, represented by black and white arrows.

Brewing operations have traditionally been carried out as batch processes, with fermentation times of typically 6-7 days. Thus economical, smaller scale, continuous processes for the primary and secondary fermentation stages are being examined. As alginate may be unstable in a wort-based medium, carrageenan has been considered as an alternative since potassium replaces calcium as the gelling ion. From a practical point of view, immobilized cells must be produced on a large scale and be stable for relatively long operational times, in the order of weeks or months. Mass transfer limitations are of critical interest [9], as it impacts on the flavour profile of the product.

Brewer's yeast was encapsulated in gelled κ-carrageenan microspheres [10] as illustrated in the process flowsheet shown in Figure 5. Carrageenan polymer and canola oil were pre-sterilized by heating, followed by cooling and holding at 40°C. Yeast was mixed with the carrageenan sol through the first static mixer, and the sol-yeast mixture was dispersed into warm oil *via* a second static mixer. The second mixer contained 12 mixing elements with a diameter of 12.7 mm. The resulting emulsion was rapidly chilled to 5°C, initiating gelation of the carrageenan droplets. Carrageen gel microspheres were then partitioned into cold sterile KCl solution (22 g L^{-1}). Process oil was then recycled back to the feed tank for reuse.

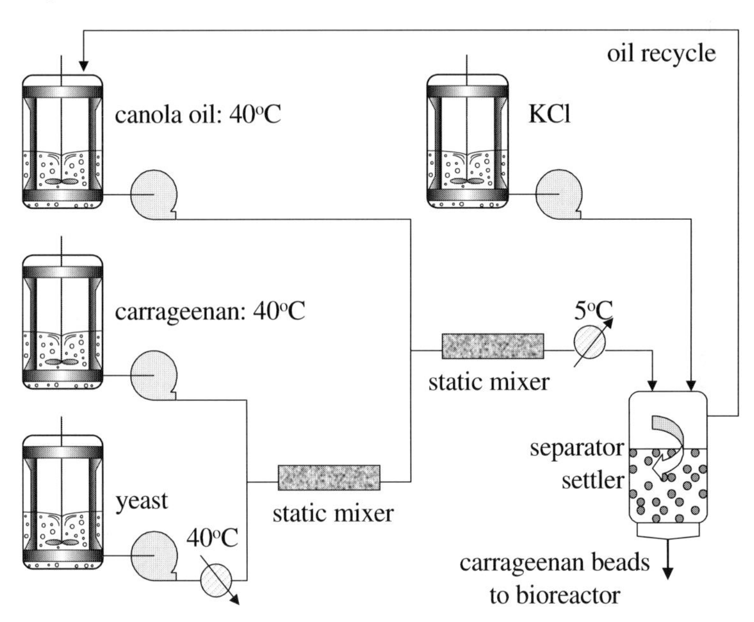

Figure 5. Flowsheet for the production of carrageenan microspheres containing brewer's yeast, in continuous process. Note the use of static mixers for both the dispersion of yeast into carrageenan sol, and for the emulsion/dispersion of the yeast carrageenan suspension into warm canola oil. Emulsified carrageenan sol is gelled by rapid cooling and partitioning into KCl solution.

Yeast loaded, carrageenan beads with a mean diameter of 1 mm and a coefficient of variability (standard deviation/mean diameter) of 40%, were then loaded into a 50 L airlift bioreactor as illustrated in Figure 6 for carrying out a primary beer fermentation. A gas mixture of CO_2:air (98:2) was sparged to the reactor at a volumetric rate of 0.3 vvm. Beer wort sugars were consumed in 22 h, compared to a conventional fermentation with free cells, requiring 6 to 7 days. Measured flavour constituents and taste panel analysis found the flavour comparable to a conventional brew.

5. Emulsification/ internal gelation of alginate, forming membrane-coated microspheres in the encapsulation of *Lactococcus lactis*

Many biological encapsulants including living cells cannot tolerate the elevated temperatures typical of thermal gelation systems (40 to 60°C). Alginate is by far the most widely used encapsulation matrix, for several reasons. Gels can be formed at ambient temperatures and alginate polymer is inexpensive, widely available, biocompatible and forms reversible gels in the presence of multivalent cations such as Ca^{++} under gentle formulation conditions. Over the last decade, more suppliers of alginates are appearing in the market place, the quality of the polymer is improving, and alginates are now being sold partially or fully characterized in terms of its chemical and physicochemical properties and are available as food or medical grade material. The conventional method for formulating alginate beads containing entrapped cells is to extrude alginate sol dropwise, from a syringe needle into a gelation bath, containing soluble calcium salt ($CaCl_2$). Beads of uniform size are produced, however reduction in bead diameter is limited by the viscosity of the alginate sol and as a result, beads less than 500 μm are difficult to produce, on anything but a lab scale. Various simple modifications to this basic procedure have been introduced, such as the use of multiple needles, electrostatics, vibration, droplet propulsion from the needle tip by concentric airflow, and liquid jet cutters. Commercial encapsulators have now appeared on the market and appear popular amongst those working with these various droplet extrusion technologies. For beads in the millimetre size range, production rates in the order of 1 to 2 L h^{-1} are possible, and multiple needles permit small-scale industrial production, ranging to 10 L capacity. The most important limitation to these techniques is that they are still not suited to industrial scale application, particularly when beads within a sub-millimetre size range are desired.

Various attempts to use emulsion techniques to form ionic polysaccharide gel beads have been awkward or still require single droplet extrusion methods. For example, a hot carrageenan/oil emulsion was dropped into cold water [11] and an oil/alginic acid emulsion was extruded dropwise into $CaCl_2$ solution to encapsulate oil droplets in alginate [12].

Pelaez and Karel [13] proposed a method to form alginate gel slabs, in a procedure termed "internal gelation". The alginate sol was placed in a mold and gelled through the internal liberation of calcium ions from insoluble citrate complex *via* spontaneous breakdown of gluconolactone. The pH of the alginate would drop due to the slow formation of gluconic acid, leading to the liberation of soluble calcium from the insoluble salt. Alginate molds would set in 5 min. The extension of this technology to the formation of alginate beads or microspheres in an emulsion system is difficult, because of the inability to trigger gelation on command, the slow gelation rate leads to aggregation due to slow gelation of the emulsified gel droplets, and the lack of control over pH in the gel.

Through a number of refinements to the internal gelation concept, adapted toward the production of gel microbeads, a novel procedure - termed "emulsification/internal gelation" was developed [14, 15, 16]. This procedure is illustrated in Figure 7 and involves the dispersion of cell encapsulant into alginate sol at a ratio of up to 1:1. Insoluble calcium salt - typically $CaCO_3$ - is then mixed into the alginate sol (500 mM Ca^{2+} equivalent, or 25 mM calcium salt). Various calcium salts may be used, although carbonate provides for near instantaneous release of soluble calcium with a slight pH adjustment in a neutral pH range, as shown in Figure 8, a range best suited to active biologicals. The alginate/calcium carbonate mixture containing cells is then dispersed into an oil phase (vegetable, mineral or silicone oil). Various mixing devices, typical of those used in food and chemical processing (see Figure 9), including static mixers, provide considerable level of control over droplet (microsphere) mean diameter and size distribution [15, 16]. Once the emulsion is formed, gelation is initiated by pH adjustment with an oil soluble acid, such as acetic acid. Approximately 0.66 mL of acetic acid is required per L of emulsion, assuming an alginate sol concentration of about 2% (pH change from 7.5 to 6). Gelation should be essentially instantaneous, otherwise aggregation of the gel microspheres may result. Following about 5 min of gelation, the emulsion is mixed gently with an aqueous phase, and microspheres partitioned from the oil dispersion. This is followed by filtration and washing with mild surfactant to remove residual oil. Recently, direct filtration of microspheres from oil was successfully attempted using fine wire mesh.

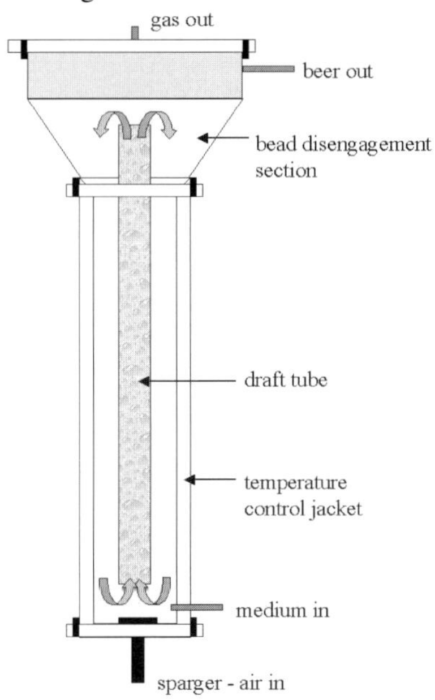

Figure 6. Gas-lift 50L bioreactor for continuous brewing process using carrageenan immobilized yeast. Figure adapted from Figure 3 in Reference [10].

An important aspect of this technique involves the need for calcium carbonate microcrystals, approaching a few microns in diameter. The smaller the calcium carbonate grain size, the more grains can be introduced within smaller and smaller microdroplets, and the faster is their dissolution rate. Resulting microspheres are thus more spherical, and less likely to aggregate. There should be no residual carbonate grains in the microsphere once gelation is complete.

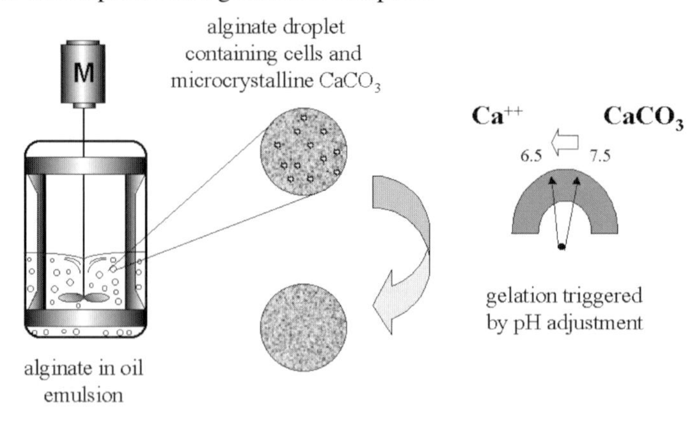

Figure 7. Cell encapsulation by emulsification/internal gelation of alginate. Alginate containing cells and calcium as insoluble microcrystalline salt is dispersed into vegetable oil. Calcium - Ca^{++} is released in situ by pH adjustment with an oil soluble acid - acetic acid, triggering instantaneous gelation, forming alginate microspheres.

Smooth, spherical microspheres with the narrowest size dispersion were produced using a low-guluronic acid and low-viscosity alginate, and carbonate as the calcium release vector. Microsphere mean diameters ranging from 50 to 1000 μm were produced, with standard deviations ranging from 35 to 45% of the mean [16].

The following is provided as an example of cell encapsulation using emulsification/internal gelation of alginate. *Lactococcus lactis* was encapsulated [17] in alginate microspheres for use in the diary industry for the manufacture of yogurt, fermented creams, and cottage cheese dressings. Over acidification can be avoided by removal of the encapsulated cells, facilitating their reuse, and extending the shelf life of the dairy product. Microspheres containing entrapped cells were spherical, with diameters extending down to 50 μm, depending on the formulation conditions. Resulting alginate microspheres were coated with poly-L-lysine, a polycationic polymer that forms a membrane coat around polyanionic alginate. Microspheres were immersed in 0.02% poly-L-lysine solution for 20 min to permit membrane formation. The purpose of the membrane was to limit the rate of release of cells from the encapsulation matrix. At high concentrations of encapsulated cells, approaching 10^9 cells mL^{-1} milk, similar rates of lactic acid production were observed between encapsulated cells, and controls consisting of an equivalent level of free cells.

This procedure has only to a limited extent been tested on a continuous basis, involving the use of static mixers. One difficulty, which would need to be overcome, requires the need to recondition the slightly acidified oil, prior to re-use. It may be that residual acetic acid has been fully removed through aqueous phase partitioning during the microsphere separation step, but this has not as yet been tested. It is unlikely that the oil has been hydrolyzed to any extent by the gentle acid treatment, but again this question has not as yet been addressed.

In the application described above (encapsulation of lactic cultures within poly-L-lysine coated alginate microspheres), the alginate microsphere served as a core matrix around which a thin membrane coat was applied. In this section, a microencapsulation approach will be described, by which a membrane can be formed by the cross-linking of pre-formed polymer through an interfacial reaction. The resulting microcapsule consists of a liquid core, contained within a thin, semi-permeable membrane coat. Individual microcapsules are formed by the emulsion dispersion of liquid biocatalyst suspension into an appropriate solvent or oil phase, followed my membrane formation around the liquid microdroplets. Following membrane formation, the microcapsules are separated from oil suspension through aqueous phase partitioning, or filtration.

The key in encapsulating living cells by interfacial cross-linking, is to maintain a separation between the cells and the toxic cross-linking reagent. Glutaraldehyde is an obvious choice of cross-linker because it is oil soluble, but since it is also water soluble, it is difficult to exclude from the liquid microcapsule core. The following examples involve similar applications, but in both cases include the use of strictly oil soluble cross-linkers. The examples are interesting to examine, because in one case, it involves a cross-linking reaction with a protein (gelatin) and in the other case, it involves the cross-linking of a polysaccharide (chitosan). Both of these materials are of interest for use in food and pharmaceutical products. The technique is easily adapted for use in other cell immobilization applications.

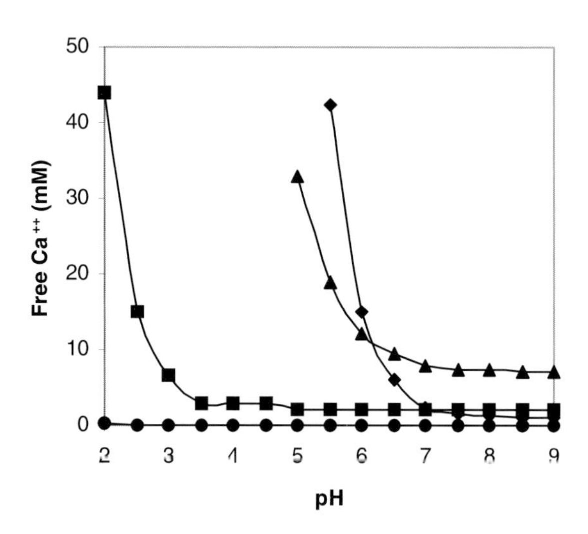

Figure 8. Free calcium concentration at different pH levels for various calcium salts. Redrawn from Figure 6 in Reference [16]

6. Microencapsulation of lactic cultures by interfacial cross-linking reaction involving pre-formed biopolymer

The bioencapsulation of *Lactococcus lactis* within cross-linked gelatin membranes was achieved as follows. Culture, suspended in 24% gelatin solution, was emulsified into either sunflower oil or silicone oil, facilitate by a surfactant [18]. Mixing was provided by a sheet lattice impeller as illustrated in Figure 9. The ratio of gelatin to oil was 1:10. The cross-linker, toluene-2,4-diisocyanate, was added to yield a final concentration of 34 mmol dm^{-3} in the oil. After 15 min reaction, the oil was decanted, and the microcapsules filtered and rinsed with water. Microcapsule diameters were 124 ± 74 and 271 ± 168 µm when formulated in sunflower and silicone oil respectively. Activity of the microencapsulated cells was determined by pH reduction in milk in repetitive sequential fermentations. Loss of activity, in comparison to the free cell controls, was realized immediately after microencapsulation, but was quickly recovered by cell growth within the microcapsules. Acidification rates similar to those achieved with cells immobilized in alginate beads, were measured by the third sequential fermentation.

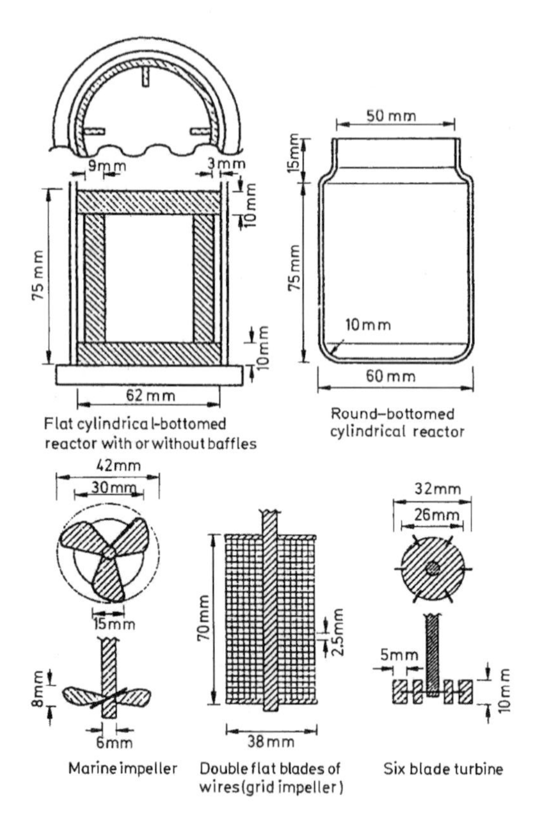

Figure 9. Design of baffled vessel and impellers used to emulsify/disperse aqueous polymer solution containing the biocatalyst, into an oil phase.

In a related study [19], *Lactococcus lactis* was microencapsulated within chitosan membranes, cross-linked with a variety of oil soluble cross-linkers. The procedure involved suspending cells into a 4% chitosan solution, followed by emulsion/dispersion into an organic phase in a ratio of 1:5, facilitated by means of Span 85 surfactant. Various organic phases were tested, including cyclohexane, or sunflower, canola or mineral oil. The emulsion was generated using a frame lattice impeller for 1 to 5 min, followed by the addition of terephthaloyl chloride, glutaraldehyde or hexamethylene diisocyanate as cross-linkers. Following membrane formation, microcapsules were separated by settling and decantation of the oil, and filtered and washed with Tween 20 solution. Microcapsule diameters depended strongly on the choice of oil used, with the oil viscosity being the determining parameter. As a result, mean diameters covered a broad range, from 100 μm to 1 mm. Lower viscosity solvents facilitated microcapsule recovery, while the more viscous oils were more problematic, particularly for the smaller diameter microcapsules. Chitosan cross-linked with hexamethylene diisocyanate or glutaraldehyde formed strong membranes, and resulting microcapsules had a narrow size distribution with a mean diameter of 150 μm. Some loss of acidification activity was observed upon microencapsulation due to the toxicity of the cross-linker, but activity recovered during sequential batch fermentations to levels equivalent to that observed with free cells.

Detailed procedures and a description of the chemistry behind the cross-linking reactions involving chitosan and various oil soluble cross-linkers, can be found in Reference [20]. The recovery and reuse of the oil phase in these procedures may be achieved by minimizing residual cross-linker. Given the high reactivity of these reagents, it is assumed that the cross-linker has been fully consumed by the reaction, but this would need to be carefully verified.

7. Microencapsulation by emulsification and interfacial polymerization reactions

In the procedure described above, pre-formed polymer is cross-linked at the droplet surface or interface, giving rise to a cross-linked membrane coat. An alternative is to carry out an interfacial polymerization reaction in an emulsion system, whereby one of the polymerization monomers is water soluble, and thus present in the aqueous phase microdroplet, and the other polymerization monomer is oil soluble, and thus present in the oil dispersed phase. Interfacial polymerization is initiated by addition of the oil soluble polymer, or by addition of a polymerization catalyst or initiator. In this case, a synthetic polymer coat is formed around the emulsified droplet, which upon separation from the oil phase, gives rise to membrane bound microcapsules. Strong and stable membranes are formed, which because of the wide range of possible polymers that can be used, provide for considerable range of compatibilities and permeabilities.

A well-known example of an interfacial polymerization reaction involves the formation of nylon or polyamide membrane microcapsules. The procedure involves the use of a diamine base (*e.g.* 1,6-hexamethylene diamine) solubilized in the aqueous internal phase, which is dispersed into an oil or solvent phase, to which is added a dichloride (e.g. sebacoyl chloride). The resulting polymerization reaction gives rise to a thin polyamide film (nylon-6,10), coating the liquid microdroplet [21].

The difficulty with this procedure lies in the extremes of pH that are encountered, solvent toxicity if an organic solvent is used as the continuous emulsion phase, and potential toxicity of the monomers themselves. Various modifications to the procedures have been introduced to minimize toxic effects such as through improved control of pH, the use of alternative monomers or various fillers, and non-toxic vegetable oils.

In our experience, and for the reasons described, interfacial polymerization reactions are problematic when the objective is to microencapsulate living cells. The potential is interesting however, but the procedures should be approached with caution. It should be mentioned that this procedure has been successfully applied to the microencapsulation of active and stable enzyme preparations such as urease [22,23].

8. Summary and conclusions

The objective of this review has been to provide a perspective on emulsification/dispersion technologies which are available to immobilize living cells and other biocatalysts through entrapment or microencapsulation protocols. The resulting product is a gel bead or microsphere, which may or may not be membrane coated, or a membrane coated microcapsule containing a cell suspension in a liquid core. The objective is to formulate these active cell preparations on an industrial scale, with a careful control over size - with mean diameters potentially extending into a micron size range. Enormous flexibility in the selection and control of mean microsphere or microcapsule diameters is exercised through the control of emulsification parameters. However, emulsions lead to size dispersions, so attention must be paid to the manner in which the emulsions are generated so as to exercise as high a level of control over the size dispersion as is possible. The key benefit then is in the ability to generate microspheres down into a micron size range, on virtually any industrial scale that could be contemplated. In fact, emulsification/dispersion systems are the only entrapment or microencapsulation protocols available for industrial scale application, in a microsphere or microcapsule size range under 500 μm. The methodologies are also readily extended to continuous processing, under aseptic conditions.

Several examples have been provided involving the encapsulation of living cells. These procedures are adaptable to the encapsulation of other cell lines.

References

[1] Poncelet, D.; Poncelet De Smet, B.; Beaulieu, C. and Neufeld, R.J. (1993) Scale-up of gel bead and microcapsule production in cell immobilization. In: Goosen, M.F.A. (Ed.) Fundamentals of Animal Cell Encapsulation and Immobilization. CRC Press; pp. 113-142.

[2] Neufeld, R.J.; Y. Peleg, J.S.; Rokem, O.; Pines and Goldberg, I. (1991) L-Malic acid formation by immobilized *Saccharomyces cerevisiae* amplified for fumarase. Enz. Microb. Technol. 13: 991-996.

[3] Sanderson, G.R.; Bell, V.L. and Ortega, D. (1989) A comparison of gellan gum, agar, κ-carrageenan, and algin. Cereal Foods World 34: 991-998.

[4] Camelin, I.; Lacroix, C.; Paquin, C.; Prevost, H.; Cachon, R. and Divies, C. (1993) Effect of chelants on gellan gum rheological properties and setting temperature for immobilization of living *Bifidobacteria*. Biotechnol. Tech. 9: 291-297.

[5] Moorehouse, R.; Colegrove, G.T.; Stanford, P.A.; Baird, J.K. and Kang, K.S. (1981) PS-60: a new fel-forming polysaccharide. In: Brant, D.A. (Ed.) Solution Properties of Polysaccharides. American Chemical Society, Washington DC; pp. 111-124.

[6] Moslemy, P.; Guiot, S.R. and Neufeld, R.J. (2002) Production of size-controlled gellan gum microbeads encapsulating gasoline-degrading bacteria. Enz. Microb. Technol. 30: 10-18.

[7] Baveye, P.; Vandevivere, P; Hoyle, B.L.; DeLeo, P.C. and Sanchez de Lozada, D. (1998) Environmental impact and mechanisms of the biological clogging of saturated soils and aquifer materials. Crit. Rev. Environ. Sci. Technol. 28: 123-191.

[8] Mutsakis, M. and Robert, R. (1986) Static mixers bring benefits to water/wastewater operations. Water Engineering and Management, November, 1986.

[9] Mensour, N.A.; Margaritis, A.; Briens, C.L.; Pilkington, H. and Russell, I. (1996) Application of immobilized yeast cells in the brewing industry. In: Wijffels, R.H.; Buitelaar, R.M.; Bucke, C. and Tramper, J. (Eds.) Immobilized Cells: Basics and Applications. Elsevier Science, Amsterdam; pp. 661-671.

[10] Neufeld, R.J.; Poncelet, D.J.C.M. and Norton, S.D.J.M. (1999) Immobilized-Cell carrageenan bead production and a brewing process utilizing carrageenan bead immobilized yeast cells. US Patent 5,869,117. Assignee: Labatt Brewing Company Ltd.

[11] Lacroix, C.; Paquin, C. and Arnaud, J.-P (1990) Batch fermentation with entrapped growing cells of *Lactobacillus casei*; optimization of the rheological properties of the entrapment gel matrix. Appl. Microbiol. Biotechnol. 32: 403-408.

[12] Lim, F. and Sun, A.M. (1980) Microencapsulated islets as bioartificial endocrine pancreas. Science 210: 908-910.

[13] Pelaez, C. and Karel, M. (1981) Improved method for preparation of fruit-simulating alginate gels. J. Food Process Preserv. 5: 63-81.

[14] Neufeld, R.J.; Lencki, R.W.J. and Spinney, T. (1989) Polysaccharide Microspheres and Method of Producing Same. US Patent 4,822,534.

[15] Poncelet, D.; Lencki, R.; Beaulieu, C.; Halle, J.P.; Neufeld, R.J. and Fournier, A. (1992) Production of alginate beads by emulsification/internal gelation. I. Methodology. Appl. Microbiol. Biotechnol. 38: 39-45.

[16] Poncelet, D.; Poncelet De Smet, B.; Beaulieu, C.; Huguet, M.L.; Fournier, A. and Neufeld, R.J. (1995) Production of alginate beads by emulsification/internal gelation. II. Physicochemistry. Appl. Microbiol. Biotechnol. 43: 644-650.

[17] Larisch, B.C.; Poncelet, D.; Champagne, C.P. and Neufeld, R.J. (1994) Microencapsulation of *Lactococcus lactis* subsp. *cremoris*. J. Microncapsulation 11: 189-195.

[18] Hyndman, C.L.; Groboillot, A.F.; Poncelet, D.; Champagne, C.P. and Neufeld, R.J. (1993) Microencapsulation of *Lactococcus lactis* within cross-linked gelatin membranes. J. Chem. Tech. Biotechnol. 56: 259-263.

[19] Groboillot, A.F.; Champagne, C.P.; Darling, G.D.; Poncelet, D. and Neufeld, R.J. (1993) Membrane formation by interfacial cross-linking of chitosan for microencapsulation of *Lactococcus lactis*. Biotechnol. Bioeng. 42: 1157-1163.

[20] Quong, D.; Groboillot, A.; Darling, G.D.; Poncelet, D. and Neufeld, R.J. (1997) Microencapsulation within cross-linked chitosan membranes. In: Muzzarelli, R.A.A. and Peters, M.G. (Eds.) Chitin Handbook. European Chitin Society; pp. 405-410.

[21] Chang, T.M.S. (1964) Semipermeable microcapsules. Science 1: 524-525.

[22] Monshipouri, M. and Neufeld, R.J. (1991) Activity and distribution of urease following microencapsulation within polyamide membranes. Enz. Microb. Technol. 13: 309-313.

[23] Monshipouri, M. and Neufeld, R.J. (1992) Kinetics and activity distribution of urease co-encapsulated with hemoglobin within polyamide membranes. Appl. Biochem. Biotechnol. 32: 111-126.

ATOMISATION TECHNIQUES FOR IMMOBILISATION OF CELLS IN MICRO GEL BEADS

JAMES C. OGBONNA
Department of Biochemistry and Biotechnology, Ebonyi State University, Abakaliki, Ebonyi State, Nigeria – Email: jcogbonna@hotmail.com

1. Introduction

Entrapment of cells in natural polymer gel beads is one of the most extensively studied methods of cell immobilisation. A mixture of pre-cultured cells and polymer solution is extruded through a nozzle into a stirred gelling agent solution, resulting in formation of gel beads entrapping the cells. The diameters of the resulting gel beads depend on the diameter of the nozzle and also on the physical properties (mainly the viscosity and surface tension) of the polymer. Gel beads entrapping microbial cells are used in either fluidised or packed bed bioreactors for various bioprocesses such as production of useful metabolites, waste water treatment and in biofiltration systems. However, most of the gel beads currently used for the various processes are more than 2.0 mm in diameter. Mass transfer limitation is often a major problem encountered in bioprocesses using cells entrapped in large diameter gel beads [1]. Oxygen and nutrients have to diffuse from the broth through the gel matrix before reaching the immobilised cells. Furthermore, as oxygen and nutrients diffuse from the broth into the gel beads, they are consumed by the cells at and near the surface of the gel beads, leading to oxygen and nutrient concentration gradients inside the gel beads. These gradients depend on the specific rates of nutrient and oxygen consumption and on the concentration of the immobilised cells. When the concentration of the immobilised cells is very high, oxygen and nutrients are completely consumed within the periphery of the gel beads. Thus, it is difficult to maintain active cells at the centre of large diameter gel beads due to oxygen and nutrient limitations. Furthermore, the produced metabolites and gasses must also diffuse out from the beads into the broth. For products with feed back inhibition, diffusion of the produced metabolites from the large diameter gel beads into the broth may be the rate-limiting step in the overall process. When the product is acidic, the pH inside the large diameter gel beads may be much lower than that in the bulk medium, making it difficult to maintain the optimum pH for the cells inside the gel beads [2,3]. Also, when the rate of gas (carbon dioxide) evolution by the immobilised cells is higher than the rate of its diffusion out of the gel beads, the gel beads crack due to increase in the internal pressure. Depending on the fluid dynamics in the bioreactor, cracked beads float with resultant decrease in the productivity of the process.

V. Nedović and R. Willaert (eds.), Fundamentals of Cell Immobilisation Biotechnology, 327-341.

In this section, the need to immobilise cells in small diameter gel beads is highlighted by discussing the effectiveness factors for oxygen and nutrient uptake by cells immobilised in gel beads of various diameters. Oxygen and nutrient concentration profiles in gel beads were also compared and immobilisation of cells in micro gel beads by rotating disk atomisation technique was discussed.

2. Need to immobilise cells in small diameter gel beads

Performance of immobilised cells in terms of mass transfer characteristics can be assessed by effectiveness factors. In this sense, the effectiveness factor for immobilised cells is defined as the ratio of nutrient uptake by the immobilised cells to that obtainable when there is no intra-particle resistance to diffusion of the nutrients into and out of the beads. In other words, when there is no mass transfer limitation, the effectiveness factor is 1.0 while low value of effectiveness factor indicates acute limitation of the nutrient or oxygen inside the gel bead. The effectiveness factors for oxygen uptake by immobilised *Corynebacterium glutamicum* cells are shown in Table1.

Table 1. Effects of gel bead diameter on the effectiveness factors for oxygen uptake by immobilised cells of Corynebacterium glutamicum. The maximum oxygen uptake rate and the Km values used for the calculation were 5.25 mM/g-cell.h and 0.0015 mM/L, respectively. The method used for the calculation is described elsewhere [4].

Gel bead diameter (mm)	Immobilised cell conc. (g/L)	Dissolved oxygen conc. (ppm)	Effectiveness factor (-)
3.0	14.0	3.0	0.232
	14.0	7.0	0.348
	85.0	7.0	0.126
2.0	14.0	3.0	0.340
	14.0	7.0	0.502
	85.0	7.0	0.216
1.0	14.0	3.0	0.620
	14.0	7.0	0.844
	85.0	7.0	0.416
0.5	14.0	3.0	0.955
	14.0	7.0	0.998
	85.0	7.0	0.753
0.25	14.0	3.0	0.998
	14.0	7.0	1.000
	85.0	7.0	0.996

With gel beads of the same diameter, the effectiveness factor can be increased by increasing the dissolved oxygen concentration in the broth and by reducing the concentration of cells immobilised in the gel beads. However, it is desirable to use high cell concentration in order to increase the overall productivity of the system. Furthermore, it is very difficult to maintain high dissolved oxygen concentration in the broth especially when the immobilised cell concentration is high. This is because the conventional methods of using high aeration rate and increasing the agitation speed generate high hydrodynamic stress that leads to breakage of the gel beads. Even when pure oxygen is used for aeration of a system with high immobilised cell concentration, the dissolved oxygen concentration would still be low due to rapid oxygen uptake by the cells within the periphery of the gel beads. Moreover, as shown in Table 1, with dissolved oxygen concentration of 7 ppm in the broth, the effectiveness factor for oxygen uptake by the cells immobilised in 3.0-mm gel beads is still as low as 0.348 even when a low cell concentration of 14 g/L is used. Reducing the diameter of the gel beads is the most effective method of increasing the effectiveness factor. With a low cell concentration of 14 g/L and bulk dissolved oxygen concentration of 7 ppm, there is no oxygen limitation in the 0.5-mm gel beads as reflected by the high value of the effectiveness factor (0.998). When the cell concentration in the 0.5 mm diameter gel beads was increased to 85 g/L, the effectiveness factor was still as high as 0.753.

Table 2. Effects of gel bead diameter on the effectiveness factors for glucose uptake by immobilised cells of Saccharomyces cerevisae. The effectiveness factors were calculated using a maximum glucose uptake rate of 2.78 g-glucose/g-cell.h and a Km value of 0.476 g/L. The effectiveness factors were calculated as described elsewhere [4].

Gel bead diameter (mm)	Immobilised cell conc. (g/L)	Glucose conc. (g/L)	Effectiveness factor (-)
3.0	14.0	1.0	0.551
	14.0	10.0	0.983
	85.0	1.0	0.253
2.0	14.0	1.0	0.728
	14.0	10.0	0.994
	85.0	1.0	0.366
1.0	14.0	1.0	0.931
	14.0	10.0	0.999
	85.0	1.0	0.636
0.5	14.0	1.0	0.984
	14.0	10.0	1.000
	85.0	1.0	0.892
0.25	14.0	1.0	0.996
	14.0	10.0	1.000
	85.0	1.0	0.977

For anaerobic processes that are not limited by oxygen supply, the effectiveness factor for uptake of the organic carbon source is a better parameter for assessing the performance of the immobilised cells. The effectiveness factors for glucose uptake by immobilised *Saccharomyces cerevisiae* cells are shown in Table 2. For the same cell concentration and gel bead diameters, the effectiveness factors for glucose uptake are generally higher than those for oxygen uptake. With bulk glucose concentration of 10 g/L and low cell concentration of 14 g/L, 3.0-mm gel beads can be used without glucose diffusion limitation inside the gel bead. However, for continuous processes, high cell concentration is used and the substrate concentration in the broth is maintained low in order to obtain high productivity and substrate conversion efficiency. As shown in Table 2, when the glucose concentration in the broth is maintained at 1.0 g/L and the cell concentration is increased to 85 g/L, the effectiveness factor for glucose uptake by the cells immobilised in 3.0-mm gel beads is only 0.253. Thus although in comparison with aerobic processes, larger diameter gel beads can be used for anaerobic processes without substrate limitation in the centre, it is still necessary to immobilise the cells in gel beads of small diameters.

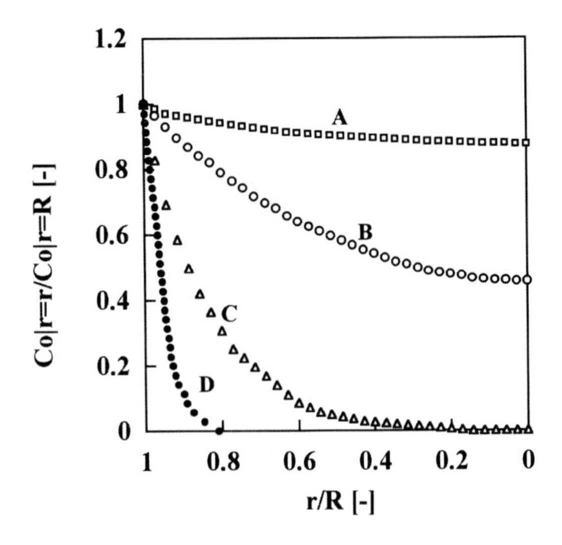

Figure 1. Effects of gel bead diameter on oxygen and glucose concentration profiles in gel beads. The cell concentration was 14 g/L while the diffusion coefficients for oxygen and glucose in the gel beads were 6.4 x 10^{-6} m^2/h and 6.2 x 10^{-10} m^2/s, respectively. The dissolved oxygen and glucose concentrations in the medium were 7 ppm and 1 g/L, respectively. A: Glucose concentration profile in 0.5-mm gel beads immobilising Saccharomyces cerevisae *cells. B: Oxygen concentration profile in 0.5-mm gel beads immobilising* Corynebacterium glutamicum *cells. C: Glucose concentration profile in 3.0-mm gel beads immobilising* Saccharomyces cerevisae *cells. D: Oxygen concentration profile in 3.0-mm gel beads immobilising* Corynebacterium glutamicum *cells.*

The effectiveness of immobilising cells in small diameter gel beads for improved mass transfer into, and out of the gel beads is also demonstrated by the oxygen and substrate concentration profiles inside the gel beads. As shown in Figure 1, with a low cell concentration of 14 g/L, and high dissolved oxygen concentration of 7ppm, oxygen is almost completely consumed within the outer 0.3 mm (r/R = 0.8) of the 3.0-mm diameter gel beads. On the other hand, in the case of 0.5 mm diameter gel bead, the cells at the centre of the beads ($r/R \sim 0$) are not oxygen-limited since the oxygen concentration at the centre of the bead is as high as 3 ppm [($Co|r=r/Co|r=R$) = 0.43]. Also there is almost no difference in glucose concentration between the surface and centre of the 0.5-mm diameter gel beads but in the case of 3.0-mm gel beads, glucose is completely consumed within the outer 1.2 mm (r/R = 0.2) of the gel bead. Because of the low nutrient and oxygen concentrations at the centre of large diameter gel beads, cells grow mainly at the periphery of the beads. Depending on the cell strain, cell concentration and the nutrient concentration in the broth, less than 10% of the large diameter gel beads may contain active cells. However, for small diameter gel beads, active cells are uniformly distributed within the beads since both the nutrients and oxygen concentrations at the centre of gel beads are still high [5].

Reducing the gel bead diameter improves mass transfer by increasing the diffusion surface area (contact area between the bead and the broth) per unit volume of the bead, and by reducing the mean distance of diffusion so that oxygen transfer from the broth to the cells is controlled mainly by molecular diffusion. Another advantage of using small diameter gel beads is that they are generally more resistant to hydrodynamic stress than large diameter gel beads. They can be used even in stirred tank bioreactors for a long period of time without mechanical damage [5].

3. Production of small diameter gel beads

The simplest method of producing small diameter gel beads is to extrude polymers with low viscosity and low surface tension slowly through nozzles of very small diameters into the gelling agent solution. However, since the viscosity of the polymer is proportional to its concentration, there is a limit to which the viscosity of the polymer can be reduced. Gel beads produced from very low concentrations of polymers are usually very fragile and are not suitable for long term processes. On the other hand, when very low flow rates are used, very long time will be required for production of sufficient amount of gel beads even for laboratory experiments. Furthermore, when the nozzle is very narrow, it is easily plugged especially when the viscosity of the polymer is high. By using multi-nozzle system and nozzles of appropriate diameters, about 1.0-mm gel beads can be produced in sufficient quantities for laboratory experiments. However, this method is not suitable for production of large amounts of micro gel beads in the order of 500 μm or below.

Various methods have been reported for immobilisation of cells in gel beads of small diameters. These include vibration of liquid (polymer-cell mixture) jets which are produced by pumping a polymer-cell mixture through a nozzle [6,7]; use of a concentric air stream to shear the droplets of polymer-cell mixture off the nozzle before the droplets grow to large diameters [8,9,10]; use of an electrostatic generator to disintegrate polymer-cell mixture into droplets [7,11,12,13]; emulsification of polymer-cell mixture in a non aqueous phase followed by addition of gelling agent solution [14,15,16]; use of jet cutting techniques [17]; and atomisation of the polymer-cell mixture on a rotating disk. Only immobilisation of cells in micro gel beads by rotating disk atomisation will be discussed here.

4. Principles of liquid atomisation on rotating disks

Atomisation of liquids on rotating disks is widely used for production of fine particles of liquids in gas-liquid contact processes such as spray drying, cooling, scrubbing, combustion, promotion of chemical reactions and in humidifying equipment.

When a liquid is dropped on top of a rotating disk, it is brought to a high velocity by centrifugal force and spreads out as a very thin film. At a relatively low volumetric flow rate of the liquid, the centrifugal force generated by the rotating disks overcomes the surface tension holding the film to the disk and it breaks up into droplets, resulting in direct drop formation at the tip of the rotating disk. If the volumetric flow rate is high, the surface tension works on the liquid film edge under the state of gas-liquid-gas system as a result of which the film becomes unstable. It then departs from the edge of the disk and changes into several ligamentary streams. Droplets are released from the tips of the ligaments due to the growth of the disturbance. The lengths of the ligaments depend on the rotation speed of the disk, the viscosity and surface tension of the liquid as well as on the volumetric flow rate of the liquid.

5. Construction of a rotating disk atomiser for immobilisation of cells in micro gel beads

Cells can be immobilized in micro gel beads by atomisation of a polymer-cell mixture and collecting the droplets in a gelling agent. The system developed for atomisation of liquids can be used for immobilisation of cells in micro gel beads with some modifications. A container equipped with a mixing device should be provided for homogeneous mixing of the cells with the polymer solution during atomisation. When the viscosity of the polymer solution is high, the cells do not sediment easily. In such a case, the cells and the polymer solution are homogeneously mixed only at the initial stage and there is no need to continue mixing during atomisation of small volume of polymer-cell mixture. However, for low viscous polymers, continuous mixing is necessary to keep the cells in suspension during atomisation. In the case of agar, carrageenan and other polymers that gel at low temperatures, the container should be temperature controlled to avoid gelling inside the container. However, when the room temperature is relatively high and the volume of the polymer-cell mixture is small, atomisation can be completed before the temperature drops to the gelling temperature. Reducing the gelling temperature of the polymer can also be used to avoid this problem. By decreasing the gelling temperature, the problem of cell deactivation due to over exposure of cells to high temperature is also reduced. As shown in Figure 2, the gelling temperature of 2% κ-carrageenan is 42ºC but by mixing it with locust bean gum (LBG), the gelling temperature can be reduced to below 30ºC. Addition of LBG also improves the physical characteristics of κ-carrageenan. The micro gel beads produced from 2% κ-carrageenan were deformed in shape but spherical micro gel beads could be produced from a mixture of 0.75% LBG and 2% κ-carrageenan.

Another major difference between a rotating disk atomiser used in gas-liquid contact processes and a system for micro gel bead production is that the later has to be equipped with a vessel for collecting and gelling the produced droplets. The polymer-cell droplets should fall directly into a continuously stirred gelling agent solution. Usually, calcium chloride solution is used for gelling alginate gel beads and cold potassium chloride solution is used for gelling κ-carrageenan gel beads. However, it has been shown that using strontium chloride for immobilisation of cells in alginate and κ-carrageenan results in gel beads that are chemically and physically more stable than the calcium alginate gel beads or carrageenan gel beads gelled in potassium chloride solution [18].

Most of the droplets released from the edge of the rotating disks or ligaments travel perpendicular to the walls of the vessel. Thus the vessel has to be rotated to create a vortex of the gelling agent solution. The distance between the tip of the rotating disk and the vortex is very important. The lengths of the ligaments formed at high polymer-cell mixture flow rates depend on the viscosity of the polymer, the rotation speed of the disk and the flow rate of the polymer-cell mixture. Thus, it is necessary to use large diameter vessels so that the distance between the edge of the rotating disk and the vortex is large enough to avoid direct contact between the gelling agent solution and the ligaments. If the ligaments make direct contact with the gelling agent solution, spaghetti-like gels are formed rather than beads. The thickness and height of the vortex as well as the distance between the disk and the vortex can be controlled by adjusting the rotation speed of the vessel.

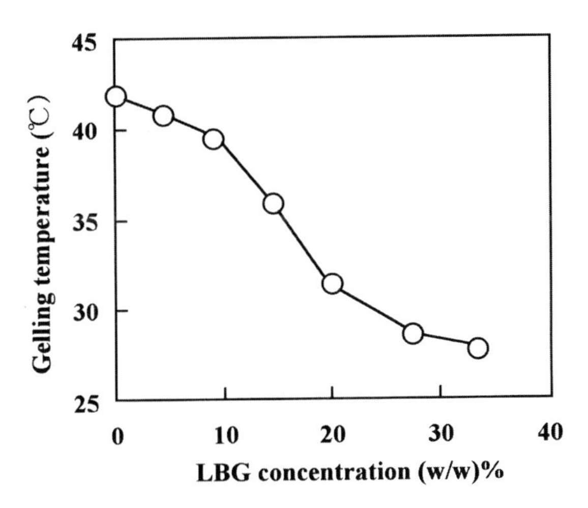

Figure 2. Effect of locust bean gum (LBG) concentration on the gelling temperature of κ–carrageenan.

An example of a rotating disk atomisation system for immobilisation of cells in micro gel beads is schematically represented in Figure 3. A vessel (V) containing the gelling agent solution is fix to a magnetic disk (H) and rotated by means of a motor. The rotating disks (D) are attached to a spindle and rotated by another motor on top of the spindle. The disks can be veined, curved at the centre, or just flat. The optimal diameter of the disk depends on the diameter of the vortex and on the scale of production. The disks can easily be replaced but for immobilisation of cells in calcium alginate of various concentrations, both 3.95 cm and 6 cm diameter disks with small curvature at the centre produced satisfactory results. The disks and vessels are housed inside a bigger vessel that acts as a water bath for temperature control during immobilisation.

The whole system is fixed to a stable metallic frame for minimal vibration during the rotation of the vessel and the disks. Such vibration would affect the uniformity of the resulting beads. The polymer solution and the cells are mixed inside the vessel(s) (A) and the mixture is pumped onto the top of the rotating disks through the nozzle (N) by the pumps (B). The generated droplets fall into the vortex of the gelling agent solution (that is formed by rotating the vessel (V)) and are subsequently gelled. The beads are left inside the gelling agent solution with gentle stirring for curing. The curing time depends on the diameter of the beads, the viscosity of the polymer (this affects the rate of diffusion of the gelling agent into the gel beads), and the concentration of the gelling agent. Usually two hours of curing is sufficient [19] but longer period of curing will ensure that the gelling is completed and that the beads shrink to a stationary state before they are used for the desired process. If long period of curing time is desired, nutrients can be added to the gelling agent during curing to avoid possible cell deactivation. Alternatively, short curing time is used but low concentration of the gelling agent is added into the culture broth during the process.

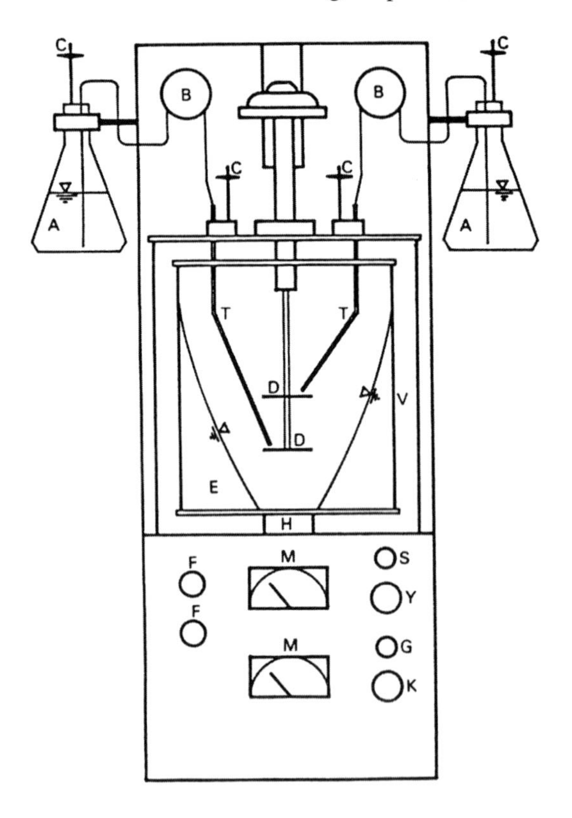

Figure 3. Schematic diagram of the rotating disk atomiser for immobilisation of viable cells in micro gel beads. A = vessel for mixing the cells and the polymer; B = pumps; C = air filter; N = feed nozzle; D = rotating disk; V = rotating vessel; E = vortex of the gelling agent solution; H = magnetic disk; F = power switch for the pumps; M = speedometer; S = power switch for the disks; G = power switch for the rotating vessel; Y = speed controller for the disk; K = speed controller for the rotating vessel.

6. Factors affecting the size, uniformity and shape of the micro gel beads produced by rotating disk atomisation

6.1. THE SIZE OF THE MICRO GEL BEADS

The diameters of the droplets formed at the tip of ligaments are proportional to the thickness of the ligaments. Thus any factor that affects the thickness of ligaments on the atomising disk also affects the diameter of the resulting gel beads. The rotation speed of the disk has an overwhelming influence on the diameter of the gel beads. At low rotation speed of the disk, a film of polymer-cell mixture is formed on the disk but as the rotation speed is increased, the centrifugal force overcomes the surface tension holding the film to the disk. Thus the film is broken down into ligaments. An increase in the rotation speed of the disk results in an increase in the centrifugal force and thus an increase in the number of the ligaments. An increase in the number of ligaments at a constant volumetric flow rate of the polymer-cell mixture means that each ligament becomes thinner and the diameter of the droplets from their tips becomes smaller. Generally, the diameters of the beads are inversely correlated with the rotation speed of the disk. The centrifugal force generated by the rotating disk is also a function of the disk diameter. Under the same rotation speed, the centrifugal force increases with increase in the disk diameters. Thus under the same rotation speed and volumetric flow rate of the polymer-cell mixture, the diameter of the gel beads is negatively correlated with the diameter of the disk. The thickness of the ligaments increases with increase in the volumetric flow rate of the polymer-cell mixture. Thus, at a constant rotation speed of the disk, the diameter of the produced gel bead is positively correlated with the volumetric flow rate of the polymer-cell mixture.

The physical properties such as the surface tension, viscosity and density of the polymer also have significant effects on the diameters of the gel beads. High viscous fluids with high surface tension are more resistant to the centrifugal force. Thus, under the same rotation speed and volumetric flow rate of the polymer-cell mixture, high viscous polymers produce fewer but thicker ligaments, leading to formation of bigger droplets (larger diameter of the gel beads).

6.2. UNIFORMITY OF THE MICRO GEL BEADS

Depending on the atomisation conditions, plots of the numerical frequency against the mean gel bead diameters yield histograms with bimodal distribution [4]. Some gel beads with very small diameters (satellite gel beads) are produced but the total volume of such satellite gel beads usually make up only a very small percentage of the total volume of the gel beads produced. The formation of satellite gel beads may be due to the effect of kinetic energy that travels at a high speed from the polymer-cell mixture into the droplets before the droplets are released. The maximum kinetic energy delivered to a liquid (solution) leaving a rotating disk varies directly with the square of the flow rate. Thus, the number of the satellite gel beads increases with increase in the volumetric flow rate of the polymer–cell mixture. Air friction can also result in formation of satellite gel beads. At high rotation speed of the disk, the centrifugal acceleration is much higher than the gravitational acceleration ($Nr^2 > 10$ g) so that the effect of gravitational effect is negligible. Thus, the number of satellite gel beads can be reduced by increasing the rotation speed of the disk. However, very high rotation speed of the disk may result in vibration, which increases the number of the satellite gel beads. Large diameter disks tend to generate vibration more than smaller disks. Thus, more satellite gel beads are produced when disks of large diameters are used. The physical properties of the carrier also affect the number of satellite gel beads. When perfluorocarbon was added to sodium alginate solution, the viscosity increased but the surface tension decreased. Beads produced from the perfluorocarbon-sodium alginate mixture were more uniform than those produced from sodium alginate of the same concentration. In summary, formation of satellite gel beads can be reduced by using high rotation speed of the disk, avoiding vibration during rotation of the disk, and by ensuring smooth and uniform flow of the polymer-cell mixture.

6.3. THE SHAPE OF THE MICRO GEL BEADS

When a gelling agent solution with high surface tension is used, the polymer droplets are deformed as they fall into the gelling agent solution. This results in gel beads of irregular shapes. In such a case, there is a need to reduce the surface tension of the gelling agent solution. This can be achieved by adding small amounts of surface-active agents such as Tween 20 into the gelling agent solution. The choice of the surface-active agent depends on the cells because some of the surface-active agents may have inhibitory effects on the cells.

However, usually, only very small amount of the surface-active agent is required to substantially reduce the surface tension of the gelling agent solution. An example of the effect of surface-active agent concentration on the surface tension of the gelling agent solution is shown in Figure 4. Addition of 0.5% Tween 20 reduces the surface tension of 0.2M calcium chloride solution from 72 mN/m to 36.8 mN/m. Gel beads produced by using 0.2M calcium chloride solution containing 0.5% Tween 20 were all spherical in shape.

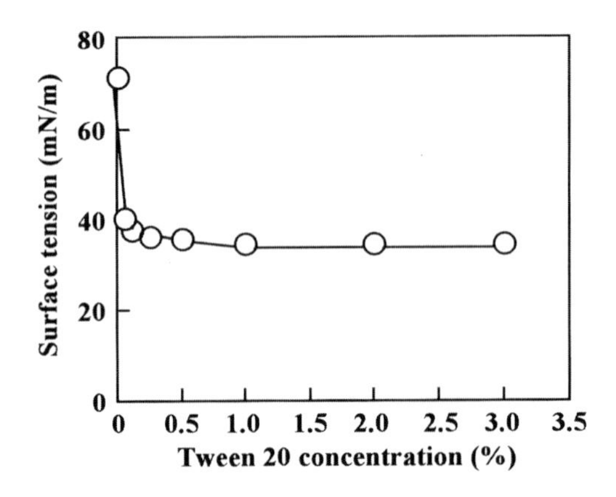

Figure 4. Effect of Tween 20 concentration on the surface tension of 0.2 M calcium chloride solution.

7. Optimum gel bead diameters

In order to produce micro gel beads with uniformly distributed active cells, the oxygen or limiting substrate concentration at the centre of the gel bead should be maintained above certain threshold values referred to as the critical centre concentrations. The critical centre concentration of oxygen or the rate-limiting substrate can be defined as the concentration above which the cell growth or production of the desired metabolites is not limited by oxygen or the substrate. A critical gel bead diameter may thus be defined as a gel bead diameter above which oxygen or the rate-limiting substrate concentration at the centre of the bead becomes less than the critical centre concentration for that process under defined culture conditions (immobilised cell concentration, rate-limiting substrate concentration in the culture broth, and other conditions that affect the substrate diffusion into and out of the bead).

The critical gel bead diameter is not necessarily the optimal gel bead diameter. A compromise has to be made between the critical gel bead diameter as defined above and the feasibility of immobilising high cell concentrations inside the gel beads. The amount of cells or enzymes immobilised per unit volume of the polymer tends to decrease as the diameter of the gel bead is reduced [7,12]. When the gel beads are too small, some of the important advantages of cell immobilisation may be lost. Very small gel beads tend to float, especially in processes that produce a lot of foams, and this would in turn reduce the contact surface area between the gel beads and the culture broth with consequent decrease in mass transfer from the broth into the gel beads. In the case of plant cells that form large aggregates, immobilisation of cells in very small diameter gel beads may not be practical. Incidentally, however, the rates of substrate uptake by plant cells are relatively low so that the critical gel bead diameters for plant cell immobilisation are relatively bigger than those for microbial cells. For the same reason, animal cells can be immobilised in relatively large diameter gel beads without mass transfer limitation [20].

The feasibility of retaining the gel beads inside the bioreactor is also an important factor to be considered while deciding the optimum gel bead diameter. Very small gel beads are washed out of the reactor together with the effluent during continuous processes. The cells can be retained inside the reactor by fixing a membrane to the effluent port or by using a settler such as the type of system shown in Figure 5. This consists of a glass cylinder (N) housed inside an outer glass jacket (J). During a continuous process, the effluent is continuously pumped out of the bioreactor through the settler at a low rate so that the rate at which the effluent moves up the settler (linear velocity) is lower than the sedimentation rate of the gel beads. The maximum effluent flow rate therefore depends on the diameter of the beads (rate of bead sedimentation). Since the effluent linear velocity is negatively correlated with the diameter of the glass cylinder (N), the effluent flow rate can be increased if a settler with larger inner glass cylinder is used. The micro gel beads sediment into the outer jacket through the opening (C) and finally back into the bioreactor through the openings (D) and (E). The liquid level inside the bioreactor is controlled by the opening (B). When the liquid level decreases below (B), air moves in from the opening (A) through the opening (B) and breaks up the effluent flow. The external glass jacket (J) protects the sedimentation process from disturbances due to aeration and agitation inside the bioreactor.

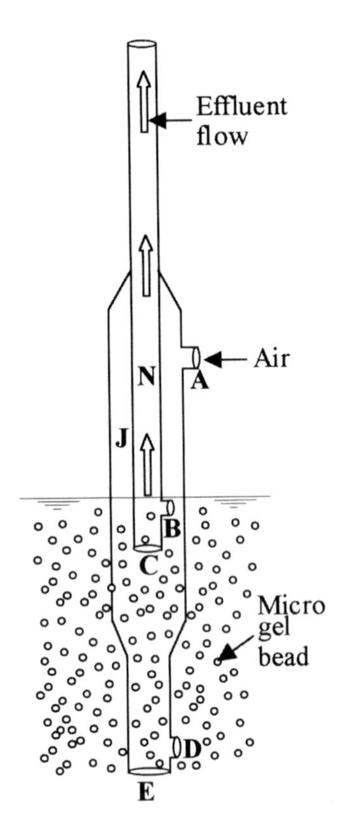

Figure 5. Schematic diagram of a settler for retaining the micro gel beads inside the bioreactor during continuous process with micro gel bead-immobilised cells.

8. Immobilisation of cells in micro gel beads of desired diameters

Cells can be immobilised in gel beads of desired diameters by varying the atomisation conditions. A correlation equation relating the diameter of the micro gel beads to the atomisation conditions has to be derived. The surface area of the droplets are very important for gas-liquid contact processes for which atomisation has been extensively studied. Thus most of the correlation equations for diameters of droplets produced by liquid atomisation are based on surface area per unit volume diameter (d_{sv}) expressed by equation 1.

$$d_{sv} = \sum n_i d_i^3 / \sum n_i d_i^2 \qquad (1)$$

Here, n_i is the number of gel beads with diameter, d_i. In the case of immobilising cells in micro gel beads, the surface area of the gel beads is also important as nutrient and gas exchange point between the immobilised cells and the bulk medium. However, the volume of the gel beads is more important when discussing the productivities and efficiency of immobilised bioreactors. The productivities of immobilised bioreactors are usually expressed in terms of production rate per unit volume of the bioreactor (or per unit volume of the gel beads). Thus the mean volume diameter (d_v), expressed by equation 2, is a more useful parameter for deriving correlation equations between the diameters of the gel beads and the operation parameters for atomisation.

$$d_v = (\sum n_i d_i^3 / \sum n_i)^{1/3} \tag{2}$$

For gel beads produced by atomisation of various concentrations of sodium alginates, and mixtures of sodium alginates and perfluorocarbons, using calcium chloride as the gelling agent [4], the mean volume diameters of the micro gel beads was expressed by equation 3.

$$d_v/D = 22.4 N_{We}^{-0.43}\, N_{Re}^{0.08}\, N_Z^{0.07} \tag{3}$$

The Reynolds number (N_{Re}) was defined as $\rho Q/\mu D$; the Ohnesorge number (N_Z) as $\mu/\sqrt{\rho\sigma D}$ and the Weber number (N_{we}) as $\rho N^2 D^3/\sigma$. Here, ρ = density of the polymer (g/cm^3), μ = viscosity of the polymer (g/cm.s) σ = surface tension of the polymer (g/s^2), Q = flow rate of the polymer (cm^3/s), D = diameter of the disk (cm), and N = rotation speed of the disk (s^{-1}). Within the range of variables tested [densities of the carriers = 1.007 ~ 1.068 (g/cm^3); viscosity of the carrier = 0.633 ~ 2.572 (Pa.s); surface tension of the carrier = 32 ~ 65.5 (mN/m); flow rate of the polymer = 0.017 ~ 0.4 (cm^3/s); rotation speed of the disk = 18.5 ~ 75.7 (s^{-1}); and diameters of the disk = 3.95 and 6.0 (cm)], this equation provided a fit to within ±15%. Uniform gel beads within the range of 200 to 1200 μm can easily be produced by selecting the carrier flow rate, the rotation speed of the disk and the diameter of the disk when the physical properties of the polymer are known.

9. Possible methods of scaling up the rotating disk atomiser for large scale immobilisation of cells in micro gel beads

A simple method of increasing the rate of micro gel bead production by rotating disk atomisation is to increase the volumetric flow rate of the polymer-cell mixture and correspondingly increase the rotation speed of the disk (equation 3). However, the length of the ligament formed at the edge of the rotating disk increases with increase in the volumetric flow rate of the polymer-cell mixture. At very high flow rates, the ligaments can be so long that they enter the gelling agent solution before the droplets are released from their tips. Thus, the distance between the edge of the disk and the vortex of the gelling agent solution is a limiting factor to the volumetric flow rate of the polymer-cell mixture. One method of scaling up this system is therefore to increase the distance between the edge of the disk and the surface of the vortex, which can be achieved by simply increasing the diameter of the vessel. For a given polymer with known physical properties, the length of the ligament (and thus the required distance between the edge of the disk and the surface of the vortex) can be calculated for a desired volumetric flow rate, using an equation proposed by Kamiya and Kayano [21].

The number of rotating disks is another very important parameter for scaling up the system. The rate of micro gel bead production is directly proportional to the number of disks. Thus, the system can easily be scaled up by increasing the number of atomising disks. A tall vessel is rotated at high rotation speed to create tall vortex so that many disks can be mounted on the same spindle with optimal spacing distance.

10. Conclusion

Mass transfer limitation is a major problem especially in aerobic processes with cells immobilised by entrapping in polymer gel beads. Many mathematical models have been used to demonstrate that immobilising the cells in micro gel beads can substantially reduce the mass transfer problem. However, most processes are still done with cells immobilised in large diameter gel beads and there are relatively very few detailed experimental studies to demonstrate the usefulness of immobilising the cells in micro gel beads. Various methods of immobilising cells in micro gel beads have been reported but whether or not such methods can be used for industrial application would depend on their simplicity, costs and scale up potentials. Furthermore, the immobilisation condition must be mild to avoid excessive stress on the cells.

Atomisation by rotating disk is mild because high pressure is not used. The stress exerted on the cell is similar to that of centrifugation. Thus, most microbial cells can be immobilised by rotating disk atomisation without significant decrease in cell viability. Also, the diameters of the beads do not depend on the diameters of the feed nozzles. This means that the polymer-cell mixture can be pumped through a large diameter nozzle to the top of the rotating disk and thus avoid the problem of nozzle plugging even when high viscous polymers are used. The power requirement is also comparably low and the process is flexible. In other words, gel beads of the same diameters can be produced from polymers of various physical characteristics by adjusting the volumetric flow rate of the polymer-cell mixture, the disk diameter and/or the rotation speed of the disk. The rotating disk atomisation system described in this section is very simple and a laboratory scale system that can produce up to 3L of micro gel beads per hour is already commercially available. By controlling the atomisation conditions, uniform gel beads of any desired sizes ranging from 200 to 1200 μm can easily be produced. Furthermore, it can be scaled up easily by increasing the sizes and number of the disks, and enlarging the diameter of the vessel.

To use cells immobilised in micro gel beads for industrial processes, there is a need for development of suitable bioreactors with appropriate fluid dynamics. Aside from the low mass transfer efficiency in packed or expanded bed systems, the inter-bead void volume in beds of micro gel beads is very small. Thus they are easily plugged, leading to nutrient channelling with consequent decrease in mass transfer and productivity. However, since micro gel beads can tolerate higher hydrodynamic stress than the conventional large diameter gel beads, long time process stability in fluidised bed reactors can be achieved. However, there is still a need to practically demonstrate this by more detailed case by case optimisation of processes with cells immobilised in micro gel beads. In this regard, more detailed studies on reactor designs, fluid dynamics, bubble column vs. airlift systems, gel bead loading ratios *etc.* are required before micro gel beads can be widely used for industrial processes.

References

[1] Ogbonna, J.C.; Matsumura, M. and Kataoka, H. (1991) Effective oxygenation of immobilised cells through reduction in bead diameters: A review. Process Biochem. 26: 109–121.

[2] Wang, H.; Seki, M. and Furusaki, S. (1995) Mathematical model for analysis ofmass transfer for immobilised cells in lactic acid fermentation. Biotechnol. Progress 11: 558-564.

[3] Cachon, R.; Lacroix, C. and Divies, C. (1997) Mass transfer analysis for immobilised cells of *Lactococcus lactis* sp. using both simulations and in-situ pH measurements. Biotechnol. Tech. 11: 251-255.

[4] Ogbonna, J.C.; Matsumura, M.; Yamagata, T.; Sakuma, H. and Kataoka, H. (1989) Production of micro gel beads by a rotating disk atomiser. J. Ferment. Bioeng. 68: 40–48.

[5] Ogbonna, J.C.; Matsumura, M. and Kataoka, H. (1991) Production of glutamine by micro gel bead-immobilised *Corynebacterium glutamicum* 9703-T cells in a stirred tank reactor. Bioproc. Eng. 7: 11–18.

[6] Hunik, J.H. and Tramper, J. (1993) Large-scale production of kappa-carrageenan droplets for gel-bead production: theoretical and practical limitations of size and production rate. Biotechnol. Progress 9: 186-192.

[7] Serp, D.; Cantana, E.; Heinzen, C.; von Stockar, U. and Marison, I.W. (2000) Characterisation of an encapsulation device for the production of monodisperse alginate beads for cell immobilisation. Biotechnol. Bioeng. 70: 41-53.

[8] Sirirote, P.; Yamane, T. and Shimizu, S. (1988) L-serine production from methanol and glycine with an immobilised methylotroph. J. Ferment. Technol. 66: 291-297.

[9] Eikmeier, H. and Rehm, H.J. (1987) Stability of calcium alginate during citric acid production of immobilised *Aspergillus niger*. Appl. Microbiol. Biotechnol. 26: 105-111.

[10] Rehg, T.; Dorger, C. and Chau, P.C. (1986) Application of an atomiser in producing small alginate gel beads for cell immobilisation. Biotechnol. Lett. 8: 111–118.

[11] Poncelet, D.; Bugarski, B.; Amsden, B.G.; Zhu, J.; Neufeld, R. and Goosen, M.F.A. (1994) A parallel plate electrostatic droplet generator: Parameters affecting microbead size. Appl. Microbiol. Biotechnol. 42: 251-255.

[12] Watanabe, H.; Matsuyama, T. and Yamamoto, H. (2001) Preparation of immobilised enzyme gel particles using an electrostatic atomisation technique. Biochem. Eng. J. 8: 171-174.

[13] Nedovic, V.A.; Obradovic, B.; Leskosek-Cukalovic, I.; Trifunovic, O.; Pesic, R. and Bugarski, B. (2001) Electrostatic generation of alginate microbeads loaded with brewing yeast. Process Biochem. 37: 17-22.

[14] Poncelet, D.; Lencki, R.; Beaulieu, C.; Halle, J.P.; Neufeld, R.J. and Fournier, A. (1992) Production of alginate beads by emulsification/internal gelation: I. Methodology. Appl. Microbiol. Biotechnol. 38: 39-45.

[15] Sheu, T.Y. and Marshall, R.T. (1993) Microentrapment of *Lactobacilli* in calcium alginate gels. J. Food Sci.58: 557-561.

[16] Mofidi, N.; Aghai-Moghadam, M. and Sarbolouki, M.N. (2000) Mass preparation and characterisation of alginate microspheres. Process Biochem.35: 885-888.

[17] Pruesse, U.; Fox, B.; Kirchhoff, M.; Bruske, F.; Breford, J. and Vorlop, K.D. (1998) The jet cutting method as a new immobilisation technique. Biotechnol. Tech. 12: 105-108

[18] Ogbonna, J.C.; Chay, B. Pham; Matsumura, M. and Kataoka, H. (1989) Evaluation of some gelling agents for immobilisation of aerobic microbial cells in alginate and carrageenan gel beads. Biotechnol. Tech. 3: 421–424.

[19] Ogbonna, J.C.; Amano, Y. and Nakamura, K. (1989) Elucidation of optimum conditions for immobilisation of viable cells by using calcium alginate. J. Ferment. Bioeng. 67: 92-96.

[20] Cristina, A.M.; Castelli, G.; Conti, F.; De Virgiliis, L.C.; Marrelli, L.; Miccheli, A. and Satori, E. (2000) Transport and consumption rate of O_2 in alginate gel beads entrapping hepatocytes. Biotechnol. Lett. 22: 865-870.

[21] Kamiya, T and Kayano, A (1975) Calculation of the continuous length of the ligamentary flow generated by a rotating disk. J. Chem. Eng. Japan 8: 72-74.

SPRAY COATING AND DRYING PROCESSES

MURIEL JACQUOT[1] AND MIMMA PERNETTI[2]
[1]LPGA, ENSAIA-INPL, 2 avenue de la Forêt de Haye, 5400 Vandoeuvre lès Nancy, France – Fax: 33 3 83 59 58 04 – E-mail: Muriel.Jacquot@ensaia.inpl-nancy.fr
[2]Department of Chemical Engineering, University of Rome "La Sapienza", via Eudossiana 18, 00184 Rome, Italy – Fax: 39 06 44 58 56 22 – Email: pernetti@tin.it

1. Introduction

Spray coating and drying are two technologies widely employed in biotechnology respectively to entrap different substances and to protect particles. Though presenting different problems of operating conditions, experimental set-up and product applications, they are both included in the same class of physical procedures for encapsulation, relying on heat exchange and phase transition.

2. Spray coating processes

The spray-coating method comes from the pharmaceutical field. Since the fifties, this technique was used to modify surface properties of drugs or active substances in order to mask undesirable taste, to control the delivery or to stabilize the molecule. Spraying devices have then been upgraded, in order to improve capsules properties, to obtain larger batch volumes and higher throughput. Today, spray-coating methods find new applications in different domains as food or cosmetics industries [1].

Spray coating consists in dispersing molten droplets of a coating material onto a surface, to produce a homogeneous membrane. The sprayed liquid, also referred as shell, wall or coat material, can be a solution, a suspension, an emulsion or a melt. Almost any material with a stable molten phase can be sprayed, allowing coatings with a thickness of 100 µm up to 10 mm, at high deposition rates. For a food industry application, the properties have been reviewed for edible films and coatings [2].

Spray coating is not strictly speaking a method of particles production because a solid support is needed and a shell-core structure is obtained. For example, cells can be immobilized on microcarriers or in polymeric beads; the resulting particles may then be spray-coated in order to provide the required properties. Nevertheless, in other cases, it is possible to apply the coat directly on the cell surface [3].

V. Nedović and R. Willaert (eds.), Fundamentals of Cell Immobilisation Biotechnology, 343-356.

Major interests of using spray coating are the following:

- delayed /controlled release of encapsulated agent
- molecules stabilization
- protection against oxygen, water, light or temperature
- masking of undesirable taste, odour or colour
- improvement of density and flow behaviour of particles
- dust reduction

It is of fundamental importance to focus the application of the desired capsules, in order to decide the suitable coating material, particle size, size distribution and wall thickness, which influence the choice of coating technique.

Different spray coating procedures are available, depending on the coating material, the set-up and the operating conditions. The main techniques are listed below:

- Fluid bed coating
- Hot melt coating
- Spray chilling
- Spray cooling

The following lines will present each technique more in details.

2.1. FLUID BED COATING

This technique relies upon a nozzle spraying the coating material into a fluidized bed of core particles in a hot environment. Effective evaporation occurs after contact between spray droplets and the surface of the particle to coat. This process allows high coating rates and it is suitable for particles with a diameter from 50 µm to 5 mm.

Fluid bed spray coating is a three-step process. First, the particles to coat are fluidized in the hot atmosphere of the coating chamber. Then, the coating material is sprayed through a nozzle on the particles and the film formation begins with a succession of wetting and drying stages.

It is possible to choose between three types of coating processes depending on the nozzle position: i) the top spray, ii) the bottom spray and iii) the tangential or rotary spray coating. However, for each type of coating device, the principle of particle coating is the same and can be schematized as follows (Figure 1).

The small droplets of the sprayed liquid contact the particle surface, spread on the surface and coalesce. The solvent or the moisture is then evaporated by the hot air and the coating material adheres on the particle.

Product quality characteristics depend on numerous variables, which affect the different steps of the process. Fluidization air velocity, temperature and humidity affect the evaporation rate and consequently the film characteristics. The successful spreading of droplets on the particle surface depends on the stickiness of the coating material, on the wettability of particles by the coating liquid and on the operating conditions. Collision between particles leading to agglomeration is generally prevented by controlling the coating material characteristics, such as viscosity and hygroscopicity, increasing the particles velocity and decreasing the bed moisture content. The thickness of the final film is determined by the number of passages of the particles in the coating zone. Coalescence of the coating material can be enhanced with a thermal treatment after the spray-coating process [4]. The properties of the coating, related to the operating conditions and the liquid characteristics, determine moisture content and permeation rates of the particles. During spraying process bubble might form due to shear and they might be trapped during droplet collision on particle surface; these holes in the coating film determine a porous structure, showing high permeability, brittleness and unstable mechanical properties [4].

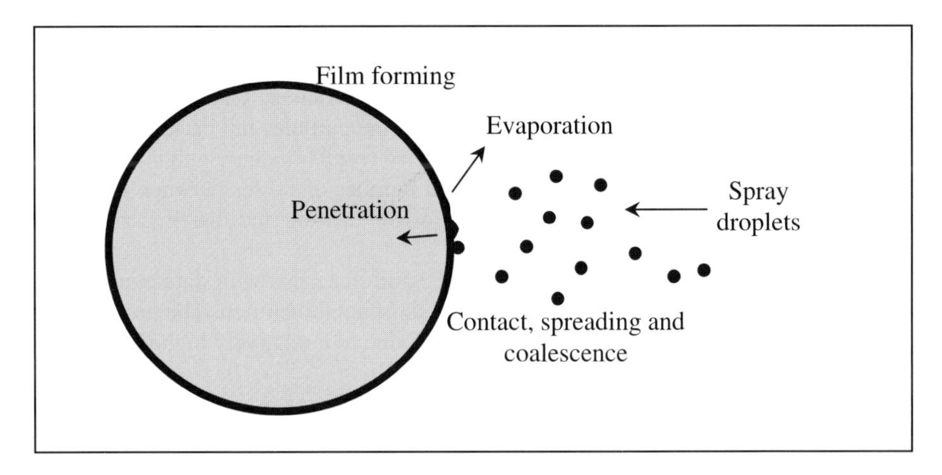

Figure 1. Schematic approach of film formation with the fluid bed coating device.

Usually, air-atomized spray nozzles are used, as the atomization air allows better control of droplet size compared with airless nozzles. Atomization pressure influences droplet size, droplet velocity and fluid bed temperature [5].

Different spray-coating fluidized bed have been realized, where the nozzle is positioned either in the bottom, in the top or on one side of the set-up. Common target of all these devices is to minimize droplet travel distance, in order to obtain optimal spreadability and coalescence upon contact with particles [6].

The next parts of the text are devoted to the presentation of each spray coating device with a quick outline of their advantages and disadvantages.

2.1.1. Top spray coating

The top spray coating is the base of spray coating technology. It has been used for more than 20 years for the lipid coating of vitamins. In this technology, the spray liquid and the air are in counter current (Figure 2). Therefore, pulverization occurs at the top of the fluidized bed, involving a high risk of droplets drying. To prevent agglomeration, the particles should travel fast trough the coating zone and the droplets should be small enough to simply deposit the coating material on the surface; the finer the droplets are, the denser the final film. Moreover, in this configuration the motions of fluidization are random, resulting in a nonuniform coating [7]. Nevertheless, for taste masking or enteric coating applications, which do not require a perfect coat, satisfying results are attained with this device [8].

2.1.2. Bottom spray coating and Wurster process

In the bottom spray coating, the liquid is pulverized concurrently with the air, straight inside the bed of particles, at the bottom of the coating chamber [9]. Collision between droplets and particles is thus increased, resulting in higher coating efficiency and lower droplets drying. In 1950, Prof. Dale Wurster improved the device by adding a cylindrical partition in the coating chamber and designing an air distribution plate to control the particles motions [10]. The largest free area is just below the central partition allowing a high-velocity air stream and as a consequence the rising of the particles. In this zone, the spreading of the droplets is the most efficient. Outside the partition, gravity overcomes the force of the fluidizing air and the particles fall back, suspended in a slowly moving “down bed” (Figure 3). This device results in dense and homogeneous coating and minimal risk of agglomeration. The number of cycles between the drying and the coating zone will decide the thickness of the coating and the final size of the particles.

This device can be employed to encapsulate solid materials with diameters ranging from 50 µm up to several centimetres, without risk of agglomeration. The process has a greater drying capacity than other coating systems, due to a relatively high fluidizing air velocity.

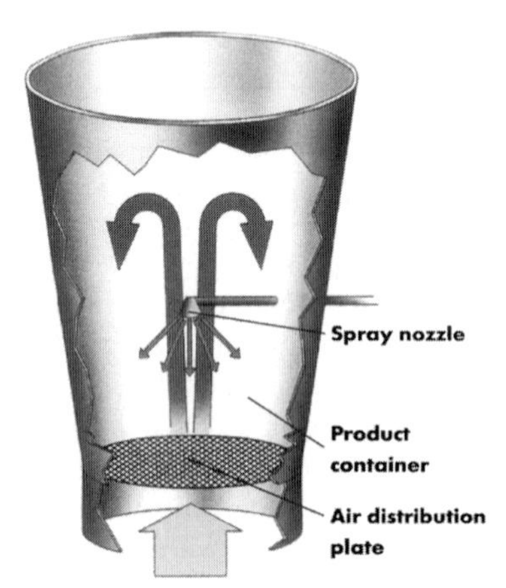

Figure 2. Top spray coating device (GLATT GmbH).

2.1.3. Tangential spray coating

The last type of spray coating device is the tangential spray coating also called rotary spray coating. This is a relatively new technology combining a fluid bed and a rotary disc in the bottom of the product container. The spray liquid is directly released tangentially inside the fluidized bed. The fluidization of solid particles is the result of an airflow through the slit between the edge of the rotor disk and the inside chamber (Figure 4). The fluidization pattern in the rotor processor can be described as a spiralling helix due to a combination of centrifugal force, air stream and gravity.

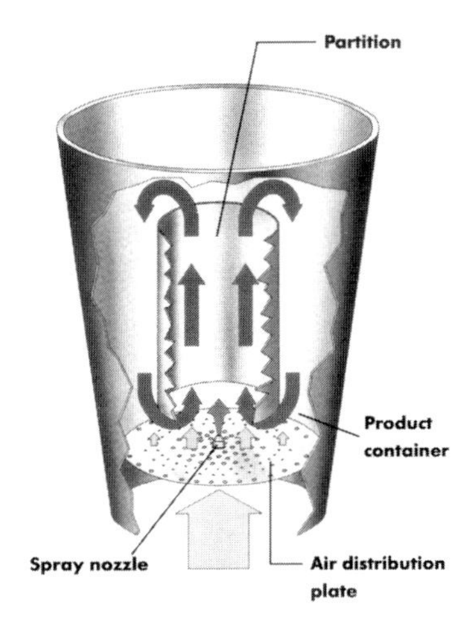

Figure 3. Bottom spray coating with the Wurster device (GLATT GmbH).

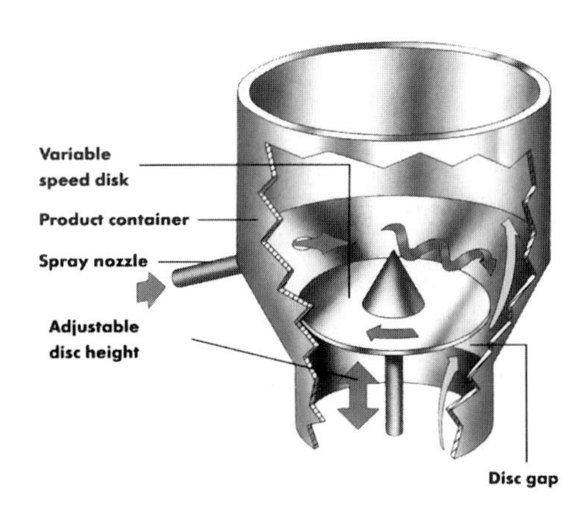

Figure 4. Rotary spray coating device (GLATT GmbH).

For the film coating application, the tangential spray system equals the bottom spray system. The coating is continuous and homogenous. However, the biggest drawback of this technique is the high shear stress applied to the particles, which does not allow its application to fragile or friable material [7].

In conclusion, these spray-coating devices have some common features and variables, but each one has unique advantages and limitations. The choice of the coating device depends on the final application and on other different criteria such as economic, product and process.

The fluid bed process has been improved in order to optimise residence time distribution in the coating chamber and film homogeneity. Some efforts are carried on modelling the fluid bed coating in order to achieve the process optimization [10,11,12].

It is still a batch process, it is expensive, and time consuming. However, continuous fluidized bed coaters have been developed, which should decrease the prices and allow a higher throughput [14]. This technique is already widely employed already in the pharmaceutical and cosmetic industry, which disposes of a higher budget, but applications in the food industry to encapsulate enzymes, proteins, aromas, bacteria and yeasts have also been studied [15,16].

2.2. HOT MELT COATING

Hot melt coating is very similar to the fluid bed technique because it is a lipid spray coating. The coating material should have a melting point lower 100°C and so can be applied as a molten solution to the fluidized particles. The solidification of droplets results of the decrease of temperature by the ambient cold air [17].

A top spray, bottom spray or rotary spray device can be employed to spread the coating material. Nevertheless, a careful control of the liquid, the product bed and the atomizing air temperatures is required.

This procedure reduces production time and energy costs, eliminates the risk of droplet drying and the use of solvent. On the other hand, final capsules are not perfectly homogeneous. This technique may be applied only to molten coating materials such as wax, and hydrogenated oils therefore it is not suitable to heat sensitive particles [18].

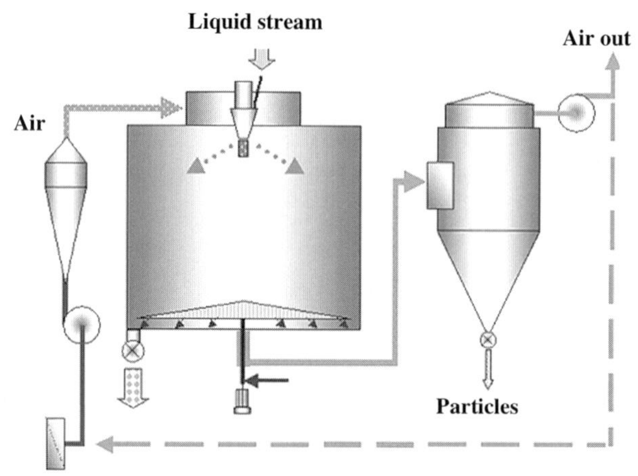

Figure 5. Spray cooler one-stage (GEA Niro Inc.).

2.3. SPRAY CHILLING

Similarly, in the spray chilling technique, the coating material is melted and atomised with a pneumatic nozzle in a vessel generally containing a carbon-dioxide ice bath (temperature -50°C) as in a hot-melt fluidized bed. Thus, droplets adhere on particles

and solidify forming a coat film. The formed capsules move countercurrent to a flow of tempered air, the coating quickly solidifies, and the capsules are collected below the spray nozzle and dried in a vacuum oven (Figure 5). Spray chilling may be applied to thermosensible substances, as the coating material does not need high temperatures.
This is also an immobilization technique, as it allows to solidify dispersion of melted material and active substances. As a matter of fact, spray chilling is often employed in the pharmaceutical industry to form a solid dispersion of drug and lipophilic excipients. The encapsulating material is generally a fractionated vegetable oil or hydrogenated vegetable oil with a relatively low melting point (32-42°C) [19]. Spray chilling (also called spray congealing) is used primarily for the encapsulation of solid food additives such as ferrous sulphate, acidulants, vitamins, and solid flavours, as well as for sensitive materials or those that are not soluble in "usual" solvents [20,21]. Liquid solution may also be encapsulated after freezing. Because of the controlled release properties of the particles, the process is suitable for protecting many water-soluble materials such as spray dried flavours, which may otherwise be volatilized from a product during thermal processing. In pharmacy, spray-chilling is employed to realize sustained-release products [21,22,23]. Spray chilled products have applications in bakery products, dry soup mixes, and foods containing a high level of fat.

2.4. SPRAY COOLING

This technique is similar to spray chilling, the only difference is the temperature of the reactor in which the coating material is sprayed. A molten matrix material containing minute droplets of the core materials may be spray cooled. As spray chilling, this technique may also be employed to immobilize substances in solid dispersion. Different coating substances may be employed, such as some derivative of vegetable oil and stearine as well as hard mono- and di-glycerides [20,21].

3. Drying processes

Drying is an encapsulation technique successfully employed when the substance to be immobilized is dispersed in the encapsulating agent, forming an emulsion or a suspension. The "solvent" is generally a food-grade hydrocolloid such as gelatine, vegetable gum, modified starch, dextrin, or non-gelling protein [24,25] and it is immiscible with the substance to be immobilized. The solution thus obtained is dried, providing a barrier to oxygen and aggressive agents.

This procedure is advantageous when the encapsulated substance is a small fraction of the whole particle. On the other side, only very small particles (up to 100 μm) can be encapsulated with this technique; the resulting beads are heterogeneous and part of the encapsulated substance may not be protected from external agents. However, the formed particles may be further coated with spray-coating techniques.

The three main drying techniques are spray-drying, fluid-bed drying and freeze drying, and the resulting particles are either in form of powder, granulates or grounded solids.

3.1. SPRAY DRYING

This process is similar to spray cooling or spray chilling, the main difference lies in the air temperature and the coating material. The suspension is atomised in a hot drying chamber, allowing the evaporation of moisture and formation of solid particles (Fig.6). A double nozzle is employed, allowing air from an annular geometry to atomize the liquid stream in order to form fine particles carrying the microencapsulated product in a dispersed state. The particles are then conveyed in a drying chamber where the solvent or aqueous media flash-evaporates. The resulting particles pass through a cyclone and are collected into a holding chamber. The size of a spray-dried bead depends both on physical and operating parameters, such as: solution viscosity, density, surface tension and atomisation pressure, nozzle type. The solid concentration influences the physical properties of the suspension and consequently the droplet size. Sometimes both the encapsulated substance and the encapsulating agent may be solid, thus creating heterogeneous suspensions, with a tendency to segregation [26].

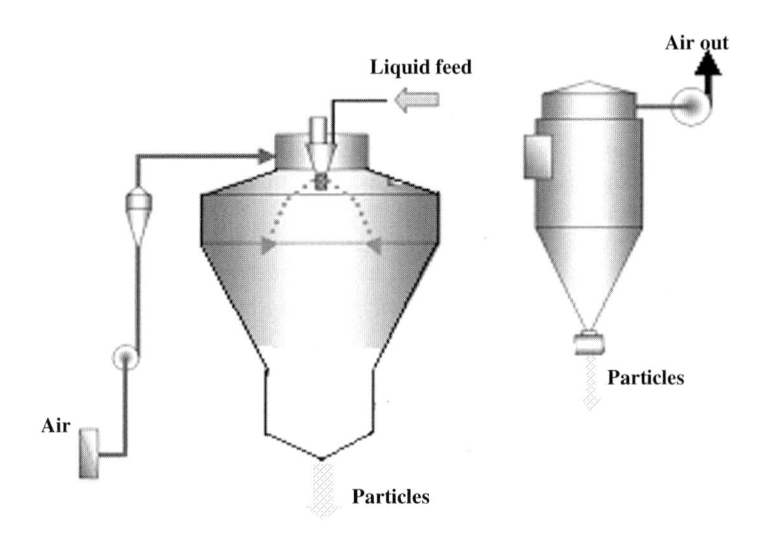

Figure 6. Spray drier.

The temperature of the sprayed liquid must be optimised in order to avoid solidificatio prior to atomisation on one side and late solidification, resulting in odd-shaped stick particles, on the other side. As a matter of fact, as the moisture content in the particle drops, the diffusion rate of water to the surface also decreases and the outlet temperatur must be high enough to continue the drying process. This can be avoided by adding fluid bed after the dryer. In order to conduct the process in an economical an environment-friendly manner, the powder should be removed from drying air, which then recycled. Fine powder is generally removed from the air with cyclones; bag filter electrostatic precipitators or scrubbers may also be employed. Fines are then returned to an agglomeration process and air is recycled to the system.

The main advantages of spray drying are the large scale production in continuous mode, with simple and cost-effective set-up and the processing of heat-sensitive materials at atmospheric pressure and low temperature. Spray-drying techniques are applied to pharmaceuticals, flavours, dairy products, blood plasma, detergents, rubber latex, ceramic powders and other chemicals [27] and can be studied through computational fluid dynamics [28].

Spray chilling and spray cooling employed for encapsulation may be considered as two variants of spray drying process, as cold air instead of hot air is employed to solidify the particles.

3.2. FLUID BED DRYING

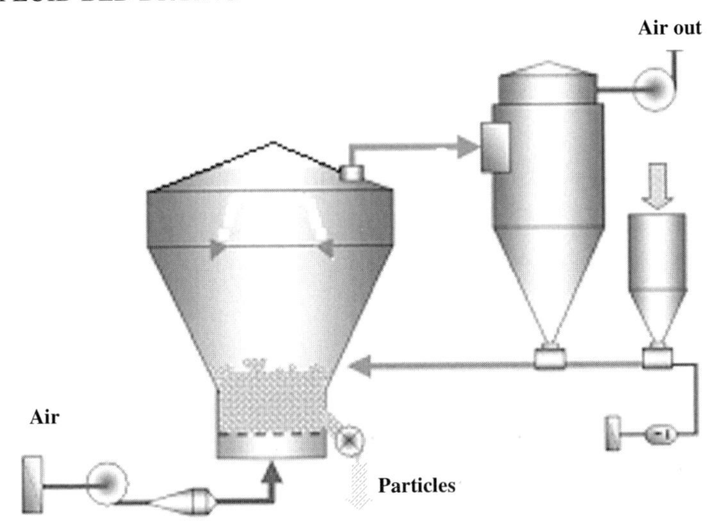

Figure 7. Fluidised bed drier.

This process requires a solid support, which is fluidised at the bottom of the set-up through a fluidization gas, usually hot air. Heat is provided by air and surfaces, panels or tubes, immersed in the fluidized layer (Figure 7). Depending on the process, the solid feed may be powder produced in spray drying or recycled fines from exhaust air. In order to prevent particle sticking to the walls of the dryer, inert pellets, larger than the dried particles, can be co-fluidised, then separated through a metallic sieve.

In a fluidised bed dryer the residence time can be fixed independently from the inlet air temperature, so that particle overheating can be avoided. Fluid bed processing offers important advantages over other methods of drying particulate materials: particle fluidization gives easy material transport, uniform temperature distribution, high rates of heat exchange and mass transfer. The properties of resulting particles are determined by operating conditions such as drying rate, fluidization gas velocity, fluidization point (*i.e.* the volatile content below which fluidization without mechanical agitation or vibration is possible), equilibrium volatile content, and heat transfer coefficient for immersed heating surfaces [29].

The process can be carried out in two stages, in order to control the final moisture content and to avoid excessive dehydration. For industrial application a continuous fluidised-bed dryer would be cost-effective. Nevertheless, there would be a distribution in residence time, damaging the heat sensitive products and inactivating encapsulated cells. Series of batch fluidised beds are therefore more used, especially in food industry.

In addition, spraying devices may be connected in sequence with fluid bed, as it is already realized in milk powder production [30]; the solution is dried to a relatively high water content in the spray dryer, then the powder is cooled and dried in a fluidised bed, allowing agglomeration of the powders, formation of homogeneous particles and dust reduction (Figure 8). With this design, the first stage realizes the intensive evaporation and the second stage allows a rigorous control of the temperature and of the final water content.

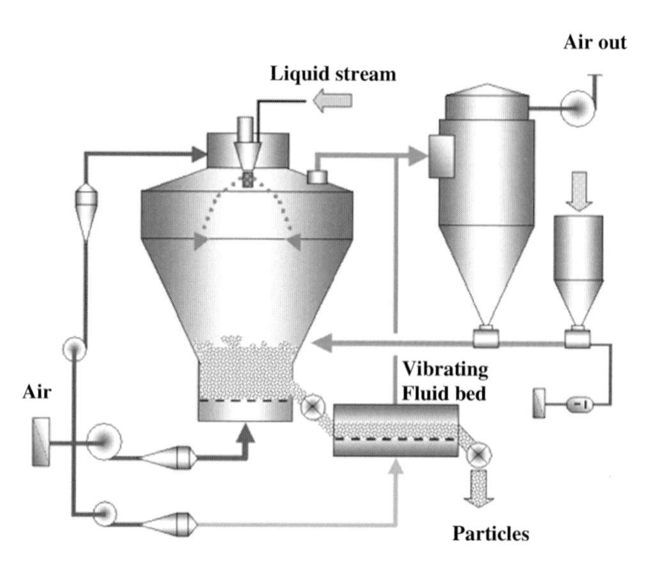

Figure 8. Spray drier with fluid bed.

A vacuum fluid bed can be employed for the processing of heat sensitive products and explosive powders and it allows solvent recovery. The particles may then be cooled using a fluid bed with cold gas, ambient or conditioned air, or cooling surfaces. Special fluidized bed for granulating, drying and coating may be integrated in the same device. These are quite expensive processes, which can be applied specially in pharmaceutical industries, where the high price of the final product justifies high running costs.

3.3. FREEZE-DRYING

Freeze-drying is often employed in the food industry. This process occurs at low temperatures and under vacuum, avoiding any water phase transition and any oxidation. The dried mixture thus obtained must be grounded, resulting in heterogeneous particles with a low surface area, if compared to spray-dried particles. The costs of freeze-drying are up to 50 times higher than spray-drying [31] and also storage and transport of produced particles are extremely expensive and make this drying technique less attractive than the others.

4. Coating and drying technologies for cell encapsulation

Both coating and drying techniques illustrated above can be applied to cell encapsulation, in order to provide protection from external agents in fermentation processes and bioremediation, to improve the storage and preservation of cultures, to enhance cell usage in food and pharmaceutical industry.

Coating techniques can be applied to cell encapsulation either to enhance the bead resistance by adding a polymeric membrane or to deposit cell suspension on inert supports. Ceramic pellets show high porosity, suitable surface structure and high stability; the coating allows a stable immobilization, reduces cell release and diffusion limitation. Yeasts have been immobilized within binder polymers on γ-alumina particles by spray-coating process, showing high activity, especially with a pre-soaking treatment of ceramic beads [32]. Bacteria have been immobilized within a PVA membrane coating expanded clay pellets [33].

Drying techniques applied to cell immobilization enhance cell preservation and allow their usage in different fields such as the production of nutraceuticals and probiotics and for the so-called biocontrol preservation, with antagonistic activity against pathogens [34,35,36]. Dried cultures are employed as bulk starters in order to reduce the risk of contamination and to guarantee constant characteristics.

Freeze-drying is often used for cell storage; providing protective additives such as skim milk, glycerol, lactose and optimising process conditions, this process can guarantee a survival rate up to 80% [37]. Nevertheless, it is a complex and expensive process, unsuitable for large production and requiring high storage and transport costs.

Spray-drying and fluid-bed drying present low costs in storage and transport and the suitability to large scale production. Nevertheless, operating conditions, such as temperature, moisture content and drying time, must be rigorously controlled in order to prevent thermal inactivation and dehydration of encapsulated cells.

Spray-drying is already widely employed in dairy industry and can be used to produce large amounts of dried viable cultures, with reduced costs for production, storage and transport [38]. Strains should be selected with regard to their resistance to heat treatment and spray-drying. Protectants and antioxidants should improve viability during powder storage, although they may have contrasting effects on stability [39]; colloids can be added to protect cells during spraying [40]. Thermal inactivation can be avoided by optimising the mass ratio of drying air to liquid feed. Dehydration can be prevented by controlling the water content of the particles. In fact, the modelling of spray-drying process and the control of operating parameters are very complex, thus hindering the optimisation. An advantage of spray-drying is the possibility to dry the cells as a suspension, without need of a support. On the other side the feed must be pumped to the drier, thus limiting cell concentration and increasing the amount of water to be evaporated [37].

Lactobacillus strains suspended in reconstituted skim milk were spray-dried, maintaining good viability and activity [39]. Entomopathogenic fungi suspended in a solution of skim milk and PVP were also spray-dried, resulting in concentrated powders with high viability [41].

Fluid-bed drying allows to use relatively low air temperature, thus minimizing thermal inactivation; moreover, with a two-stage process it is possible to control the water content of the particles, in order to prevent dehydration. Fluid-bed process can be easy modelled, allowing insight into the influence of the different parameters. The main drawbacks of this process are the low porosity of resulting particles, which require a long rehydration time [37] and the need of a support to dry the cells. Generally starch, wheat bran, xantan gum, alginate, k-carrageenan are employed in different applications; milk powder, lactose, maltodextrine are used in food industry.

A β-carotene producing microalgae has been immobilized in Ca-alginate beads, then dried in a fluid-bed dryer, enhancing the product stability [42,43]. Lactic bacteria have been encapsulated in starch then dried in a fluidised bed [44]; after eight months storage under refrigeration, particles could be rehydrated and recover their viability and fermentation activity. Some experiments were carried out with Lactic bacteria and yeasts immobilized in Ca-alginate and k-carrageenan beads then dried, showing that storage in immobilized dried state is feasible if drying times and temperature are rigorously controlled and cells are activated prior to drying [45].

In conclusion, spray-drying and fluid-bed drying are two promising immobilization techniques for large-scale production of culture-containing powders to be applied in food and pharmaceutical industry, for cell preservation and cell encapsulation for environmental applications.

References

[1] Risch, S.J., (1995) Encapsulation: overview of uses and techniques. In: Encapsulation and Controlled Release of Food Ingredients, ACS Symposium Series, 590: 2-7.

[2] Kester, J.J. and Fennema, O.R. (1986) Edible films and coating: a review. Food Tech. 40: 47-59

[3] Kampf, N.; Zohar, C. and Nussinocitch A. (2000) Hydrocolloid coating of *Xenopus laevis* embryos. Biotech. Progr.16: 480-487.

[4] Sun, Y.; Huang, W. and Chang, C. (1999) Spray-coated and solution-cast ethylcellulose pseudolatex membranes. J. Membr. Sci.157: 159-170.

[5] Dewettinck, K. and Huyghebaert, A. (1998) Top-spray fluidized bed coating: effect of process variables on coating efficiency. Liebensmittel Wissenchaft und Technologie 31: 568-575.

[6] Eichler, K. (2002) Fluid bed film coating. In: Proceedings of the TTC Workshop n°53 on "Nutraceutical and Probiotics", Binzen (Germany), 26-28 June, 12; pp. 1-9.

[7] Dewettinck, K.; Messens, W.; Deroo, L. and Huyghebaert, A. (1999) Agglomeration tendency during top-spray fluidized bed coating with gelatin and starch hydrolyzate. Lebensmittel-Wissenschaft und - Technologie 32: 102-106.

[8] Dewettinck, K. and Huyghebaert, A. (1999) Fluidized bed coating in food technology. Trends in Food Sci. and Tech. 10: 163-168.

[9] Shelukar, S.; Ho, J.; Zega, J.; Roland, E.; Yeh, N.; Quiram, D.; Nole, A.; Katdare, A. and Reynolds S. (2000) Identification and characterization of factors controlling tablet coating uniformity in a Wurster coating process. Powder Tech. 110: 29-36.
[10] Wurster, D.E. (1950) Means for applying coatings to tablets or like. J. Am. Pharmac. Assoc. 48(8): 451-460.
[11] Dewettinck, K.; Visscher, A.; Deroo, L. and Huyghebaert, A. (1999) Modeling the steady state thermodynamic operation point of top-spray fluidised bed. J. Food Eng. 39: 131-143.
[12] Fyhr, C. and Kemp, C. (1999) Mathematical modelling of batch and continuous well-mixed fluidised bed dryiers. Chem. Eng. and Processing 38: 11-18.
[13] Guignon, B.; Duquenoy, A. and Dumoulin, E. (2000) Modelling of coating of solid particles in fluid bed process. In: Proceedings of the 12th Int. Drying Symposium, Noordwijkerhout (The Netherlands), 2000, 28-31 August; p. 46.
[14] Teunou, E. and Poncelet, D. (2002) Batch and continuous fluid bed coating – review and state of the art. J. Food Eng. 53: 325-340.
[15] Dezarn, T.J. (1998) Food ingredient encapsulation. In: Risch, S.J. and Reineccius, G.A. (Eds.) Encapsulation and Controlled Release of Food Ingredients. ACS Symposium Series 590; pp. 74-86.
[16] Lee, S.Y. and Krochta J.M. (2002) Accelerated shelf life testing of whey protein coated peanuts analysed by static gaz chromatography. J. Agric. Food Chem. 50: 2022-2028.
[17] Jozwiakowski, M.J.; Franz, R.M. and Jones, D.M. (1990) Characterization of a hot-melt fluid bed coating process for the granules. Pharmac. Res. 7: 1119-1126.
[18] Emas, M. and Nyqvist, H. (2000) Methods of studying aging and stabilization of spray-congealed solid dispersions with carnauba wax.1.Microcalorimetric investigation. Int. J. Pharmaceutics 197: 117-127.
[19] Uhlemann, J.; Schleifenbaum, B. and Bertram, H-J. (2002) Flavor encapsulation technologies: an overview including recent developments. Perfumer & Flavorist 27: 54-61.
[20] Gibbs, B.F.; Kermasha, S.; Alli, I. and Mulligan, C.N. (1999) Encapsulation in the food industry: a review. Int. J. Food Sci. Nutrition 50: 213-224.
[21] Savolainen, M.; Khoo, C.; Glad, H.; Dahlquist, C. and Juppo, A.M. (2002) Evaluation of controlled-release polar lipid microparticles. Int. J. Pharmaceutics 244: 151-161.
[22] Rodriguez, L.; Passerini, N.; Cavallari, C.; Cini, M.; Sancin, P. and Fini, A. (1999). Description and preliminary evaluation of a new ultrasonic atomizer for spray-congealing processes. Int. J. Pharm. 183: 133-143.
[23] Passerini, N.; Perissutti, B.; Albertini, B.; Voinovich, D.; Meneghini, M. and Rodriguez, L. (2003) Controlled release from waxy microparticles prepared by spray-congealing. J. Controlled Release 88: 263-275.
[24] Apintanapong, M. and Noomhorm, A. (2003) The use of spray drying to microencapsulate 2-acetyl-1-pyrroline, a major flavour component of aromatic rice. Int. J. Food Sci. Technol. 38: 95-102.
[25] Balassa, L.L. and Fanger, G.O. (1971) Microencapsulation in the food industry. In: CRC Critical reviews in food technology. CRC Press, Boca Ration (FL); pp. 245-265.
[26] Diosady, L.L.; Alberti, J.O. and Venkatesh, M.M.G. (2002) Microencapsulation for iodine stability in salt fortified with ferrous fumarate and potassium iodide. Food Res. Int. 35: 635-642.
[27] Benczedi, D. (2002) Flavor encapsulation using polymer-based delivery systems. In: Taylor, A.J. (Ed.) Food Flavour Technology. Sheffield Academic Press, Sheffield, UK; pp. 153-166.
[28] Langrish, T.A.G. and Fletcher, D.F. (2001) Spray drying of food ingredients and applications of CFD in spray drying. Chem. Eng. Proc. 40: 345-354.
[29] Bayrock, D. and Ingledew, W.M. (1997) Fluidized bed drying of baker's yeast: moisture levels, drying rates and viability changes during drying. Food Res. Int. 30(6): 407-415.
[30] Bylund, G. (1995) Dairy Processing handbook. Tetra Pak Processing Systems AB ed. 1995, Lund, Sweden; p. 436.
[31] Desobry, S.; Netto, F. and Labuza, T. (1997) Comparison of spray-drying, drum-drying and freeze-drying for β–carotene encapsulation and preservation. J. Food Sci. 62: 1158-1162.
[32] Isono, Y.; Araya, G. and Hooshino, A. (1995) Immobilization of a *Saccharomyces cerevisiae* for ethanol fermentation on g-alumina particles using a spray-dryer. Proc. Biochem. 30: 743-746.
[33] Massart, B.; Nicolay, X.; Van Aelst, S. and Simon, J.-P. (2002) PVA-gel entrapped specific microorganisms enhancing the removal of VOCs in waste air and biofiltration systems. Landbauforschung Volkenrode SH 241: 93-98.
[34] Domingues, D.J. and Hanlin, J.H. (2000) Food preservation by using dried lactic acid bacteria as biocontrol agents. PCT Int. Appl.; pp. 68.

[35] Hunter, K.W., Jr.; Gault, R.A. and Berner, M.D. (2002) Preparation of microparticulate β-glucan from *Saccharomyces cerevisiae* for use in immune potentiation. Letters in Appl. Microbiol. 35: 267-271.
[36] Silva, J.; Carvalho, A.S.; Teixeira, P. and Gibbs, P.A. (2002) Bacteriocin production by spray-dried lactic acid bacteria. Letters in Appl. Microbiol. 34: 77-81.
[37] Lievense, L.C. and van't Riet, K. (1993) Convective drying of bacteria. I. The drying processes. Adv. Bioch. Eng. Biotech. 50: 45-63.
[38] Ré, M.I. (1998) Microencapsulation by spray drying. Drying Tech. 16: 1195-1236.
[39] Gardiner, G.E.; O'Sullivan, E.; Kelly, J.; Auty, M.A.; Fitzgerald, G.F.; Collins, J.K.; Ross, R.P. and Stanton, C. (2000) Comparative survival rates of human-derived probiotic *Lactobacillus paracasei* and *L. salivarius* strains during heat treatment and spray drying. Appl. Env. Microb. 66: 2605-2612.
[40] Millqvist-Fureby, A; Malmsten, M.; Bergenstahl, B. (1999) An aqueous polymer two-phase system as a carrier in the spray-drying of biological material. J. Coolid. Interface Sci. 225: 54-61.
[41] Horaczek, A. and Viernstein, H. (2002) Aerial conidiaa of *Metarhizium anisopliae* subjected to spray drying for encapsulation purposes. Landbauforschung Volkenrode SH 241: 93-98.
[42] Leach, G.; Oliveira, G. and Morais, R. (1998) Production of a carotenoid-rich product by alginate entrapment and fluid-bed drying of *Dunaliella salina*. J. Sci. Food Agric. 76: 298-302.
[43] Leach, G.; Oliveira, G. and Morais, R. (1998) Spray-drying of *Dunaliella salina* to produce a β-carotene-rich powder. J. Ind. Microbiol. Biotechnol. 20: 82-85.
[44] Clementi, F. and Rossi, J. (1984) Effect of drying and storage conditions on survival of *Leuconostoc Oenos*. Am. J. Enol. Vitic. 35: 183-186.
[45] Turker, N. and Hamamci, H. (1998) Storage behaviour of immobilized dried micro-organisms. Food Microbiol. 15: 3-11.

PART 3

CARRIER CHARACTERISATION AND BIOREACTOR DESIGN

DIFFUSIVE MASS TRANSFER IN IMMOBILISED CELL SYSTEMS

RONNIE WILLAERT[1], GINO V. BARON[2] AND VIKTOR NEDOVIĆ[3]

[1]*Department of Ultrastructure, Flanders Interuniversity Institute for Biotechnology, Vrije Universiteit Brussel, Pleinlaan 2, B-1050 Brussel, Belgium – Fax 32-2-6291963 – Email: Ronnie.Willaert@vub.ac.be*

[2]*Department of Chemical Engineering, Vrije Universiteit Brussel, Pleinlaan 2, B-1050 Brussel, Belgium – Fax: 32-2-6293248 – Email: gvbaron@vub.ac.be*

[3]*Department of Food Technology and Biochemistry, University of Belgrade, Nemanjina 6, PO Box 127, 11081 Belgrade-Zemun, Yugoslavia – Fax: 381-11-193659 – Email: vnedovic@EUnet.yu*

1. Introduction

The analysis of the influence of mass transfer on the reactor performance in immobilised-cell reactors is an important topic since the effectiveness of these reactors may often be reduced by the rate of transport of reactants to and products from the immobilised cell system (external mass transfer limitation), and by the rate of transport inside the immobilised cell system (internal mass transfer limitation). External mass transfer limitations can be reduced or eliminated by a proper design of the reactor and immobilised cell system. Several phenomena are involved such as axial dispersion, convective flow, and macro- and micro-mixing. More information about these topics can be found in various (bio)chemical engineering books and several review articles (*e.g.* [1-6]). Internal mass transfer limitations are often more difficult to eliminate and their knowledge is a prerequisite to analyse and optimise the performance of the immobilised cell system.

In this chapter, internal mass transport by diffusion in immobilized cells systems is discussed, since in most cases convective transport within the system is negligible. The various methods of measurement of the diffusion coefficient are reviewed. The influence of the presence and growth of living cells in the immobilisation matrix on the diffusion coefficient is discussed. Next, internal diffusion in gel systems and dense cell masses – *i.e.* biofilms, bioflocs, mammalian cell aggregates – is treated in detail. Finally, the diffusion in microcapsules is discussed.

V. Nedović and R. Willaert (eds.), Fundamentals of Cell Immobilisation Biotechnology, 359-387.

2. Definitions of diffusion coefficients

Mass transport by molecular diffusion is defined by Fick's law, i.e. the rate of transfer of the diffusing substance through a unit area is proportional to the concentration gradient measured normal to the section:

$$J = -D\frac{\partial C}{\partial x} \quad (1)$$

where J is the mass transfer rate per unit area of the section, C the concentration of diffusing substance (amount per total volume of the system), x the space co-ordinate, and D is called the diffusion coefficient. It is general practice to use an effective diffusion coefficient (D_e), which can be readily used in the expression for the Thiele modulus and for the determination of the efficiency factor of a porous biocatalyst. This effective diffusion coefficient can be defined as

$$J = -D_e\frac{\partial C_L}{\partial x} \quad (2)$$

where C_L is the amount of solute per unit volume of the liquid void phase. Concentration C may be correlated with C_L by using the void fraction (ε), which is the accessible fraction of the porous particle to the diffusion solute as

$$C = \varepsilon C_L \quad (3)$$

Hence, the relationship between the effective diffusion coefficient and the diffusion coefficient can be written as

$$D_e = \varepsilon D \quad (4)$$

In the (bio)chemical engineering literature however, alternative definitions of the term effective diffusion coefficients have been introduced (see for example [7-12]).

3. Internal diffusion in immobilised cell systems

The effective diffusion coefficient through a porous support material (matrix) is lower than the corresponding diffusion coefficient in the aqueous phase (D_a) due to the exclusion and obstruction effect. By the presence of the support, a fraction of the total volume (1-ε) is excluded for the diffusing solute. The impermeable support material obstructs the movement of the solute and results in a longer diffusional path length that can be represented by a tortuosity factor (τ), which equals the square of the tortuosity [13]. The influence of both effects on the effective diffusion coefficient can be represented by

$$D_e = \frac{\varepsilon}{\tau} D_a \tag{5}$$

This equation holds as long as there is no specific interaction of the diffusion species with the porous carrier. In the case of gel matrices, predictions using the polymer volume fractions are recommended, since ε nor τ can be measured for a gel in a simple way [14-16]:

$$D = \frac{(1-\phi_p)^2}{(1+\phi_p)^2} D_a \tag{6}$$

where ϕ_p is the polymer volume fraction. For low molecular weight solutes in cell-free gels, an approximate measure of ε can be given as

$$\varepsilon = 1 - \phi_p \tag{7}$$

D_e can also be expressed as a function ϕ_p by combining Equation (4), (6) and (7) [17-20] as

$$D_e = \frac{(1-\phi_p)^3}{(1+\phi_p)^2} D_a \tag{8}$$

3.1. METHODS OF MEASUREMENT OF THE DIFFUSION COEFFICIENT

Both steady-state and transient methods for measurements of diffusion coefficients are used, sometimes in combination as in the lag-time method. A few researchers have used indirect methods, such as the use of a reaction-diffusion model, to determine the diffusion coefficient of gel immobilised cells [21-23] or biofilms [24-29]. Westrin and co-workers [30] have reviewed in depth the methods for measuring diffusion coefficients. They compared different methods (including holographic laser interferometry and nuclear magnetic resonance) with regard to accuracy, reproducibility, time, cost and limitations.

The equation, which describes the transient diffusion, can be readily derived by writing the mass balance over the system:

$$\varepsilon \frac{\partial C_L}{\partial t} = x^{-n} \frac{\partial}{\partial x}\left(x^n D_e \frac{\partial C_L}{\partial x}\right) \tag{9}$$

where n is a shape factor which is 1 for planar, 2 for cylindrical or 3 for spherical geometry. In Table 2 the most popular methods to determine diffusion coefficients (in gels and dense cell masses) are listed. Recently, Frazier *et al.* [31] used near-infrared spectroscopy to measure nutrients within a special designed diffusion chamber that permits the non-invasive measurement of effective diffusivities in cell immobilised gels (for example D_e of glutamine in agarose gel).

Table 1. Experimental methods to determine diffusion coefficients.

Method	Reference examples
Concentration gradient methods: steady-state	
• True steady-state diaphragm cell	[32-39]
• Pseudo-steady-state diaphragm cell	[18,39,40-43]
Concentration gradient methods: transient	
• Time-lag diaphragm cell	[8,36,44-47]
• Uptake/release from particles dispersed in a stirred solution	[7,10,48-71]
• Concentration profile in a material/cell mass:	
- Electrode or mass spectrometric probe covering	[72-77]
- Fixed ultrasonic probes	[78]
- (Holographic) laser/light interferometry	[79-81]
- Magnetic resonance imaging	[82]
- Microelectrodes in:	
- Gel phase or biofilm	[19,25,73,83-86]
- Fungal pellets	[87-90]
- Near-infrared spectroscopy	[31]
- Sectioning method	[91,92]
- UV/VIS absorption scanning	[93]
- Confocal scanning microscopy	[94]
• Indirect methods	
- Chromatographic curves	[95-96]
Instrumental methods	
• Dynamic light scattering	[97]
• Fluorescence recovery after photobleaching	[98-100]
• Pulsed-gradient spin-echo (PGSE)/NMR	[20,101-106]
• Holographic relaxation spectroscopy	[107,108]

Experimentally, a concentration disturbance is applied over the system and the change of the concentration as a function of time is followed until the steady-state is reached. Diffusion coefficients are determined by fitting the theoretical curve to the experimental one, which is obtained after solving Equation (9) using the appropriate initial and boundary conditions. Analytical solutions for simple cases are described by Crank [109]. Crank gives also the solution in terms of M_t, the amount of solute in the matrix at time t, as a fraction of M_∞, the equilibrium amount, which is practically more accessible.

Steady-state measurements can be performed by using a "diaphragm-diffusion cell". Two liquid-filled well-stirred compartments are separated by a membrane under investigation, through which a steady-state diffusional flux is set up [36-39,41,42,44,45]. In the steady-state method, a constant upstream and different downstream solute concentration is employed. Whereas in the pseudo-steady-state method, a change in solute concentration in the compartments is allowed.

If the diffusion coefficient is measured by the steady-state method, the value of the equilibrium partition coefficient (K) should be known. Another approach is to determine directly the product of K and D, which equals the effective diffusion coefficient [110]. Sun and co-workers [36] used a non steady-state method to determine immediately the product of K and D_e for oxygen in cell-free and cell-containing Ca-alginate and PVA-SbQ gels. The partition coefficient is defined as

$$K = \frac{C_m}{C_L} \tag{10}$$

where C_m is the solute concentration in the matrix (amount per volume matrix) and C_L the solute concentration in the free (surrounding) liquid phase. In the absence of specific adsorption phenomena, the partition coefficient for cell-containing matrices can be predicted by the following equation [18,41,110]

$$K = 1 - (\alpha\phi_c + \phi_p) \tag{11}$$

where α represents the volume fraction of the individual cell that is not accessible to the solute and ϕ_c is the volume fraction of the cells. If the outer cell membrane totally excludes the solute, the value of α will be unity. Axelsson and Persson [41] found a value of 1 for glucose, lactose, galactose and ethanol for alginate gels containing yeast cells deactivated by iodoacetic acid. Chresand and co-workers [8] measured the partition coefficient of labelled sodium acetate between 1% (w/w) agar gel containing mammalian cells and the surrounding solution, and found a value that was unity. This result corresponds with a value of zero for α. It can also be assumed that the solute is excluded from a certain volume fraction of the individual cell: $0 < \alpha < 1$. This volume fraction may, *e.g.*, be approximately equal to the dry weight fraction of the cells, which can be assumed to be equal to 0.25 [18]. If there are no immobilised cells present, Equation (11) is reduced to $1-\phi_p$ what equals ε (*cfr.* Equation 7). Values of experimentally determined partition coefficients for cell-free gels were close to unity; *e.g.*, between 0.97 and 1.02 for glucose in 2% Ca-alginate [52]; 0.98 for lactose in 2.75%/0.25% (w/w) κ-carrageenan/locust bean gum [58]; 1.2 for lactose, 1.3 for glucose, 1.1 for galactose and 1.32 for ethanol in 2.4-2.8% (w/w) alginate, the higher value for ethanol was explained as adsorption of ethanol on the alginate gel; 1.14 for catechol, 0.99 for L-DOPA and 2.03 for dopamine in cross-linked gelatin (10% w/v) beads at pH 7.1 and 35°C [59]. For solute molecules with a large molecular weight, a low value for K was determined: 0.57 for ovalbumin (43.5 kDa), 0.37 for BSA (67 kDa) and 0.18 for IgG (155 kDa) in 3% Ca-alginate; 0.48 for ovalbumin, 0.23 for BSA and 0.17 for IgG in 4% agarose [70]. Due to the obstruction effect of cells immobilised in κ-carrageenan/locust bean gum, partition coefficients decreased from 0.89 to 0.79 and from 0.98 to 0.87 for lactose and lactic acid, respectively [60]; Øyaas *et al.* [68] found that K was constant for both lactose and lactic acid (1.00 and 1.16, respectively) with increasing cell concentration. In biofilms, the partition coefficient between the biomass and water is determined through isotherm experiments. For biofilms composed of *Xanthobacter autotrophicus* and *Pseudomonas* sp., a value of 1 was found for 1,1,2-trichloroethane [77].

In measuring solute diffusivities, it is important to ensure that the external overall mass transfer resistance is negligible. In the case of permeable spheres, the effect of the external mass transfer resistance on the overall uptake and/or release rate by the beads may be quantitatively evaluated by calculating the time constant for the external film (τ_e), and to compare it to the time constant for diffusion in the sphere (τ_i) [74]. The internal time constant can be calculated as [111]

$$\tau_i = \frac{R^2}{15D_e} \tag{12}$$

where R is the radius of the bead. Alternatively, according to the film theory, the film thickness can be estimated ([50]). The external mass transfer resistance can also be neglected if the Biot number (Bi) is much larger than one [58]. Bi for beads is defined as the ratio of the characteristic film transport rate to the characteristic intraparticle diffusion rate:

$$Bi = \frac{k_s R}{D_e} \tag{13}$$

An estimation of the external mass transfer coefficient (k_s) is required to calculate τ_e, Bi or the film thickness. The value of k_s can be calculated by a procedure recommended by Harriot [112]. Merchant and co-workers [52] determined Bi for a rotating sphere. Using the empirical correlation of Noordsij and Rotte [113], k_s could be estimated using the following equation:

$$Sh = 10 + 0.43\,\mathrm{Re}_r^{1/2}\,Sc^{1/3} \tag{14}$$

where Sh is the Sherwood number, Re_r the rotational number of Reynold and Sc the Schmidt number. In the case of diffusion through a membrane or thin disc, Bi can also be calculated. For free-moving particles k_s can be determined using the following correlations [112,114-117]:

$$Sh = \sqrt{4 + 1.21\left(Re_p Sc\right)^{0.67}} \quad \text{for } Re_p Sc > 10^4 \tag{15}$$

$$Sh = 2 + 0.6 Re_p^{0.5} Sc^{0.33} \quad \text{for } Re_p < 10^3 \tag{16}$$

where Re_p is the (particle) Reynolds number, which can be estimated using the following correlations:

$$Re_p = Gr/18 \quad \text{for } Gr < 36 \tag{17}$$

$$Re_p = 0.153 Gr^{0.71} \quad \text{for } 36 < Gr < 8\ 10^4 \tag{18}$$

$$Re_p = 1.74Gr^{0.5} \qquad \text{for } 8\ 10^4 < Gr < 3\ 10^9 \tag{19}$$

where Gr is the Grashof number. Another correlation which has been used to estimate k_s for gel beads in agitated reactors [69], is [118]:

$$Sh = 2 + 0.52\left(e_s^{1/3} d_p^{4/3} / \nu\right)^{0.59} Sc^{1/3} \tag{20}$$

where d_p is the average diameter of the particle, ν is the cinematic viscosity, e_s is the energy dissipation given as $e_s = N_p n_i^3 D_i^5 / V$ for a stirred tank (where N_p is the power number, n_i the impeller speed, D_i the impeller diameter and V the volume of the reactor). The ranges of validity for this correlation are:

$$10 < \left(e_s^{1/3} d_p^{4/3} / \nu\right) < 1500 \text{ and } 120 < Sc < 1450 \tag{21}$$

Also, a correlation, which was originally developed for fluidised particles [119], has also been recommended for agitated dispersions of small, low density solids [69,120]:

$$k_s = \frac{2D_{e0}}{d_p} + 0.31(Sc)^{-2/3}\left(\frac{\Delta\rho\nu g}{\rho_l}\right)^{1/3} \tag{22}$$

where $\Delta\rho$ is the particle/liquid density difference and ρ_l the density of the bulk liquid.

For spherical particles in a packed bed, k_s depends on the liquid velocity around the particles. For the range $10 < Re_p < 10^4$, Sherwood number has been correlated by the following equation [1]:

$$Sh = 0.95 Re_p^{0.5} Sc^{0.33} \tag{23}$$

An estimation of k_s can be calculated if the stirred chambers have the shape of flat cylinders [61] using the following correlation [112]

$$k_s = 0.62 D_a^{2/3} \nu^{-1/6} \omega^{1/2} \tag{24}$$

where ν is the cinematic viscosity and ω the rotational speed of the stirrer (in rad/s). Other correlations can be adapted from heat transfer correlations [42].

On the other hand, the external mass transfer limitation can be experimentally investigated by observing the concentration-time profile at different rotation speeds of the stirrer [36].

3.2. DIFFUSION IN CELL-CONTAINING MATRICES

The presence of immobilised cells in the immobilisation matrix can have a substantial effect on the effective diffusion coefficient, especially at high biomass concentrations.

Also, the effective diffusion coefficient decreases as a function of time in growing immobilised systems because the cell volume fraction increases. Diffusion in gels containing immobilised cells has been reviewed by Westrin and Axelsson [18]. The influence of the cells on the effective diffusion coefficient can be described as

$$D_e = f(\phi_c, D_c) D_{e0} \tag{25}$$

where D_c is the effective diffusion coefficient within the cells and D_{e0} the effective diffusion coefficient in a cell free matrix. Two theoretical approaches have been developed: (1) the cells are impermeable to the diffusing solute or (2) the solute diffuses in the cells with a very low D_e. Suitable expressions have been derived for f(ϕ_c, D_c) that were based on heterogeneous media ([121,122]). Expressions could be classified in five categories:

3.2.1. The exclusion model

This model assumes that D_c is zero. The presence of the impermeable cells reduces the volume available for diffusion. Equation (25) can now be written as

$$D_e = (1 - \phi_c) D_{e0} \tag{26}$$

Using this equation, effective diffusion coefficients have been determined for substrate diffusion in calcium alginate gel with entrapped deactivated yeast cells [41,61].

3.2.2. Models of suspended impermeable spheres

These models regard the cells as impermeable spheres (D_C equals zero), which are suspended in a continuum. The expression for this model has been adapted from the expression derived by Maxwell [123] for effective conductivity in a composite medium of periodically spaced spheres:

$$D_e = \frac{1 - \phi_c}{1 + (\phi_c / 2)} D_{e0} \tag{27}$$

This equation has been used to correlate experimental diffusion coefficients with immobilised mammalian cells in agar gel [8] and in artificial biofilms (agar containing inert polystyrene particles of the same size as bacteria) [19].

Due to limitations of Maxwell's model in other immobilised cell systems, a more successful correlation was developed by using a scaling relationship [124]:

$$D^*(\phi_c) = \frac{1 - D_e / D_{e0}}{1 - D_c / D_{e0}} \tag{28}$$

and a polynomial approximation to D^*:

$$D^*(\phi_c) = 1.7271\phi_c - 0.8177\phi_c^2 + 0.09075\phi_c^3 \quad (29)$$

Diffusivity values used for constructing the correlation were generated from Monte Carlo computer simulations.

3.2.3. Models of suspended permeable spheres

This model assumes only permeable cells ($D_c > 0$). In this case, a more complex expression (adapted from Maxwell [125]) has been derived:

$$D_e = \frac{2/D_c + 1/D_{e0} - 2\phi_c(1/D_c - 1/D_{e0})}{2/D_c + 1/D_{e0} + \phi_c(1/D_c - 1/D_{e0})} D_{e0} \quad (30)$$

Values for D_c/D_{e0} have been determined to be 0.31, 0.30 and 0.20 for fermentation media of *Saccharomyces cerevisiae*, *Escherichia coli* and *Penicillium chrysogenum*, respectively [126]. Equation (30) has been used to correlate experimental diffusion coefficients in cell-containing gels [8] and dense cell suspensions [9,126].

3.2.4. Capillary models

Capillary models, which have been developed to quantify the influence of the tortuosity on the effective diffusion coefficient, i.e. the model of Kozeny (*e.g.* [127]) and the improved random-pore model [128]. This model has been adapted to immobilised cell systems where cells are considered as impermeable (i.e. $D_c = 0$):

$$D_e = (1-\phi_c)^2 D_{e0} \quad (31)$$

Several investigators have used this equation to correlate the effective diffusion coefficient with the immobilised cell concentration [7,11,36,67,71,129,130].

This random pore approach has been adapted to obtain an improved general diffusion model, which accurately includes systems containing impermeable or permeable cells [131]:

$$\frac{D_e}{D_{e0}} = (1-\phi_c)^2 + \phi_c^2 \frac{KD_c}{D_{e0}} + 4\phi_c(1-\phi_c)\frac{KD_c}{D_{e0}}\left(1 + \frac{KD_c}{D_{e0}}\right)^{-1} \quad (32)$$

3.2.5. Empirical models

Prediction equations have been obtained by fitting some arbitrary function to experimental data [132]. Empirical models are only valid for the studied gel material and immobilised cell type; and can only be used to predict D_e between the investigated boundaries. Ogston *et al.* [133] and Cukier [134] developed an empirical equation to describe the restricted diffusion of proteins in polymer gels:

$$\frac{D_e}{D_{e0}} = \exp\left(-Br_0 C_f^{1/2}\right) \tag{33}$$

where r_0 is the interaction radius between the protein and the polymer fibres, B a proportionality constant and C_f the polymer (fibre) concentration in particles. This correlation allowed the estimation of several protein diffusivities in an agarose matrix based on the molecular weight of the protein and the polymer concentration [96]. Scott *et al.* [12] used the following correlation:

$$\frac{D_e}{D_{e0}} = 1 - 0.9\phi_c + 0.27\phi_c^2 \tag{34}$$

Korgel and co-workers [130] also developed the following empirical correlation:

$$\frac{D_e}{D_{e0}} = 1 - 2.23\phi_c + 1.40\phi_c^2 \tag{35}$$

Some researchers have reported that the diffusivity was not influenced by the presence of cells. Kurosawa and co-workers [23] found that the oxygen diffusivity in Ca- and Ba-alginate was not influenced by the presence of immobilised yeast cells up to a cell density of 30 gDW/l. Estapé *et al.* [62] studied the influence of counter-diffusion of ethanol and glucose on the diffusion coefficient of glucose and ethanol respectively: there was no significant decrease of the diffusion coefficients in the presence of yeast cells when counter-diffusion was involved; without counter-diffusion the presence of cells resulted in a significant decrease of the diffusion coefficient for ethanol, but in contrast an increase was observed for the glucose diffusion coefficient.

In the determination of the effective diffusion coefficient in cell-containing matrices by methods based solely on diffusion, the solute may not react, or erroneous results would be obtained. In the literature, different methods have been described to deactivate living cells. Cells have been deactivated using ultraviolet radiation [28]; heat [33,36,36,135-138]; organic solvents like ethanol [44,61], chloroform [138] and glutaraldehyde [8,138,139]; (organic) acids like iodoacetic acid [41] and hydrochloric acid [67]; toxins or inhibitors like mercuric chloride [28,19,86,138,140], dimethyl sulphoxide [10] and sodium azide [58,60]; detergents, *e.g.* Triton X-100 [9]. Some researchers have investigated the effect of the deactivating agents on the effective diffusion coefficient. Matson [141] found that the deactivation process by mercuric chloride or heat had no effect on the diffusion process. The results of Dibdin [135] and McNee and co-workers [139] suggest that neither heat treatment nor glutaraldehyde fixation have significant effect on the rate of diffusion through dental plaque. In contrast, Tatevossian [136,137] found a significant effect by using heat deactivation. Libicki and co-workers [9] found that the mass transport of nitrous oxide in aggregates of *E. coli* prepared from cells treated with detergents or disrupted by dehydration and grinding differed only slightly from the values obtained for aggregates formed from treated cells. Pu and Yang [10] observed that the permeabilisation of apple cells with dimethyl sulfoxide led to an increase in effective diffusivity. Beuling *et al.* [19] showed that glucose was excluded by deactivated (with mercuric chloride) *Micrococcus luteus* cells entrapped in agar, while the cells were somewhat permeable for oxygen.

Another strategy is to select a solute that will not be consumed by the micro-organism, *e.g.*, nitrous oxide [9,142], inert gases (H_2, He, CH_4, $C_2 H_2$, N_2O, $CHClF_2$, SF_6) [32], yohimbine that is a secondary metabolite of plant cells [10]; potassium chloride [38], galactose that is not metabolised by *Zymomonas mobilis* [11].

Itamunoala [92] compared three methods of determining D_e based on the shape of the matrix – designated (a) thin disc, (b) cylinder, and (c) beads types – and showed that, by using a sensitivity and error analysis, the thin disc and cylindrical techniques gave more accurate results than the bead method. Westrin and Zacchi [143] studied theoretically the method of calculating diffusion coefficients in gel beads based on transient experiments of solute uptake from a finite volume. Monte Carlo simulations were used to investigate the effect of random errors on concentration measurements. The influence of sampling frequency and diffusion direction was investigated. Systematic errors derived from an assumption of constant diffusion coefficient and of monodisperse beads were also investigated.

3.3. DIFFUSION IN GELS

A lot of research has been performed on the diffusion in polymeric gels with and without encapsulated cells. Various values of measurements of diffusion coefficients in cell-free and cell-containing gels are presented in Willaert and Baron [144].

Size-exclusion chromatography (SEC) has been used to study the porosity of alginate gels [145-147]. Smidsrød [146] studied the porosity of alginate gels using scanning electron microscopy and found that there was a very broad distribution in pore diameter, ranging from 5 to 200 nm. Globular proteins with radii of gyration of approximately 3 nm and with net negative charges should therefore not be trapped in the alginate network, but be able to diffuse at a rate dependent upon their sizes [64]. In contrast, Klein *et al.* [145] showed, using SEC and dextrans of known size, that the pore diameters in Ca-alginate gels were fairly uniform. The exclusion volumes for standard macromolecules were recorded and values of 6.8, 14.1 and 16.6 nm were estimated for 3 different alginates. SEC with 2% Ca-alginate beads using proteins of known Stokes radii indicated a pore diameter of 8-10 nm when eluents of 30 and 150 mM $CaCl_2$ were used at pH 6.2 [147]. The whey proteins α-lactoalbumine and β-lactoglobulin both easily penetrated the alginate pores and no proteins had complete access to the total internal volume of the matrix.Polakovic [148] used SEC to evaluate the pore size distribution of 5% Ca-pectate gel and predicted — using a cylindrical and slit-pore model — the pore diameters of 50 and 35 nm respectively.

From pore-size distribution studies using electron microscopy [149], it was suggested that there is a more constricted network on the bead surface than in the gel core. Skjåk-Bræk and co-workers [150] found that the alginate structure is governed not only by the concentration and chemical structure of the gel material, but also by the kinetics of gel formation. They showed that gels with varying degrees of anisotropy (heterogeneity) can be prepared by controlling the kinetics of the process. Therefore, it is difficult to compare the different diffusion coefficients reported by different workers because of the considerable variations in the experimental conditions of preparing alginate gels. When gels were formed in the presence of anti-gelling cations, such as Na^+ or Mg^{2+}, isogeneous Ca-alginate gels were obtained [151]. Such beads were mechanically stronger and had a higher porosity than those formed in the absence of anti-gelling ions. Klein and co-workers [145] have reported that glucoamylase is retained for many days in ferric-alginate gel, although it is lost from Ca-alginate or aluminium alginate beads. Gray and Dowsett [152] showed that insulin may be entrapped in zinc alginate and zinc-calcium alginate gels.

The average pore diameters of alginate and agarose gels have been estimated from the following semi-empirical expression, often used to describe the diffusion of solutes in porous materials and originally derived by Renkin [153] to characterise the diffusion of protein molecules through cellulose membranes [70]:

$$\frac{D_{gel}}{D_{water}} = \left(1 - \frac{d}{\delta_p}\right)^2 \left\{1 - 2.104\left(\frac{d}{\delta_p}\right) + 2.09\left(\frac{d}{\delta_p}\right)^3 - 0.95\left(\frac{d}{\delta_p}\right)^5\right\} \tag{36}$$

where δ_p is the average pore diameter of the gel and d is the hydrodynamic diameter of the solute. This equation accounts for steric and frictional hindrance to diffusion under Stokes flow conditions but neglects charge related interactions between solute and matrix. The effect of varying crosslinking condition, polymer concentration, and direction of diffusion on transport for alginate and agarose gels have been investigated [70]. In general, 2-4% agarose gels offered little transport resistances for solutes up to 150 kDa, while 1.5-3% alginate gels offered significant transport resistances for solutes in molecular weight range 44-155 kDa – lowering their diffusion rates from 10- to 100-fold as compared to their diffusion in water. Doubling the alginate concentration had a more significant effect on hindering diffusion of larger molecular weight species than did doubling the agarose concentration. Average pore diameters of approximately 170 and 147 Å for 1.5 and 3% agarose gels, respectively, and 480 and 360 Å for 2 and 4% agarose gels, respectively, were estimated using Equation (36).

3.3.1. Influence of the gel type and concentration

Studies on the diffusion of NAD and haemoglobin in calcium and barium alginate gels showed that NAD diffusion characteristics are unaffected by alginate (from 2.5% to 4% w/v) and calcium chloride (from 0.125 to 0.5 M) concentrations; however, haemoglobin diffusion was affected by the alginate concentration [154]. Hannoun and Stephanopoulos [44] demonstrated that D_e for ethanol and glucose decreased when the alginate concentration was increased from 1 to 4% (w/v). Itamunoala [91] showed that by increasing the concentration of either the calcium chloride (1-4% w/v) or sodium alginate (2-8% w/v) component used in calcium alginate formation substantially decreased D_e, but the effect of the alginate was the greater of the two. A slight reduction of the product of K and D_e for oxygen was found when the alginate concentration was increased from 2 to 4% (w/v) by Sun and co-workers [36]. In contrast, Tanaka *et al.* [49] demonstrated that the diffusion of solutes with a molecular weight less than 2 10^4 (glucose, *L*-tryptophan and α-lactalbumin) was not disturbed by increasing the alginate concentration (2-4% w/v) and $CaCl_2$ concentration used in the gel preparation; and diffusion coefficients comparable with those in water were found. For larger molecules such as albumin, γ-globulins and fibrinogen, the diffusion in the gel was retarded to an extent that depended upon the concentration of alginate and calcium chloride concentration. Moreover, these proteins could diffuse out of, but not into the beads. Therefore, it was suggested that the structure of Ca-alginate gel formed in the presence of large protein molecules was different from that of the gels formed in their absence. Axelsson and Persson [41] found also that the dependence of D_e on the alginate concentration (in the range 1.4-3.8% w/w) for glucose, lactose, galactose and ethanol was negligible. Hulst and co-workers [56] examined the influence of the concentration of Ca-alginate, gellan gum, κ-carrageenan, agarose and agar on D_e for oxygen. For agarose and agar, D_e decreased when the gel concentration was increased from 2 to 8% (w/v). Gellan gum and alginate showed a remarkable maximum value for the diffusion coefficient with respect to the gel concentration: at 1 and 2% (w/v) respectively. κ-Carrageenan had a maximum at 5% (w/v) in the concentration range of 1 to 5% (w/v). The effective diffusion coefficient determined in pure 2% alginate beads was 10 to 25% lower than the diffusivity in water for various solutes, and a 4% alginate concentration decreased D_e, corresponding well with increased inclusion effect due to the increase in polymer concentration [68]. Chen *et al.* [155] used a linear absorption model (LAM) and a shrinking core model (SCM) to interpret the diffusion data of Cu^{2+} in Ca-alginate over a biopolymer range from 2% to 5%. The diffusion coefficient of Cu^{2+} calculated from the LAM was independent of the biopolymer concentration. The LAM had theoretical advantages over the SCM; the latter calculated an unreasonable exponential increase in the diffusion coefficient as the density of the alginate concentration increased. The diffusivity of Cu^{2+} in Ca-alginate increased by a factor of two when the gel concentration was increased from 2 to 5% (w/v) [156,157].

The diffusion of solutes in alginate gel is also affected by the chemical composition of the alginate. Martinsen and co-workers [65] showed that the diffusion coefficient of albumin in 4% *Laminaria digitata*, *Macrocyctis pyrifera* and *L. hyperborea* Ca-alginate gels decreased with decreasing content of guluronic acid and with any decrease in the average length of the guluronic acid (GG) blocks. Also, Itamunoala [[91] found a glucose diffusion coefficient which was slightly higher for Manutex RS gel compared to Manutex RH gel.

The effective diffusivity of sucrose in agar (0.5%, 1.0% and 1.5% w/v) was similar to that in water [43]. Dextran fractions (MW range from 10.5 kDa to 1950 kDa) displayed restricted diffusion in the agar membranes. Their D_e values were a decreasing function of the agar content of the gel membrane. The diffusivity in a given membrane decreased as the molecular weight of the diffusing molecule increased. The diffusion data did not agree with the Renkin model for a hard sphere diffusing through a cylindrical pore. Beuling *et al.* [20] studied the influence of agar concentration on the diffusion coefficient of glucose by using Pulsed Field Gradient-Nuclear Magnetic Resonance. The obtained values corresponded with the ones predicted by the model of Mackie and Meares.

The diffusivity of glucose in κ-carrageenan was affected by the presence of other solutes in the glucose solution [50]: electrolytes such as ammonium sulphate, potassium chloride and calcium chloride were observed to enhance the diffusion coefficient. Estapé *et al.* [62] demonstrated that de glucose and ethanol effective diffusion coefficients were significantly reduced in the presence of counter-diffusion of ethanol and glucose, respectively.

It was found that lactic acid modified the structure of κ-carrageenan (2.75% w/w)/locust bean gum (0.25% w/w) gel [60], since lactose diffusion characteristics (D_e and K) differed significantly from other studies. Øyaas *et al.* [68] found that D_e for lactose and lactic acid in Ca-alginate were constant between pH 5.5 and 6.5, but significantly reduced at pH 4.5.

Renneberg and co-workers [73] investigated the influence of different variables on the oxygen diffusivity in gels derived from prepolymers. A series of ENT-type (prepared from hydroxyethylacrylate) hydrophilic polymers showed a steady increase in water content and diffusivity with increasing chain length of the prepolymers from 10 nm to 60 nm. The slightly higher diffusivity of the anionic ENT-type polymer compared with the cationic seemed to be due to the lower water content rather than to electronic charges. The polyvinyl-alcohol stilbazolium (PVA-SbQ) gels have an higher water content and also a higher diffusivity than the polymers of ENT-type. The polymers with 22 nm distance between the photo-functional groups showed a higher O_2 diffusion than the polymers with 6.6 and 6.5 nm distance. The PVA-SbQ 1800-100 with the highest polymerisation degree (1800) and the highest saponification (100%) had the highest water content (94.2%) and the highest O_2 diffusivity.

3.3.2. Influence of the temperature

The influence of the temperature on the glucose diffusion in 3% (w/v) κ-carrageenan has been investigated in a temperature interval between 10 and 33°C [50]. The diffusivity remained unchanged in the interval from 10 to 25°C; from 25 to 32°C, the diffusivity increased linearly from $3.73\ 10^{-10}$ to $6.10\ 10^{-10}$ m^2/s. This behaviour was explained by the fact that the diffusion in the gel system is governed by the bead pore size as well as by the viscosity of the gel. The κ-carrageenan gel viscosity remains unchanged at temperatures below 20°C, and the viscosity decreases when the temperature increases above 2°C. The D_e values of glucose and galactose in κ-carrageenan (2.5% w/v) increased when the temperature was increased from 10°C, to 25°C and 32°C [46]. The diffusion coefficient of glucose, lactate, hydroquinone and urea in 2% collagen and 1% agar increased when the temperature was raised from 25°C to 37°C [8]. The temperature dependence of glucose in 3% (w/v) Ca-alginate followed the Arrhenius relation with an activation energy of 4350 J/mol [51]. D_e of glucose increased from $6.1\ 10^{-6}$ to $7.8\ 10^{-6}$ m^2/s and the value for ethanol from $1.0\ 10^{-6}$ to $1.2\ 10^{6}$ m^2/s in 2% Ca-alginate when the temperature was increased from 22°C to 30°C [45]. Martinsen *et al.* [65] calculated the activation energy as 23.5 kJ/mol for the Arrhenius temperature dependence of the diffusion of albumin in 4% (w/v) Ca-alginate. The temperature dependency of the diffusion coefficients for lactose and lactic acid in Ca-alginate were found to be well described by an Arrhenius-type equation [68]. The activation energy was constant in the temperature range tested (10°C to 55°C) and was found to be 20.0 kJ/mol for lactose and 19.1 kJ/mol for lactic acid.

3.3.3. Influence of the pH

Because the alginate gel matrix is negatively charged, the pH influences the diffusion of charged substrates and products. Most proteins are negatively charged at pH 7 and will not easily diffuse into the gel matrix. However, when immobilised in the gel, they tend to leak out more rapidly than would be expected from their free molecular diffusion. The rate of diffusion of bovine serum albumin out of alginate beads increased with increasing pH, due to the increased negative charge of the protein [65].

3.3.4. Influence of the diffusing substance

Teo *et al.* [51] found that the effective diffusion coefficient of glucose in 3% (w/v) Ca-alginate is independent of the glucose concentration over the range from 10 to 150 g/l. Concentrations of glucose in the range 2-100 g/l and ethanol in the range of 10-80 g/l did not affect their diffusion coefficients in Ca-alginate [44]. In contrast, Itamunoala [91] showed that only at low glucose concentrations was the glucose diffusivity in Ca-alginate higher. The diffusion of glucose in κ-carrageenan decreased when the glucose concentration was increased from 6.5 to 65 g/l [50].

Most of the reported effects can be explained on the basis of: a change in tortuosity τ or porosity ε; change in partition coefficient; interaction of large molecules with the polymer network of the matrix; interaction between different diffusing solutes; and effect of the temperature.

3.4. DIFFUSION IN DENSE CELL MASSES

3.4.1. Diffusion in biofilms

Some values (and additional information) for the diffusion of various components in biofilms are given in Table 2.

Diffusion coefficients were found to be dependent on the biomass concentration, C/N ratio in the growth medium, temperature [34], and sludge age [28]. Diffusion studies of lactose in acidogenic biofilms revealed that the active biofilms had about 66% void volume made up of channels through which the lactose molecules were transported into the bacterial aggregate [29]. The decrease in lactose diffusivity was mainly caused by the biofilm's solid biomass fraction rather than by the tortuosity of the channels. The influence of the diffusion potential and the Donnan potential in grown biofilms have been evaluated by comparing the diffusion coefficients of a positively and negatively charged ions and a neutral molecule in experiments with different background electrolyte concentrations [158]. Mass transfer effects by electrostatic forces are negligible at the ionic strength of wastewater and tap water. Holden and co-workers [159] determined the diffusion coefficient of toluene as a function of the water potential through unsaturated *Pseudomonas putida* biofilms. The effective diffusion coefficient for toluene was approximately two orders of magnitude lower than the diffusivity in water and did not vary markedly with the water potential.

Table 2. Diffusion in biofilms.

Biomass type	Cell density (kg/m^3)	Component	Relative diffusivity[a]	Measurement method	Reference
Mixed-culture	94	Glucose	38 (25°C)	Reaction-diffusion model	[27]
Mixed-culture		Oxygen	39-62 (20°C)	Steady-state flux	[28]
		Glucose	27-52 (20°C)		
Mixed-culture	29-84	Oxygen	31-36 (20°C)	Steady-state flux	[140]
Nitrifier culture	42-109	Oxygen	85 (20°C)	Transient flux	[160]
		Ammonia	81-88 (20°C)		
		Nitrate	90-97 (20°C)		
		Nitrite	84 (20°C)		
Biofilm	23.3-24.5	Glucose	50 (15-25°C)	Reaction-diffusion model	[174]
	23.3-24.5	Oxygen	50 (15-25°C)		
Mixed-culture	62.1	Nitrate	46 (22°C)	Reaction-diffusion model	[175]
Mixed-culture	37	Valeric acid	34-67 (23°C)	Reaction-diffusion model	[176]
Mixed-culture	3-100	Glucose	100-87 (20°C)	Steady-state flux	[34]
		Oxygen	100-86		
Mixed-culture	19	Glucose	97 (20°C)	Steady-state flux	[158]
	26	Bromide	50 (20°C)		
	14-25	Sodium	50-70 (20°C)		
Mixed-culture	72-152	Phenol	28-10 (23°C)	Reaction-diffusion model	[177]
Mixed-culture	130-180	Phenol	39-13 (25°C)	Transient flux	[57]
Acidogenic biofilm	30 ($kg C/m^3$)	Lactose	66 (35°C)	Reaction-diffusion model	[29]
Pseudomonas putida		Toluene	0.9-2.3 (27°C)	Steady-state flux	[159]
Pseudomonas putida		Fluorescein	91[e]	FRAP[f]	[100]
		Dextran[g]	28		
		Dextran[h]	15		
		BSA[i]	54		
		Hexokinase	49		
		DNA	19		
Pseudomonas		TCE[b]	200 (30°C)[c] 30[d]	Steady-state flux	[77]
Xanthobacter autotrophus		TCE	100 (30°C)	Steady-state flux	[77]

[a] Relative diffusivity = effective diffusivity in cell aggregate/diffusivity in water (%).
[b] TCE = 1,1,2-trichloroethane
[c] Biofilm thickness < 1 mm; [d] Biofilm thickness > 1 mm.
[e] Biofilm overall thickness was 232 μm; [f] FRAP = Fluorescence Recovery After Photobleaching; [g] MW = 10000, [h] MW = 70000; [i] BSA = bovine serum albumin

Diffusion coefficients have been evaluated using only particulate biomass filtered onto a membrane [28,160-162]. The biofilm has been assumed to be a homogeneous phase in which mass transport is described by Fickian diffusion. More recently, insight into biofilm structure have led to the recognition that the biofilm is heterogeneous, and mass transfer also occurs by eddy diffusion and convection in the voids or channels and diffusion through the bacterial cells [77,84,85,158,163-171]. A 2-D model has been developed to evaluate the effect of convective and diffusive substrate transport on biofilm heterogeneity [172]. It was found that in the absence of detachment, biofilm heterogeneity is mainly determined by internal mass transfer rate and by initial percentage of carrier-surface colonisation. Model predictions showed that biofilm structures with highly irregular surface develop in the mass transfer-limited regime. As the nutrient availability increases, there is a gradual shift toward compact and smooth biofilms. A smaller fraction of colonised carrier-surface leads to a patchy biofilm. Biofilm surface irregularity and deep vertical channels are caused by the inability of the colonies to spread over the whole substratum surface. The maximum substrate flux to the biofilm was greatly influenced by both internal and external mass transfer rates, but not affected by the inoculation density.

Recently, a new technique, which is based on microinjection of fluorescent dyes and analysis of the subsequent plume formation using confocal laser microscopy, has been used to study the liquid flow in aerobic biofilms [173] and to determine the local diffusion coefficients in biofilms [94]. The diffusion coefficients of fluorescein (MW 332), TRITC-IgG (MW 150000) and phycomerythrin (MW240000) were measured in the cell clusters and interstitial voids of a heterogeneous biofilm.

Zhang and co-workers [77] used a single tube extractive membrane bioreactor (STEMB) for the determination of the effective diffusion coefficient for a non-reactive tracer (1,1,2-trichloroethane). A video imaging technique was used for the *in situ* measurement of biofilm thickness and continuous monitoring and recording of biofilm growth during experiments.

Bryers and Drummond [100] refined an alternative analytical technique to determine the local diffusion coefficients on a micro-scale to avoid the errors created by the biofilm architectural irregularities. This technique is based upon Fluorescence Return After Photobleaching (FRAP), which allows image analysis observation of the transport of fluorescently labelled macromolecules as they migrate into a micro-scale photobleaching zone. The technique allows to map the local diffusion coefficients of various solute molecules at different horizontal planes and depths in a biofilm. These mappings also directly indicate the distribution of water channels in the biofilm. Fluorescence return after photobleaching results indicate a significant reduction in the solute transport coefficients in biofilm polymer gel *vs.* the same value in water, with the reduction being dependent on solute molecule size and shape.

Beuling *et al.* [20] used Pulse Field Gradient-NMR to measure the water mobility inside complex heterogeneous systems like natural and active biofilms. Diffusion coefficients measured in both well-defined biofilms and spontaneously grown aggregates corresponded well to glucose diffusion coefficients determined in the same matrices. Diffusion coefficients of the natural biofilms could be related to their physical characteristics. The monitored PFG-NMR signal contained supplementary information on cell fraction or spatial organisation but quantitative analysis was not yet possible.

Wood and co-workers [178] developed a scheme for numerically calculating the effective diffusivity of cellular systems such as biofilms and tissues. A finite-difference model was used to predict the effective diffusivity of a cellular system on basis of the subcellular scale geometry and transport parameters. These predictions were compared to predictions from simple analytical solution [179,180] and experimental data. Their results indicate that, under many practical experimental circumstances, the simple analytical solution can be used to provide reasonable estimates of the effective diffusivity.

3.4.2. *Diffusion in bioflocs*

Microbial flocs have been reviewed by Atkinson and Daoud [181], and Kosaric and Blaszczyk [182]. Yeast flocculation has been reviewed by Speers *et al.* [183], Stratford [184], Straver *et al.* [185], Teunissen and Steensma [186], and Dengis and Rouxhet [187]. The growth and modelling of the growth of fungal pellets have been reviewed by Whitaker and Long [188], Metz and Kossen [189], Nielsen [190], Nielsen and Villadsen [4], Nielsen and Carlsen [191]. Examples of diffusion data for bioflocs are shown in Table 3.

Smith and Coackley [140] used surface area determinations to calculate the mean pore radius of activated sludge flocs. This value was found to be between 108 and 130 Å. The tortuosity of the pores in activated sludge (3-9% solids) was also calculated and found to have an average value of 2.73.

Libicki and co-workers [9] determined the effective diffusive permeability in aggregates of *E. coli* confined in hollow fiber reactors. It was found that the effective diffusive permeability decreased with increasing cell volume fraction to a value, for aggregates comprising 95% cells, of approximately 30% that obtained for cell-free buffer solution. The dependence on the cell volume fraction was described adequately by the Hashin-Shtrikman bounds for a two-phase medium [192]:

$$K_2D_2 + \frac{f_1}{\dfrac{f_2}{3K_2D_2} + \dfrac{1}{K_1D_1 - K_2D_2}} \le D_e \le K_1D_1 + \frac{f_2}{\dfrac{f_1}{3K_1D_1} + \dfrac{1}{K_2D_2 - K_1D_1}} \tag{37}$$

where f_1 and f_2 are the volume fractions; K_1 and K_2 are the partition coefficients; and D_1 and D_2 are the solute diffusion coefficients for the two components (cells and interstitial fluid), with the indices chosen such that $K_2D_2 \le K_1D_1$.

Table 3. Diffusion in bioflocs.

Biomass type	Cell density (kg/m^3)	Component	Relative diffusivity[a]	Measurement method	Reference
A. niger pellet	12-200	Oxygen	4-93 (20°C)	Reaction-diffusion model	[193]
A. niger pellet	15	Oxygen	15	Reaction-diffusion model	[87]
A. niger pellet	19	Oxygen	52.5 (30°C)	Reaction-diffusion model	[194]
Mixed-culture floc		Glucose	7-67 (20°C)	Steady-state flux	[195]
Mixed-culture floc	29-84	Oxygen	31-36 (20°C)	Steady-state flux	[140]
Zoogloea ramigera floc	400	Oxygen	8 (20-30°C)	Reaction-diffusion model	[196]
Zoogloea ramigera floc	390	Glucose	5-12(20-30°C)	Reaction-diffusion model	[197]

[a] Relative diffusivity = effective diffusivity in cell aggregate/diffusivity in water (%).

Mass transfer limitations into fungal pellets is usually due to intraparticle diffusion [4]. Only when intraparticle mass transfer is severe, external mass transport resistance should be taken into account. The average pellet density is found to increase with decreasing pellet radius, and the porosity is related to the pellet density [198]:

$$\varepsilon = 1 - \frac{\rho_{pellet}}{\rho_{hyph}(1-\omega)} \tag{38}$$

where ρ_{pellet} is the average pellet density, ρ_{hyph} the hyphal density, and ω is the thickness of the active biomass layer in a pellet. Metz [198] found that for *Penicillium chrysogenum* the average density of the pellet varies between 15 and 100 kg/m^3, and the porosity is in the range 0.70 – 0.95. Similar values for the density of *P. chrysogenum* [199-201] and *Aspergillus niger* [193] pellets have been reported. In cultures of filamentous fungi, it is typically glucose or oxygen that causes mass transfer limitations. The critical pellet radius for various concentrations of these substrates in the medium have been calculated using the appropriate reaction-diffusion model [191]. It was observed that except for low glucose concentrations, it is most often oxygen that becomes the limiting substrate in the pellet. For pellets with a radius larger than the critical radius, there is an active growth layer. The active layer thickness depends on the pellet size. Wittler *et al.* [88] recorded the oxygen concentration profile in *P. chrysogenum* pellets using a microelectrode and found that the critical radius was in the range of 100 to 400 μm depending on the pellet structure and operation conditions. It was found that – as a result of zero oxygen concentration in the centre of the pellet – cell lysis occurred, which led to the appearance of hollow pellets. Microelectrode studies by Cronenberg *et al.* [90] also showed that internal mass transport properties of *P. chrysogenum* pellets were highly affected by their morphological structure. Relative young pellets possessed a homogeneous and dense structure. These pellets were partly penetrated by oxygen at air saturated bulk conditions. Older pellets were stratified and fluffy. They were completely penetrated by oxygen due to a decreased activity and a higher diffusivity. Investigations with glucose microelectrodes revealed that glucose consumption inside pellets of all lifetimes exclusively occurred in the periphery, indicating that growth was restricted to these regions only.

Usually, it is assumed that intraparticle molecular diffusion is the only mass transfer mechanism (see for earlier work for example [87,193,194,202,203]). However, despite the low density difference between pellets and the bulk medium, some convective flow into the pellets can occur [88,204].

3.4.3. Diffusion in mammalian cell aggregates

The Krogh's diffusion constant and the diffusion coefficient of various inert gases have been determined in rat skeletal muscle [32] (see Table 4). Graham's law (inverse proportionality between the diffusion coefficient and the square root of the molecular mass) was found to represent a useful approximation for these gases. However, a better correlation between the diffusion coefficient and the molecular diameter was found.

The diffusion of ^{3}H-inulin (which does not penetrate the intact cell membrane of mammalian cells) in two types of multicellular tumour spheroids have been studied [48]. It was found that the value of the diffusion coefficient in spheroids with a smaller extracellular space was significantly larger than in spheroids with a higher extracellular space. It was suggested that the way the cells attach to each other and the extracellular matrix are responsible for this difference.

Table 4. Diffusion in mammalian cell aggregates.

Cell type	Component	Diffusion coefficient (m^2/s)	Relative diffusivity[a]	Measurement method	Reference
Rat skeletal muscle	Helium		62 (37°C)	Transient flux	[32]
	Hydrogen		49		
	Acetylene		52		
	Nitrous oxide		59		
Tumour cells:V-79	Inulin	$8.9\ 10^{-12}$	–	Transient flux	[48]
EMT6/Ro	Inulin	$3.3\ 10^{-12}$	–		
Normal rat tissue	FITC[b]-BSA[c]		1.2-1.7	Transient flux	[93]
	FITC-D20[d]		22-17		
	FITC-D40[e]		2.4-3.1		
	FITC-D70[f]		0.9-1.0		
	Na-F[g]		29-34		
Tumour tissue (VX2 carcinoma)	FITC-BSA		10	Transient flux	[93]
	FITC-D20		71		
	FITC-D40		56		
	FITC-D70		31		
	Na-F		89		

[a] Relative diffusivity = diffusivity in cell aggregate/diffusivity in water (%).
[b] FITC = fluorescein isothiocyanate; [c] BSA = bovine serum albumin; [d] D20 = dextran (MW 20500);
[e] D40 = dextran (MW 44200); [f] D70 = dextran (MW 71800); [g] Na-F = sodium fluorescein.

The diffusion of fluorescein isothiocyanate conjugated bovine serum albumin, a graded series of FITC-dextrans and sodium fluorescein in both normal tissue and tumour grown in a rabbit ear chamber have been determined [93]. Apparent interstitial diffusion coefficients showed a relationship with molecular size, which progressively deviated from that of free diffusion in a single solute, single phase system with values of albumin being significantly reduced from that for a dextran of equivalent hydrodynamic radius. Macromolecular transport in tumour tissue was hindered to a lesser extent than in normal tissue, which was consistent with reports of reduced [GAG] content of tumours.

3.5. DIFFUSION IN MICRO-CAPSULES

The permeable membrane of a micro-capsule introduces a supplementary mass transfer barrier compared to gel beads. The core of the micro-capsule can be composed of cells in a liquid solution, cells in gel phase, or a combination of both. In the latter case, three mass transfer layers are present. A model for the transient diffusion of proteins from a bulk solution into micro-capsules has been recently developed [205,206]. It was used to describe the experimental concentration profiles and to determine the membrane diffusion coefficient of alginate-PLL-alginate capsules. By introducing an effective volume fraction as parameter in the model, micro-capsules with different sizes of impermeable gel cores could be studied. It was found that the ability of proteins to diffuse into the capsule was not only controlled by the membrane diffusivity, but also by the amount of Ca-alginate core remaining in the micro-capsule.

The permeability of the capsule membrane is often characterised by the molecular weight cut-off value (MW cut-off). The application of micro-capsules for immunoisolation required a membrane with a MW cut-off of around 50000 to prevent the permeability of cytotoxic antibodies [207]. The permeability of a membrane for a certain compound is not only dependent on the pore size and structure but also on other physical characteristics such as the hydrophobicity/hydrophilicity and charge, and the nature of the solute molecule (*e.g.* electrostatic interactions between proteins and synthetic polymers).

By manipulating the parameters involved in membrane manufacturing such as the molecular weight and concentration of the membrane forming molecules, the reaction time, polymerisation conditions (temperature, pH, ionic strength, ...), addition of (an) additional layer(s) at the outside of the capsule (usually to improve the biocompatibility). Permeation can also be influenced by changing size, swelling and shape of the capsules [208].

A heterogeneous mixture of dextrans of different molecular weights (MW 10000 – 500000) has been used as a test solute to study the membrane permeability of alginate polylysine microcapsules [209]. The diffusion experiment revealed that the permeability of APL microcapsules decreased with dextrans with MW > 80000. At equilibrium, dextrans of MW 104000 – 112000 represented only approximately 50% of dextrans of equivalent MW outside of microcapsules. The microcapsule membrane does not show a distinct MW cut-off to dextran, but spans a very wide range of at least 80000 to 110000. It is suggested that this can be due to differences in permeability between microcapsules within a batch or to a non-uniform membrane.

Awrey *et al.* [210] have constructed a series of genetically modified cell lines secreting recombinant gene products (human growth hormone, rat serum albumin, human arylsulphatase A, human immunoglobulin, mouse β-hexosaminidase, mouse β-glucuronidase) of different molecular weights ranging from 21 through 150 to 300 kDa. They investigated the delivery of these products by alginate-PLL-alginate microcapsules enclosing these producer cell lines. It was found that the secretion rates of the gene product were similar between nonencapsulated and encapsulated cells with the exception of the largest molecule (the 300-kDa β-glucuronidase), which showed an eightfold reduced secretion. Increasing the thickness of the PLL-membrane did not provide a lower molecular weight cut-off; and an additional coating with alginate reduced the leakage of the larger molecular species but the effect was short lived. Hence, they concluded that immunoisolation of encapsulated cells with alginate-PLL-alginate microcapsules cannot provide a molecular weight cut-off below 300 kDa.

References

[1] Moo-Young, M. and Blanch, H.W. (1981) Design of biochemical reactors: mass transfer criteria for simple and complex systems. Adv. Biochem. Eng. 19: 1-69.

[2] Radovich, J.M. (1985) Mass transfer in fermentations using immobilized whole cells. Enzyme Microb. Technol. 7: 2-10.

[3] Bailey, J.E. and Ollis, D.F. (1986) Bioreactor engineering fundamentals, 2nd edition, McGraw-Hill, NY, USA.

[4] Nielsen, J. and Villadsen, J. (1994) Bioreaction engineering principles. Plenum Press, New York, USA.

[5] Blanch, H.W. and Clark, D.S. (1997) Biochemical engineering. Marcel Dekker, New York.

[6] Pilkington, P.H.; Margaritis, A. and Mensour, N.A. (1998) Mass transfer characteristics of immobilized cells used in fermentation processes. Crit. Rev. Biotechnol. 18: 237-255.

[7] Furusaki, S. and Seki, M. (1985) Effect of intraparticle mass transfer resistance on reactivity of immobilized yeast cells. J. Chem. Eng. Jpa. 18: 389-393.

[8] Chresand, T.J.; Dale, B.E.; Hanson, S.L. and Gillies, R.J. (1988) A stirred batch technique for diffusivity measurements in cell matrices. Biotechnol. Bioeng. 32: 1029-1036.

[9] Libicki, S.B.; Salmon, P.M. and Robertson, C.R. (1988) The effective diffusivity permeability of a nonreacting solute in microbial cell aggregates. Biotechnol. Bioeng. 32: 68-85.

[10] Pu, H.T. and Yang, R.Y.K. (1988) Diffusion of sucrose and yohimbine in calcium alginate gel beads with or without entrapped plant cells. Biotechnol. Bioeng. 32: 891-896.

[11] Sakaki, K.; Nozawa, T. and Furusaki, S. (1988) Effect of intraparticle diffusion in ethanol fermentation by immobilized *Zymomonas mobilis*. Biotechnol. Bioeng. 31: 603-606.

[12] Scott, C.D.; Woodward, C.A. and Thompson, J.E. (1989) Solute diffusion in biocatalyst gel beads containing bioctalysis and other additives. Enzyme Microb. Technol. 11: 258-263.

[13] Epstein, N. (1989) On tortuosity and the tortuosity factor in flow and diffusion through porous media. Chem. Eng. Sci. 44: 777-779.

[14] Mackie, J.S. and Meares, P. (1955) The diffusion of electrolytes in a cation-exchange resin membrane, I: Theory. Proc. Roy. Soc. Lond. A232: 498-509.

[15] Brown, W. and Johnsen, R.M. (1981) Polymer 22: 185.

[16] Muhr, A.H. and Blanshard, J.M.V. (1982) Diffusion in gels. Polymer 23: 1012-1026.

[17] Mackie, J.S. and Meares, P. (1953) Proc. Roy. Soc. Lond. 75: 5705.

[18] Westrin, B.A. and Axelsson, A. (1991) Diffusion in gels containing immobilized cells: a critical review. Biotechnol. Bioeng. 38: 439-447.

[19] Beuling, E.E.; van den Heuvel, J.C. and Ottengraf, S.P.P. (1996) Determination of biofilm diffusion coefficients using micro-electrodes. In: Wijffels, R.H.; Buitelaar, R.M.; Bucke, C. and Tramper, J. (Eds.) Immobilized cells: basic and applications. Elsevier Science, Amsterdam, The Netherlands; pp. 31-38.

[20] Beuling, E.E.; van Dusschoten, D.; Lens, P.; van den Heuvel, J.C.; Van As, H. and Ottengraf, S.P.P. (1998) Characterization of the diffusive properties of biofilms using Pulsed Field Gradient-Nuclear Magnetic Resonance. Biotechnol. Bioeng. 60: 283-281.

[21] Hiemstra, H.; Dijkhuizen, L. and Harder, W. (1983) Diffusion of oxygen in alginate gels related to the kinetics of methanol oxidation by immobilized *Hansenula polymorpha* cells. Eur. J. Microbiol. Biotechnol. 18: 189-196.

[23] Kurosawa, H.; Matsumura, M. and Tanaka, H. (1989) Oxygen diffusivity in gel beads containing viable cells. Biotechnol. Bioeng. 34: 926-932.

[24] Benefield, L. and Molz, F. (1985) Mathematical simulation of a biofilm process. Biotechnol. Bioeng. 27: 921-931.

[25] Beyenal, H.; Tanyolac, A. and Lewandowski, Z. (1998) Measurement of local effective diffusivity in heterogeneous biofilms. Water Sci. Technol. 38: 171-178.

[26] Cunningham, A.B.; Visser, E.; Lewandowski, Z. and Abrahamson, M (1995) Evaluation of a coupled mass transport-biofilm process model using dissolved-oxygen microsensors. Wat. Sci. Technol. 32, 107-114.

[27] LaMotta, E. (1976) Internal diffusion and reaction in biological films. Environ. Sci. Technol. 10: 765-769.

[28] Matson, J.V. and Characklis, W.G. (1976) Diffusion into microbial aggregates. Water Res. 10, 877-885.

[29] Yu, J.A. and Pinder, K.L. (1993) Diffusion of lactose in acidogenic biofilms. Biotechnol. Bioeng. 41, 736-744.

[30] Westrin, B.A.; Axelsson, A. and Zacchi, G. (1994) Review: diffusion measurement in gels. J. Control. Rel. 30: 189-199.

[31] Frazier, B.L.; Larmour, P. and Riley, M.R. (2001) Noninvasive measurement of effective diffusivities in cell immobilization gels through use of near-infrared spectroscopy. Biotechnol. Bioeng. 72: 364-368.

[32] Adlercreutz, P. (1986) Oxygen supply to immobilized cells: 5. theoretical calculations and experimental data for the oxidation of glycerol by immobilized *Gluconobacter oxydans* cells with oxygen or *p*-benzoquinone as electron acceptor. Biotechnol. Bioeng. 28: 223-232.

[32] Kawashiro, T.; Carles, A.C.; Perry, S.F. and Piiper, J. (1975) Diffusivity of various inert gases in rat skeletal muscle. Pflügers Arch. 359: 219-230.

[33] Matsunaga, T.; Karube, I. and Suzuki, S. (1980) Some observations on immobilized hydrogen-producing bacteria: behavior of hydrogen in gel membranes. Biotechnol. Bioeng. 22: 2607-2615.

[34] Onuma, M. and Omura, T. (1982) Mass transfer characteristics within microbial systems. Wat. Sci. Technol. 14: 553-568.

[35] Smith, B.A.H. and Sefton, M.V. (1988) Permeability of a heparin-polyvinyl alcohol hydrogel to thrombin and antithrombin III. J. Biomed. Meter. Res. 22: 673-685.

[36] Sun, Y.; Furusaki, S.; Yamauchi, A. and Ichimura, K. (1989) Diffusivity of oxygen into carriers entrapping whole cells. Biotechnol. Bioeng. 34: 55-58.
[37] Mignot, L. and Junter, G.A. (1990) Diffusion in immobilized-cell agar layers: influence of microbial burden and cell morphology on the diffusion coefficients of *L*-malic acid and glucose. Appl. Microbiol. Biotechnol. 32: 418-423.
[38] Mignot, L. and Junter, G.A. (1990) Diffusion in immobilized-cell agar layers: influence of bacterial growth on the diffusivity of potassium chloride. Appl. Microbiol. Biotechnol. 32: 167-171.
[39] Converti, A.; Casagrande, M.; De Giovanni, M.; Rovatti, M. and Del Borghi, M. (1996) Evaluation of diffusion coefficient through cell layers for the kinetic study of an immobilized cell bioreactor. Chem. Eng. Sci. 51: 1023-1026.
[40] Reinhart, C.T. and Peppas, N.A. (1984) Solute diffusion in swollen membranes. Part II: Influence of crosslinking in diffusive properties. J. Membane Sci. 18: 227-239.
[41] Axelsson, A. and Persson, B. (1988) Determination of effective diffusion coefficients in calcium alginate gel plates with varying yeast cell content. Appl. Biochem. Biotechnol. 18: 231-250.
[42] Axelsson, A. and Westrin, B. (1991) Application of the diffusion cell for the measurement of diffusion in gels. Chem. Eng. Sci. 46: 913-915.
[43] Lebrun, L. and Junter, G.-A. (1993) Diffusion of sucrose and dextran through agar gel membranes. Enzyme Microb. Technol. 15: 1057-1062.
[44] Hannoun, B.J.M. and Stephanopoulos, G. (1986) Diffusion coefficients of glucose and ethanol in cell-free and cell-accupied calcium alginate membranes. Biotechnol. Bioeng. 28: 829-835.
[45] Hannoun, B.J.M. and Stephanopoulos, G. (1990) Intrinsic growth and fermentation rates of alginate-entrapped *Saccharomyces cerevisiae*. Biotechnol. Prog. 6: 341-348.
[46] Brito, E.; Don Juan, J.; Dominguez, F. and Casas, L.T. (1990) Diffusion coefficients of carbohydrates in modified κ-carrageenan gels with and without Escherichia coli immobilized. J. Ferment. Bioeng. 69: 135-137.
[47] Park, T.G. and Hoffman, A.S. (1990) Immobilization of *Arthrobacter simplex* cells in thermally reversible hydrogel: effect of temperature cycling on steroid conversion. Biotechnol. Bioeng. 35: 152-159.
[48] Freyer, J.P. and Sutherland, R.M. (1983) Determination of diffusion constants for metabolites in multicell tumor spheroids. Adv. Exp. Med. Biol. 159: 463-475.
[49] Tanaka, H.; Matsumura, M. and Veliky, I.A. (1984) Diffusion characteristics of substrates in Ca-alginate gel beads. Biotechnol. Bioeng. 26: 53-58.
[50] Nguyen, A. and Luong J.H.T. (1986) Diffusion in κ-carrageenan in gel beads. Biotechnol. Bioeng. 28: 1261-1267.
[51] Teo, W.K.; Ti, H.C. and Tan, W.H. (1986) Sorption of glucose in calcium alginate gel. Proceedings World Congress III of Chemical Engineering, Tokyo; pp. 851-854.
[52] Merchant, F.J.A.; Margaritis, A. and Wallace, J.B. (1987) A novel technique for measuring solute diffusivities in entrapment matrices used in immobilization. Biotechnol. Bioeng. 30: 936-945.
[53] Haggerty, L.; Sugarman, J.H. and Prud'homme, R.K. (1988) Diffusion of polymers through polyacrylamide gels. Polymer 29: 1058-1063.
[54] Kou, J.H.; Amidon, G.L. and Lee, P.I. (1988) pH dependent swelling and solute diffusion characteristics of poly(hydroxyethyl methacrylate-CO-methacrylic acid) hydrogels. Pharm. Res. 5: 592-597.
[55] Miller, D.R. and Peppas, N.A. (1988) Diffusional effects during albumin adsorption on highly swollen poly(vinyl alcohol) hydrogels. Eur. Polymer J. 24: 611-615.
[56] Hulst, A.C.; Hens, H.J.H.; Buitelaar, R.M. and Tramper, J. (1989) Determination of the effective diffusion coefficient of oxygen in gel materials in relation to gel concentration. Biotechnol. Techn. 3: 199-204.
[57] Fan, L.-S.; Leyva-Ramos, R.; Wisecarver, K.D. and Zehner, B.J. (1990) Diffusion of phenol through a biofilm grown on activated carbon particles in a draft-tube three-phase fluidized-bed bioreactor. Biotechnol. Bioeng. 35: 279-286.
[58] Arnaud, J.-P. and Lacroix, C. (1991) Diffusion of lactose in κ-carrageenna/locust bean gum gel beads with or without entrapped growing lactic acid bacteria. Biotechnol. Bioeng. 38: 1041-1049.
[59] Anderson, W.A.; Reilly, P.M.; Moo-Young, M. and Legge, R.L. (1992) Application of a Bayesian regression method to the estimation of diffusivity in hydrophilic gels. Can. J. Chem. Eng. 70: 499-504.
[60] Arnaud, J.-P.; Lacroix, C. and Castaigne, F. (1992) Counterdiffusion of lactose and lactic acid in κ-carrageenan/locust bean gum gel beads with or without entrapped lactic acid bacteria. Enzyme Microb. Technol. 14: 715-724.

[61] De Backer, L.; Devleminck S.; Willaert, R. and Baron, G. (1992) Reaction and diffusion in a gel membrane reactor containing immobilized cells. Biotechnol. Bioeng. 40: 322-328.

[62] Estapé, D.; Gòdia, F. and Solà, C. (1992) Determination of glucose and ethanol effective diffusion coefficients in Ca-alginate gel. Enzyme Microb. Technol. 14: 396-401.

[63] Longo, M.A.; Novella, I.S.; García, L.A. and Díaz, M. (1992) Diffusion of proteases in calcium alginate beads. Enzyme Microb. Technol. 14: 586-590.

[64] Martinsen, A.; Skjåk-Bræk, G. and Smidsrød, O. (1989) Alginate as immobilization material: I. Correlation between chemical and physical properties of alginate beads. Biotechnol. Bioeng. 33: 79-89.

[65] Martinsen, A.; Storrø I. and Skjåk-Bræk, G. (1992) Alginate as immobilization material: III. Diffusional properties. Biotechnol. Bioeng. 39: 186-194.

[66] Takamura, A.; Ishii, F. and Hidaka, H. (1992) Drug release from poly(vinyl alcohol) gel prepared by freeze-thaw procedure. J. Control. Rel. 20: 21-28.

[67] Wu, W.; Sidhoum, M. and DeLancey, G.B. (1994) Diffusion of acetophenone and phenethyl alcohol in the calcium-alginate-bakers' yeast-hexane system. Biotechnol. Bioeng. 44: 1217-1227.

[68] Øyaas, J.; Storrø, I.; Svendsen, H. and Levine, D.W. (1995) The effective diffusion coefficient and the distribution constant for small molecules in calcium-alginate gel beads. Biotechnol. Bioeng. 47: 492-500.

[69] Øyaas, J.; Storrø, I.; Lysberg, M.; Svendsen, H. and Levine, D.W. (1995) Determination of effective diffusion coefficients and distribution constants in polysaccharide gels with non-steady-state measurements. Biotechnol. Bioeng. 47: 501-507.

[70] Li, R.H.; Altreuter, D.H. and Gentile, F.T. (1996) Transport characterization of hydrogel matrices for cell encapsulation. Biotechnol. Bioeng. 50: 365-373.

[71] Mateus, D.M.R.; Alves, S.S. and da Fonseca, M.M.R. (1999) Diffusion in cell-free and cell immobilizing κ-carrageenan gel beads with and without chemical reaction. Biotechnol. Bioeng. 49: 625-631.

[72] Ho, C.S.; Ju, L.-K. and Baddour, R.F. (1988) The anomaly of oxygen diffusion in aqueous xanthan solutions. Biotechnol. Bioeng. 32: 8-17.

[73] Renneberg, R.; Sonomoto, K.; Katoh, S. and Tanaka, A. (1988) Oxygen diffusivity of synthetic gels derived from prepolymers. Appl. Microbiol. Biotechnol. 28: 1-7.

[74] Willaert, R.G. (1993) Reaction and diffusion in immobilised cell systems. PhD dissertation, Vrije Universiteit Brussel, Brussel, Belgium.

[75] Willaert, R. (1996) Membrane mass spectrometry. In: Willaert, R.G.; Baron, G.V. and De Backer, L. (Eds.) Immobilised living cell systems: modelling and experimental methods. John Wiley & Sons, Chichester, UK; pp. 177-213.

[76] Willaert, R.G. and Baron, GV (1994) The dynamic behaviour of yeast cells immobilised in porous glass studied by membrane mass spectrometry. Appl. Microbiol. Biotechnol. 42: 664-670.

[77] Zhang, S.-F.; Splendiani, A.; Freitas dos Santos, L.M. and Livingston, A.G. (1998) Determination of pollutant diffusion coefficients in naturally formed biofilms using a single tube extractive membrane bioreactor. Biotechnol. Bioeng. 59, 80-89.

[78] Rassing, J. (1985) New ultrasonic differential method for measuring diffusion coefficients of drugs inside small gel samples. Application to the diffusion of lidocaine hydrochloride and lidocaine base inside a 25% Pluronic F-127 gel. J. Control. Rel. 1: 169-175.

[79] Korthäuer, W.; Gelléri, B. and Sernetz, M. (1987) Interferometric determination of the effective diffusion coefficient of albumin in single Sepharose beads. Ann. NY Acad. Sci. 501: 517-521.

[80] Robert, M.C. and Lefaucheux, F. (1988) Crystal growth in gels: principles and applications. J. Cryst. Growth 90: 358-367.

[81] Ruiz-Beviá, F.; Fernández-Sempere, J. and Colom-Valiente, J. (1989) Diffusivity measurement in calcium alginate gel by holographic interferometry. AIChE J. 35: 1895-1898.

[82] Prasad, P.V. and Nalcioglu, O. (1991) Application of MRI to measure inter-diffusion coefficients of paramagnetic ions. Proc. Soc. Magnetic Tesonance in Medicine Ann. Meeting; pp. 1139.

[83] Ottengraf, S.P.P. and Van den Heuvel, J.C. (1996) Microelectrodes. In: Willaert R.G., Baron G.V., De Backer L. (Eds.) Immobilised living cell systems: modelling and experimental techniques. John Wiley & Sons, Chichester, UK; pp. 147-176.

[84] Yang, S.N. and Lewandowski, Z. (1995) Measurement of local mass-transfer coefficient in biofilms; Biotechnol. Bioeng. 48: 737-744.

[85] Rasmussen, K. and Lewandowski, Z. (1998) Microelectrode measurements of local mass transport rates in heterogeneous biofilms. Biotechnol. Bioeng. 59, 302-309.

[86] Beuling, E.E.; van den Heuvel, J.C. and Ottengraf, S.P.P. (2000) Diffusion coefficients of metabolites in active biofilms. Biotechnol. Bioeng. 67: 53-60.

[87] Huang, M.Y. and Bungay, H.R. (1973) Microprobe measurements of oxygen concentration in mycelial pellets. Biotechnol. Bioeng. 15: 1193-1197.

[88] Wittler, R.; Baumgartl, H.; Lübbers, D.W. and Schügerl, K. (1986) Investigations of oxygen transfer into *Penicillium chrysogenum* pellets by microprobe measurements. Biotechnol. Bioeng. 28: 1024-1036.

[89] Michel Jr., D.C.; Grulke, E.A. and Reddy, C.A. (1992) A kinetic model for fungal pellet lifecycle. AIChE J. 38: 1449-1460.

[90] Cronenberg, C.C.H.; Ottengraf, S.P.P.; van den Heuvel, J.C.; Pottel, F.; Sziele, D.; Schügerl, K. and Bellgradt, K.H. (1994) Influence of age and structure of *Penicilium chrysogenum* pellets on the internal concentration profiles. Bioprocess Eng. 10: 209-216.

[91] Itamunoala, G.F. (1987) Effective diffusion coefficients in calcium alginate gel. Biotechnol. Prog. 3: 115-120.

[92] Itamunoala, G.F. (1988) Limitations of methods of determining effective diffusion coefficients in cell immobilization matrices. Biotechnol. Bioeng. 31: 714-717.

[93] Nugent, L.J. and Jain, R.K. (1983) Interstitial diffusion of macromolecules in normal and tumor capillary beds. AIChE Symposium Series 227: 1-10.

[94] de Beer, D.; Stoodley, P. and Lewandowski, Z. (1997) Measurement of local diffusion coefficients in biofilms by microinjection and confocal microscopy. Biotechnol. Bioeng. 53: 151-158.

[95] Davies, P.A. (1989) Determination of diffusion coefficients of proteins in beaded agarose gel filtration. J. Chromat. 483: 221-252.

[96] Boyer, P.M. and Hsu, J.T. (1992) Experimental studies of restricted protein diffusion in an agarose matrix. AIChE J. 38: 259-272.

[97] Fang, L. and Brown, W. (1990) Decay time distributions from dynamic light scattering for aqueous poly(vinyl alcohol) gels and semi-dilute solutions. Macromol. 23: 3284-3290.

[98] Poitevin, E. and Wahl, P. (1988) Study of the translational diffusion of macromolecules in beads of gel chromatography by the FRAP method. Biophys. Chem. 31: 247-258.

[99] Scalettar, B.A.; Selvin, P.R.; Axelrod, D.; Klein, M.P. and Hearst, J.E. (1990) A polarized photobleaching study of DNA reorientation in agarose gels. Biochemistry 29: 4790-4798.

[100] Bryers, J.D. and Drummond, F. (1998) Local macromolecule diffusion coefficients in structurally non-uniform bacterial biofilms using fluorescence recovery after photobleaching (FRAP). Biotechnol. Bioeng. 60: 462-473.

[101] Sellen, D.B. (1987) Laser light scattering studies of polyacrylamide gels. J. Polymer Sci. B25: 699-716.

[102] Nyström, B.; Moseley, M.E.; Brown, W. and Roots, J. (1981) Molecular motion of small molecules in cellulosic gels studied by NMR. J. Appl. Polymer Sci. 26: 2285-2294.

[103] Everhart, C.H. and Johnson Jr., C.S. (1982) The determination of tracer diffusion coefficients for proteins by means of pulsed field gradient NMR with applications to hemoglobin. J. Magn. Reson. 48: 466-474.

[104] Brown, W.; Stilbs, P. and Lindström, T. (1984) Self-diffusion of small molecules in cellulose gels using FT-pulsed field gradient NMR. J. Appl. Polymer Sci. 29: 823-827.

[105] Stilbs, P. (1987) Fourier transform pulsed-gradient spin-echo studies of molecular diffusion. Progr. NMR Spectr. 19: 1-45.

[106] Lundberg, P. and Kuchel, P.W. (1997) Diffusion of solutes in agarose and alginate gels: ^{1}H and ^{23}Na PFGSE and ^{23}Na TQF NMR studies. Magn. Reson. Med. 37: 44-52.

[107] Stewart, U.A.; Bradley, M.S.; Johnson Jr., C.S. and Gabriel, D.A. (1988) Transport of probe molecules through fibrin gels as observed by means of holographic relaxation methods. Biopolymers 27: 173-185.

[108] Park, I.H.; Johnson Jr., C.S. and Gabriel, D.A. (1990) Probe diffusion in polyacrylamide gels as observed by means of holographic relaxation methods: search for a universal equation. Macromol. 23: 1548-1553.

[109] Crank, J. (1975) The mathematics of diffusion, 2nd edition, Clarendon Press, Oxford, UK.

[110] Westrin, B.A. (1990) Diffusion in immobilized-cell gels. Appl. Microbiol. Biotechnol. 34: 189-190.

[111] Glueckauf, E. (1955) Theory of chromatography: part 10. formulae for diffusion into spheres and their application to chromatography. Trans. Farad. Soc. 51: 1540-1551.

[112] Sherwood, T.K.A.; Pigford, R.L. and Wilke, C.R. (1975) Mass transfer, McGraw-Hill Book Company, London, UK.

[113] Noordsij P., Rotte J.W. (1967) Mass transfer coefficients to a rotating and a vibrating sphere. Chem. Eng. Sci. 22: 1475-1481.

[114] van 't Riet, K. and Tramper, J. (1991) Basic Bioreactor Design. Marcel Dekker, New York, USA.

[115] McCabe, W.L. and Smith, J.C. (1976) Unit operations of chemical engineering. McGraw-Hill, New York, USA.

[116] Brian, P.L.T. and Hales, H.B. (1969) Effects of transpiration and changing diameter on heat and mass transfer to spheres. AIChE J. 15: 419-425.
[117] Ranz, W.E. and Marshall, W.R. (1952) Evaporation from drops. Chem. Eng. Prog. 48: 141-146, 173-180.
[118] Kikuchi, K.I.; Sugarawa, T. and Ohashi, H. (1988) Correlation of liquid-side mass transfer coefficient based on the new concept of specific power group. Chem. Eng. Sci 43: 2533-2540.
[119] Calderbank, P.H. and Moo-Young, M.G. (1961) The continuous phase heat and mass-transfer properties of dispersions. Chem. Eng. Sci. 16: 39-54.
[120] Horstmann, B.J. and Chase, H.A. (1989) Modelling the affinity adsorption of immunoglobulin G to protein A immobilised to agarose matrices. Chem. Res. Des. 67: 243-254.
[121] Neale, G.H. and Nader, W.K. (1973) Prediction of transport processes within porous media: diffusive flow processes within an homogeneous swarm of spherical particles. AIChE J. 19: 112-119.
[122] Akanni, K.A.; Evans, J.W. and Abramson, I.S. (1987) Effective transport coefficients in heterogeneous media. Chem. Eng. Sci. 42: 1945-1954.
[123] Maxwell J.C. (1892) Electricity and magnetism, 2nd edition, Clarendon Press, Oxford, UK.
[124] Riley, M.R.; Muzzio, F.J.; Buettner, H.M. and Reyes, S.C. (1996) A simple correlation for predicting effective diffusivities in immobilized cell systems. Biotechnol. Bioeng. 49: 223-227.
[125] Maxwell J.C. (1881) A treatise on electricity and magnetism, 2nd edition, Clarendon Press, Oxford, UK.
[126] Ho, C.S. and Ju, L.-K. (1988) Effects of microorganisms on effective oxygen diffusion coefficients and solubilities in fermentation media. Biotechnol. Bioeng. 32: 313-325.
[127] Coulson, J.M. and Richardson, J.F. (1978) Chemical engineering, Vol. 2, 3rd edition, Pergamon Press, Oxford, UK.
[128] Wakao, N. and Smith, J.M. (1962) Diffusion in catalyst pellets. Chem. Eng. Sci. 17: 825-834.
[129] Klein, J. and Schara, P. (1981) Immobilization of living microbial cells in covalent polymeric networks. II. A quantitative study on the kinetics of oxidative phenol degradation by entrapped *Candida tropicalis* cells. Appl. Biochem. Biotechnol. 6: 91-117.
[130] Korgel, B.A.; Rotem, A. and Monbouquette, H.G. (1992) Effective diffusivity of galactose in calcium alginate gels containing immobilized *Zymomonas mobilis*. Biotechnol. Prog. 8: 111-117.
[131] Zhao, Y. and DeLancey, G.B. (2000) A diffusion model and optimal cell loading for immobilized cell biocatalysts. Biotechnol. Bioeng. 69: 639-647.
[132] Klein J., Manecke G. (1982) In: Chibata, I; Fukui, S. and Wingard Jr., L.B. (Eds.), Enzyme engineering, vol. 6, Plenum Press, New York, USA.
[133] Ogston, A.G.; Preston, B.N. and Wells, J.D. (1973) On the transport of compact particles through solutions of chain-polymers. Proc. R. Soc. Lond. 333 A: 297.
[134] Cukier, R.I. (1984) Diffusion of Brownian spheres in semidilute polymer solutions. Macromol. 17: 252.
[135] Dibdin G.H. (1981) Arch. Oral Biol. 26: 515
[136] Tatevossian A. (1985) Some factors affecting the diffusion of [14C]-lactate in human dental plaque. Arch. Oral Biol. 30: 141-146.
[137] Tatevossian, A. (1985) The effect of heat inactivation, tortuosity, extracellular polyglucan and on-exchange sites of the diffusion of [^{14}C]-sucrose in human dental plaque residue *in vitro*. Arch. Oral Biol. 30: 365-371.
[138] Cronenberg C.C.H. (1994) Biofilms investigated with needle-type glucose sensors. PhD Thesis, University of Amsterdam, The Netherlands.
[139] McNee S.G., Geddes D.A.M., Weetman D.A. (1982) Diffusion of sugars and acids in human dental plaque *in vitro*. Arch. Oral Biol. 27: 975-979
[140] Smith, P.G. and Coackley, P. (1984) Diffusivity, tortuosity and pore structure of activated sludge. Water Res. 18: 117-122.
[141] Matson, J.V. (1975) Diffusion through microbial aggregates. Doctoral dissertation, Rice University, U.S.A.
[142] Libicki, S.B.; Salmon, P.M. and Robertson, C.R. (1986) Measurement of inert gas permeabilities in compact bacterial cell aggregates using an annular reactor. Ann. N.Y. Acad. Sci. 469: 145-151.
[143] Westrin, B.A. and Zacchi, G. (1991) Maesurement of diffusion coefficients in gel beads: random and systematic errors. Chem. Eng. Sci. 46: 1911-1916.
[144] Willaert RG, Baron G.V. (1996) Gel entrapment and micro-encapsulation: methods, applications and engineering principles. Rev. Chem. Eng. 12, 1-205.
[145] Klein, J.; Stock, J. and Vorlop, K.-D. (1983) Pore size and properties of spherical Ca-alginate biocatalysts. Eur. J. Appl. Microbiol. Biotechnol. 18: 86-91.
[146] Smidsrød, O. (1974) Faraday Discuss. Chem. Soc. 57: 263-274.

[147] Stewart, W.W. and Swaisgood, H.E. (1993) Characterization of calcium alginate pore diameter by size-exclusion chromatography using protein standards. Enzyme Microb. Technol. 15: 922-927.
[148] Polakovic, M. (1996) Evaluation of density function of pore size distribution of calcium pectate hydrogel. In: Wijffels, R.H.; Buitelaar, R.M.; Bucke, C. and Tramper, J. (Eds.) Immobilized cells: basic and applications. Elsevier Science, Amsterdam, The Netherlands; pp. 62-69.
[149] Andresen, I.-L.; Skipnes, O.; Smidsrød, O.; Østgaard, K. and Hemmer, P.C. (1977) ASC Symp. Ser. 48: 361-381.
[150] Skjåk-Bræk, G.; Grasdalen, H. and Smidsrød, O. (1989) Inhomogeneous polysaccharide ionic gels. Carbohydr. Polym. 10: 31-54.
[151] Skjåk-Bræk, G.; Grasdalen, H. and Larsen, B. (1986) Monomer sequence and acetylation pattern in some bacterial alginates. Carbohydr. Res. 154: 239-250.
[152] Gray, C.J. and Dowsett, J. (1988) Retention of insulin in alginate gel beads. Biotechnol. Bioeng. 31: 607-612.
[153] Renkin, E.M. (1954) Filtration, diffusion, and molecular sieving through porous cellulose membranes. J. Gen. Physiol. 38: 225-243.
[154] Kierstan, M.; Darcy, G. and Reilly, J. (1982) Studies on the characteristics of alginate gels in relation to their use in separation and immobilization applications. Biotechnol. Bioeng. 24: 1507-1517.
[155] Chen, D.; Lewandowski, Z.; Roe, F. and Surapaneni, P. (1993) Diffusivity of Cu^{2+} in calcium alginate gel beads. Biotechnol. Bioeng. 41: 755-760.
[156] Lewandowski, Z. and Roe, F. (1994) Diffusivity of Cu^{2+} in calcium alginate gel beads: recalculation. Biotechnol. Bioeng. 43: 186-187.
[157] Jang, L.K. (1994) Diffusivity of Cu^{2+} in calcium alginate gel beads. Biotechnol. Bioeng. 43: 183-185.
[158] Siegrist, H. and Gujer, W. (1985) Mass transfer mechanisms in a heterotrophic biofilm. Water Res. 19:1369-1378.
[159] Holden, P.A.; Hunt, J.R. and Firestone, M.K. (1997) Toluene diffusion and reaction in unsaturated *Pseudomonas putida* biofilms. Biotechnol. Bioeng. 56: 656-670.
[160] Williamson, K. and McCarty, P.L. (1976) Verification studies of the biofilm model for bacterial substrate utilization. J. Wat. Pollut. Control Fed. 48: 281-296.
[161] Pipes, D.M. (1974) Variations in glucose diffusion coefficients through biological flocs. M.S. thesis. Rice University, Houston, USA.
[162] Onuma, M.; Omura, T.; Umita, T. and Aizawa, J. (1985) Diffusion coefficients and its dependency on some biochemical factors. Biotechnol. Bioeng. 27: 1533-1539.
[163] San, H.A., Tanik, A. and Orhon, D. (1993) Micro-scale modelling of substrate removal kinetics in multicomponent fixed-film systems. J. Chem. Technol. Biotechnol. 58: 39-48.
[164] Tanik, A.; Orhon, D. and San, H.A. (1993) Micro-scale modelling of substrate removal kinetics in multicomponent fixed-film systems. J. Chem. Technol. Biotechnol. 58: 49-55.
[165] Lawrence, J.R.; Wolfaardt, G.M. and Korber, D.R. (1994) Determination of diffusion-coefficients in biofilms by confocal laser microscopy. Appl. Environ. Microbiol. 60: 1166-1173.
[166] Zhang, T.C. and Bishop, P.L. (1994) Evaluation of tortuosity factors and effective diffusivities in biofilms. Water Res. 28: 2279-2287.
[167] Bishop, P.L.; Zhang, T.C. and Fu, Y.C. (1995) Effects of biofilm structure, microbial distributions and mass-transport on biodegradetion processes. Wat. Sci. Technol. 31: 143-152.
[168] de Beer, D. and Stoodley, P. (1995) Relation between the structure of an aerobic biofilm and transport phenomena. Wat. Sci. Technol. 32: 11-18.
[169] Lewandowski, Z.; Stoodley, P. and Altobelli, S. (1995) Experimental and conceptual studies on mass-transport in biofilms. Wat. Sci. Technol. 31: 153-162.
[170] van Loosdrecht, M.C.M. and Heijnen, J.J. (1996) Biofilm processes. In: Willaert, R.G.; Baron, G.V. and De Backer, L. (Eds.) Immobilised living cell systems: modelling and experimental methods. John Wiley & Sons, Chichester, UK; pp. 255-271.
[171] Wanner, O. and Reichert, P. (1996) Mathematical-modelling of mixed-culture biofilms. Biotechnol. Bioeng. 49: 172-184.
[172] Picioreanu, C.; van Loosdrecht, M.C.M. and Heijnen, J.J. (2000) Effect of diffusive and convective substrate transport on biofilm structure formation: a two-dimensional modelling study. Biotechnol. Bioeng. 69: 504-515.
[173] de Beer, D.; Stoodley, P. and Lewandowski, Z. (1994) Liquid flow in heterogeneous biofilms. Biotechnol. Bioeng. 44: 636-641.
[174] Fujie, K.; Tsukamoto, T. and Kubota, H. (1979) J. Ferment. Technol. 57: 539.
[175] Mulcahy, L.T.; Shieh, W.K. and LaMotta, E.J. (1981) Biotechnol. Bioeng. 23: 2403.

[176] Wang, S.-C.P. and Tien, C. (1984) Bilayer film model for the interaction between adsorption and bacterial activity in granular activated carbon columns. Part I: Formulation of equations and their numerical solutions. AIChE J. 30: 786-794.
[177] Tang, W.-T. and Fan, L.-S. (1987) Steady state phenol degradation in a draft-tube, gas-liquid-solid fluidized-bed bioreactor. AIChE J. 33: 239-249.
[178] Wood, B.D.; Quintard, M. and Whitaker, S. (2002) Calculation of effective diffusivities for biofilms and tissues. Biotechnol. Bioeng. 77: 495-516.
[179] Ochoa-Tapia, J.A.; Stroeve, P. and Whitaker, S. (1994) Diffusive transport in two-phase media: spatially periodic models and Maxwell's theory for isotropic and anisotropic systems. Chem. Eng. Sci. 49: 709-726.
[180] Wood, B.D. and Whitaker, S. (2000) Multi-species diffusion and reaction in biofilms and cellular media. Chem. Eng. Sci. 55: 3397-3418.
[181] Atkinson, B. and Daoud, I.S. (1976) Microbial flocs and flocculation in fermentation process engineering. Adv. Biochem. Eng. 4: 41-124.
[182] Kosaric, N. and Blaszczyk, R. (1990) Microbial aggregates in anaerobic wastewater treatment. Adv. Biochem. Eng./ Biotechnol. 42: 27-62.
[183] Speers, R.A.; Tung, M.A.;Durance, T.D. and Stewart, G.G. (1992) Biochemical aspects of yeast flocculation and its measurement: a review. J. Inst. Brew. 98: 293-300.
[184] Stratford, M. (1992) Yeast flocculation: a new perspective. Adv. Microb. Physiol. 33: 2-71.
[185] Straver, M.H.; Kijne, J.W. and Smit, G. (1993) Cause and control of flocculation in yeast. Tibtech 11: 228-232.
[186] Teunissen, A.W. and Steensma, H.Y. (1995) Review: the dominant flocculation genes of *Saccharomyces cerevisiae* constitute a new subtelomeric gene family. Yeast 11: 1001-1013.
[187] Dengis, P.B. and Rouxhet, P.G. (1997) Flocculation mechanisms of top and bottom fermenting brewing yeast. J. Inst. Brew. 103: 257-261.
[188] Whitaker, A. and Long, P.A. (1973) Fungal pelleting. Process Biochem. 8: 27-31.
[189] Metz, B and Kossen, N.W.F. (1977) The growth of molds in the form of pellets – a literature review. Biotechnol. Bioeng. 19: 781-799.
[190] Nielsen, J. (1992) Modelling the growth of filamentous fungi. Adv. Biochem. Eng./Biotechnol. 46: 187-223.
[191] Nielsen, J. and Carlsen, M. (1996) Fungal pellets. In: Willaert, R.G.; Baron, G.V. and De Backer, L. (Eds.) Immobilised living cell systems: modelling and experimental methods. John Wiley & Sons, Chichester, UK; pp. 273-293.
[192] Hashin, Z. and Shtrikman, J. (1962) A variational approach to the theory of the effective magnetic permeability of multiphase materials. J. Appl. Phys. 33: 3125-3131.
[193] Yano, T.; Kodama, T. and Yamada, K. (1961) Fundamental studies on the aerobic fermentation: Part VIII. Oxygen transfer within a mold pellet. Agr. Biol. Chem. 25: 580-584.
[194] Ngian, K.F. and Lin, S.H. (1976) Diffusion coefficient of oxygen in microbial aggregates. Biotechnol. Bioeng. 18: 1623-1627.
[195] Pipes, D.M.; Characklis, W.G. and Matson, J.V. (1974) Discussion on substrate removal mechanism of trickling filters. J. Environ. Eng. Div., ASCE 100: 225-226.
[196] Mueller, J.A.; Boyle, W.C. and Lightfoot, E.N. (1968) Oxygen diffusion through zoogloeal flocs. Biotechnol. Bioeng. 10: 331-358.
[197] Baillod, C.R. and Boyle, W.C. (1970) Mass transfer limitations in substrate removal. J. Sanit. Eng. Div. ASCE 96: 525-545.
[198] Metz, B. (1976) From pulp to pellets. Ph.D. thesis, Technical University of Delft, Delft, The Netherlands.
[199] Phillips, D.H. and Johnson, M.J. (1961) Aeration in fermentations. J. Biochem. Microb. Technol. Eng. 3: 277-309.
[200] Pirt, S.J. (1966) A theory of mode of growth of fungi in the form of pellets in submerged culture. Proc. Roy. Soc. B. 166: 369-373.
[201] Trinci, A.P.J. (1970) Kinetics of the growth of mycelial pellets of *Aspergillus nidulans*. Arch. Mikrobiol. 73: 353-348.
[202] Phillips, D.H. (1966) Oxygen transfer into mycelial pellets. Biotechnol. Bioeng. 8: 456-460.
[203] Kobayashi, T.; van Dedem, G. and Moo-Young, M. (1973) Oxygen transfer into mycelial pellets. Biotechnol. Bioeng. 15: 27-45.
[204] Stephanopoulos, G. and Tsiveriotis, K. (1989) The effect of intraparticle convection on nutrient transport in porous biological pellets. Chem. Eng. Sci. 44: 2031-2039.

[205] Kwok, W.Y.; Kiparissides,C.; Yuet P.; Harris, T.J. and Goosen, M.F.A. (1991) Mathematical modelling of protein diffusion in microcapsules: a comparison with experimental results. Can. J. Chem. Eng. 69: 361-370.

[206] Yuet, P.K.; Kwok, W.; Harris, T.J. and Goosen, M.F.A. (1993) Mathematical modelling of protein diffusion and cell growth in microcapsules. In: Goosen, M.F.A. (Ed.) Fundamentals of animal cell encapsulation and immobilisation. CRC Press, Boca Raton, Florida, USA; pp. 79-111.

[207] Christenson, L., Dionne, K.E. and Lysaght, M.J. (1993) Biomedical applications of immobilized cells. In: Goosen M.F.A. (Ed.) Fundamentals of animal cell encapsulation and immobilisation. CRC Press, Boca Raton, Florida, USA; pp. 7-41.

[208] Okhamafe A.O., Goosen M.F.A. (1993) Control of membrane permeability in microcapsules. In: Goosen M.F.A. (Ed.) Fundamentals of animal cell encapsulation and immobilization. CRC Press, Boca Raton, USA; pp. 55-78.

[209] Coromili, V. and Chang, T.M.S. (1993) Polydisperse dextran as a diffusing test solute to study the membrane permeability of alginate polylysine microcapsules. Biomat. Art. Cells & Immob. Biotech. 21: 427-444.

[210] Awrey, D.E.; Tse, M.; Hortelano, G. and Chang, P.L. (1996) Permeability of alginate microcapsules to secretory recombinant gene products. Biotechnol. Bioeng. 52: 472-484.

CHARACTERIZATION OF MICROCAPSULES

DAVID HUNKELER[1], CHRISTINE WANDREY[2],
STEFAN ROSINSKI[3], DOROTA LEWINSKA[3]
AND ANDRZEJ WERYNSKI[3]
[1]*AQUA+TECH Specialties S.A., 4 Chemin du Chalet-du-bac, CP 28 La Plaine, CH-1283, Geneva, Switzerland – Fax: +41 22 756 09 23 – Email: david.hunkeler@AQUAplusTECH.ch*
[2]*Swiss Federal Institute of Technology Lausanne, Laboratory of Chemical Biotechnology, CH-1015 Lausanne, Switzerland – Fax: +41 21 693 60 30 – Email: christine.wandrey@epfl.ch*
[3]*Institute of Biocybernetics and Biomedical Engineering, Trojdena Str 4, 02-109 Warsaw, Poland – Fax: +4822 6582872 – Email: bioenc@ibib.waw.pl*

1. The problem of encapsulation

The encapsulation of mammalian cells has been a topic of intensive research, if not since Chang's pioneering work [1], certainly after Sun showed long-term normoglycaemia in diabetic rats [2] by immunoisolating xenotransplanted islets in a semipermeable polymeric membrane. Therefore, since the early 1980s, two dozen academic groups and approximately the same number of, generally venture capital funded, private firms, have attempted to move the field into larger animal trials, and the clinic, all the while with the ambition of increasing the period of transplanted cell function. However, despite the field's potential, and the outstanding groups working therein, progress has been very slow. This can be explained, to a large extent, by the inability to consistently isolate, and disseminate, technologies for cell isolation. Primary cell lines are also lacking for the great majority of hormone deficient diseases, which would require such a therapy. Concomitant with the lack of tissue supply is the inability of all but the most selective groups, to be able to cryopreserve or "bank" cell lines, limiting the number of pre-clinical trials. A final difficulty has been the lack of any batches of biocompatible materials, even for the relatively simply alginate bead-based capsules. The longstanding goal of having firms, or laboratories, provide clinically pre-certified lots, therefore, seems unrealizable.

V. Nedović and R. Willaert (eds.), Fundamentals of Cell Immobilisation Biotechnology, 389-409.

1.1. THE ROLE OF CHARACTERIZATION IN MOVING FORWARD

The present chapter examines, within the xeno-*versus*-allotransplant debate, means by which multicomponent microcapsules can be characterized. The latter are certainly needed to block both the direct and indirect mechanisms that reject xenografts. Rather than overview a complete list of techniques available, which has been recently carried out and is beyond the scope of this contribution, two emerging techniques will be highlighted in detail. Specifically, a summary of the requirements to measure the ingress and egress of small and macromolecules through the membrane will be discussed, along with the relevant theory. This section on mass transport will be preceded by a discussion of the analytical ultracentrifuge, as applied to characterize the kinetics and quality of polysymplexes formed by reacting oppositely charged polyelectrolytes. Prior to this a review of the relevant factors to consider in microcapsule design will be presented, along with a brief overview of recent techniques for micromechanical characterization.

It is clear that the authors have been selective. We will not provide extensive explanations of mechanical techniques, which enable the determination of the yield and deformability of beads and capsules, since most of such "devices" are suitably strong to withstand long term implantation in all sites other than possible internal organs such as the liver. Nevertheless, we will discuss mechanical properties as they relate to material purity as well as the durability of the microcapsules. We also omit to summary techniques for estimation of the hydration of the beads or capsules, nor will we hark on specific chemistries, as these have been the topic of recent reports by EC-sponsored working groups. This chapter, therefore, seeks to define the analytical tools, which are necessary to provide feedback to a designer of microcapsules, which would be used to block both ingress and egress of immunostimulating molecules. One would require, in any case, biocompatible materials, which themselves may necessitate genetic modification, as Skjåk-Bræk has recently done for alginate [3]. We therefore cannot avoid the debate as to if any polycation at all is sufficiently benign to avoid an immune response. However, given that the authors subscribe to a philosophy that xenotransplantation, which will be preceded by publicly funded allotransplantation, will require short term immunosuppression, the choice of capsules, over homogeneous beads, is clear. This chapter summarizes the two techniques that would be required to develop, and eventually patent, a microcapsule as a product based on existing materials and cells. While modest, and perhaps minimalist, it is the best that can be done with the existing state of development, and technology.

1.2. CUSTOMIZED MICROCAPSULES

One should note that, encapsulation, particularly of living cells, requires a balance of several biological, micromechanical and materials science related parameters. This includes morphological issues such as the size, shape and permeability of the capsule. Mechanical properties including the deformability, relaxation and bursting force, many of which are strain dependent, also must be considered. In light of the application at hand (*e.g.* encapsulated islets for a bioartificial pancreas transplanted intraportally in the liver) the durability, elasticity and retrievability should also be optimized. From a polymeric perspective, the membrane chemistry must be chosen to block the indirect pathway, at least for xenografts, as well as to provide a suitable cut-off. Therefore, the membrane thickness and structure, as well as the homogeneity of the polymer network play a role. One could question, in light of the length of the preceding list, how one embarks on a rational design of a microcapsule. The authors believe that the building blocks must be the existence of an appropriate polyanion which is compatible with the cell to be immunoisolated and can be purified and depyrogenated to the extend needed, without sacrificing mechanical and transport properties. Examples of this will be included in Section 1.1 for alginate based materials.

It is important to note that this chapter generally discusses microcapsules, which can be defined as sub-mm semipermeable polymeric spheres, generally based on hydrogels, which include a defined membrane. In contrast a microbead, or bead if the size is above 1mm, would be based on a hydrogel network, in the absence of a polycation coating. Such a hydrogel could be either homogeneous in composition, across the radius, or heterogeneous, depending on the polysaccharide employed, and the gelation method with mono- and divalent cations.

2. Mechanical characterization of microcapsules

A variety of methods have been published describing different possibilities for the mechanical characterizations of microcapsules. These include both qualitative and quantitative techniques applicable to different microcapsule types [4-7]. The "explosion assay" may be considered as a qualitative method [8] applicable specifically to membrane capsules. Lowering the ionic strength by transferring the capsules from a saline solution into water the osmotic conditions causes the capsules to swell and finally to break. Some mechanical techniques measure the breakage of microcapsules under various conditions, for example, in a turbine reactor [9], in a bubble column [10,11], while shaking [12-14], or in a cone-and-plate flow device [15]. All these procedures permit the qualitative estimation of the capsule's resistance as a function of the shear forces. However, quantification remains difficult.

A highly sensitive technique is to measure the resistance or deformation of beads/capsule under compression. Such measurements enable conclusions concerning the mechanical stability, the deformability/rigidity, and deliver quantitatively the bursting force. Various techniques have been developed and successfully applied to different kinds of beads/capsules by several authors [10,16,17]. A micromanipulation technique introduced by Zhang *et al.* [18] makes the compression method applicable to very small capsules even below 1 μm. Though the bursting force can be considered as a precise parameter it does not deliver sufficient information for the applicability of capsules in dynamic systems such as exposition to agitation or shear forces. Therefore, often the combination of various characterization methods is required.

Figure 1 demonstrates the principle of a force-displacement curve and typical characteristic values obtained by uniaxial compression of multicomponent microcapsules [19]. The characteristic values are the force at 60% displacement, the displacement at bursting and the force at bursting. Applying this method to capsules of different size a correlation between the force at 60% displacement and the membrane volume has been identified, as it is shown in Figure 2 [19]. This is critical, since it implies that the membrane chemistry and the capsule size are secondary parameters, with mechanical properties governed by the volume of the membrane. Therefore, if capsules are constructed in two steps, with a membrane following as a coating on the gel, one can imagine that mechanical properties and transport rates can be completely decoupled. This is the current state of the art in multicomponent polysymplex based microcapsules [20].

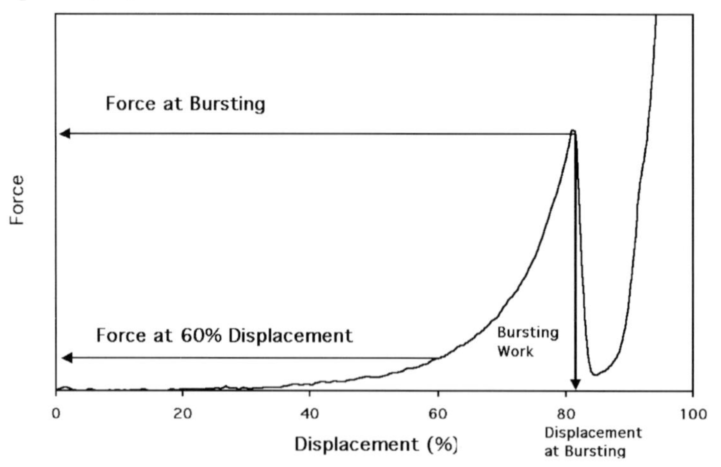

Figure 1. Principle of a force-displacement curve and characteristic values.

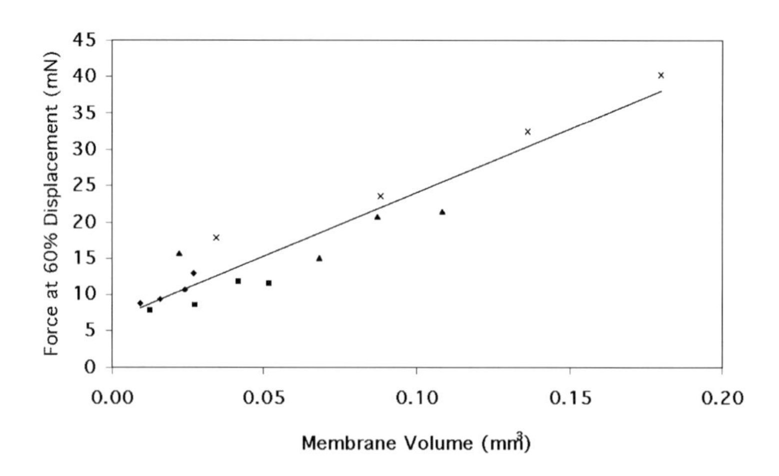

Figure 2. Correlation between the force necessary for the compression of microcapsules (60% displacement) and the membrane volume for various diameters and membrane thickness. Capsule diameter: ◆ 400 μm, ■ 600 μm, ▲ 800 μm, × 1000 μm.

The influence of the alginate microstructure and molar mass has also been identified for the same type of multicomponent microcapsules (1:1 blend of sodium alginate and sodium cellulose sulfate pre-gelled with calcium chloride before the membrane formation with poly(methylene-co-guanidine)) using the compression method.

Figure 3 summarizes the experimental results of appropriate studies showing the force at which the various capsules break. The highest mechanical resistance has been found for capsules produced from high molar mass high-G alginate. The purification of this alginate type did not negatively influence the capsule properties as it is also demonstrated in Figure 3.

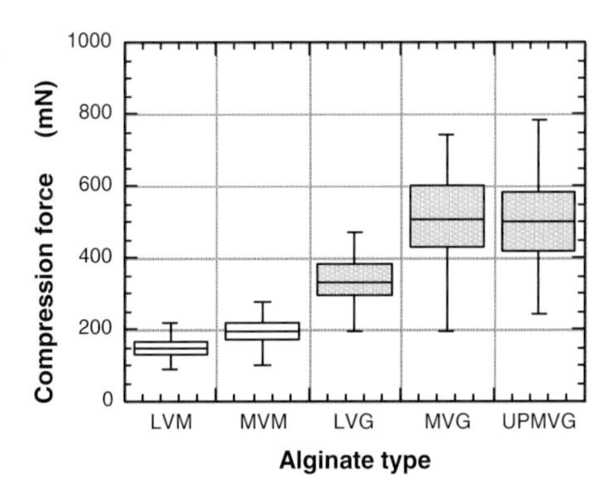

Figure 3. Influence of the alginate quality on the mechanical stability of multi-component microcapsules. Bursting force as a function of the alginate type. LVM: low viscous high-M, MVM: medium viscous high-M, LVG: low viscous high-G, MVG: medium viscous high-G, UPMVG: highly purified medium viscous high-G [18].

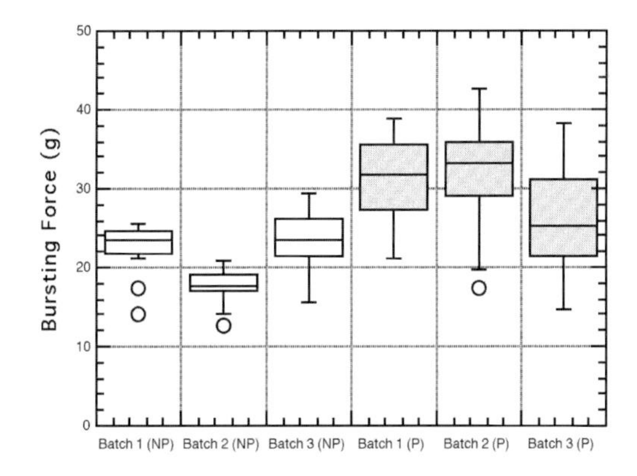

Figure 4. Influence of the purification on the mechanical properties of multicomponent microcapsules prepared from non-purified (NP) and purified polyanions (P).

For more general conclusions concerning the correlation of mechanical resistance and capsule parameters some modelling has been performed for both beads and capsules based on different physical approaches [21,22] considering them, for example, as elastic spheres, which are completely gelled or full of liquid [21], or model them as an empty sphere with a thin wall [23]. Figure 4 reveals that the depyrogenation of alginates does not influence the mechanical properties of the resulting beads or capsules.

3. On-line control of membrane formation

The prediction of polyelectrolyte complex networks/membranes, as a function of macromolecular, physicochemical, structural, and environmental conditions, has not been elaborated to date. A multitude of combinations have been screened empirically to find optimum pairs [24]. However, no approach exists which tries to find general correlation between the physicochemical parameters and the membrane properties. Such screening suffers from the fact that no experimental technique was used that can follow the membrane formation on-line. Recently the synthetic boundary experiment and the optical system of an analytical ultracentrifuge (AUC) have been proposed to study the membrane formation on-line [25-28]. Sodium alginate and chitosan were employed as a first model system [25-27] but also other polycations have been screened such as, for example, polyvinylamine and poly(L-lysine)[28].

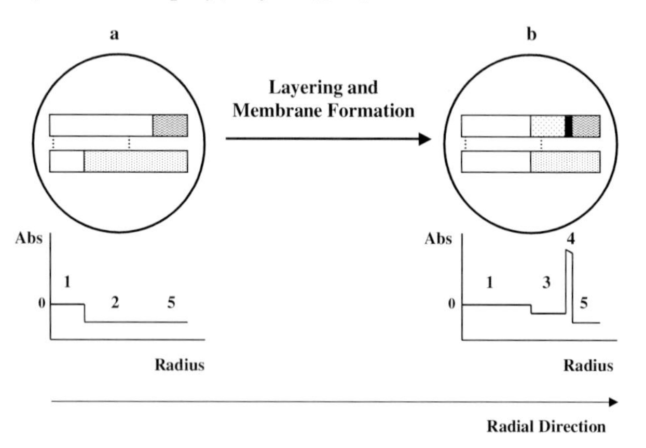

Figure 5. Principle of the membrane formation experiment in a synthetic boundary cell if a polyanion solution, in the solution sector (upper sector), is layered with a polycation solution, in the solvent sector (lower sector). The position of the components and the absorption scan are shown for the case before the layering has been started (a) and when the layering is completed (b). The five ranges monitor the absorbance difference signals (solution sector/solvent sector) of air/air (1), air/polycation solution (2), polycation solution which passed the capillary/polycation solution (3), polyelectrolyte complex membrane/polycation solution (4), polyanion solution/polycation solution (5). Here, the polycation in the solvent sector absorbs, whereas the polyanion in the solution sector does not. (This holds for the case of modified chitosan as the polycation. For other polycations forming complexes with alginate the membrane signal may result only from the turbidity.)

Polyelectrolyte membranes require for their formation that the contact surface of the polyelectrolyte solutions is well defined, permitting the undisturbed interpenetration of the reaction partners. The contact surface may be flat (layering of two solutions), or spherical (dropping one solution into a second one). To investigate the polyelectrolyte complex membrane formation one of the principal techniques of AUC, the synthetic boundary experiment, was modified. The less viscous polyion solution possessing, in addition, a lower density, was filled in the solvent sector and layered over the higher viscous polyion solution in the solution sector. The horizontal projection (view from the above) of a synthetic boundary centrepiece and the principle of the experiment are illustrated in Figure 5. Additionally, the appropriate absorbance plot is presented schematically in this figure. The membrane signal results from the absorption when the light beam passes the polyelectrolyte complex network. It has to be mentioned that the absorbance signal can be caused by the membrane turbidity, the absorption of the polymer components, or a combination of both. The absorbance signals as shown in Figure 6 for the membrane formation of alginate with poly(vinylamine) and poly(L-lysine), in combination with the interference signals (not shown herein), finally permit to evaluate the membrane in regards to membrane growth, thickness, compactness, homogeneity, and symmetry.

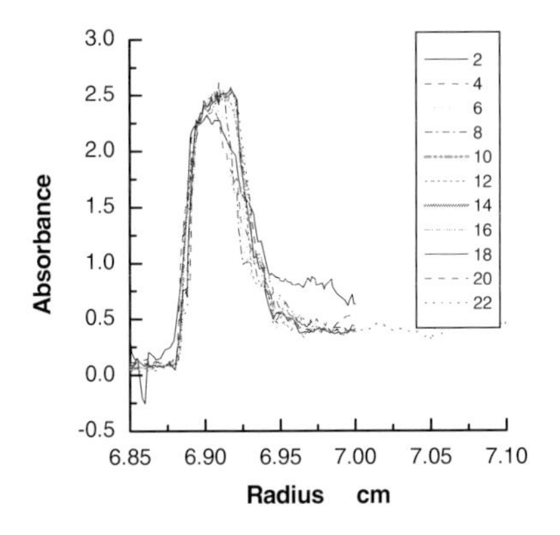

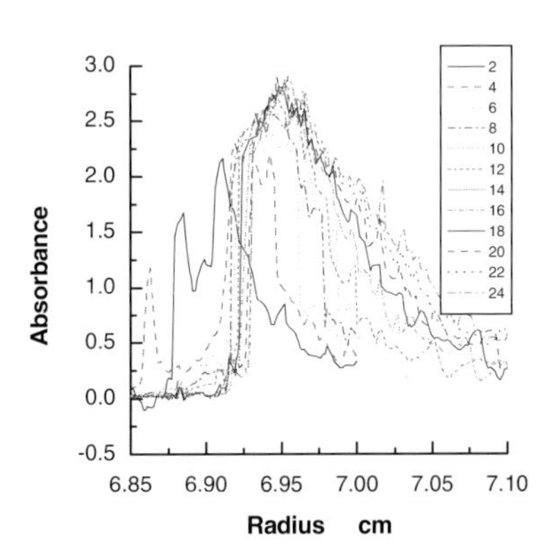

Figure 6. Detection of the membrane formation by absorbance scans for various capsule chemistries under otherwise identical conditions. Left: Poly(vinylamine)/calcium chloride layered onto alginate. Right: Poly(L-lysine)/calcium chloride layered onto alginate. (Scans were taken every two minutes). [28]

As it is clearly visible the membrane formation with poly(vinylamine) reveals, under the selected experimental conditions, more homogeneous, symmetrical and thinner membranes than with poly(L-lysine). The information content of the absorbance scans from the AUC experiment have been confirmed by microscopy as visualized in Figure 7 where the heterogeneous structure of the poly(L-lysine) membrane removed from the synthetic boundary cell after finishing the experiment is detected.

As a conclusion, the AUC method permits to evaluate in model experiments the influence of molecular characteristics such as, for example, charge density, molar mass/chain length, chemical structure, type of the ionic group, chain architecture, polydispersity, and medium conditions such as, for example, concentration, pH, ionic strength, temperature. As such the method provides a useful tool in capsule development. The authors believe that methods which probe the membrane structure, during formation, along with those which can both accurately and precisely assess transport properties (Section 4) are the principal requirements in rationalizing capsule design based on pre-existing biocompatible materials.

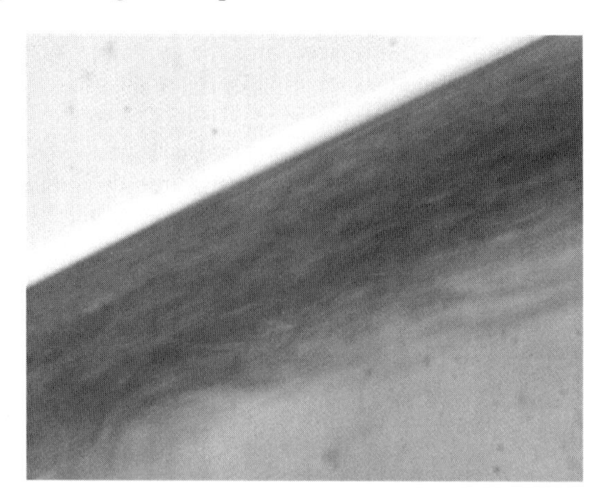

Figure 7. Photomicrograph of a poly(L-lysine)/alginate membrane on a calcium alginate gel. The image corresponds to the last scan in Figure 5 (right) taken after 24 minutes. [26]

4. Mass transport properties of microcapsules

Microcapsule used in the final application comprises of the encapsulated, active material (cells, macromolecules or any other objects) entrapped in the permeable matrix, coated (or not) by a semipermeable membrane or other protective layers. It is a crucial point to know how the encapsulation matrix behaves relative to the chemical and biological substances of interest; *e.g.* what is the mass transfer rate and the extent of penetration of the matrix by molecules of different molecular mass and structure. In the case of cells entrapped inside the gel matrix the non-restricted inflow of nutrients (ingress of glucose, oxygen, amino-acids *etc.*) and outflow of metabolites and cell-released factors (egress of *e.g.* insulin, neuro-transmitters, hormones *etc.*) is necessary for cell survival, proliferation and fulfilling of desired function in bioencapsulation. Therefore, first the selected methods of the experimental determination of mass transport properties, as applied to microcapsules, will be summarized below. Furthermore, some examples of mass transport characteristics of materials used in cell encapsulation will be given.

4.1. EXPERIMENTAL METHODS OF MASS TRANSPORT EVALUATION

As it was already described in this book, microcapsules are made mostly from water-soluble polymers and comprise of gel matrix (core), surrounded (or not) by a distinct membrane layer (shell), both core and shell being of hydrophilic nature. Thus, diffusion of substances inside a hydrophilic, loose and porous microcapsule gel structure is the most important process in evaluation of mass transport. For characterization of mass transport properties, diffusion in microcapsules is usually investigated in simple experimental systems. The closed system, comprising of known volumes of microcapsules and solution of molecular probe (marker) with its predefined initial concentration, is often used. After contacting microcapsule sample with marker solution, the process of mass transport is observed, mostly by measuring marker concentration in the solution in definite time intervals and, thus, by recording experimental points of a kinetic curve of concentration in time of the process. A schematic illustration of the typical experimental system for evaluation of mass transport during ingress process in two phases of experiment: in initial state and in final, equilibrium state, is presented in Figure 8.

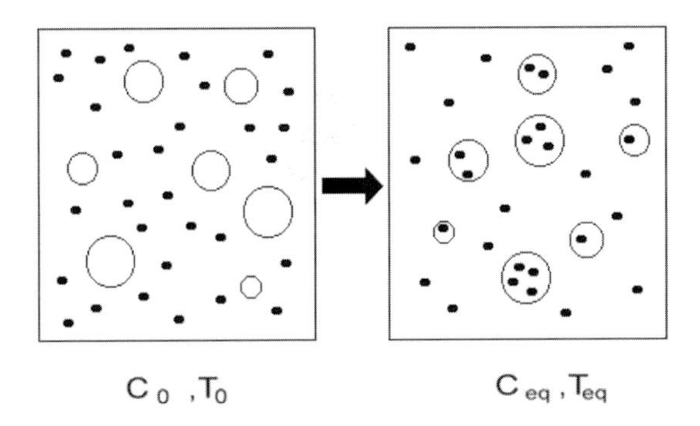

Figure 8. A schematic illustration of a closed system for the evaluation of mass transport during mass ingress experiment in two situations: in the initial state (t=0) with initial marker concentration in the solution C_0 (left) and in the equilibrium state ($t \to \infty$) with the marker molecules distributed between microcapsules and solution in the equilibrium concentration C_{eq}.

The analysis of kinetic curves of substance ingress and egress into and out off microcapsules provides the fundamental information needed for the characterization of their mass transport properties. Kinetic curves for various alginate microbeads, presented as the dimensionless substance concentration (relative to its initial or final value) in function of process time, are exemplified in Figure 9. Slow ingress of immunoglobulin G (157 kDa) from 0.2 M Tris buffer, pH=7.4, into alginate beads with the equilibration time longer than 20 hours can be observed (left part of Figure 9, upper curve) [29]. On the other hand, the low molecular mass marker vitamin B_{12} (1355 Da), attains an equilibrium distribution in less than 30 min (left part of Figure 9, lower curve). This result is obtained in the system comprised of vitamin B_{12} saline solution, pH=7.0, and alginate microbeads with the volume of solution V and volume of microbeads V_S in 1:1 ratio [30]. For such a system a rapid decrease in marker concentration is observed, with 50% of initial marker mass diffused into microbeads. Thus, the free diffusion of this low molecular mass marker inside gel volume and the total accessibility of gel interior for marker molecules can be expected in this case. These assumptions are confirmed in egress experiment (right part of Figure 9, upper curve) where vitamin B_{12}, diffusing out of alginate microbeads by the same $V{:}V_S$=1:1 volume ratio, reaches the predicted for equidistribution equilibrium concentration value C_{eq} (50% of initial concentration) in, practically, the same time. A different behaviour (slow diffusion process) is observed in the same system for the substance with larger molecular mass, interleukin 2 (15 kDa) - after two and half hour the egress process is still far from being completed (right part of Figure 9, lower curve).

The information, collected during the measurements of kinetic curves for various marker substances, is consecutively used for the determination of mass transport indexes, based on the assumption that diffusion is a dominating process in ingress and egress experiments. There are three common indexes of microcapsule mass transport properties: molecular mass cut-off value (MMCO), diffusion coefficient and mass transfer coefficient. The first index is usually based on a part of information gathered in marker kinetic experiments, namely on marker equilibrium concentration and total amount of mass diffused into the microcapsules. In the evaluation of diffusion and mass transfer coefficients, on the other hand, the marker concentration course in time is important for the determination of their values.

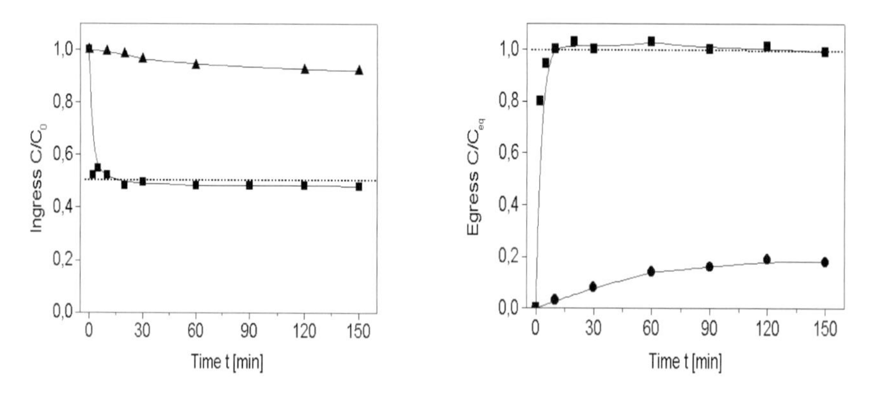

Figure 9. Examples of kinetic curves for various alginate microbeads and marker substances. Left: ingress of immunoglobulin G (157 kDa)-upper curve, ingress of vitamin B_{12} (1355 Da). Right: Egress of vitamin B_{12} (upper curve) and interleukin-2 (lower curve).

To assess to what extent different molecules can penetrate the microcapsule matrix and to estimate its exclusion limit, the MMCO characteristics should be evaluated i.e. amounts of mass of different markers present inside microcapsules in function of marker molecular mass must be known. The most common method is the investigation of diffusive mass ingress (and egress) from solution of the given substance, treated as a molecular probe, into the microcapsules [31]. In this case the amount of substance, which diffused into microcapsules, until an equilibrium state in the system has been attained, is evaluated. In the simple experimental setup, comprising of the known volumes of microcapsules and marker solution (usually in 1:1 volume ratio), the equilibrium concentration C_{eq} of a marker is determined. If the initial marker concentration C_0 is known, the mass ingress relative to the initial mass in the system

Ing, expressed in percents (percent mass ingress), can be evaluated from the expression (1):

$$Ing = (C_o - C_{eq}) / C_o x100\% \qquad (1)$$

A similar equation can be adapted for the investigation of marker egress when the substance applied as a marker is concentrated inside microcapsules at the beginning of the diffusion process. To characterize extensively the diffusive properties of microcapsule matrix a wide spectrum of markers should be used. Various synthetic and natural substances are applied for this purposes, including copper ions [32,33], low molecular mass markers like glucose [29,34], urea, creatinine, vitamin B_{12}, synthetic polysaccharides like dextrans of different molecular mass (from 10 kDa up to 220 kDa), pullulans [30], polyethylene glycols, cytokines like interleukin-1 [35], natural proteins like ovalbumin, albumin, gamma immunoglobulins [29,35] *etc.* Collection of mass ingress data for markers of different molecular mass enables one to obtain the Molecular Mass Cut-Off (MMCO) curve for an investigated microcapsule type, which represents the dependence between percent mass ingress and molecular mass of a marker. Such curves are exemplified in Figure 10.

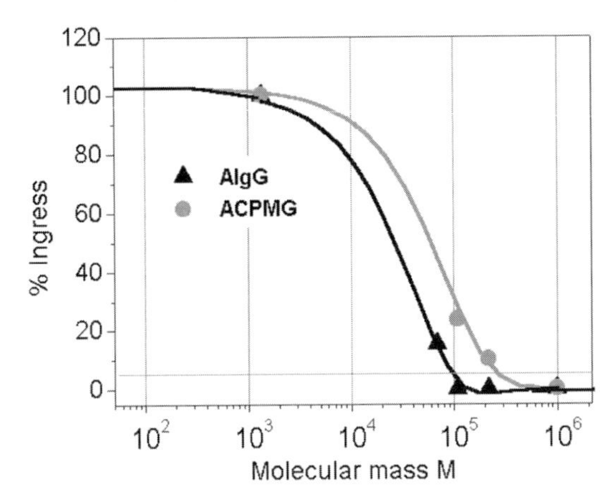

Figure 10. The example of estimation of molecular mass cut-off curves for two different microcapsule types: high-guluronic content alginate microbeads (AlgG, left curve) and membrane coated microcapsules with poly(methylene-co-guanidine) based shell and mixed alginate-cellulose sulphate core (ACPMG, right curve) [36].

The expected value of the marker molecular mass at which the percent mass ingress decreases below 5% (MMCO value) is commonly assumed to be an index of microcapsule permeability, pointing to its exclusion limit. Concerning the example in Figure 10, MMCO for alginate microbeads of high guluronic acid content (AlgG) can be estimated to be around 100 kDa while for microcapsules comprised of mixed alginate-cellulose sulphate gel core and poly(methylene-co-guanidine) based shell (ACPMG the respective value is 210 kDa approximately. The experimental results were obtained using vitamin B_{12} and a mixture of dextran standards (70, 110 and 220 kDa)

[36]. A similar investigation of MMCO characteristics for the same ACPMG microcapsules, conducted using pullulan standards (50, 110, 210, 400 and 790 kDa), yielded MMCO value around 180 kDa [30].

A more fundamental approach to the problem of microcapsule permeability is based on the investigation of diffusional properties of the matrix for a given substance or marker. As diffusion is a time-dependent process, the values of marker ingress (or egress) in function of process time are investigated as a rule. The starting equation is the second Fick's law for the diffusion inside a sphere:

$$\frac{\partial C}{\partial t} = D_{eff}\left(\frac{\partial^2 C}{\partial r^2} + \frac{2}{r}\frac{\partial C}{\partial r}\right) \tag{2}$$

where C is the marker concentration inside a sphere, r is the radial coordinate and D_{eff} is the diffusion coefficient averaged over entire sphere volume (a possible spatial variability of diffusion coefficient is not taken into account). Various solutions of this equation for different sets of initial and boundary conditions are known [37]. Diffusion profiles inside microcapsule (*i.e.* concentration in function of microcapsule radius and time) can be determined. More important, on assumptions of good mixing in the solution phase and of negligible mass transfer resistance in the boundary layer around microcapsule, the theoretical concentration course in time in the solution during diffusion process can be determined and fitted to the experimental data. The solution of Equation 2, often used in the characterization of microcapsules by ingress method (the corresponding solution for egress of marker can be found elsewhere [34,37]), is represented by the following equation:

$$C_S / C_0 = \frac{\alpha}{1+\alpha}\left(1 + \sum_{n=1}^{\infty} \frac{6(6+\alpha)}{9+9\alpha+\alpha^2 q_n^2} \exp\left(-D_{eff}\frac{q_n^2}{R^2}t\right)\right) \tag{3}$$

The value of C_S is the marker concentration in the solution phase relative to its initial concentration C_0, α is the constant parameter specified only by the volumes of solution and microcapsules V and V_S, respectively, and the partition coefficient K ($K=(C_{eq})_{microcpsule}/(C_{eq})_{solution}$). Assuming that the investigated sample can be represented by N equal spheres with radius R, parameter α can be expressed as:

$$\alpha = \frac{V}{V_S}\frac{1}{K} = \frac{3V}{4\pi R^3 NK} \tag{4}$$

whereas q_n are the consecutive positive roots of the following functional equation:

$$\tan(X) = \frac{3X}{3+\alpha X^2} \tag{5}$$

where X is the variable and root values are tabularized for different α values in [37].

The value of the effective diffusion coefficient D_{eff}, which can be determined, using Equation 3, is a measure of the diffusive properties of the microcapsule matrix for a marker, especially when compared with (usually known) the diffusion coefficient for a free diffusion of the marker in a solution. The experimentally determined diffusion coefficient values decrease, as a rule, with increasing molecular mass of a marker. For cations like calcium and copper D_{eff} values for diffusion in typical alginate microbeads (alginate concentrations in the range of 0.5-2%) are only slightly smaller than the respective values in water that are about 10^{-5} cm^2/s [33,38]. In the same type of gels for the low molecular mass marker, vitamin B_{12}, the values of D_{eff} smaller than for a free diffusion in saline were found [39]. Other authors [40] report the values below 3.8×10^{-6} cm^2/s (diffusion in water) for vitamin B_{12} diffusion in polyvinyl alcohol (PVA) and polyacrylate (PAA) gels, with the partition coefficient K bigger than 1. For high molecular mass substances like immunoglobulin G (IgG), diffusing into alginate beads, D_{eff} values of an order of magnitude smaller than for IgG diffusion in buffers (about 10^{-7} cm^2/s) can be evaluated (based on data from [29]).

The diffusion coefficient is an established index of microcapsule permeability and can be used widely for its characterization. Nevertheless, the reported diffusion coefficient values should be always treated with caution, as there are different experimental systems used for D_{eff} evaluation. In principle, the above described method should be used only when diffusion is the most important process and the influence of other phenomena (*e.g.* adsorption) can be neglected. There are some empirical means to check whether diffusion is a dominating process. The first rough test is to see, using a neutral, low molecular mass marker, whether the equidistribution (a half of marker mass in capsules and a half in solution) is attained in the system with 1:1 volume ratio in the equilibrium state. Another method is to check the identity of the determined diffusion coefficient values for selected marker when examining mass ingress and egress. The third method is based on the thermodynamic description of diffusion and adsorption processes and evaluation of the activation energy. If the ingress experiment is conducted twice in two different temperatures T_1 and T_2 ($T_1 < 298$ K $< T_2$) and two corresponding values of diffusion coefficient D_1 and D_2 are determined, the activation energy E_A can be calculated using the following formula:

$$E_A = \mathrm{R}\frac{T_1 T_2}{T_1 - T_2}\ln\left(\frac{D_2}{D_1}\right) \qquad (6)$$

where R is the gas constant. The activation energy E_A for pure diffusion process should lay in the range of 16-20 kJ/mol in temperature 298 K [41], otherwise other phenomena can accompany diffusion (*e.g.* adsorption).

Another method of characterizing the mass transport, though less often used in the case of microcapsules, is the determination of the value of mass transfer coefficient h. In contrast to the diffusion coefficient determination, this method does not need to be limited to the pure diffusion processes. For a general description of processes, including adsorption from dilute solutions, diffusion, first order chemical reaction *etc.*, where the concentration difference between a capsule surrounding and capsule interior close to interphase boundary is the predominant driving force the following equation of mass transport can be used [42]:

$$V\frac{dC_S}{dt} = -hS(C_S - C) \quad (7)$$

where V is the volume of solution phase, S is the surface area of interphase boundary, C_S is the marker concentration in the solution and C is the concentration inside microcapsule close to an interphase boundary while h designates the mass transfer coefficient. As C is generally unknown during the process, Equation 7 can be solved in some cases with various simplifying assumptions [36]. Particularly for low molecular mass markers (well under microcapsule MMCO value), on assumption of full accessibility of microcapsule interior, the following solution can be given:

$$C_S / C_0 = C_{eq} / C_0 + (1 - C_{eq} / C_0)\exp(-\frac{1}{V} + \frac{1}{V_S} hSt) \quad (8)$$

where V_S is the total volume of microcapsules and C_{eq} is the respective equilibrium concentration in the experimental system. Determination of h gives an idea about how fast the molecules of a given substance cross an interphase boundary and, thus, about an effectiveness of mass transport for the investigated microcapsules. To give an example, determination of h values for vitamin B_{12} in the series of investigated calcium alginate microbeads, manufactured from alginates with low and high guluronic acid content and with average bead diameters in the range of 0.5-2.5 mm, yielded results between $5x10^{-5}$ and $7x10^{-4}$ cm/s [36]. Mass transfer coefficient for immunoglobulin G ingress into alginate beads was of order of 10^{-6} cm/s and decreased down to about $0.3x10^{-6}$ cm/s after putting an immunoisolation aminopropyl-silicate membrane around alginate (unpublished calculations based on results of Sakai *et al* [29]).

In comparison with D_{eff} method, determination of mass transfer coefficient h is straightforward and can be easily implemented for quantitative calculations. The method can also serve as the screening test for the choice of proper marker substances. Whereas the diffusion coefficient is in principle a local parameter, mass transfer coefficient characterizes mass transport in general terms. Both are fundamental indexes for the characterization of mass transport properties of microcapsules.

4.2. SUMMARY OF MASS TRANSPORT PROPERTIES OF MICROCAPSULES

As the majority of cell encapsulation matrices is still based on various alginate gel formulations, the following review will concentrate on this material. There are several factors influencing mass transport properties of microcapsules, including:

- microcapsule material (core), including the role of concentration and chemistry of the polymer used, crosslinking agent, homogeneity *vs* nonhomogeneity of the capsules,
- microcapsule membrane (outer layer)
- microcapsule size

In addition to the aforementioned effects, the type of selected marker can influence the results and their interpretation.

It is known, for many years, that natural polysaccharide gels demonstrate enhanced, in comparison with porous synthetic polymer matrices, permeability for low and high molecular mass substances, even up to several hundred kilo Daltons. Tanaka *et al* [32] have shown that low molecular mass substances (glucose, tryptophan and α-lactalbumin), below 20 kDa, diffuse freely in 2% alginate beads. They have not observed any significant diffusion of albumin (69 kDa), γ-globulin (154 kDa) and fibrinogen (341 kDa) into alginate beads but were able to determine diffusion coefficients for these markers in egress diffusion. The determined values were: $3.5\text{x}10^{-7}$ cm^2/s, $2\text{x}10^{-7}$ cm^2/s and $3.3\text{x}10^{-8}$ cm^2/s respectively and amounted to about 50%, 45% and 17% of diffusion in water. They were also able to show that diffusion of these markers decreases sharply in more concentrated gels, made from 4% alginate. This influence of alginate concentration on molecular diffusion was further confirmed for many markers, including small substances like copper ions. For the series of concentrated alginate gels, with concentration ranging between 1.5% and 21%, the diffusion coefficient D_{eff} for Cu^{2+} decreased from 10^{-5} cm^2/s down to $0.5\text{x}10^{-5}$ cm^2/s [33] in egress experiments.

In addition to the polymer concentration, the polymer chemistry influences mass transport properties, though to a lesser extent. The polymer chain in alginate is built from guluronic acid (G) and mannuronic acid (M) monomers, occurring in different sequences and with different relative contents, depending on alginate source and type. This variety of alginate chain structures results in different alginate gels with various porosities and pore size distributions and, thus, different diffusive properties [43]. Although there is an evidence that the porosity of alginate gels increases with the content of guluronic acid in the polymer and, especially, with G-block length [43,44], other reports point to a larger permeability of high-M alginates in comparison with high-G alginates [20]. Referring to the role of polymer chemistry in the formation of the final internal structure of a gel bead, the role of crosslinking agent should be also stressed. Although the concentration of calcium during external gelation does not seem to be as important as polymer concentration [34], the type of crosslinker is important. Barium alginate (or mixed barium-calcium) capsules are known to offer a better mechanical resistance but on cost of diffusive properties [45].

It is not true that the same kind of alginate, at a given concentration, would always give the same kind of gel [45]. Much depends on kind of gelation process and its kinetics. An example is the difference between homogeneous and nonhomogeneous gels [46]. Homogeneous gels can be obtained by dispersion of an inactive crosslinking agent inside hydrophilic polymer sol droplets, suspended in an oil phase, and subsequent activation of the agent [47]. Nonhomogeneous gels, on the other hand, are formed in the process of external gelation and have the polymer concentration decreasing inwards (denser shell with less dense core). Such radial concentration profiles were calculated theoretically [38] and evidenced experimentally using fluorescence markers and laser confocal microscopy [48]. It has been demonstrated, by release of bovine serum albumin (BSA) from both types of gels, that BSA diffusion was faster for beads with uniform alginate distribution than for beads where alginate was concentrated near the outer surface [44]. Better permeability for homogeneous capsules was also confirmed by results of measurements of haemoglobin diffusion from erythrocytes encapsulated in both types of gels and subjected to osmotic shock [45].

The most important factor, concerning microcapsule applications, is the membrane layer formed around an internal gel core. The primary role of the membrane is to form a selective shell, restricting ingress of high molecular mass substances into microcapsule interior. Whereas the MMCO value for typical alginates can be on the order of 230 kDa [49], placing e.g. poly(L-lysine) (PLL) membrane around alginate beads can reduce it to approximately 150 kDa [50]. To highlight the influence of the presence of membrane on global mass transport properties of a microcapsule, the examples will be presented concerning the following topics:

- microcapsule with polyanion (alginate) core and polycation (PLL) membrane,
- microcapsule with chemically modified polyanion (alginate) and polycation (PLL) membrane,
- microcapsule with polyanion (alginate) core and synthetic polymer membrane,
- microcapsule with mixed polymer core (alginate + cellulose) and polycation (poly(methylene-co-guanidine)) membrane.

Among microcapsules, based on alginate gel core, the most widely investigated formulation is the alginate-PLL-alginate (APA) microcapsule. APA manufacturing according to the classical procedure of Sun *et al.* [3,51] comprises of the following main steps: formation of alginate gel core, formation of PLL membrane during reaction between alginate outer layer and PLL in a solution bath, formation of external alginate layer (for biocompatibility purposes) by washing in dilute alginate solution and, optionally, liquefying of internal alginate core by washing capsules in the solution of cation sequestering agent (*e.g.* sodium citrate for complexing calcium ions). Considering the PLL membrane, *e.g.* Goosen *et al.* [52] were able to show that MMCO values of the resultant microcapsules decrease with increasing PLL concentration and PLL-alginate reaction time. On the other hand, MMCO was found to increase with increasing average molecular mass of PLL, used for membrane formation. The relationship of such type between the molecular mass of membrane forming substance and a global microcapsule permeability was reported in many publications. For example, Thu *et al.* [45] have found that leaching of unbound, low molecular mass fragments of alginate (below 100 kDa) from APA microcapsules is much more marked for membrane made of 38 kDa PLL than for 18 kDa PLL membrane. The same authors evidenced experimentally, by investigation of kinetics of BSA and haemoglobin release, that, even with the same PLL based membrane, microcapsules with liquid core offer a better permeability than hard core ones.

The significant influence of microcapsule core content and its structure on membrane and global mass transport properties of microcapsule can be illustrated *e.g.* by results of investigation of the influence of chemical modifications of alginate core on capsule permeability, obtained by Lee and Chu [53]. For APA-type microcapsules made of LV (low viscosity) alginate and polydisperse PLL (molecular mass in the range of 30-70 kDa) the diffusion coefficient of BSA was found to be about $7x10^{-8}$ cm^2/s in ingress experiments (in comparison with $6.8x10^{-7}$ cm^2/s BSA diffusion in water). Microcapsules with the same membrane and different gel core, modified by inclusion of polyethylenimine (PEI) into alginate matrix, demonstrated lower permeability, with the D_{eff} value decreasing to $5x10^{-8}$ cm^2/s. On the other hand, increased permeability to BSA was observed when, in the first step of microcapsule formulation, the core was formed from carboxyl methylo cellulose (CMC) and alginate in 6:1 weight ratio. The authors found that the core was liquid, due to low content of alginate, and D_{eff} of BSA increased up to about $15x10^{-8}$ cm^2/s (more than in a typical ACA microcapsule). These and other findings strongly suggest that a microcapsule membrane, surrounding an internal gel core, can not be considered in separation from the core material.

Problems with biocompatibility (*e.g.* fibrotic overgrowth after transplantation on PLL containing microcapsules [54]) and demands concerning permeability, to be satisfied in various applications, motivate to search for other, alternative to polycation PLL, membranes. Different aminopropyl-silicate membranes [29], synthesized from two precursors at different molar ratios and formed on calcium alginate micro beads yielded an average MMCO value of about 60 kDa (investigated using marker series: glucose, myoglobin, ovalbumin, BSA and γ-globulin) for different capsule formulations. By varying the membrane precursors ratio is was possible to modulate microcapsule permeability for γ-globulin from total exclusion to a restricted permeability with D_{eff} circa 40 times smaller than the respective D_{eff} value for uncoated alginate beads (about 10^{-7} cm^2/s, calculations based on results in [29]).

The search for new microcapsules resulted also in introduction of a capsule with mixed alginate-cellulose sulphate core and poly(methylene-co-guanidine) based membrane (ACPMG) [55] as an alternative to a classical polyanion-polycation APA capsule. Generally, the capsule permeability can be controlled by changing the polyanion concentration and polycation composition, or via additional treatment with a secondary polycation. Taking advantage of these possibilities, a series of capsules with different MMCO values in the range from 6.5 kDa up to 1770 kDa (as determined using protein standards) was manufactured [35]. Selecting two types of ACPMG microcapsules, with 3.2 kDa and 230 kDa exclusion limits, the authors were able to demonstrate restricted and free ingress of interleukin-1 to both types of capsules respectively. Interleukin-1 and immunoglobulin G ingresses were analysed by enclosing Protein A-bound antibodies inside capsules and applying size exclusion chromatography for determination of marker concentrations. Thus, in this type of capsule it is possible to influence to a great extent the resultant permeability of entire microcapsule by modulating membrane properties together with the properties of polymer core.

An important factor also influencing the assessment of microcapsule permeability, and often not taken into account, is the microcapsule size (one should note, in reference to Section 1.1, that capsule size was not significant in the determination of mechanical properties. Dependence of various microcapsule properties, including mass transport indexes, on its size was the topic of interlaboratory study in the frames of the European project COST 840 [28]. For the series of alginate microbeads with different G and M content and ACPMG microcapsules in the range of sizes between 0.25 mm and 2.0 mm, it was possible to find a positive correlation between diffusion coefficients for vitamin B_{12} and capsule size. Furthermore, no significant differences were observed between microbeads and microcapsules, made from various materials [39]. An additional information can be obtained by the application of the method of determination of mass transfer coefficient h for vitamin B_{12}. The method not only enabled one to reveal the increase of this mass transport index with capsule size but also to differentiate the capsules (Figure 11), made from different materials and by different methods, in separate groups according to their particular h/D index [36] (D is the capsule diameter).

Application of this method should offer a possibility of convenient capsule classification and comparison of samples of different sizes.

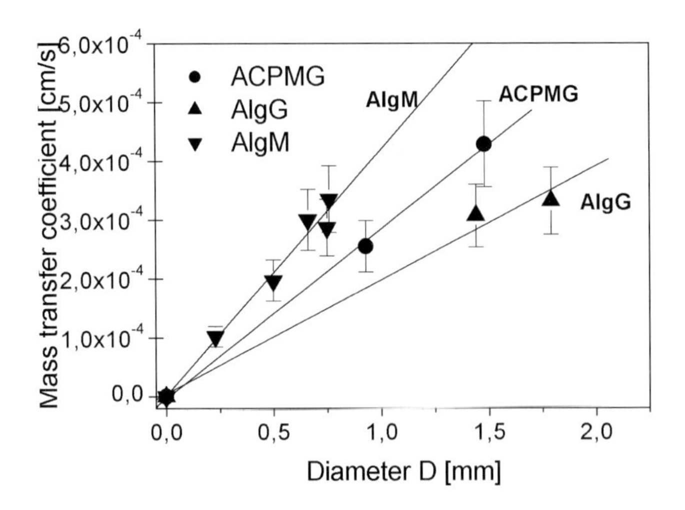

Figure 11. Relationship between mass transfer coefficient h values and capsule size. Vitamin B_{12} diffusion into high-M alginate beads (AlgM, ▼), high-G alginate beads (AlgG, ▲) and mixed alginate-cellulose sulphate core poly(methylene-co-guanidine) coated microcapsules (ACPMG, ●) [36].

A final, though still critical, factor in permeability assessment is the choice of marker substances. Application of different markers can give contradictory results. Briefly, for the same microcapsule an application of the series of *e.g.* dextran molecular mass standards to the evaluation of MMCO can give the result 100% different from the result obtained using protein molecular mass standards. This topic, concerning the problem of marker standardization, is highlighted by Powers, Brissova, Lacik *et al.* in [35]. They have found a correspondence between different dextran standards and protein standards with different molecular masses, based on comparison of their molecular 3D-conformations especially when comparing viscosity radiuses R_η of different molecules. In the light of these results, *e.g.* a dextran sample with molecular mass of 44 kDa is equivalent in diffusion experiment to protein sample with molecular mass of 200 kDa due to the similar value of R_η on the order of 5.1 nm. Thus, concerning investigations of microcapsule diffusivity for a selected marker, the markers should be compared to each other with respect to their equivalent dimensions in solution rather than with respect to their molecular masses.

The following conclusions are based on various cooperative experimental projects, as well as an analysis of the literature.

- There are established methods of evaluation of microcapsule mass transport properties – molecular mass cut-off and diffusion coefficient determination.
- Mass transfer coefficient, related to a capsule size, is a valuable index in characterization of microcapsules with different sizes.

- A broad spectrum of markers or at least multimolecular mass standard mixtures should be used for microcapsule characterization; care must always be taken concerning the marker mass, size and conformation in the interpretation of results.
- Single and multi-membrane layers can be formed to control the entire microcapsule permeability, which is also influenced by modification to the internal capsule core.
- There is still a need for a simple and convenient method of mass transport characterization, which would enable one to consider membrane and core in more details – present methods treat a microcapsule as a single entity with averaged properties (*e.g.* diffusion coefficient).
- Development of new imaging markers and procedures to reveal microcapsule structure would offer new possibilities when integrated with the evaluation of mass transport indexes.
- Analytical ultracentrifugation has emerged as a technique to examine, on-line during membrane formation, the kinetics as well as structure of the semipermeable barrier.
- Challenges remaining in encapsulation include the need for standardized polymeric materials as well as reproducible protocols for cell isolation.

References

[1] Chang, T.M.S. (1964) Semipermeable microcapsules. Science 146: 524-525.

[2] Lim, F. and Sun, A.M. (1980), Microencapsulated iselts as bioartificial pancreas. Science 210: 908-910.

[3] Strand, B.L.; March, Y.A.; Rokstad, A.M.; Kulseng, B.; Espevik, T. and Skjåk-Bræk, G. (2001) Improved capsule properties by the use of epimerised alginate. In: Proceedings of the IX International BRG workshop, Warsaw, 11-13 May 2001; S.I-1 (copies available at bioenc@ibib.waw.pl).

[4] Gaserød, O. (1998) Microcapsules of alginate-chitosan: A study of capsule formation and functional properties. Ph.D. Thesis, Norwegian University of Science and Technology, Trondheim, Norway.

[5] Ma, X.; Vacek, I. and Sun, A. (1994) Generation of alginate-polylysine-alginate (APA) biomicrocapsules: The relationship between the membrane strength and the reaction conditions. Art. Cells, Blood Subst., and Biotechn. 22: 43-69.

[6] Renken, A. (2000) Formation of microcapsules by reaction of polyanion blends with divalent cations and oligocations. Ph.D. Thesis, Swiss Federal Institute of Technology, Lausanne, Switzerland.

[7] Gugerli, R. (2003) Polyelectrolyte-complex and covalent-complex microcapsules for encapsulation of mammalian cells: potential and limitations. Ph.D. Thesis, Swiss Federal Institute of Technology, Lausanne, Switzerland.

[8] Thu, B.; Bruheim, P.; Espevik, T.; Smidsrød, O.; Soon-Shiong, P. and Skjåk-Bræk, G. (1996) Alginate polycation microcapsules: I. Interaction between alginate and polycation. Biomaterials 17: 1031-1040.

[9] Poncelet, D. and Neufeld, RT. (1989) Shear breakage of nylon membrane microcapsules in a turbine reactor. Biotechnol. Bioeng. 33: 95-103.

[10] dos Santos, V.; Leenen, E. and Rippoll, M. (1997) Relevance of rheological properties of gel beads for their mechanical stability in bioreactors. Biotechnol. Bioeng. 56: 517-529.

[11] Lu, G.Z.; Thompson, F.G. and Gray, M.R. (1992) Physical modelling of animal cell damage by hydrodynamic forces in suspension cultures. Biotechnol. Bioeng. 40: 1277-1281.

[12] Chen, J. and Park, S. (1995) Polysaccharide hydrogels for protein drug delivery. Carbohydrate Polym. 28: 69-76.

[13] Uludag, H.; De Vos, P. and Tresco, P. (2000) Technology of mammalian cell encapsulation. Adv. Drug Delivery Rev. 42: 29-64.

[14] Wang, J. (2000) Development of new polycations for cell encapsulation with alginate. Mater. Sci. Eng. C: 59-63.

[15] Peirone, M.; Ross, C.; Hortelano, G. and Brash, J. (1998) Encapsulation of various recombinant mammalian cell types in different alginate microcapsules. J. Biomed. Mater. Res. 42: 587-596.
[16] Martinsen, A.; Skjåk-Bræk, G. and Smidsrød, O. (1989) Alginate as immobilization material: I. Correlation between chemical and physical properties of alginate gel beads. Biotechnol. Bioeng. 33: 79-89.
[17] Schoichet, M.; Li, R.; White, M. and Winn, S. (1995) Stability of hydrogels used in cell encapsulation: An *in vitro* comparison of alginate and agarose. Biotechnol. Bioeng. 50: 374-381.
[18] Zhang, Z.; Saunders, R. and Thomas, C.R. (1999) Mechanical strength of single microcapsules determined by a novel micromanipulation technique. J. Microencapsulation 16: 117-124.
[19] Rehor, A.; Canaple, L.; Zhang, Z. and Hunkeler, D. (2001) The compression deformation of multicomponent microcapsules: Influence of size, membrane thickness and compression speed. J. Biomater. Sci., Polym. Ed. 12: 157-170.
[20] Wandrey, C.; Espinosa, D.; Rehor, A. and Hunkeler, D. (2002) Influence of alginate characteristics on the properties of multi-component microcapsules. J. Microencapsulation. In press.
[21] Andrei, D.; Briscoe, B.J. and Williams, D.R. (1996) The deformation of microscopic gel particles. J Chim. Phys. 93: 960-976.
[22] Liu, K.K.; Williams, D.R. and Briscoe, B.J. (1996) Compressive deformation of a single microcapsule. Phys. Rev. 54: 6673-6680.
[23] Ohtsubo, T.; Tsuda, S. and Tsuji, K (1990) A study of the physical strength of microcapsules. Polymer 32: 2395-2399.
[24] Prokop, A.; Hunkeler, D.; Haralson, M.; Dimari, S. and Wang. T.G. (1998) Water soluble polymers for immunoisolation I: Complex coacervation and cytotoxicity. Adv. Polym. Sci. 136: 1-54.
[25] Wandrey, C.; Bartkowiak, A. and Hunkeler, D. (2000) Development of biomembranes by utilizing analytical ultracentrifugation techniques. In: Transact. 6th World Biomat. Congr. Vol II; p. 893.
[26] Wandrey, C. (2000) Study of polyelectrolyte complex formation by analytical ultracentrifugation. Polym. News 25: 299-301.
[27] Wandrey, C. and Bartkowiak, A. (2001) Membrane formation at interfaces examined by analytical ultracentrifugation techniques. Colloids Surf. A 180: 141-153.
[28] Wandrey, C.; Grigorescu, G. and Hunkeler, D. (2002) Study of polyelectrolyte complex formation applying the synthetic boundary technique of analytical ultracentrifugation. Progr. Colloid Polym. Sci 119: 84-91.
[29] Sakai, S.; Ono, T.; Ijima, H. and Kawakami, K. (2001) Newly developed aminopropyl-silicate immunoisolation membrane for a microcapsule-shaped bioartificial pancreas. In: Hunkeler, D.; Cherrington, A.; Prokop, A. and Rajotte, R. (Eds.) Bioartifical Organs: Tissue Sourcing, Immunoisolation and Clinical Trials. Annals of the NY Academy of Sciences, Vol. 944; pp. 278-283.
[30] Rosinski, S.; Grigorescu, G.; Lewinska, D.; Ritzen, LG.; Viernstein, H.; Tenou, E.; Poncelet, D.; Zhang, Z.; Fan, X.; Serp, D.; Marison, I. and Hunkeler, D. (2002) Characterization of microcapsules: recommended methods based on round-robin testing. J. Microencapsulation 19: 641-659.
[31] Schuldt, U. and Hunkeler, D. (2000) Characterization methods for microcapsules. Minerva Biotec. 12: 249-264.
[32] Jang, L.K. (1994) Diffusivity of Cu^{2+} in calcium alginate gel beads. Biotechnol. Bioeng. 43: 183-185.
[33] Kwiatkowska, S. and Wojcik, M. (2000) Diffusion of copper (II) chloride in concentrate alginate gel. In: Proceedings of COST 840 Workshop "Structure-function properties of biopolymers in relation with bioencapsulation", December 8-10 2000, Espoo, Finland; pp. 1-4.
[34] Tanaka, H.; Matsumara, M. and Veliky, I.A. (1984) Diffusion characteristics of substrates in Ca-alginate gel beads. Biotechnol. Bioeng. 26: 53-58.
[35] Powers, A.C.; Brissova, M.; Lacik, I.; Anilkumar, A.V.; Shahrokhi, K. and Wang, T.G. (1997) Permeability assessment of capsules for islet transplantation. In: Prokop, A.; Hunkeler, D. and Cherrington, A.D. (Eds.) Bioartifical Organs: Science, Medicine and Technology. Annals of the NY Academy of Sciences, Vol. 931; pp. 208-217.
[36] Lewinska, D.; Rosinski, S.; Hunkeler, D.; Poncelet, D. and Werynski, A. (2002) Mass transfer coefficient in characterization of gel beads and microcapsules. J. Membrane Sci. 209: 533-540.
[37] Crank, J. (1975) The Mathematics of Diffusion. Clarendon Press, Oxford.
[38] Mikkelsen, A. and Elgsæther, A. (1995) Density distribution of calcium induced alginate gels. Biopolymers 36: 17-41.
[39] Rosinski, S. and Lewinska, D. (2002) Mass transfer coefficient and diffusion coefficient in characterisation of microcapsules for bioencapsulation. In: Proceedings of COST 840 Workshop "Formulation and Characterisation of Biocompatible Capsules", 30 August-1 September 2002, University of Birmingham, UK; T13, pp. 1-4.

[40] Peppas, N.A. and Wright, S. (1996) Solute diffusion in poly(vinyl alcohol)/poly(acrylic acid) interpenetrating networks. Macromolecules 29: 8798-8804.
[41] Wojcik, M. (1991) The methods of determination of the effective diffusion coefficient in biocatalyser carriers (in Polish). Biotechnologia 2: 20-27.
[42] Radcliffe, D.F. and Gaylor, J.D.S. (1981) Sorption kinetics in haemoperfusion columns. Part I: Estimation of mass transfer parameters. Med. and Biol. Eng. Comp. 19: 617-627.
[43] Martinsen, A.; Skjåk-Bræk, G. and Smisrød, O. (1989) Alginate as immobilization material: I. Correlation between chemical and physical properties of alginate beads. Biotechnol. Bioeng. 39: 186-194.
[44] Martinsen, A.; Stodrø, I. and Skjåk-Bræk, G. (1992) Alginate as immobilization material: III. Diffusional properties. Biotechnol. Bioeng. 39: 186-194.
[45] Thu, B.; Bruheim, P.; Espevik, T.; Smisrød, O.; Soon-Shiong, P. and Skjåk-Bræk, G. (1996) Alginate polycation microcapsules: II. Some functional properties. Biomaterials 17: 1069-1079.
[46] Skjåk-Bræk, G.; Smisrød, O. and Grasdalen, H. (1989) Inhomogeneous polysaccharide ionic gels. Carbohydr. Polym. 10: 31-54.
[47] Poncelet, D. (2001) Production of alginate beads by emulsification/internal gelation. In: Hunkeler, D.; Cherrington, A.; Prokop, A. and Rajotte, R. (Eds.) Bioartifical Organs: Tissue Sourcing, Immunoisolation and Clinical Trials. Annals of the NY Academy of Sciences, Vol. 944; pp. 74-82.
[48] Strand, B.L.; Mørch, Y.A.; Espevik, T. and Skjåk-Bræk, G. (2003) Visualization of alginate-poly-*L*-lysine-alginate microcapsules by confocal laser scanning microscopy. Biotechnol. Bioeng. (in press).
[49] Brissova, M.; Petro, M.; Lacik, I.; Powers, A. and Wang, T. (1996) Evaluation of microcapsule permeability *via* inverse size exclusion. Analytical Biochemistry 242: 104-111.
[50] Awrey, D.E.; Tse, M.; Hortelano, G. and Chang, P.L. (1996) Permeability of alginate microcapsules to secretory recombinant gene products. Biotechnol. Bioeng. 52: 472-484.
[51] Sun, A.M. (1997) Microencapsulation of cells. In: Prokop, A.; Hunkeler, D. and Cherrington, A.D. (Eds.) Bioartifical Organs: Science, Medicine and Technology. Annals of the NY Academy of Sciences, Vol. 931; pp. 271-279.
[52] Goosen, M.F.A.; O'Shea, G.M.; Gharapetian, H.M.; Chou, S. and Sun, A.M. (1984) Optimization of microencapsulation parameters. Semipermeable microcapsules as a bioartificial pancreas. Biotechnol. Bioeng. 27: 146-150.
[53] Lee, C.-S. and Chu, I.-M. (1997) Characterization of modified alginate-poly-*L*-lysine microcapsules. Artif. Organs 21: 1002-1006.
[54] Strand, B.L.; Rayn, L.; In't Veld, P.; Kulseng, B.; Rokstad, A.M.; Skjåk-Bræk, G. and Espevik, T. (2001) Poly-*L*-lysine induced fibrosis on alginate microcapsules *via* the induction of cytokines. Cell Transplant. 10: 263-275.
[55] Wang, T.G.; Lacik, I.; Brissova, M.; Prokop, A.; Hunkeler, D.; Anilkumar, A.V.; Green, R.; Shahrokhi, K. and Powers, A.C. (1997) An encapsulation system for the immunoisolation of pancreatic islets. Nature Biotechnol. 15: 358-362.

IMMOBILISED CELL BIOREACTORS

BOJANA OBRADOVIC[1], VIKTOR A. NEDOVIĆ[2], BRANKO BUGARSKI[1], RONNIE G. WILLAERT[3] AND GORDANA VUNJAK-NOVAKOVIC[4]

[1]Department of Chemical Engineering, Faculty of Technology and Metallurgy, University of Belgrade, Karnegijeva 4, PO Box 3503, 1120 Belgrade, Yugoslavia – Fax: 381-11-337038 – Email: bojana@elab.tmf.bg.ac.yu

[2]Department of Food Technology and Biochemistry, University of Belgrade, Nemanjina 6, PO Box 127, 11081 Belgrade-Zemun, Yugoslavia – Fax: 381-11-193659 – Email: vnedovic@EUnet.yu

[3]Department of Ultrastructure, Flanders Interuniversity Institute for Biotechnology, Vrije Universiteit Brussel, Pleinlaan 2, B-1050 Brussels, Belgium – Fax: 32-2-6291963 – Email: Ronnie.Willaert@vub.ac.be

[4]Harvard – MIT Division of Health Sciences and Technology, Massachusetts Institute of Technology, E25-330, 77 Massachusetts Avenue, Cambridge MA 02139, USA – Fax: 1-617-2588827 – Email:gordana@mit.edu

1. Introduction

Over the last decades, immobilised cell systems were widely investigated and used in different fields of biotechnology such as pharmacy, biomedicine, food and environmental technologies. Immobilised cell technology is successfully established at the industrial scale in wastewater treatment and production of biopharmaceuticals and fermented beverages. Cell immobilisation provides several advantages over the conventional free cell systems including higher cell concentrations, higher volumetric productivities, cell protection required for shear sensitive cells, and easy separation of cells and products. In addition, immobilised cell systems can be operated in continuous mode at higher dilution rates without the risk of cell washout. Especially attractive are the options for co-immobilisation of different cell types for the simultaneous implementation of consecutive reactions.

The key elements of an immobilised cell system are: cell type, immobilisation material and method, and bioreactor design. The cell type determines the desirable conditions in the cell environment such as the optimal pH, oxygen concentration and permissible level of shear, which are then set as guidelines in the design of the immobilised system. The choice of the immobilisation material and method determines

V. Nedović and R. Willaert (eds.), Fundamentals of Cell Immobilisation Biotechnology, 411-436.

the local cell environment within a biocatalyst, while the bioreactor design governs the overall fluid dynamic, mass and heat transfer conditions at the biocatalyst surface. Selections of immobilisation methods and bioreactor type are often interrelated and various potential combinations are possible.

This chapter describes the main features and strategies in bioreactor design for use with immobilised cell systems. General types of immobilisation methods are briefly described and related to the major bioreactor types. Fluid dynamics and mass transfer properties are discussed in more detail for selected applications.

1.1. IMMOBILISATION METHODS

A variety of immobilisation methods and materials are available for different cell types and applications. These methods can generally be classified into 4 major groups: self-aggregation, adsorption onto pre-formed carriers, entrapment into carrier matrices and containment in or behind a barrier (Figure 1).

Self-aggregation of cells (Figure 1A) can be natural or artificially induced by cross-linking agents. This is the simplest and least expensive immobilisation method but the most sensitive to changes in the operating conditions. In addition, there is a high risk of cell washout. This cell immobilisation method provides higher cell densities as compared to free cell systems, but no protection of cells.

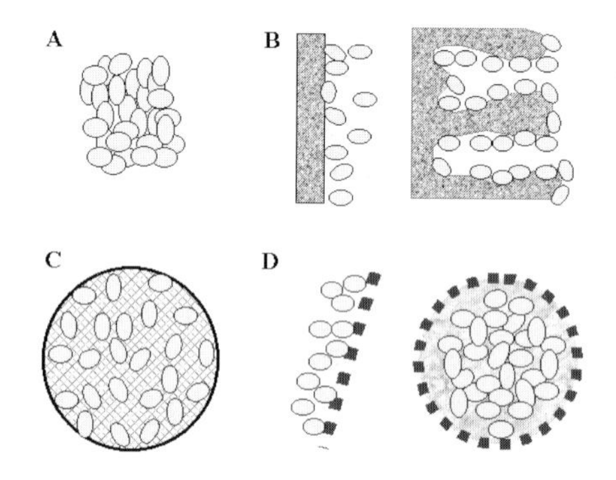

Figure 1. Major methods for cell immobilisation: A) self-aggregation of cells. B) adsorption onto pre-formed carriers. C) entrapment into carrier matrix. D) containment of cells in or behind a barrier. Figure adapted from Figure 1 in Reference [1].

Adsorption onto pre-formed carriers is the oldest immobilisation method, and is based on cell adsorption onto external surfaces of solid carriers (Figure 1B). Cells can be attached by van der Waals forces, electrostatic interactions, covalent bonding, and physical entrapment in the pores. Furthermore, cells can be restricted to external surfaces only, or adsorbed inside the pores of carriers. Various types of materials are suitable for cell adsorption such as sand, glass, ceramic, wood, and synthetic materials. The immobilisation procedure is gentle, carried out by recirculation of cell suspension through a packed bed of carriers, or by mixing a suspension of cells and carriers. An additional advantage of this immobilisation method is the carrier regeneration and reuse.

This immobilisation method provides direct contact of cells and substrate, which results in the enhancement of mass and heat transfer rates, but gives no cell protection. The second drawback could be the high cell loss by washout coupled with the relatively low cell loadings in these systems.

Cell entrapment into carrier matrices is based on the use of low porosity matrices such as hydrogels, which provide cell retainment while allowing metabolite diffusion (Figure 1C). Alginate, κ-carrageenan, agar, polyvinyl alcohol and pectate gels are some of the commonly used gel materials for cell entrapment. The main advantage of this method is the attainment of very high cell loadings. However, in some cases cell proliferation and activity can be limited by low mass transfer rates within the matrices. In addition, relatively weak mechanical properties of gels used for cell entrapment can limit the use of this method.

Containment of cells in or behind a barrier includes cell immobilisation in semi-permeable microcapsules as well as immobilisation in or behind membranes incorporated in membrane bioreactors (Figure 1D). This method can provide high cell concentrations, cell protection, and simultaneous product separation. The mass transfer is mainly governed by the membrane properties. A drawback can be the higher cost as compared to other immobilisation methods.

2. Bioreactor design and mass transport phenomena

A proper bioreactor design needs to provide appropriate fluid dynamic conditions and an efficient mass and heat transfer to and from the biocatalyst surface. The level of shear stress imposed on biocatalysts, flow and mixing patterns are particularly important in immobilised cell systems. The level of shear stress is examined from different perspectives in several immobilised cell applications. Shear sensitive cells, like mammalian, insect or plant cells, require a low shear environment so that the protection by entrapment or a membrane is favoured. In addition, matrices used for cell entrapment are often mechanically weak and require gentle mixing conditions. In systems with adsorbed cells, high shear may induce cell leakage, which can lead to cell depletion and washout. Nevertheless, in some systems partial cell leakage from immobilisation supports can be beneficial for providing a balance between the amount of immobilised and suspended cells resulting in a continuous long-term stable operation [*e.g.* 2,3]. In biofilm bioreactors, an optimal level of shear stress is desired in order to maintain a constant biofilm thickness and compact structure [4,5,6].

The flow and mixing conditions in a bioreactor determine the overall distribution of substrates and products, and affect the reaction rates. Generally, good local mixing is preferred while the most favourable overall flow pattern depends on the reaction kinetics. For single positive order kinetics or product inhibited reactions, plug flow is more efficient than ideal mixing. However, for complex reactions multistage reactor systems may be needed for optimal results. In biofilm systems, flow patterns were reported to largely influence microbial population and their kinetics [4]. In addition, the existence of stagnant zones, channelling, and bypassing are common problems that can decrease the overall reactor performance.

Heat and mass transport to the immobilised cells are usually studied in terms of internal and external transfer rates. Internal transfer designates the transport of heat, nutrients, gases and metabolites within the biocatalyst. It depends on the biocatalyst properties and it usually occurs by molecular diffusion. External transfer indicates transport of heat and mass between the bulk fluid and the biocatalyst surface. It depends on the fluid dynamic properties in the reactor and the activity of the biocatalyst particle. The external liquid-solid mass transfer is usually described by the film theory, which assumes that the resistance to mass transport is concentrated in a thin film adherent to the solid surface (Figure 2A) [7].

The relative importance of the external and internal mass transfer rates can be determined by the dimensionless Biot number (Bi):

$$Bi = \frac{k_{ls} L}{D_s} \tag{1}$$

where k_{ls} is the liquid-solid mass transfer coefficient, D_s is the diffusivity of the transferred species in the solid phase and L is the characteristic length.

In 3-phase systems, mass transfer modelling is usually complicated by the addition of one more step, namely gas-liquid transport (Figure 2B). Especially in aerobic applications, oxygen is often a limiting substrate and the rate of oxygen transport through the gas-liquid interface is often the rate controlling step. Thus, the volumetric gas-liquid mass transfer coefficient ($k_L a$) is one of the main design and scale-up parameters for three-phase bioreactors.

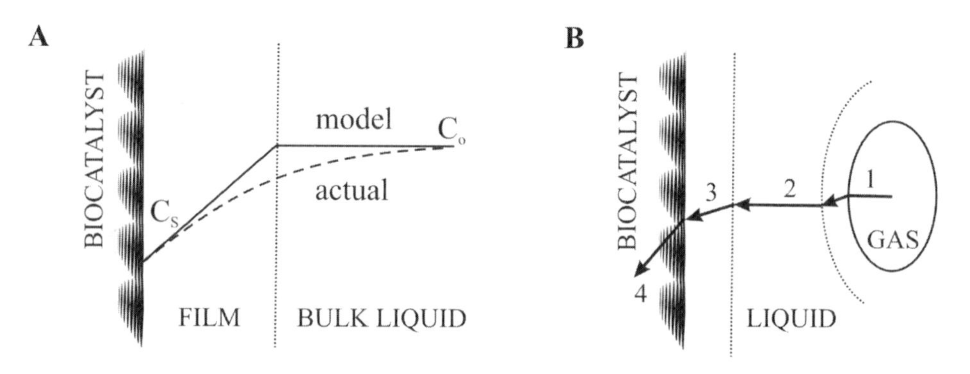

Figure 2. Mass transfer phenomena in heterogeneous systems. A) the film theory model presumes that the concentration change from bulk liquid (C_o) to the biocatalyst surface (C_s) occurs linearly within a thin film; B) mass transport of a gaseous component in 3-phase systems: 1 - gas-liquid transport; 2 - transport through the bulk liquid phase; 3 - liquid-solid transport; 4 – diffusion inside the biocatalyst and reaction.

2.1. STIRRED TANK BIOREACTOR

The stirred tank bioreactor was one of the first reactor types used in biotechnology and it is still widely used for various biotechnological processes. It consists of a vessel with one or more stirrers with a variety of designs to satisfy specific operation conditions

(Figure 3). A detailed description of common stirrer types can be found in the (bio)chemical engineering literature [*e.g.* 8-11]. Vessels are usually equipped with baffles, which facilitate mixing and prevent undesirable bulk rotation of liquid since it causes high mechanical stresses in the stirrer shaft, bearings and seal. The liquid phase can be supplied in batch or continuous mode, while the gas phase can be introduced optionally.

The main characteristic of this bioreactor type is the almost ideal mixing of the fluid phase. The fluid flow is generally turbulent with a non-uniform distribution of shear, being the highest near the stirrer and lowest near the reactor walls. As energy dissipation in turbulent flow can be the cause of cell death, low power inputs are preferred for shear sensitive cells. In addition, regions of close clearances should be avoided [12]. As a consequence, stirrers such as helical ribbon, screw or anchor are preferred in immobilised cell systems for their gentler stirring than turbines or propellers [13]. Modified impellers such as cell-lift and centrifugal impellers were also proposed for the reduction of shear stress on animal cells [14,15]. Correspondingly, support matrices for cell immobilisation should have good mechanical properties and be resistant to abrasion. Alginate gel was shown to disintegrate quickly in a bioreactor stirred with a 6-blade impeller [16] while cell carriers made of polyvinyl alcohol (PVA) [16], PVA with stilbazolium groups (PVA-SbQ) [17] and κ-carrageenan-locust bean gum [18] where shown to have better mechanical stability in similar bioreactor systems.

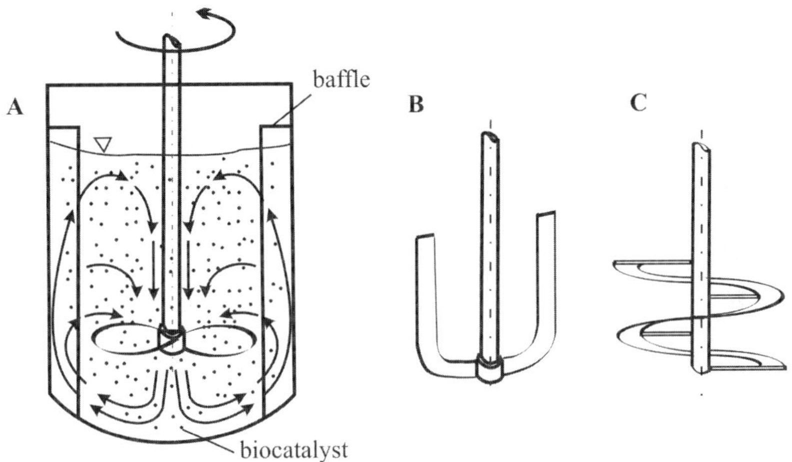

Figure 3. Stirred tank reactor and some stirrer types. A) Flow patterns in a baffled tank with an axial-flow propeller. B) Anchor stirrer. C) Helical ribbon stirrer.

The main operating parameter in a stirred tank bioreactor is the rotational speed of the stirrer. Generally, it is required to determine the "critical" rotational speed at which all solid particles are suspended in liquid at the minimum power input. The critical rotational speed depends on the stirrer diameter and properties of particle suspension (*e.g.* diameter, loading, density of particles, viscosity and density of the liquid). Power absorption in mechanically agitated vessels is related to the flow regime in terms of the dimensionless quantities, the power number (N_P) and Reynolds number (Re), which are defined in this system as:

$$N_p = \frac{P}{n^3 d^5 \rho} \quad (2)$$

$$Re = \frac{nd^2}{\nu} \quad (3)$$

where P is the external power from the agitator, n is the stirrer rotational speed, d is the stirrer diameter, ρ is the density, and ν is the kinematic viscosity of the medium. The power characteristic of a stirrer correlates the power number to the Reynolds number for each stirrer type. For laminar flow, where $Re \leq 10$ except for stirrers with very small wall clearances such as anchor or helical ribbon mixer where $Re \leq 100$, it is expressed by the general form [9,19]:

$$N_p = KRe^{-1} \quad (4)$$

where K is a constant. In the transient regime, which arises only in vessels without baffles and for Re in the range from 10 to 5×10^4, there is no general relationship. In the turbulent flow the power number is constant independent of Re. Power absorption is reduced in gassed systems as compared to two-phase solid-liquid systems.

Turbulent flow and good mixing in stirred tank bioreactors provide efficient mass transfer between the liquid phase and solid particles. However, in gassed systems such as aerobic fomenters, gas-liquid mass transfer of oxygen can be rate limiting. The gas-liquid mass transfer coefficient, $k_L a$, in this case, is correlated to the volumetric gassed power input, P_g/V_L, and superficial gas velocity, u_G, [19]:

$$k_L a = K \left(\frac{P_g}{V_L} \right)^{\alpha} u_G^{\beta} \quad (5)$$

where K, α and β are constants.

The basis for the scale-up of stirred tank reactors can be the constant volumetric oxygen transfer coefficient, constant power per unit volume, constant shear or constant mixing time. Usually, a similar geometry is not as important to scale-up as $k_L a$ and shear. Mass transfer of oxygen to the cells is usually the controlling step in fermentation systems, and therefore used as a basis for scale-up. The current practice is to maintain a constant $k_L a$ and impeller tip speed and adjust the geometry [19].

Some laboratory, pilot and industrial scale applications of stirred tank bioreactors with immobilised cells include: ethanol production by *Saccharomyces* species and *Zymomonas mobilis* [20]; ethanol production from starch by a co-immobilised system of *Aspergillus awamori* and *Saccharomyces cerevisiae* [21]; the continuous yoghurt production with *Lactobacillus bulgaricus* and *Streptococcus thermophilus* entrapped in

Ca-alginate [22]; cheese production using a continuous prefermentation process of milk [23]; the production of cephalosporin C by immobilised *Cephalosporium acremonium* in Ba-alginate beads [24]; hepatitis A vaccine production using human MRC-5 diploid fibroblasts [25].

2.2. PACKED BED BIOREACTORS

Packed bed bioreactors are characterised by a simple design, consisting of a column, which is packed with biocatalysts and is continuously perfused by the liquid phase (Figure 4A).

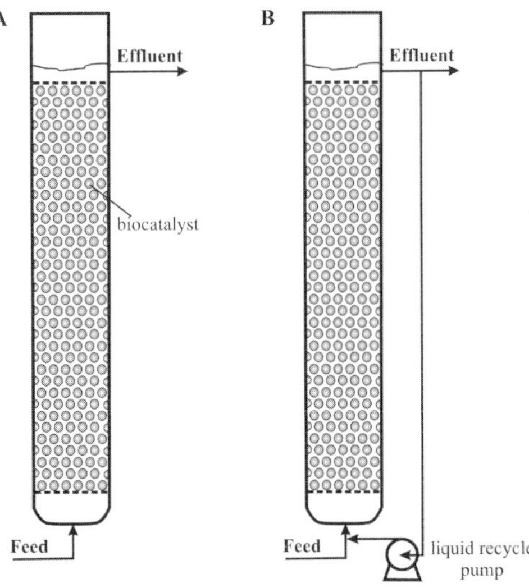

Figure 4. Packed bed reactor. A) Liquid is pumped through a bed of packed biocatalyst particles. B) Packed bed reactor with liquid recirculation.

In this way, close to plug flow of liquid is established at low shear rates. As a consequence, a variety of materials for cell adsorption can be used in this reactor type, including those that are fragile and susceptible to abrasion as well as adsorbents with weakly adsorbed cells. Some common materials include sand as a cheap adsorbent for growth of bacterial biofilms suitable for water treatment applications [26], DEAE-cellulose and wood chips used for brewing yeast immobilisation in the beer industry [27,28], and different forms of glass, silica, ceramics, and synthetic materials used for immobilisation of various cell types. Cells immobilised by entrapment in porous matrices such as alginate or agar can also be used in packed bed reactors [29-32]. However, the use of mechanically week materials such as hydrogels can be limited to lower bed heights and liquid flow rates due to possible compression of beads [32]. Chien and Sofer [33] proposed a modified packed bed reactor in which layers of alginate beads are separated by screens. In this way, a more uniform flow distribution was provided, which avoided the formation of densely packed beads at the top of reactor and the compression of beads at the bottom.

The main operating parameter in this reactor type is the liquid flow rate. Uniform liquid distribution through narrow channels of packing material should provide liquid velocities high enough to generate efficient mass transfer. However, both external and internal mass transfer limitations can be present. General correlations for the Sherwood number for particulate immobilised cell systems are presented in Box 1.

Non-uniform flow distribution with flow channelling causing concentration and temperature gradients are frequent problems in packed bed reactors. One of the attempts to improve the performance of this reactor type was to introduce forced liquid recirculation (Figure 4B), which resulted in better mass and heat transfer rates [34,35]. Plugging of the reactor is also a common problem. This can be solved by occasional gas purging or by increasing the liquid flow rate.

Nearly plug flow conditions in packed bed reactors make them attractive for product inhibited reactions such as the production of ethanol [2,29,30,32,33,36], butanol [37] or propionic acid [38,39]. Generally, higher productivities can be achieved with plug flow as compared to ideally mixed conditions. However, in all investigated packed bed systems for ethanol production the main problem was the build-up of CO_2, which resulted in a decreased productivity. Pulsation in the liquid flow was proposed as a method for degassing the bed and providing a renewal of the interfacial area. When the pulsing frequency was matched to the gas production rate, the productivity of ethanol by alginate immobilised yeast cells in a 2.1-L packed bed reactor was increased [30]. Chotani and Constantinides [29] proposed a cross-flow reactor – similar to a shell and tube heat exchanger – for ethanol production by yeast cells immobilised in silica reinforced alginate beads. The variation in the number of baffles in this reactor design could provide different types of fluid flow, ranging from flow with a significant back mixing to almost ideal plug flow when a large number of baffles was used. Alternatively, fibrous bed bioreactors based on spirally wound cotton matrix supported by stainless steel mesh were proposed for cell immobilisation and propionic acid production [38-40]. Open spaces between wound layers of the matrix provided the escape of the produced CO_2, low pressure drop through the bed, and an easy fall off of the accumulated biomass to the reactor bottom, which prevented bed clogging. However, relatively high liquid flow rates were required to reduce the mass transport limitations. As an upgrade, a centrifugal fibrous-bed bioreactor was designed for xanthan gum fermentation in which the fibrous matrix was rotated, providing a radial liquid flow through the matrix and xanthan gum separation by the centrifugal force [41].

Some pilot and industrial scale applications of packed bed reactors include: water treatment such as denitrification using sand immobilised heterotrophic bacteria [26,42] or sulphur/limestone immobilised autotrophic bacteria [43,44], and sulphate removal using bacteria immobilised on polyurethane foam [45]. In addition, packed bed reactors are used at an industrial scale for beer maturation and low alcohol beer production [46,47].

2.3. FLUIDIZED BED BIOREACTORS

Fluidized bed bioreactors provide solutions for some of the problems inherent to the use of packed bed reactors. In this reactor type, particles with immobilised cells are fluidized in the liquid up-flow while gas can be optionally supplied (Figure 5A). For some applications such as reactions with a gaseous product, a tapered bed configuration can be beneficial (Figure 5B), in which liquid velocity gradients along the reactor height are obtained. In this way, higher liquid velocities are established in the tapered zone without the risk of particle washout. The enlargement of the reactor diameter in the upper zone results in a decrease in the local liquid velocity. However, the top of the reactor is usually provided with a settling zone in order to assure particle retention within the reactor. Alternative configurations utilize particles lighter than the liquid phase, which can be fluidized in liquid down-flow (inverse fluidization, Figure 5C) or under an up-flow of gas only (pseudo-fluidization, Figure 5D) [48]. As a consequence of particle fluidization, moderate local mixing is established and the liquid flow is generally described as plug flow with axial dispersion [5,49-51]. However, in systems with a high production of gas or at high gas inputs, the degree of mixing increases towards well mixed systems. Generally, fluidized bed reactors provide better mass and heat distribution with more uniform liquid flow throughout the reactor volume as compared to packed bed reactors. The main operating parameter is the liquid flow rate in 2-phase systems together with the gas flow rate in 3-phase systems.

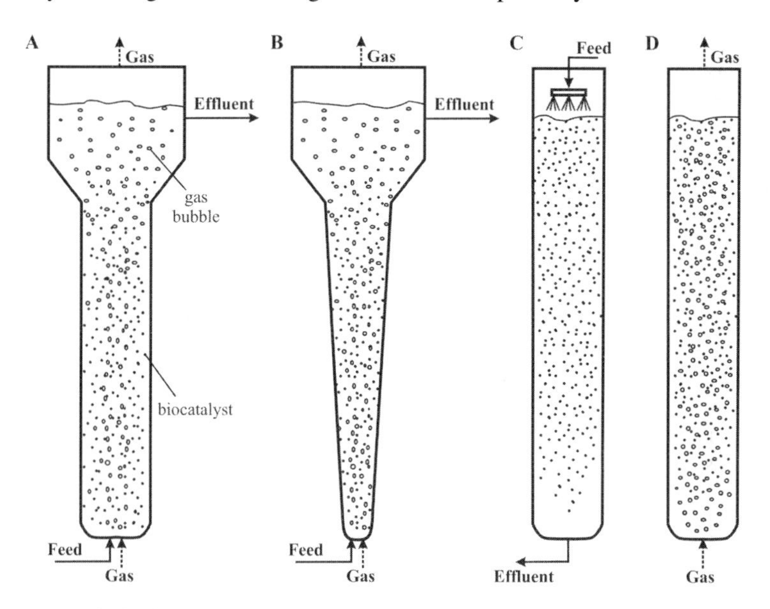

Figure 5. Fluidized bed reactors. Biocatalyst particles are fluidized in liquid and optionally gas up-flow: A) straight bed configuration; B) tapered bed configuration; C) inverse fluidized bed: biocatalyst particles are fluidized in liquid down-flow; D) pseudo-fluidized bed: biocatalyst particles are fluidized by gas up-flow.

Particle movements and collisions in the fluidized state result in shear stresses and abrasion, creating a need for mechanically stable supports in this type of reactors. In addition, low density particles require very low liquid flow rates for fluidization, which can be difficult to maintain. As a result, particle-based biofilms were found to be a favourable option for applications in fluidized bed reactors. Cell aggregates in the form of sludge granules and microbial biofilms grown on sand (particle size in range

0.2 -0.8 mm) are effectively used in several types of fluidized bed reactors for wastewater treatment [*e.g.* 5,6].

Upflow sludge blanket (USB), expanded granular sludge blanket (EGSB) and biofilm fluidized bed (BFB) reactors are used at an industrial scale for the treatment of wastes. The USB reactor was designed first and it exhibits somewhat different hydrodynamic properties than a classical fluidized bed reactor. Dense sludge granules are fluidized in the up-flow of the wastewater and as the conversion of COD to biogas proceeds, gas-borne granules are carried to the top of the reactor. At this point, degasification takes place so that granules fall back to the bottom and a continuous recirculation of sludge granules is established. As a consequence, a higher degree of mixing is obtained as compared to the conventional fluidized bed reactor [5]. In the USB reactor, accumulation of solids is the major problem.

The EGSB reactor development was based on the same concept as the USB reactor with an addition of a specially designed three phase separator. In this way, the operation at much higher liquid flow rates was allowed as compared to the USB reactor and hence solid accumulation in the reactor was eliminated. This reactor type was used at a large scale for industrial wastewater treatment [52-54].

The BFB reactor was also designed in an attempt to overcome problems intrinsic to USB reactors. Sand particles are used for cell adsorption and biofilm growth. Consequently, higher liquid circulation rates are utilized for particle fluidization and thus solid accumulation within the reactor is reduced. However, the control of the biofilm structure and growth is difficult such that the washout of particles with overgrown biofilms is a major problem in this system [6].

Inverse fluidization (Figure 5C and D) could be an interesting solution for biofilm overgrowth since over coated particles sink to the bottom of the reactor where they can be collected and recovered. In addition, inverse fluidization can be enhanced by the up-flow of gas, resulting in turbulence and mixing, while providing better control of the bed height [48].

Plug flow with axial dispersion in fluidized bed reactors has attracted the attention for product inhibited reactions similarly to packed bed reactors. High ethanol productivities were obtained by κ-carrageenan immobilised bacterial cells in straight and tapered fluidized bed reactors [49,55].

2.4. AIR-LIFT BIOREACTORS

Air-lift reactors are especially attractive for applications in biotechnology because they can provide an almost ideal mixing comparable to that in stirred tanks but at a much lower and evenly distributed shear stress [56]. Efficient mixing is achieved by liquid and solid circulation due to the pressure difference in the gassed and non-gassed sections of the reactor. Air-lift reactors can be constructed as internal loop or external loop configurations, generally consisting of a base, riser, down-comer and a top section (Figure 6). Gas is usually introduced at the bottom of the riser and partially or completely disengaged in the top section. In this way, different gas hold-ups are established in the riser and down-comer, resulting in different bulk densities in these two sections and hence inducing a liquid and solid circulation. In aerobic applications, air is commonly used as a gas phase, while in anaerobic processes the circulation is generated by gas produced in the reaction [5] or alternatively by the introduction of an inert gas [1,57]. The top section of the reactor serves also for the retention of carriers with immobilised cells within the reactor while permitting the washout of suspended free cells and cell debris.

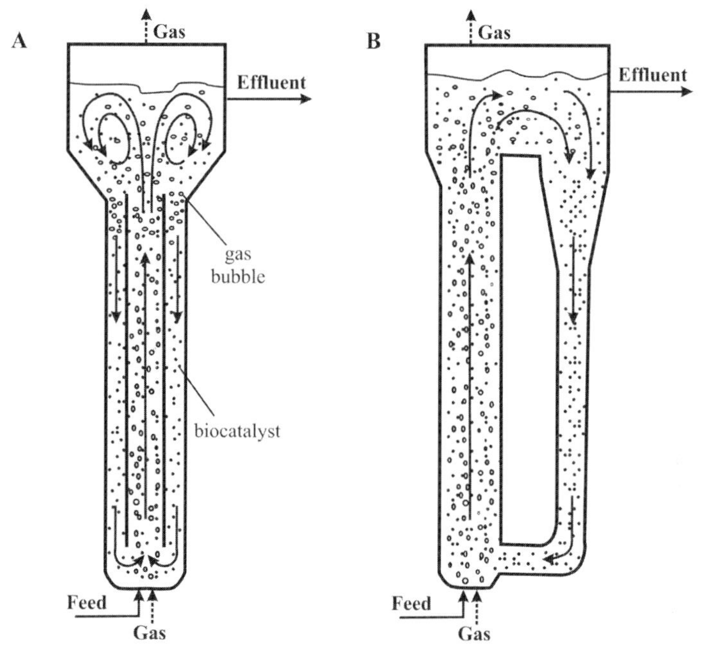

Figure 6. Air-lift reactors: A) internal loop configuration; B) external loop configuration.

Internal loop air-lift reactors can be operated at three flow regimes. At low gas velocities and low liquid circulation rates all gas is disengaged in the top section. In a transitional regime some gas bubbles are entrained in the down-comer. At high liquid circulation rates gas bubbles that are entrained in the down-comer, are carried with the liquid flow and full gas circulation in the reactor is established. Gas circulation is especially favourable in aerobic processes for providing higher gas hold-ups and gas-liquid mass transfer rates as compared to the first two regimes. Full scale internal loop air-lift reactors always operate at full bubble circulation [58].

The overall mixing in air-lift reactors is reported to be close to ideal [59-63] achieved at all scales [58]. However, liquid flow patterns and mixing properties were shown to vary in individual reactor sections [61,64,65]. In general, the flow in the riser and down-comer can be described as plug flow with axial dispersion while the top section is well mixed. The mixing efficiency in air-lift reactors is thus a consequence of high liquid and solid recirculation, and high degree of back mixing in the top section [63,66].

Efficient mixing and low shear rates make air-lift reactors as suitable for all types of particulate immobilised cell systems. Especially advantageous are applications with low density and mechanically weak carriers such as hydrogels, which are easily compressed in packed beds, on one side, and require unfeasible low fluidization rates, on the other.

Air-lift reactors are also superior to fluidized beds in applications based on the use of particulate biofilms, due to better control of the biofilm thickness and structure [5,6]. At optimal operating conditions and initial carrier concentration, thin, dense biofilms can be maintained [67].

The use of cell carriers lighter than the liquid phase was also investigated in air-lift reactors, which are then operated in the inverse mode similarly to inverse fluidized beds. In sparged systems, cell contact with the gas-liquid interface and bubble bursting at the suspension interface were reported to cause cell damage [12,68]. Inversed fluidized bed air-lift reactors provide the separation of the gas and solid phases such that direct contact of gas bubbles with cell carriers is avoided. The gas phase is introduced in the riser, inducing a liquid circulation and thus providing the fluidization of the solid cell carriers in the down-comer by a downward liquid flow. The external loop reactor configuration provides a more stable operation and ensures the prevention of three phase contacts as compared to an internal loop configuration. In addition, the installation of a valve in the external loop configuration for the regulation of the liquid circulation rate provided the control of the gas hold-up independently of the gas flow rate and solid loading [69].

The main design parameters for air-lift reactors are down-comer to riser area ratio (A_d/A_r), reactor height, and top section height. The superficial gas velocity is the main operating parameter. These parameters determine the liquid circulation rate, gas hold-up, energy dissipation and as a result, the mass transfer and mixing properties in the reactor.

Gas-liquid mass transfer has been often investigated for aerobic applications in air-lift reactors as the rate limiting step for oxygen supply and cell growth. The volumetric gas-liquid mass transfer coefficient, k_La, in 2-phase systems was generally related to the superficial gas velocity, u_G, using the empirical correlation [70]:

$$k_L a = K u_G^m \tag{6}$$

where K and m are constants. Miron *et al.* [71] derived and experimentally verified a theoretical relationship:

$$k_L a = \frac{\Phi}{u_G^y - 1} \tag{7}$$

where Φ and y are constants dependent on the reactor geometry and the liquid properties. However, in three phase systems no general correlation for k_La was derived. Bugarski *et al.* [72-74] reported that the k_La in external loop air-lift reactors loaded with alginate beads or microcapsules was a function of the superficial gas velocity while Nicollela *et al.* [75] found a linear dependence of k_La on gas hold-up in an biofilm internal loop air-lift reactor. The oxygen mass transfer per power input, a measure of aeration efficiency, is generally higher for air-lift reactors than for other reactor configurations [55].

Liquid-solid mass transfer is generally efficient in air-lift reactors such that it is often neglected as compared to the internal diffusion, which is considered as rate limiting. However, in some immobilised cell systems such as thin bacterial biofilms, internal diffusion limitations can be minimized and the rate of external mass transfer can become comparable to the utilization rate of a limiting substrate [76]. General approaches to estimate the liquid-solid mass transfer coefficient in air-lift reactors are presented in Box 1.

The scale-up of this reactor type is usually based on the maintenance of similar flow regimes and mass transfer rates. Since the gas hold-up, liquid circulation rate, energy dissipation rate, and hence the gas-liquid and liquid-solid transport are largely determined by the superficial gas velocity, the maintenance of the same superficial gas velocity is the main scale-up criterion [58]. In addition, identical reactor heights and superficial liquid velocities should also be established, in order to provide similar flow patterns and mixing properties [58]. Another scale-up issue is the geometry and hydraulic load of the settler in order to provide an effective washout of cell debris while retaining the cell carriers [58,67]. However, the top section can be specially designed for each system or, in the case of light particles such as hydrogels, the installation of a mesh can be sufficient to keep the particles in the reactor while free cells can be washed out [77].

Internal loop air-lift reactors are used at an industrial scale for aerobic wastewater treatments [5,6]. Another commercial application is the production of single cell protein [78].

The attractive fluid dynamic features of air-lift reactors have been utilized for several alternative applications. A four-phase external-loop gas-lift reactor, which provided integrated production and extraction of extracellular metabolites, has been described [79,80]. In this system, alginate immobilised plant cells were suspended in a continuous aqueous phase, dispersed liquid solvent, and dispersed gas phase. Simultaneous reaction and extraction provided a 30-50% higher yield of antraquinones as compared to shaked flask cultures.

The multiple air-lift loop reactor was designed to approximate aerated plug flow and incorporate serial bioreactors within one vessel [81]. Multiple coaxial concentric tubes provide alternate arrangement of several risers and down comers such that medium is flowing through the cascade *via* overflows. This reactor can be used instead of several bioreactors in series, for a single reaction or potentially for consecutive reactions since subsequent reactor compartments can be supplied with different gases [82].

Box 1. Liquid-solid mass transfer in particulate immobilised cell systems

Liquid-solid mass transfer in turbulent flow is typically described by film theory, which predicts that the resistance to mass transfer is concentrated in a thin liquid film adherent to the solid surface [7]. Since the film thickness is generally unknown, the liquid-solid mass transfer coefficient, k_{ls}, can be estimated using the dimensionless Sherwood number (Sh), which presents the ratio of convective to diffusive transport rates:

$$Sh = \frac{k_{ls} d_p}{D_l} \qquad (1.1)$$

where d_p is particle diameter and D_l is the diffusivity of the transferred species in the liquid phase. The Sherwood number is related to the flow rate, geometry, fluid and particle properties. For a spherical particle in a stagnant liquid of infinite volume Sh = 2.0. As the liquid starts to flow, the thickness of the stagnant film around the particle decreases, inducing an increase in k_{ls} and hence in the Sh number, which is generally represented by an equation proposed by Ranz and Marshall [83]:

$$Sh = 2 + 0.6Sc^{1/3}Re^{1/2} \qquad (1.2)$$

where Sc is the Schmidt number ($Sc = \nu/D_l$) and Re is the Reynolds number ($Re = ud_p/\nu$). In the definitions of these dimensionless numbers ν is the kinematic viscosity of the liquid and u is the liquid velocity.

The general form of Eq. (1.2.) is commonly accepted for the estimation of Sh numbers in packed bed and 2-phase fluidized bed reactors with experimentally determined constants. In three-phase bubble columns, fluidized beds and air-lift reactors, the Reynolds number is frequently defined according to Kolmogoroff's theory of turbulence:

$$Re \sim \frac{\varepsilon d_p^4}{\nu^3} \qquad (1.3)$$

where ε is the energy dissipation rate [5,76]. Assuming a uniform energy dissipation rate throughout the reactor, it can be estimated as [70]:

$$\varepsilon = gu_G \qquad (1.4.)$$

where u_G is the superficial gas velocity. The general form of correlations for the Sh number in this case becomes:

$$Sh = 2 + aSc^b \left(\frac{\varepsilon d_p^4}{\nu^3} \right)^c \qquad (1.5.)$$

The exponent of Sc (constant b) is mainly determined as 1/3 while the values for constants a and c vary in ranges from 0.265 to 1.01 and from 0.2 to 0.274, respectively [5,76].

2.5. MEMBRANE BIOREACTORS

The main advantage of membrane bioreactors is that they provide simultaneous bioconversion and product separation, which is especially attractive for the production of high value biological molecules. As compared to conventional reactor types, the design of membrane reactors is relatively more complex and more expensive (due mainly to the high cost of the membrane material). However, since the attainment of highly concentrated products could eliminate the need for some steps of costly product purification, the utilisation of these reactors can be favourable. For low value biological products, conventional bioreactors are more appropriate [84].

Membrane reactors can be configured as flat sheet or hollow fibre modules (Figure 7). Hollow fibre modules provide a higher surface to volume ratio without the need for membrane support. However, the geometry of flat sheet modules is simpler, providing an accurate regulation of the distances between the membranes. Additionally, these modules can be easily disassembled, providing an easy access to module compartments and options for membrane cleaning and replacement [84].

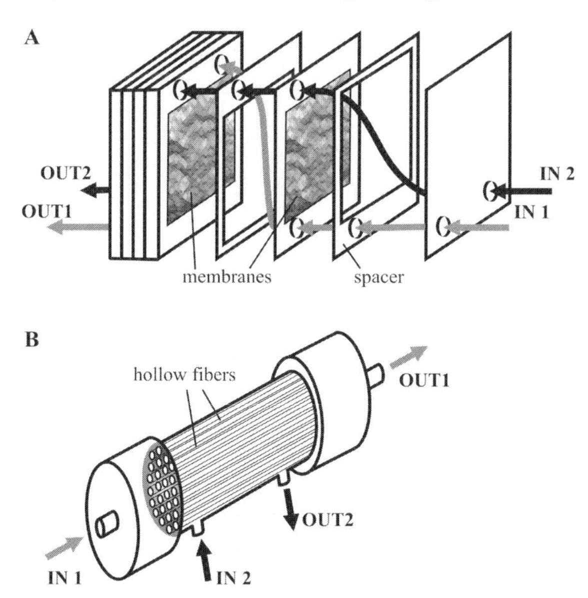

Figure 7. Membrane reactors. Generally, membranes separate two compartments in these reactors through which two independent inlet streams can be transported in co-current (shown) or counter-current mode. A) Flat sheet module. B) Hollow fibre module.

In membrane reactors, three types of cell immobilisation are possible (Figure 8):

- cell immobilisation on the membrane where a biofilm is formed,
- cell immobilisation within the membrane,
- cell immobilisation in a cell compartment separated by the membrane.

In all cases, the membrane provides the shear protection of cells and controlled supply of reactants. The contact of gas bubbles with cells and bubble bursting at the cell surface are generally avoided. This is particularly relevant for mammalian cells. Drawbacks of the use of membrane reactors can be a reduced mass transfer rate depending on the membrane properties, available surface area and external flow conditions.

Biofilms grown on membranes can be applied for wastewater treatment where membrane permselectivity is used for the extraction of targeted components or a controlled supply of oxygen (Figure 8A). Industrial waste streams are often at extreme pH or contain high concentrations of salts or carried-over catalysts. They decrease or diminish the activity of microbial cultures aimed for the biodegradation of pollutants.

The use of membranes, which are permselective only to the particular compound, provides biodegradation without exposure of biomedium to the toxic environment and prevents microbial contamination of the treated water [85,86]. In addition, the separation of the biomedium from the wastewater in a membrane reactor provides an independent control of the biofilm growth and hydraulic residence time of the wastewater [87]. A pilot plant based on an extractive membrane bioreactor was used for benzene removal from an aqueous solution of aluminium trichloride with an average efficiency of 98% over a 6-month period [88].

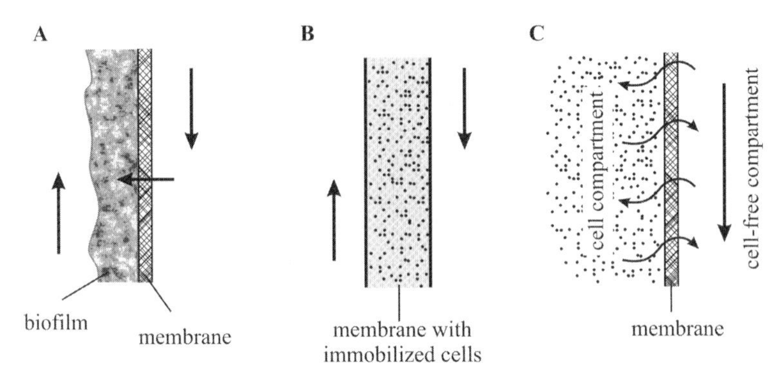

Figure 8. Cell immobilisation in membrane reactors. A) Immobilisation on the membrane. B) Immobilisation within the membrane. C) Immobilisation behind the membrane in a cell compartment.

Membrane-aerated biofilm reactors provide controlled supply of oxygen to the biofilm through the membrane. The oxygen penetration depth is easily regulated by its partial pressure in the gas phase. As compared to three-phase contactors, membrane reactors are advantageous in providing a constant area for oxygen transfer, minimization of the stripping of volatile compounds and avoidance of foaming [89]. Controlled supply of oxygen through the membrane can be beneficial for biofilms with high oxygen demand such as immobilised nitrifying bacteria, or biofilms composed of mixed cultures where oxygen can be supplied to slower growing organisms in deeper layers. Furthermore, treatments of high strength wastewater could need enhanced oxygen transfer, which is achieved in this reactor type [89].

The immobilisation of cells within a membrane (Figure 8B) can provide a constrained arrangement of biomass within a defined membrane geometry preventing cell leakage and allowing for controlled supply of reactants [90]. In addition, this type of membrane reactors is especially attractive for the simultaneous implementation of two consecutive reactions by co-immobilisation of two cell types. Such a process is found in the biological removal of nitrogen from wastewater, where ammonia is converted into nitrate and nitrite in a nitrification step, followed by the conversion of nitrate and nitrite ions into nitrogen gas in a denitrification step. Nitrification is performed by aerobic nitrifying bacteria and denitrification by anaerobic denitrifying bacteria. Nitrifying and denitrifying bacteria were co-immobilised in photo-cross-linkable PVA-SbQ gel membranes, which could be moulded in plates or tubes. The exposure of one membrane surface to ammonia-containing wastewater and the other to an electron donor (hydrogen gas, ethanol or lactate solutions) resulted in the redistribution of co-immobilised bacteria and a steady conversion of ammonia to nitrogen without the accumulation of the produced gas [91-94].

Cell immobilisation in a compartment separated by a membrane (Figure 8C) is especially attractive for mammalian cell cultures for the attainment of high cell densities (10^8 to 10^9 cells/ml) and hence an increased productivity, as well as for providing product concentration and separation. In addition, in these reactors the risk of contamination is decreased, while the cell growth and product formation can be decoupled [25,84,95,96].

The low shear environment in membrane reactors is also favourable for plant cells, which have been successfully cultured in hollow fibre and flat membrane modules with increased productivities as compared to suspension cultures. Nevertheless, the immobilisation of microbial cells in this type of membrane reactors can be limited by the oxygen supply, excessive cell growth or gas production, which can lead to membrane rupture [84].

One of the main design parameters in membrane reactors is the membrane type. There is a wide variety of commercially available membranes, which facilitate the selection of a proper membrane for a particular application [97]. The main properties of a membrane are the pore size, structure and material. Microfiltration membranes with pores of 0.01 to 10 μm or ultrafiltration membranes with a molecular weight cut-off of 10^3 to 10^6 are typically used for cell immobilisation [84]. Microfiltration membranes allow transmembrane fluxes of proteins, providing the separation of products from the cells. Ultrafiltration membranes on the contrary, retain proteins and growth factors within the cell compartment enhancing in some cases cell growth but necessitating the separation of the products from cells. Regarding the structure, membranes can be isotropic with homogenous composition and anisotropic with a thin, dense layer supported by a porous understructure. The membrane material determines the adsorption properties such that the membrane surface has often to be treated in order to prevent the loss of medium proteins and membrane fouling [98].

The performance of a membrane reactor depends on the operation mode, fluid dynamic conditions, and mass transfer properties. General approaches to the mathematical modelling of immobilised cell membrane reactors are presented in Box 2. The investigation and modelling of conventional axial-flow hollow fibre reactors have shown that mass transport by diffusion only, results in concentration gradients and an heterogeneous cell distribution [98,99]. The periodic reversal of the medium flow direction can decrease concentration gradients in the reactor but they are unavoidable without convection in the extracapillary space (ECS).

Convective flow can be increased by forcing some part of the flow through the ECS compartment. The operation in transverse mode was shown to improve Tc(VII) reduction by immobilised cells of *Escherichia coli* as compared to counter current flow along the fibres probably due to an improved mass transfer [100]. Several additional reactor designs were proposed to enhance convective flow. In one specific hollow fibre reactor fluid is forced from one set of fibres to another [101]. In this way, mass transfer limitations were decreased and a more uniform cell distribution was obtained as compared to the axial-flow hollow fibre reactor. Multi-coaxial hollow fibre reactors can also be operated under radial flow through the cell region between the fibres, proposed to mimic liver acinar structure and hepatocyte cultivation [102]. However, in radial-flow hollow fibre reactors membrane fouling may present a problem.

Other approaches focused on separating the supply of oxygen, which is often the limiting substrate, from medium flow. Some of the solutions are to deliver oxygen through an external membrane enveloping the ECS with immobilised cells [103] or reactor designs utilizing different arrangements of fibres within fibres with some of them dedicated to gas flow [98].

Box 2. Mass transfer in membrane reactors

The mass transfer of a certain component in a membrane reactor generally includes 3 steps: from the bulk fluid in the cell free compartment to the membrane surface, transfer through the membrane, and mass transfer coupled with the reaction in the cell compartment or biofilm. The mathematical modelling of the reactor performance depends on the overall fluid dynamic conditions, operation mode, reactor geometry, membrane properties and the reaction kinetics.

One of the simplest cases is a steady state operation of a flat sheet, non-porous membrane reactor with convective flow in the cell-free compartment at one membrane side, and a biofilm (Figure 8A) or stagnant cell compartment (Figure 8C), on the other. Examples for this situation can be membrane-aerated biofilm reactors and extractive membrane bioreactors. In these reactors, operating under convective flow of bulk fluid, mass transfer to the membrane is usually described by the film theory (Figure 2A):

$$J_i = k_{ls}(C_b - C_{in}) \tag{2.1}$$

where J_i is the molar flux of the transferred component, k_{ls} is the transfer coefficient through the fluid film, C_b and C_{in} are concentrations of the transferred component in the bulk fluid and at the membrane surface, respectively.

For describing the mass transfer through membranes, various models have been proposed, depending on the membrane and fluid properties [104]. For non-porous membranes, the solution-diffusion theory is typically applied, which for planar geometry yields:

$$J_i = D_{mi} k_i \frac{(C_{in} - C_{out})}{l} \tag{2.2}$$

where D_{mi} is the diffusivity in the membrane, k_i is the partition coefficient of the transferred component in the membrane, l is the membrane thickness, and C_{out} is the component concentration at the membrane surface facing the cell compartment. The fluid film and the membrane in this case present resistances in series, included in the overall expression obtained by combining equations (2.1) and (2.2):

$$J_i = K_o(C_b - C_{out}) \tag{2.3}$$

K_o is the overall mass transfer coefficient defined as:

$$\frac{1}{K_o} = \frac{1}{k_{ls}} + \frac{l}{D_{mi}k_i} \qquad (2.4)$$

Mass transfer through the cell compartment is generally described by diffusion, which coupled with the reaction at steady state results in:

$$D_{ieff} \frac{\partial^2 C_i}{\partial x^2} = Q \qquad (2.5)$$

where D_{ieff} is the effective diffusivity in the cell compartment, C_i is the component concentration, x is distance, and Q is the component consumption rate. Finally, equations (2.3) and (2.5) are coupled by the condition that the flux of the relevant component through the membrane has to be equal to the flux entering the cell compartment at the respective membrane interface:

$$J_i = -D_i \left(\frac{\partial C_i}{\partial x} \right) \qquad (2.6)$$

In the case of a porous membrane, there is a bulk flow of fluid through the membrane such that the flow of each component is a combination of diffusive and convective transport through the pores:

$$J_i = -D_i \frac{dC_i}{dx} + C_i J_V \qquad (2.7)$$

where D_i is the diffusivity of a component i in the fluid and J_v is the total transmembrane volume flux, which can be described by Poiseuille flow:

$$J_V = \frac{\varepsilon r_p^2 (\Delta P - \Delta \Pi)}{8 \mu \tau l} \qquad (2.8)$$

where ε is the fraction of the open-pore area, r_p is the equivalent pore radius, ΔP and $\Delta \Pi$ are the transmembrane hydrostatic and effective osmotic pressure differences, respectively, μ is the viscosity of the fluid in the pores, τ is the pore tortuosity and l is the membrane thickness.

On the other hand, the modelling of hollow fibre reactors is complicated by their complex geometry and fluid dynamics. Hollow fibre reactors are typically operated as an axial-flow set-up where the cells are entrapped in the extracapillary space (ECS) and the medium flows through the lumen [98]. The mass transfer of substrates from the lumen to the ECS is a combination of diffusive and convective transport through the fibre pores (Eq. 2.7) [105-112]. The convective transport depends on the membrane properties and transmembrane pressure (Eq. 2.8.). Generally, in the entrance half of the reactor, pressure is higher in the lumen and some of the fluid enters the ECS. Conversely, in the end half of the reactor, the pressure is higher in the ECS and fluid returns to the lumen. This type of flow is known as Starling (or toroidal) flow [113]. The modelling of axial-flow hollow fibre reactors is reviewed by *e.g.* Heath [98] and Brotherton and Chau [99]. In general, the fluid flow and mass transfer are two-dimensionally described by the conservation equations:

• continuity equation:

$$\frac{1}{r}\frac{\partial}{\partial r}(r\upsilon_r) + \frac{\partial \upsilon_z}{\partial z} = 0 \tag{2.9}$$

• momentum equations:

$$\rho\left(\frac{\partial \upsilon_z}{\partial r}\upsilon_r + \frac{\partial \upsilon_z}{\partial z}\upsilon_z\right) = -\frac{\partial p}{\partial z} + \mu\left[\frac{1}{r}\frac{\partial}{\partial r}\left(r\frac{\partial \upsilon_z}{\partial r}\right) + \frac{\partial^2 \upsilon_z}{\partial z^2}\right] \tag{2.10}$$

$$\rho\left(\frac{\partial \upsilon_r}{\partial r}\upsilon_r + \frac{\partial \upsilon_r}{\partial z}\upsilon_z\right) = -\frac{\partial p}{\partial r} + \mu\left[\frac{1}{r}\frac{\partial}{\partial r}\left(r\frac{\partial \upsilon_r}{\partial r}\right) + \frac{\partial^2 \upsilon_r}{\partial z^2}\right] \tag{2.11}$$

• substrate mass balance:

$$\frac{\partial C}{\partial t} + \upsilon_r\frac{\partial C}{\partial r} + \upsilon_z\frac{\partial C}{\partial z} = D\left[\frac{1}{r}\frac{\partial}{\partial r}\left(r\frac{\partial C}{\partial r}\right) + \frac{\partial^2 C}{\partial z^2}\right] - Q \tag{2.12}$$

where r and z are radial and axial distances, respectively, υ_r and υ_z are fluid velocities in radial and axial directions, respectively, p is the pressure, C is the concentration of the limiting substrate, D is the substrate diffusivity, and Q is the substrate consumption rate. The set of equations (2.9) to (2.12) and the appropriate boundary conditions are applied for all three reactor compartments: the lumen, membrane and ECS with a note that consumption terms, Q, in the lumen and the membrane are zero. In certain situations these general equations can be simplified such that for low porosity membranes, low lumen flow or high concentrations of cells in ECS, primary mass transport to the ECS is by diffusion [98].

Most models of hollow fibre reactors focus on the transport of limiting substrates, typically relatively small molecules in dilute solutions. The transport of macromolecules and concentration polarization were addressed in studies of protein transport in axial-flow hollow fibre reactors [111,114,115]. However, the effect of the transport and distribution of macromolecules, secondary flows in the ECS, cell sedimentation and migration still need to be further investigated.

3. Conclusion

Immobilised cell systems are progressively being developed, resulting in applications for the production of specialty and commodity chemicals in pharmaceutical, biomedical, and food technologies, as well as for waste treatments in environmental technologies. The design of these systems is mainly concerned with providing adequate cell environment including physical signals, nutrients and gases necessary for cell survival, and chemical factors inducing desired bioconversion. Immobilisation methods and support materials determine the local cell environment within the biocatalyst, while bioreactors govern overall fluid dynamic conditions and heat and mass transfer rates to and from the biocatalyst surface.

This chapter addressed immobilised cell systems with a particular emphasis on the bioreactor design. The major bioreactor types were presented, and the fluid dynamic and mass transfer properties were discussed. The design and operating parameters and approaches to mathematical modelling of different bioreactor types were outlined. An effort was made to refer to applications that can take advantage of specific bioreactor properties. In addition, alternative bioreactor designs intended to change or improve fluid dynamic conditions and mass transfer properties were briefly discussed.

References

[1] Nedovic, V.A.; Obradovic, B.; Leskosek-Cukalovic, I. and Vunjak-Novakovic, G. (2001) Immobilized yeast bioreactor systems for brewing - recent achievements. In: Thonart, Ph. and Hofman, M. (Eds.) Focus on Biotechnology, Volume 4: Engineering and Manufacturing for Biotechnology. Kluwer Academic Publishers, Dordrecht; pp. 277-292.

[2] Gil, G.H.; Jones, W.J. and Tornabene, T.G. (1991) Continuous ethanol production in a two-stage, immobilised/suspended-cell bioreactor. Enzyme Microb. Technol. 13: 390-399.

[3] Masschelein, C.A. and Andries, M. (1997) The Meura-Delta immobilised yeast fermenter for the continuous production of beer. Cerevisia 21(4): 28-31.

[4] Cao, Y.S. and Alaerts, G.J. (1995) Influence of reactor type and shear stress on aerobic biofilm morphology, population and kinetics. Wat. Res. 29: 107-118.

[5] Nicolella, C.; van Loosdrecht, M.C.M. and Heijnen, S.J. (2000) Wastewater treatment with particulate biofilm reactors. J. Biotechnol. 80: 1-33.

[6] Nicolella, C.; van Loosdrecht, M.C.M. and Heijnen, S.J. (2000) Particle-based biofilm reactor technology. Trends Biotechnol. 18: 312-320.

[7] Characklis, W.G.; Turukhia, M.H. and Zelver, N. (1990) Transport and interfacial transfer phenomena. In: Characklis, W.G. and Marshall, K.C. (Eds.) Biofilms. Willey Intersci., New York; pp. 265-340.

[8] Zlokarnik, M. (1972) Ruhrtechnik. In: Ullmann Encyklopaedie der Technischen Chemie, Band II, Verlag Chemie, Weinheim, pp 259.

[9] Zlokarnik, M. and Judat, H. (1987) Mixing, Bayer, Leverkusen, pp 1.

[10] Doran, P.M. (1995) Bioprocess engineering principles. Academic Press, London.

[11] Perry, R.H. and Green, D.W. (1997) Perry's chemical engineers' handbook. 7th edition, McGraw-Hill, New York.

[12] van der Pol, L. and Tramper, J. (1998) Shear sensitivity of animal cells from a culture-medium perspective. Trends Biotechnol. 16: 323-328.

[13] Baron, G.V.; Willaert, R.G. and De Backer, L. (1996) Immobilised cell reactors. In: Willaert, R.G.; Baron, G.V. and De Backer, L. (Eds.) Immobilised living cell systems: Modelling and experimental methods. John Wiley & Sons Ltd., New York; pp. 67-95.

[14] Wang, S.-J. and Zhong, J.-J. (1996) A novel centrifugal impeller bioreactor. I. Fluid circulation, mixing, and liquid velocity profiles. Biotechnol. Bioeng. 51: 511-519.

[15] Wang, S.-J. and Zhong, J.-J. (1996) A novel centrifugal impeller bioreactor. II. Oxygen transfer and power consumption. Biotechnol. Bioeng. 51: 520-527.

[16] Ting, Y.-P. and Sun, G. (2000) Use of polyvinil alcohol as a cell immobilisation matrix for copper biosorption by yeast cells. J. Chem. Technol. Biotechnol. 75: 541-546.
[17] Vogelsang, C.; Husby, A. and Ostgaard, K. (1997) Functional stability of temperature-compensated nitrification in domestic wastewater treatment obtained with PVA-SbQ/alginate gel entrapment. Wat. Res. 31: 1659-1664.
[18] Lamboley, L.; Lacroix, C.; Champagne C. P. and Vuillemard J. C. (1997) Continuous mixed strain mesophilic lactic starter production in supplemented whey permeate medium using immobilised cell technology. Biotechnol. Bioeng. 56: 502-516.
[19] Wang, D.I.C.; Cooney, C.L.; Demain, A.L.; Dunnill, P.; Humphrey, A.E. and Lilly, M.D. (1979) Fermentation and enzyme technology, John Wiley & Sons, Inc., New York, pp. 157.
[20] Godia, F.; Casas, C. and Sola, C. (1987) A survey of continuous ethanol fermentation systems using immobilised cells. Process Biochem. April: 43-48.
[21] Kurosawa, H.; Nomura, N. and Tanaka, H. (1989) Ethanol production from starch by a coimmobilised mixed culture system of *Aspergillus awamori* and *Saccharomyces cerevisiae*. Biotechnol. Bioeng. 33: 716-723.
[22] Prévost, H.; Diviès, C. and Rousseau, E. (1985) Continuous yoghurt production with *Lactobacillus bulgaricus* and *Streptococcus thermophilus* entrapped in Ca-alginate. Biotechnol. Lett. 7: 247.
[23] Sodini, I.; Corrieu, G. and Lacroix, C. (1996) Practical use of an immobilised cell bioreactor for continuous prefermentation of milk. In: Wijffels, R.H.; Buitelaar, R.M.; Bucke, C. and Tramper, J. (Eds.) Immobilised cells: basic and applications. Elsevier, Amsterdam; pp. 687-692.
[24] Park, H.-J. and Khang, Y.-H. (1995) Production of cephalosporin C by immobilised *Cephalosporium acremanium* in polyethyleneimine-modified barium alginate. Enzyme Microb. Technol. 17: 408-412.
[25] Chu, L. and Robinson, D.K. (2001) Industrial choices for protein production by large-scale cell culture. Curr. Opin. Biotechnol. 12: 180-187.
[26] Kappelhof, J.W.N.M.; Hoek, J.P. van der and Hijnen, W.A.M. (1991) Experiences with fixed bed denitrification using ethanol as substrate for nitrate removal from ground water. In: Proc. IWSA International Workshop Inorganic Nitrogen Compounds and Water Supply, Hamburg, Germany, November 27-29, 1991; pp. 101-112.
[27] Pajunen, E. (1996) Immobilized yeast lager beer maturation: DEAE-cellulose at Sinebrychoff. In: Monograph XXIV, Eur. Brew. Conv., Verlag Hans Carl Getranke-Fachverlag, Nurnberg (Germany); pp. 24-34.
[28] Kronlof, J. and Virkajarvi, I. (1999) Primary fermentation with immobilized yeast. In: Proc. 27th Congr. Eur. Brew. Conv., Cannes (France); The European Brewery Convention, Zoeterwoude (Netherlands); pp. 761-771.
[29] Chotani, C.K. and Constantinides, A. (1984) Immobilised cell cross-flow reactor. Biotechnol. Bioeng. 26: 217-220.
[30] Roca, E.; Flores, J.; Nunez, M.J. and Lema, J.M. (1996) Ethanolic fermentation by immobilised *Sacharomyces cerevisiae* in a semipilot pulsing packed-bed bioreactor. Enzyme Microb. Technol. 19: 132-139.
[31] Khattar, J.I.S.; Sarma, T.A. and Singh, D.P. (1999) Removal of chromium ions by agar immobilised cells of the cyanobacterium *Anacystis nidulans* in a continuous flow bioreactor. Enzyme Microb. Technol. 25: 564-568.
[32] Goksungur, Y. and Zorlu, N. (2001) Production of ethanol from beet molasses by Ca-alginate immobilised yeast cells in a packed-bed bioreactor. Turk. J. Biol. 25: 265-275.
[33] Chien, N.K. and Sofer, S.S. (1985) Flow rate and bead size as critical parameters for immobilised-yeast reactors. Enzyme Microb. Technol. 7: 538-542.
[34] Andersen, K.; Bergin, J.; Ranta, B.; and Viljava, T. (1999) New process for continuous fermentation of beer. In: Proc. 27th Congr. Eur. Brew. Conv, Cannes (France); IRL Press, Oxford; pp. 771-778.
[35] Tata, M.; Bower, P.; Bromberg, S.; Duncombe, D.; Fehring, J.; Lau, V.; Ryder, D. and Stassi, P. (1999) Immobilized yeast bioreactor systems for continuous beer fermentation. Biotechnol. Prog. 15: 105-113.
[36] Vega, J.L.; Clausen, E.C. and Gaddy, J.L. (1987) Maximizing productivity in an immobilised cell reactor. In: Shuler, M.L. and Weigand, W.A. (Eds.) Biochemical Engineering V. Annals of New York Academy of Sciences, New York; vol 506; pp. 208-228.
[37] Lienhardt, J.; Schripsema, J.; Qureshi, N. and Blaschek, H. (2002) Butanol production by *Clostridium beijerinckii* BA101 in an immobilised cell biofilm reactor. Appl. Biochem. Biotechnol. 98-100: 591-598.
[38] Lewis, V.P. and Yang, S.-T. (1992) Continuous propionic acid fermentation by immobilised *Propionibacterium acidipropionici* in a novel packed-bed bioreactor. Biotechnol. Bioeng. 40: 465-474.

[39] Yang, S.-T.; Zhu, H.; Li, Y. and Hong, G. (1994) Continuous propionate production from whey permeate using a novel fibrous bed bioreactor. Biotechnol. Bioeng. 43: 1124-1130.
[40] Yang, S.-T.; Huang, Y. and Hong, G. (1995) A novel recycle batch immobilised cell bioreactor for propionate production from whey lactose. Biotechnol. Bioeng. 45: 379-386.
[41] Yang, S.-T.; Lo, Y.-M. and Min, D.B. (1996) Xanthan gum fermentation by *Xanthomonas campestris* immobilised in a novel centrifugal fibrous-bed bioreactor. Biotechnol. Prog. 12: 630-637.
[42] van der Hoek, J.P.; Jong, R.C.M.; Kappelhof, J.W.N.M.; Hijnen, W.A.M.; Creusen, A.J.H.F.; Bekkers, A.J.M.E. and Feij, L.A.C. (1993) Nitrate removal from ground water by biological filtration using the fixed bed/ethanol process. In: Proc. European Water Filtration Congress, Ostende, Belgium, 15-17 March 1993, vol. 17; pp. 2.55-2.66.
[43] van der Hoek, J.P.; Kruithof, J.C. and Schippers, J.C. (1991) Design, operation and maintenance of a $35 m^3/h$ sulphur/limestone demonstration plant for nitrate removal from groundwater. In: Proc. 18th International Water Supply Congress, Copenhagen, Denmark, 25-31 May, 1991; pp. SS1-4 - SS1-10.
[44] Shan, J. and Zhang, T.C. (1998) Septic tank effluent denitrification with sulphur/limestone process. In: Proceedings of the 1998 Conference on Hazardous Waste Research, Snowbird, Utah, 18-21 May, 1998; pp. 348-362.
[45] Silva, A.J.; Varesche, M.B.; Foresti, E. and Zaiat, M. (2002) Sulphate removal from industrial wastewater using a packed-anaerobic reactor. Proc. Biochem. 37: 927–935.
[46] Lommi, H. (1990) Immobilized yeast for maturation and alcohol-free beer, Brew. Dist. Int. 5: 22-23.
[47] Pajunen, E. and Jaaskelainen, K. (1993) Sinebrychoff Kerava – a brewery for 90s, In: Proc. 24th Congr. Eur. Brew. Conv., Oslo (Norway); IRL Press, Oxford (UK); pp. 559-567.
[48] Buffiere, P.; Bergeon, J-P. and Moletta, R. (2000) The inverse turbulent bed: a novel bioreactor for anaerobic treatment. Wat. Res. 34: 673-677.
[49] Petersen, J.N. and Davison, B.H. (1991) Modelling of an immobilised-cell three-phase fluidized-bed bioreactor. Appl. Biochem. Biotechnol. 28: 685-698.
[50] Petersen, J.N. and Davison, B.H. (1995) Development of a predictive description of an immobilised-cell, three-phase, fluidized-bed bioreactor. Biotechnol. Bioeng. 48: 139-146.
[51] Buffiere, P.; Fonade, C. and Moletta, R. (1998) Mixing and phase hold-ups variations due to gas production in anaerobic fluidized-bed digesters: Influence on reactor performance. Biotechnol. Bioeng. 60: 36-43.
[52] Frankin, R.; Koevoetes, W.A.A.; van Gils, W.M.A. and van der Pas, A. (1992) Application of the Biobed upflow fluidized-bed process for anaerobic waste water treatment. Water Sci. Technol. 25: 373-382.
[53] Zoutberg, G.R. and Frankin, R. (1996) Anaerobic treatment of chemical and brewery waste water with a new type of anaerobic reactor: the Biobed EGSB reactor. Water Sci. Technol. 34: 375-381.
[54] Gonzales-Gil, G.; Lens, P.N.L.; Van Aelst, A.; Van As, H.; Versprille, A.I. and Lettinga, G. (2001) Cluster structure of anaerobic aggregates of an expanded granular bed reactor. Appl. Environ. Microbiol. 67: 3683-3692.
[55] Webb, O.F.; Davison, B.H.; Scott, T.C. and Scott, C.D. (1995) Design and demonstration of an immobilised-cell fluidized-bed- reactor for the efficient production of ethanol. Appl. Biochem. Biotechnol.; 51/52: 559-568.
[56] Merchuk, J. (1990) Why use air-lift bioreactors? Trends Biotechnol. 8: 66-71.
[57] Nedovic, V.A.; Leskosek-Cukalovic, I.; Milosevic, V. and Vunjak-Novakovic, G. (1997) Flavour formation during beer fermentation with immobilized *Saccharomyces cerevisiae* in a gas-lift bioreactor. In: Conference proceedings, Godia, F. and Poncelet, D. (Eds.) Int. Workshop Bioencapsulation VI "From fundamentals to industrial applications", Barcelona (Spain), August 30–September 1, 1997; UAB, Barcelona (Spain); T5.3, pp. 1-4.
[58] Heijnen, J.J. (1996) Scale-up aspects of immobilised cell reactors. In: Wijffels, R.H.; Buitelaar, R.M.; Bucke C. and Tramper, J. (Eds.) Immobilised cells: basics and applications. Elsevier, Amsterdam; pp. 497-504.
[59] Fields, P.R. and Slater, N.H.K. (1983) Tracer dispersion in a laboratory air-lift reactor. Chem. Emg. Sci. 38: 647-653.
[60] Verlaan, P.; Tramper, J.; van't Riet, K. and Luyben, K.Ch.A.M. (1986) Hydrodynamics and axial dispersion in an air-lift loop bioreactor with two and three-phase flow. In: Proc. Int. Conf. on Bioreactor Fluid Dynamics, Cambridge, England; pp. 15-17.
[61] Verlaan, P.; van Eijs, A.M.M.; Tramper, J.; van't Riet, K. and Luyben, K.Ch.A.M. (1986) Estimation of axial dispersion in individual sections in an airlift-loop reactor. Chem. Eng. Sci. 44: 1139-1146.
[62] Vunjak-Novakovic, G.; Jovanovic, G.; Kundakovic, Lj. and Obradovic, B. (1992) Flow regimes and liquid mixing in a draft tube gas-liquid-solid fluidized bed. Chem.Eng. Sci. 47: 3451-3458.

[114] Taylor, D.G.; Piret, J.M. and Bowen, B.D. (1994) Protein polarization in isotropic membrane hollow-fibre bioreactors. AIChE J. 40: 321-333.
[115] Labecki, M.; Weber, I.; Dudal, Y.; Koska, J.; Piret, J.M. and Bowen, B.D. (1998) Hindered transmembrane protein transport in hollow-fibre devices. J. Memb. Sci. 146: 197-216.

PART 4

PHYSIOLOGY OF IMMOBILISED CELLS: EXPERIMENTAL CHARACTERISATION AND MATHEMATICAL MODELLING

nucleus with a non-zero magnetic moment can be detected with NMR spectroscopy. The simplest example of an unpaired spin is the hydrogen nucleus, which is a single proton and has a spin of 1/2. The overall spin-state of a nucleus is quantised and only a limited number of energy states exist. For the hydrogen nucleus, the spin state can be either +1/2 or -1/2. When a hydrogen nucleus is placed in a static magnetic field (B_o), the spins will align so that their magnetisation is either parallel (+1/2) or anti-parallel (-1/2) to the field. At equilibrium, a net excess of spins is in the +1/2 state. Externally applied energy at the resonant frequency of the nucleus can be used to transfer spins from the low-energy state to the high-energy state. The frequency required for this transition can be calculated from the gyromagnetic ratio of the nucleus and the magnitude of the static magnetic field, $\omega = \gamma B_o$, and is generally in the radio frequency (RF) range. As the nuclear spins relax back to equilibrium, they give off energy at the same frequency as the excitation frequency, which is detected and recorded by the NMR spectrometer.

Individual spins precess about the main axis (generally called the z-axis) of the static magnetic field in a circular motion, similar to a spinning top. With a slight excess of spins in the lower energy state, the net magnetisation of the sample will be parallel to the main axis of the magnet. Manipulation of the net magnetic moment with RF pulses can be readily described with classical mechanics. If the spins are pulsed with energy at the resonant frequency, the net magnetisation can be tipped into a plane that is orthogonal to the static magnetic field (the x-y plane). The rotating x-y magnetisation produces a small sinusoidal voltage in the receiver coil that decays exponentially with time as the spins relax back to their equilibrium state. This signal is called a free induction decay (FID). The magnetisation in the x-y plane decays with a characteristic relaxation time T2, while the magnetisation returns to the z-axis with a characteristic relaxation time T1. In pure liquids, the two relaxation times are nearly equal, while in biological samples T2 is generally much shorter than T1. T2 relaxation is caused by many mechanisms including the interactions between spins in the sample and is called spin-spin relaxation. T1 relaxation results from interactions between the relaxing nucleus and the surrounding environment and is called spin-lattice relaxation.

The frequency components of the time domain signal generated with an RF pulse can be determined by Fourier transforming the digitised FID. The result is a spectrum of resonances that correspond to the various chemical environments of the nucleus being detected. The offset of each resonance from the reference frequency of a nucleus is termed the chemical shift and is usually expressed in parts per million (ppm) of the reference frequency. NMR spectra are generally rich with information about the chemical group in which the nucleus exists, as well as through bond and through space coupling to other nuclei with magnetic moments.

If following a 90° pulse, the x-y magnetisation of a sample is allowed to dephase due to inhomogeneities in the static magnetic field, the signal can be refocused with a 180° pulse applied about either the x or y axis. Either type of pulse will reverse the orientation of the spins in such a way that they will rephase and generate an "echo", which is an exponentially increasing and then exponentially decreasing signal. Spin-echoes are extremely useful in imaging and spatially resolved spectroscopy. Magnetic field gradients applied during RF pulses, during the echo development time or during the echo acquisition time can be used to spatially encode the NMR signal in two or three dimensions. A complete image can be generated from a series of echoes after appropriate Fourier transformation. A rigorous discussion of MRI fundamentals is given by Callaghan [5]. More simplified descriptions are also available [6]. Anatomical images are usually generated from the hydrogen signals of water and lipids. Metabolic maps can also be produced from a series of spatially encoded spectra. Finally, convection and diffusion can be mapped with techniques that use magnetic field gradients to selectively dephase signals of nuclei undergoing translational motion.

2.2. DETECTABLE NUCLEI AND THEIR USES

A list of biologically important nuclei that can be detected with NMR spectroscopy is given in Table 1, which lists the spin quantum number, the relative gyromagnetic ratio and the resonant frequency at 11.7 Tesla. The relative sensitivity of a nucleus varies strongly with its gyromagnetic ratio; for a given magnetic field strength, the relative sensitivity is proportional to $\gamma^{11/4}$ [5]. Of all nuclei, tritium is the most sensitive but it is not commonly used because it is unstable. The hydrogen nucleus, commonly referred to as "proton", is the most sensitive nucleus found in nature and is widely used for clinical MR spectroscopy and imaging. For *ex vivo* work, proton imaging has been used to study flow in hollow fibre bioreactors (HFBRs) [7, 8, 9] and packed beds of porous cellulose sponge [10]. Diffusion-weighting has been used to determine the distribution of cells in alginate beads [11] and HFBRs [12]. With the development of methods to selectively suppress the proton signal from water and extracellular metabolites, intracellular levels of amino acids, lipids and lactate have been determined in agarose embedded cells [13].

The predominant isotope of phosphorous in nature is ^{31}P, which has a moderately high magnetic moment. ^{31}P has been used extensively for monitoring intracellular levels of high-energy phosphate metabolites such as nucleoside triphosphates (NTP) and phosphocreatine. Phospholipids can also be detected with ^{31}P NMR, including phosphomonoesters (such as phosphocholine and phosphoethanolamine) and phosphodiesters (such as glycerol 3-phosphocholine and glycerol 3-phosphoethanolamine). Many tumour cells have significantly elevated levels of phosphomonesters and with some types of chemo- and radiotherapy, their magnitude can be reduced [14]. The reduction is currently being evaluated in clinical trials as an indicator of response to therapy.

^{13}C has also been used extensively for biological studies. Since essentially all carbon in nature is ^{12}C, ^{13}C-enriched compounds can be used to trace metabolic pathways [15] and determine flux rates [16]. NMR spectroscopy is especially useful for determining rates of cyclical pathways that involve metabolites that do not cross the cell membrane. For example, TCA cycle flux can be estimated from relative labelling patterns in glutamate, which in some tissues is in rapid equilibrium with α-ketoglutarate. Such pathways can not be studied with standard extracellular biochemical engineering approaches (*e.g.* mass balances). Moreover, experiments done with ^{13}C may yield unique information for identifying previously unknown metabolic pathways. ^{13}C is a much less sensitive nucleus than proton, but many methods have been developed to improve its detectability. These generally involve manipulation of the magnetisation for hydrogen atoms directly bonded to the carbon atom. The most common of these methods is the nuclear Overhauser enhancement (NOE), which can improve sensitivity by a factor of 3. Other methods include polarisation transfer and indirect detection (where ^{13}C magnetisation is "read" through the attached protons), which can be combined to give a 16-fold enhancement [3].

The predominant isotope of sodium found in nature is ^{23}Na. It is present at high levels in living tissue and has been used extensively in biological studies [17]. This nucleus has an intrinsic sensitivity similar to carbon. However because it has a quadrapolar magnetic moment and consequently a very high T1 relaxation rate, the sensitivity per unit time is also very high. Specialised techniques have been developed to discriminate between intracellular and extracellular sodium. These include triple-quantum filtration, which differentiates the two signals based on their relaxation rates [18], and shift reagents, which are excluded from the intracellular space and alter the chemical shift of extracellular sodium [19].

Both ^{15}N and ^{2}H are fairly insensitive nuclei but because they do not exist commonly in nature, both can be used for tracing metabolic pathways [20, 21]. Similar to ^{13}C, NOE, polarisation transfer and indirect detection can be used to enhance ^{15}N signals very significantly [3]. ^{35}Cl and ^{39}K are both insensitive nuclei but they have been used for biological studies because their concentration in tissue is fairly high [17]. Also, their T1 relaxation rates are comparable to that for sodium, whereby extensive signal averaging can be accomplished in a relatively short period of time.

^{17}O has gained recent popularity as a marker for aerobic metabolism. Because it is an insensitive nucleus, it generally is not detected directly. Rather it is often detected indirectly by its affect of water proton relaxation [22, 23]. It has not yet been used with immobilised cells but it could potentially be a very useful tool.

Table 1. NMR properties of biologically important nuclei.

Nucleus	*Spin Quantum Number*	*Natural Abundance (%)*	*Relative sensitivity* $(\gamma_x/\gamma_{^1H})^{11/4}$	*NMR Frequency at 11.7 Tesla (MHz)*
^{1}H	1/2	99.985	100.0	500.0
^{2}H	1	0.015	0.579	76.8
^{3}H	1/2	~0	119.4	533.3
^{7}Li	3/2	92.58	7.432	194.3
^{13}C	1/2	1.108	2.244	125.7
^{14}N	1	99.63	0.073	36.1
^{15}N	1/2	0.37	0.185	50.7
^{17}O	5/2	0.037	0.411	67.8
^{19}F	1/2	100	84.55	470.4
^{23}Na	3/2	100	2.583	132.3
^{31}P	1/2	100	8.316	202.4
^{35}Cl	3/2	75.5	0.168	49.0
^{39}K	3/2	93.1	0.022	23.3

Neither ^{19}F nor ^{7}Li naturally exist at very high levels in living tissue, which allows pharmacological agents containing these nuclei to be detected without interference from background signals. For example, ^{19}F has been used to monitor the pharmacodynamics of the chemotherapeutic agent 5-fluorouracil [24, 25]. Because ^{19}F is a very sensitive nucleus, it can be detected at very low levels (~1 mM). ^{19}F labelled compounds have also been used for detecting extracellular oxygen levels [26, 27, 28] and pH [29]. Lithium is an important pharmacological agent for the treatment of a number of psychiatric disorders and ^{7}Li NMR has been used to study pharmacodynamics in the brain [30, 31]. ^{7}Li has also been used to study lithium transport in cultured neuroblastoma cells, which could further the understanding of how lithium affects normal neurons [32].

2.3. NMR HARDWARE

2.3.1. Overview

There are four basic components to an NMR spectrometer. These include a primary magnet (generally a liquid helium cooled super-conducting coil), a radio frequency (RF) probe, an RF console that is used to transmit and receive radio frequency pulses, and a computer that controls the RF console and digitises the NMR signals. Many modern spectrometers are also equipped with micro-imaging gradient coils. These coils are used to create relatively small magnetic field gradients within the sensitive volume of the spectrometer and can be used to "localise" the NMR signal; that is, determine the spatial distribution of the detected nuclei. In this way, they perform the same function as the gradient coils of a clinical MR imager.

As the technology for constructing super-conducting magnets has advanced, so has the static field strength for NMR spectrometers. Narrow-bore (54 mm) spectrometers with 21.1 Tesla (900 MHz for proton) magnets are now commercially available. Wide-bore spectrometers (89 mm) with 14.1 Tesla magnets (600 MHz) are also available with micro-imaging gradients. Horizontal bore animal imaging systems (30 cm) are available with 11.7 Tesla (500 MHz) magnets. In general, the signal-to-noise from a biological sample (i.e. an aqueous sample that contains physiological levels of cations and anions) scales linearly with field strength. Because noise levels vary indirectly with the square root of acquisition time, doubling the field strength of the static magnet essentially quadruples the temporal resolution of an NMR spectrometer. Therefore, with improvements in super-conducting materials, even higher field strength spectrometers will likely be constructed in the future.

2.3.2. Magnetic field gradients coils

As mentioned in the previous paragraph, micro gradient coils are now commonly included in high-resolution NMR spectrometers. These gradients are generally used for micro-imaging [5] and for determining the spatial distribution of metabolites. They can also be used to suppress the water signal in a biological sample to allow other proton metabolites to be examined with a standard 16-bit analog-to-digital converter. Gradient coils can also be used to selectively suppress signals from moving spins, which allows diffusion and convection in biological samples to be studied non-invasively. These capabilities are very powerful for studying immobilised cells, where measurements of convection and diffusion of metabolites would be difficult to measure by any other means. These methods have very recently been adopted clinically for characterising pathology, especially in the brain [33]. They can also be used to discriminate between the intra and extracellular compartments *in vitro* [13].

2.3.3. RF probes

A critical component of any NMR spectrometer is the sample probe, which contains one or more resonant (inductive/capacitive) circuits that can be used to transmit and receive radio frequency radiation. These probes are commonly purchased from commercial manufacturers but they can also be constructed in laboratories with fairly good results. The noise level in biological NMR spectroscopy is generally determined by the characteristics of the sample more than those of the RF coil or the signal amplifying circuitry, whereby a relatively simple RF probe can perform nearly as well as a commercial RF probe [34]. The advantage of constructing customised probes is that the geometry can be optimised for the sample of interest. For example, probes have been constructed for HFBRs with atypical geometries (see below) [35, 36].

3. Methods commonly used for studying cells with NMR spectroscopy

3.1. HYDROGELS

3.1.1. Hydogel threads

Immobilisation in agarose threads (or filaments) is a widely used method for examining living mammalian cells with NMR spectroscopy. Agarose is a thermoset polysaccharide, which was originally used by Foxall et al. for Chinese hamster lung fibroblasts [37]. It has been used to study a variety of anchorage-dependent and anchorage-independent cell types including breast cancer cells [38, 39] erythrocytes [40], hepatocytes [41], lymphocytes [42] and cerebral astrocytes and neurons [43]. To immobilise cells, they are mixed with a liquid agarose solution, and the mixture is extruded through small diameter tubing into long filaments. As the filaments are formed, the agarose cools and forms a gel. This method is relatively simple to implement because the agarose filaments are commonly perfused in standard NMR tubes. The filaments are held in the tube with either a perforated Teflon disc or a polyethylene filter. Medium is usually recirculated through the cell mass since the consumption of metabolites other than oxygen per pass is minimal. Cells can be immobilised at high densities for short (~1-2 days) experiments. The total number of cells in the sensitive volume (approximately 1 ml in a 10-mm RF probe) of the spectrometer is typically 1×10^8 [39]. With a 9.4 Tesla magnet, quantifiable ^{31}P spectra, with signal/noise > 5:1, can be obtained in 10 minutes.

Agarose is a good immobilisation support for cells that grow in suspension and cells without fastidious surface-attachment requirements. Some anchorage dependent cells can be sustained in a non-proliferating state in agarose. For example, MDA-MB 231 breast cancer cells do not proliferate in agarose. For such cells, a gel made of basement membrane proteins (*e.g.* Matrigel®) can be used; extensive growth of MDA-MB 231 cells was observed over an 8-day period [44]. Primary astrocytes and neurons from rat embryos [45] and neonatal rat cerebral cortex [43] have also been immobilised in Matrigel® for NMR studies. The disadvantages of using Matrigel® are that it is very expensive and diffusion of macromolecules such as transferrin may be hindered.

Problems with cell outgrowth from gel filaments are not commonly reported, but would likely be important with anchorage independent cells. Long term mechanical stability (~weeks) of the filaments has also not been discussed. In early studies, filament diameters on the order of 0.5 mm were used, which would very likely result in diffusion limited metabolism for threads containing high cell numbers. More recently, 0.3-mm threads have been used, which would have a maximum diffusion distance of 150-μm. However, this is still nearly twice the normal diffusion distance observed for vascularised tissue *in vivo* and much larger than the diffusion distance for the HFBR described below.

A major advantage of using gels to study tumour cells is that they are well suited for monitoring metabolic response to therapy. Many cancer therapeutics induce apoptosis, which leads to the formation of cellular "blebs" that can fall away from the primary cell mass. With gels, the blebs can be retained so that their intracellular constituents remain in the NMR detectable region.

3.1.2. Alginate beads

A commonly used immobilisation method for NMR studies of mammalian cells is entrapment in calcium alginate beads. This method has been used to study normal erythrocytes [46], lymphocytes [47, 48], Burkitt lymphoma cells [47], metastatic human prostate carcinoma cells[47], mouse fibrosarcoma cells [49], and mouse mammary adenocarcinomas [26]. Insulinomas have also been studied extensively in alginate gels and usually the gels are coated with an immunoprotective layer such as poly-lysine [50]. The immunoprotective layer is necessary since the ultimate goal is to use the beads as therapeutic implants for type I diabetics.

For NMR studies, alginate beads are usually packed into a standard NMR tube and perfused in a manner similar to agarose threads. In theory, the beads should allow for better mass transfer than threads, since the surface area to volume ratio for a sphere is much higher than for an equivalent diameter cylinder. However, bead sizes used in early studies were large, 2.6-mm in diameter [49, 26], which would result in an excessively long maximum diffusion distance of 1,300 microns. With such beads, cells can only reach high densities at the outer-most region and metabolic gradients within them are necessarily large. Using improved bead formation technology, smaller calcium alginate beads 1.0 mm in diameter (and coated with polylysine) have been used to study insulinoma metabolism [51] with NMR spectroscopy. Smaller beads can also be made [52] but when used in a packed bed inside a spectrometer, the axial pressure drop is very large at the flow rates necessary for adequate for oxygen delivery [53]. In a 22-ml packed bed reactor, 1.5×10^9 cells could be sustained in 1-mm beads that contained 7×10^7 cells/ml of gel [53]. With such a large number of cells, the recirculated medium must be reoxygenated and depleted of CO_2 in a gas exchange module. With a horizontal animal imager operating at 4.7 Tesla, ^{31}P spectra with good signal to noise (~ 10:1) were acquired in 27 minutes.

Some anchorage independent cells, such as hybridomas, can grow out of calcium alginate gels [54]. Metabolically active cells flowing freely outside of the NMR detected volume would significantly complicate the interpretation of NMR observations. For example, in kinetic experiments of ^{13}C-labeled metabolite uptake, some consumption would occur in the NMR detectable volume and some would not.

3.2. MICROCARRIERS

Both porous and non-porous microcarriers have been used to study cells with NMR spectroscopy. Non-porous microcarriers that have been used include dextran (Cytodex-1 and Cytodex-2, Amersham Biosciences, Upsala, Sweden) and collagen-coated dextran (Cytodex-3, Amersham), solid polystyrene (Biosilon, Nunc, Roskilde, Denmark), solid plastic (unspecified composition) with a collagen coating (Solohill, Ann Arbor, MI, USA), and polyacrolein-agarose (Galistar, no longer commercially available). Cell types that have been studied on non-porous microcarriers include: mouse embryo fibroblasts [55], cerebral cortical neurons [56], rat glioma cells and astrocytes [57], human prostate cancer cells [58], porcine vascular endothelial cells [59], and human mammary adenocarcinomas [58, 60]. Non-porous microcarriers are generally used in packed beds inside standard NMR tubes. Medium is continuously pumped through the bed and is recycled, since the consumption of metabolites per pass is very small. Flow rates used are on the order of 1-2 ml/min and are usually limited by the pressure drop through the bed. The beads are generally retained within the NMR tube with porous plastic filters (frequently polyethylene) that are precisely cut to match the inside diameter of the NMR tube. Of the non-porous microcarriers, those manufactured by Nunc (Biosilon) and Solohill are the best suited for perfused NMR studies because they are essentially incompressible. This characteristic allows fairly high flow rates of medium through the tube, without compression of the fixed bed. With compressible microcarriers such as Cytodex-3, the pressure drop limits the flow rate and hence the supply of oxygen, which is generally the first metabolite to become rate limiting.

The primary advantage of using non-porous microcarriers is that the cells grow in thin layers on the surface of the beads. The diffusion distance is very short and metabolite consumption is not diffusion limited. The disadvantages of using non-porous microcarriers are: (1) no cells grow in the core of the beads and much of the NMR detectable volume is wasted; (2) at high medium flow rates, the cells are subjected to high shear forces; and (3) with growing cells the interstitial space between the beads shrinks with time, which will increase the pressure drop through the bed. However, with quiescent or slowly growing cells, long-term experiments should be possible. Not all organisms grow well on non-porous microcarriers and in some cases the surface must be coated with proteins, such as fibronectin or poly-lysine, to affect adequate cell attachment.

For 10-mm NMR tubes, the total number of cells is approximately 3-5 x 10^7 [61]. Much higher cell numbers can be sustained in a 20-mm NMR tube, since the detected volume is approximately 10 times larger. However, the sensitivity of a 20-mm NMR probe per nucleus is only ~1/2 that for a 10-mm probe (because of its larger diameter) and therefore switching to the larger probe gives a signal enhancement of approximately only 3-4 fold. Nevertheless, this significantly reduces the scan time since the signal to noise in NMR spectroscopy increases with the square root of the acquisition time. In a recent study, 8 x 10^8 EMT6/SF cells were examined on collagen-coated microcarriers in a 20-mm NMR tube and quantifiable ^{31}P spectra were obtained in less than 60 seconds, with a 9.4 Tesla magnet [62]. This is comparable to the best signal to noise obtained in HFBRs.

Porous microcarriers have not been used for many NMR studies. The primary reason for this is that they are compressible and when they are used in a packed bed, they can only be perfused at very low flow rates. Porous collagen microcarriers have been used successfully in loosely packed beds to culture RIF-1 mouse fibrosarcomas [63, 64]. In a 10-mm NMR tube, the total cell number was ~1 x10^8. These microcarriers have also been used for primary cerebellar neurones but very low cell densities were observed [43].

3.3. HOLLOW FIBRE BIOREACTORS

3.3.1. Overview

HFBRs have proven very useful for cultivating many mammalian cell types, both anchorage-dependent and anchorage-independent, at high cell concentrations. They have been used for the commercial manufacture of high-value proteins in the pharmaceutical industry and they may have use clinically as artificial organs, such as pancreas and liver. For example, an HFBR containing hepatocytes was recently tested in patients with severe liver dysfunction [65].

HFBRs have been used for NMR studies of intracellular metabolites in hybridomas [66, 67, 68, 69], melanomas [70], Chinese hamster ovary cells [71], chondrocytes [72], chicken embryo fibroblasts [73], and hepatocytes [74]. Total cell numbers as high as 10^9 have been sustained in the sensitive volume of spectrometers.

Porous collagen microcarriers have also been injected into the extracapillary space (ECS) to increase the surface area available for anchorage-dependent cells [12, 75]. While the additional surface area would support a higher cell density, the microcarriers would probably hinder diffusion and convection in the cell mass. This may result in fairly large gradients of metabolites, unless the organisms being studied had especially low metabolic rates.

One problem with trying to use HFBRs for NMR spectroscopy is that they are generally not designed for use inside a standard probe. NMR samples are usually analysed in precisely manufactured cylindrical tubes; bioreactors typically have side ports for accessing the ECS. One solution to this problem is to construct customised NMR probes, with RF coils that fit tightly around the reactor for optimal signal to noise [35, 36, 54]. Another solution to this problem is to custom manufacture the HFBR so that all of the ports are at one end and there are no projections from the sides. Units have been built that fit into 10-mm [73] and 25-mm NMR probes [69].

HFBRs are currently used for the commercial manufacture of monoclonal antibodies with hybridomas (Cellex, Minneapolis, MN, USA). To reduce product recovery costs, these bioreactors are often designed so that the antibodies remain on the shell side. This design is not optimal for producing a homogeneous environment in the cell mass and nutrient gradients probably exist. Metabolic rates determined with these reactors will likely be mass transfer limited. Much has been written about metabolic gradients in HFBRs [76, 77] and their suitability for NMR spectroscopy studies of hybridoma metabolism has been questioned. However, HFBRs designed specifically for improved homogeneity have been shown to produce more uniform environments than many of the methods described above [36, 54]. To minimise axial gradients, short reactors should be used with high intra-luminal flow rates. For optimal mass transfer in the radial direction (from the fibre lumen to the shell space) the fibres should have thin walls, high porosity, and a large pore size. In addition, the fibres should have a small diameter and the number of fibres should be high to maximise the surface area available for mass transfer.

For NMR studies, the fibre type that has been used to obtain the highest cell densities is made of a composite of cellulose acetate and cellulose nitrate [12, 36, 69]. These fibres have an 80% porosity, a 0.2 μm pore size and an 80-μm wall thickness. Reactors manufactured with these fibres by Microgon Inc. (currently part of Spectrapor Laboratories, Rancho Domingues, CA, USA) contained 1050-1900 fibres with an outer diameter of 360 μm. The inner diameter of these bioreactors was 2.2 cm and the maximum diffusion distance in the cell mass is only 60 microns with 1900 fibres. This is comparable to the extra-luminal diffusion distance in mammals. From a culture of hybridomas grown to approximately 1×10^9 cells, quantifiable NTP signals were acquired in just 80 seconds with a 9.4 Tesla system [36]. With improved spectral line shape, even higher signal to noise was subsequently demonstrated by Gillies et al. [69] at the same frequency. In addition, very long-term studies are possible with HFBRs and experiments several weeks in duration have been reported [66, 69]. These devices are generally operated as recycle reactors, since the conversion per pass of all nutrients other than oxygen is negligible [36]. An artificial lung is used for medium re-oxygenation. The medium pH can also be controlled by adjusting the lung CO_2 level.

3.3.2. Experimental examination of oxygen transport limitations

Oxygen is only sparingly soluble in aqueous media and gradients in its concentration in HFBRs can be problematic. In a study with 4A2 hybridomas, an 8-cm long, 2.2-cm diameter HFBR, containing 1600 fibres (MC1600) with the characteristics described in the previous section, was used with a lumenal flow rate of ~200 ml/min [78]. The fibre volume was 42% of the reactor volume and the maximum diffusion distance in the shell space was approximately 80 microns. Oxygen probes, located at the inlet and the outlet of the reactor, indicated that the axial drop of oxygen could be limited to just 0.04 mM (20% of saturation with air), when the reactor contained approximately 1×10^9 cells. Therefore, if the oxygen concentration at the inlet of the reactor is maintained near 0.20 mM (100% of saturation with air), the reactor lumen concentration will remain above levels that are metabolically limiting. In addition, the 0.04 mM axial drop should not significantly affect the cells in the reactor, since hybridoma metabolism is not sensitive to oxygen concentration over a wide range [79].

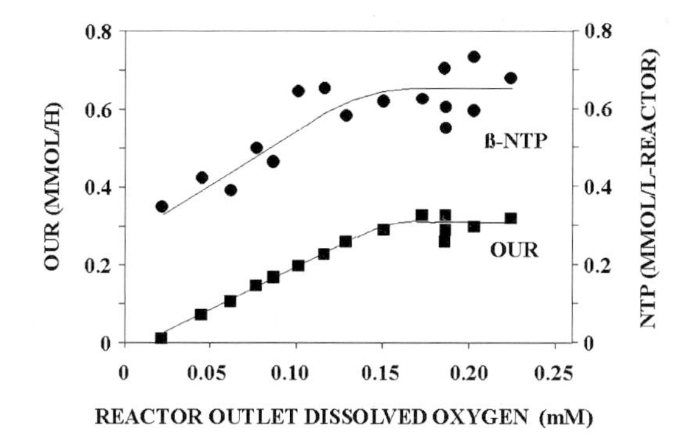

Figure 1. Effect of HFBR outlet dissolved oxygen concentration on total nucleoside triphosphates (β-NTP) and oxygen utilisation rate. The reactor contained 1600 fibres and a cell concentration of approximately 1.6 x 10^8 cells/ml of ECS [78].

While axial gradients can be well controlled by using high lumen flow rates, radial gradients can not. Lumen flow rates can impact the extent of Starling flow [8] but when the shell side is filled with cells, little or no convection is possible. This is especially true with hybridomas, which are not contact inhibited and tend to grow in dense masses that are difficult to perfuse. The extent of oxygen mass transfer limitations can be evaluated experimentally by varying the inlet oxygen concentration and monitoring the change in oxygen consumption and ATP level. An overall assessment of the extent of mass transfer limitations in HFBRs was made for 4A2 hybridomas grown in an MC1600 HFBR [78]. The cells were grown in a reactor for approximately 10 days and reached a concentration of 1.6 x 10^8 cells/ml of ECS. By adjusting the oxygen concentration in the system lung, the outlet oxygen concentration was varied from 0.23 to 0.03 mM while the oxygen consumption rate and ATP levels were monitored continuously (Figure 1). These parameters were relatively unaffected by reductions in the reactor oxygen concentration, until the outlet concentration was below ~0.12 mM. If the culture had been severely oxygen limited, reduction of the inlet oxygen concentration would have resulted in reduced ATP and OUR. Despite their high rates of lactate formation, hybridomas derive a significant portion of their metabolic energy from aerobic metabolism of glucose and glutamine [68].

3.3.3. Calculated radial distributions of metabolites

3.3.3.1. Oxygen. Calculated concentration profiles for a number of metabolites have been presented for 4A2 hybridomas grown in Microgon HFBRs [54]. A brief description of the most important findings is presented below. For these calculations, all shell-side mass transport was assumed to be diffusive for the reasons given in the previous section. Radial gradients within the fibre lumen were assumed to be negligible. The accuracy of this assumption is discussed below. Cells were assumed to be spherical with an average diameter of 15 microns. Because hybridomas are not subject to contact inhibition of growth, the diffusivities for metabolites in the cell mass were assumed to be the same as those observed for the interstitial spaces of tumours [80, 81, 82]. Diffusivities used for the calculations are summarised in Table 2.

Table 2. Diffusivities used to calculate radial concentration profiles. w = water, used for diffusivity in the fibre lumen, fm = fibre membrane, c = cell mass.

Diffusivities at 37°C	D_w (cm^2/s)	D_{fm} (cm^2/s)	D_c (cm^2/s)
Oxygen	3.0×10^{-5}	1.7×10^{-5}	1.5×10^{-5}
Glutamine	1.0×10^{-5}	3.5×10^{-6}	3.0×10^{-6}

Metabolite consumption rates were assumed to be zero order in metabolite concentration. In chemostatic studies, Miller et al. [79] and Ozturk [83] both found that variations in oxygen concentration had little affect on metabolism above 0.02 mM. Also, the K_s values for metabolites such as glucose and glutamine are generally very low. The oxygen consumption rate was assumed to be 0.20 mmol/10^9 cells/hr [84]. The cell density was assumed to be 1.6×10^8/ml of ECS.

The governing equation for diffusion in the fibre membrane is:

$$D_{fm}\left(\frac{1}{r}\frac{d}{dr}\left(r\frac{d[C]}{dr}\right)\right)=0 \tag{1}$$

and for the cell mass is:

$$D_C\left(\frac{1}{r}\frac{d}{dr}\left(r\frac{d[C]}{dr}\right)\right)=q[X] \tag{2}$$

In these equations, r is the distance from the centre of the fibre lumen, $[C]$ is the concentration of the metabolite of interest, $[X]$ is the concentration of cells in the ECS and q is the consumption rate per cell of the metabolite of interest. These equations were solved analytically with the boundary conditions detailed previously [54]. The calculated profiles indicate that the oxygen concentration in the cell mass would not be metabolically limiting unless its concentration at the reactor outlet was below 0.07 mM. This is significantly lower than the experimentally observed value of 0.12 mM. Therefore, the model over estimates the level of oxygen in the cell mass. To correct this, the effect of diffusion in the fibre lumen was included in the concentration profile calculations. The Reynolds number for flow rate in the fibres was on the order of 10, so the flow is laminar. The governing equation for radial diffusion in the presence of laminar axial flow in a cylinder is:

$$v_z\frac{\partial C}{\partial z}=\frac{D_W}{r}\frac{\partial}{\partial r}\left(r\frac{\partial C}{\partial r}\right) \tag{3}$$

where, v_z is the velocity parallel to the long axis of the fibre. This equation was solved analytically and combined with the solutions to the first two equations [54]. The resulting calculated oxygen profile is shown below in Figure 2. The data indicate that oxygen limitation will occur when the average outlet oxygen concentration is < 0.10 mM. This is nearly equal to the experimentally determined value of 0.12 mM.

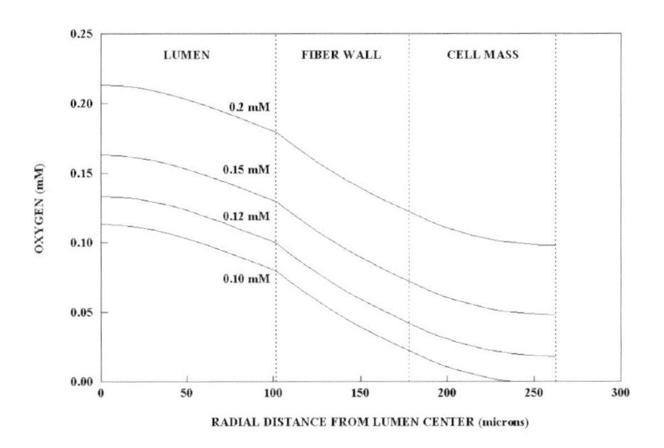

Figure 2. Calculated oxygen concentration profiles for a MC-HFBR containing 1600 fibres and 1.6 x 10^8 cells/ml. The curves are labelled with their average concentrations.

The results shown in Figures 1 and 2 clearly indicate that if the inlet oxygen concentration is maintained near 0.2 mM (saturation with air), the concentration in the cell mass will be above 0.1 mM. This level is much higher than the level associated with oxygen limited growth. Miller et al. [79] found that glucose and glutamine consumption rates were not affected by oxygen concentration, if it was maintained above 0.02 mM. Therefore, MC1600 HFBRs are more than adequate for culturing hybridomas without oxygen mass transfer limitations.

However, these reactors are not well suited for studying the effect of very low oxygen concentrations on metabolism. A thorough examination of the metabolic effects of oxygen would require that cultures be examined at concentrations near the K_s value (oxygen concentration for half maximal consumption rate). For murine hybridomas, this value is probably on the order of 0.001 mM [85]. Oxygen concentration profiles in the cell mass for MC1600 are very non-uniform at this level.

Mass transfer in the radial direction can be improved by increasing the number of fibres in the reactor. In 1991, Microgon succeeded in constructing reactors with 1900 fibres (MC1900) of the same type and size as the MC1600 HFBR. Calculated oxygen profiles for an MC1900 reactor with 1.6 x 10^8 cells/ml demonstrated that oxygen transfer rates would not be metabolically limiting above a reactor outlet concentration of 0.08 mM. The effect of oxygen concentration on NTP levels, determined with ^{31}P NMR, was evaluated experimentally with a culture grown in an MC1900 HFBR for 15 days. This culture was grown slightly past saturation; a significant amount of dead cell mass was present in the extra-capillary space. If dead cell mass packs tightly in the reactor, it may adversely affect mass transfer rates. The inlet oxygen concentration was varied from 0.16 to 0.28 mM and had no impact on NTP levels [54]. Therefore, these reactors were not oxygen transfer limited under normal operating conditions (0.20 – 0.24 mM oxygen at the inlet).

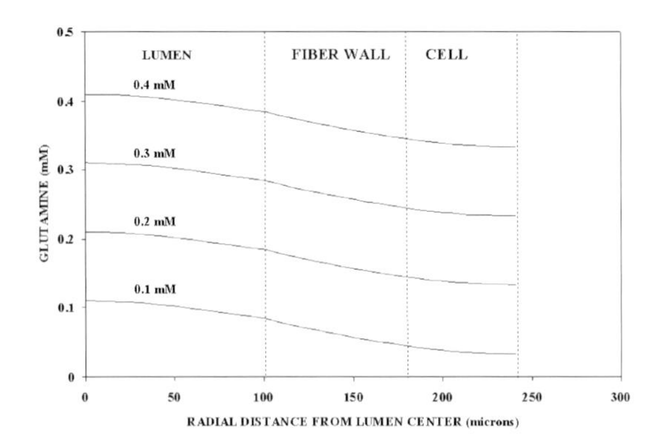

Figure 3. Calculated glutamine concentration profiles for a MC1900 HFBR containing 1.6 x 10^8 cells/ml. The concentrations shown near the curves represent the average concentrations in the fibre lumina for each curve.

3.3.3.2. Glutamine. Additional calculations were performed to determine if other metabolites in HFBRs would limit hybridoma metabolism in MC1900 HFBRs [54]. The calculations indicate, that at concentrations typically used in experiments (>1 mM), gradients of glucose and glutamine were much smaller than they were for oxygen. The primary reason for this is that the lumen concentration is much higher. To determine whether or not MC1900 HFBRs are useful for examining the effect of low glutamine concentrations on hybridoma metabolism, concentration profiles were calculated assuming a cell density of 1.6 x 10^8 cells/ml and a glutamine consumption rate of 0.042 mmol/10^9 cell-hr [68]. The results are shown in Figure 3. The gradients in glutamine concentration are not nearly as steep as those for oxygen. Therefore, these devices could be used to study the effects of low glutamine concentration on hybridoma metabolism.

4. Applications of NMR spectroscopy to the study of cellular physiology

4.1. HYBRIDOMAS

Steady-state ^{13}C NMR spectroscopy studies have been conducted with 4A2 hybridomas cultured in MC1900 HFBRs. Cultures grown to 1-2 x 10^8 cells/ml were infused with [1-^{13}C] glucose, [2-^{13}C] glucose and [3-^{13}C] glutamine to examine the pathways of primary and secondary metabolism [66, 67]. NMR spectra were acquired continuously of intracellular metabolites and at selected times for extracellular metabolites. A typical spectrum acquired during [1-^{13}C] glucose infusion is shown in Figure 4. The metabolic map in Figure 5 shows the pathways responsible for the labelling patterns observed.

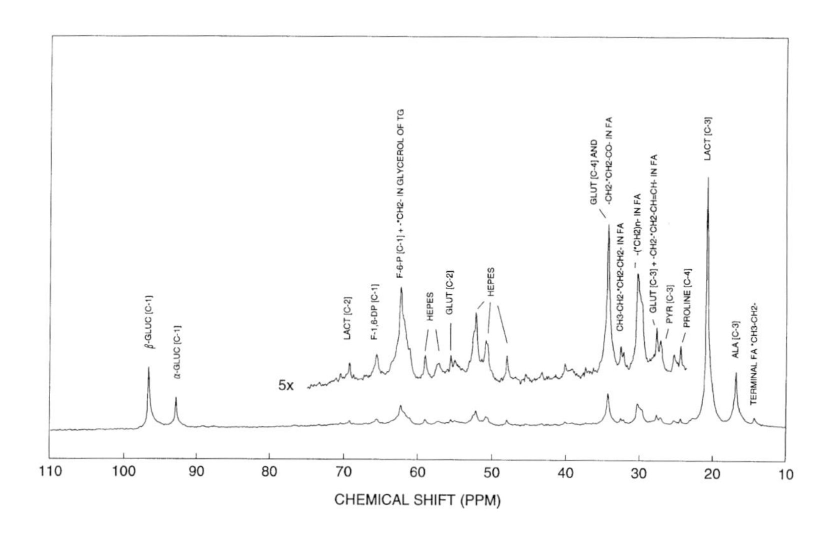

Figure 4. ^{13}*C NMR spectrum of 4A2 hybridomas grown in an HFBR during infusion of [1-*^{13}*C] glucose [66].*

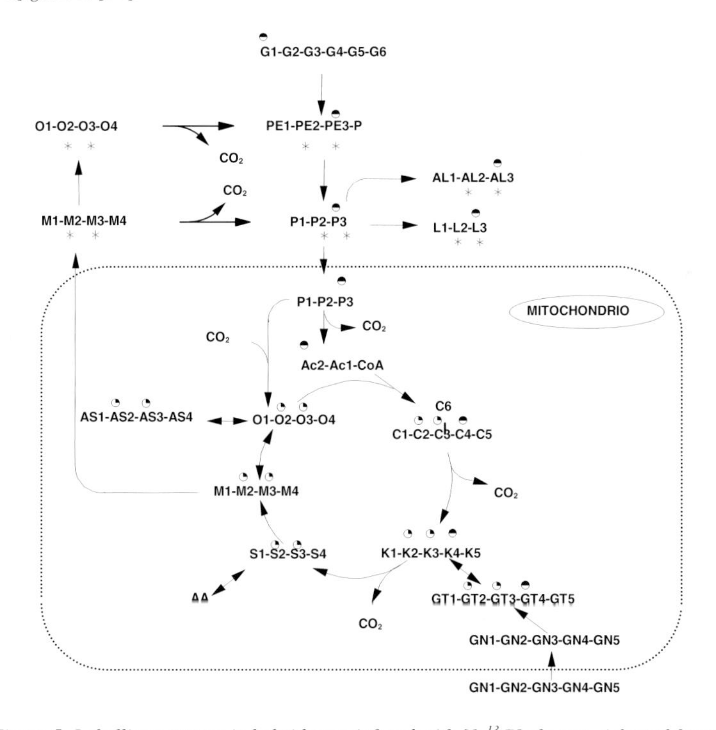

Figure 5. Labelling patterns in hybridomas infused with [1-^{13}*C] glucose. Adapted from [66].*

Addition details were published previously [66]. Labelling was observed in glutamate-2, 3, and 4, which is believed to reflect labelling in α-ketoglutarate-2, -3 and -4 [16]. The incorporation of labelled pyruvate into the tricarboxylic acid (TCA) cycle appears to occur through only pyruvate dehydrogenase and not pyruvate carboxylase as indicated by strong labelling in glutamate-4 and very weak labelling in glutamate-2 and 3. Preliminary calculations based on labelling of lactate and alanine showed that 76% of pyruvate was derived directly from glycolysis. Some was also derived from the malate shunt, the pyruvate/malate shuttle associated with lipid synthesis, and the pentose phosphate pathway. The rate of formation of pyruvate from the pentose phosphate pathway was at least 4% of that from glycolysis. Thc malate shunt rate was approximately equal to the rate of glutamine uptake. The TCA cycle rate between isocitrate and alpha-ketoglutarate was 110% of the glutamine uptake rate. Significant labelling was also observed in many mobile lipid groups. The rate of incorporation of glucose-derived acetyl-CoA into lipids was 4% of the glucose uptake rate. An extensive stoichiometric model was developed to try to calculate many additional fluxes [67]. However, subsequent data and analysis demonstrated that the model inaccurately overestimated the rate of lipid formation [68].

Glutamine is a very important nutrient for hybridomas that is used both catabolically and anabolically. Abrupt decreases in glutamine level have been shown to cause a dramatic increase in antibody production [67,68]. Also, minimizing its use in culture medium may be important since by-product ammonia formation can be inhibitory [85].

^{13}C NMR spectroscopy has been used to investigate the effects of extracellular glutamine levels on primary metabolism of 4A2 hybridomas [67,68]. Intracellular metabolites were labelled by infusing a culture inside an MC1900 HFBR with a constant level of [1-^{13}C] glucose. During the infusion, the glutamine level was varied. The effects of reduced glutamine concentration are illustrated in Figures 6 and 7.

When the glutamine level in the recirculating medium was abruptly reduced from 0.67 mM to approximately 0 mM, a significant increase was observed in the rate of antibody secretion and glucose consumption. Ammonia and alanine secretion were also increased but no significant change was observed in glutamate ^{13}C labelling.

When the extracellular glutamine level was reduced from ~0.3 mM to less than 0.1 mM and then kept constant for approximately 10 hours, glucose consumption and lactate formation were surprisingly reduced (Figure 7). Alanine and ammonia formation were also reduced and a slight reduction in glutamate-4 occurred. If glutamine was being used primarily for energy production and was rate limiting for that purpose, then reducing its extracellular concentration should have resulted in an increase in glucose consumption and glutamate-4 labelling. The results suggest that under the conditions studied, significant futile cycling occurs in energy producing pathways when excess glucose and glutamine are available. At 0.08 mM extracellular glutamine, some anabolic pathways were also affected as indicated by changes in metabolism involving amine groups [68].

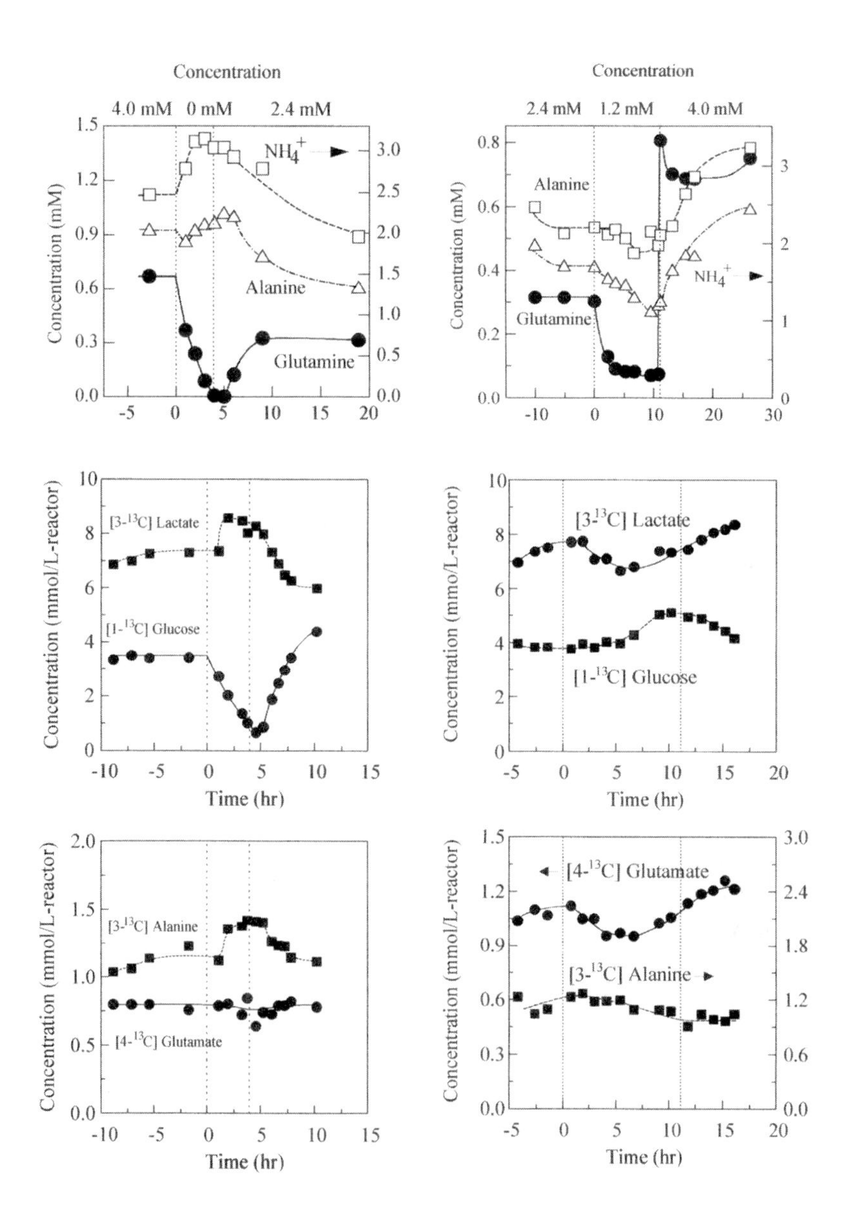

Figure 6. Changes in ^{13}C labelling from [1-^{13}C] glucose during a rapid reduction in glutamine concentration [68].

Figure 7. Changes in ^{13}C labelling form [1-^{13}C] glucose during step changes in glutamine concentration [68].

4.2. CANCER CELL METABOLISM

4.2.1. Differences between MCF7 wild-type and drug resistant cells

One potentially important application of *in vivo* NMR spectroscopy is in distinguishing chemotherapy resistant cells from chemotherapy responsive cells before therapy is initiated. This information would be very valuable in designing therapeutic approaches to individual cancers. A number of ^{13}C and ^{31}P NMR experiments have been conducted with the oestrogen-responsive human mammary adenocarcinoma, MCF7. The parental or wild-type (WT) MCF7 cells have been compared to a drug resistant sub-line, which has markedly reduced sensitivity to adriamycin (ADR). ADR cells express the MDR (multiple drug resistance) gene and can pump cytotoxic drugs from the intracellular space to the extracellular space. Because the process requires energy, ADR cells were hypothesised to have different catabolic characteristics than wild-type cells. In initial studies, ^{31}P spectra of both cell lines were acquired during growth in agarose threads (see Figure 8) [86]. The resistant cells had significantly higher levels of phosphocreatine (peak 4) and significantly lower levels of glyceryl-phosphoethanolamine (peak 3a) and glyceryl-phosphocholine (3b) and uridine diphosphoglucose (peak 7).

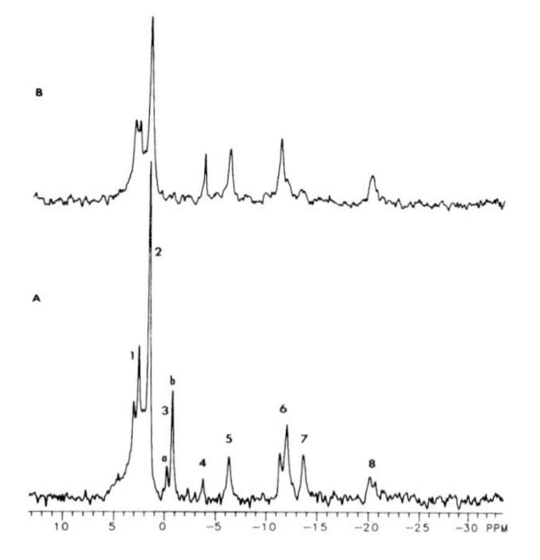

Figure 8. ^{31}P spectra of wild type (bottom) and adriamycin resistant MCF-7 cells [86]. Ordinate, chemical shift (ppm)

Another important characteristic that distinguished the ADR cells from the WT cells is that the glucose consumption rate of the ADR cells is much higher. This was demonstrated clearly during consumption of the glucose analogue, 2-deoxyglucose [87]. This compound enters cells through glucose transporters but is then trapped inside after it is phosphorylated and cannot be metabolised further. ^{31}P studies clearly showed that the rate of accumulation of phosphorylated 2-DG in ADR cells was much higher than it was in the WT cells. These results were confirmed with ^{13}C NMR experiments that were conducted with 2-DG labelled in the 6 position. Similar results were observed when both types of cells were allowed to grow sub-cutaneously in nude mice, indicating that the *in vitro* observations were relevant to *in vivo* metabolism.

Another study, which showed a clear distinction between the ADR cells and their wild type parental line, was done with [1-^{13}C] glucose. Cells immobilised in agarose threads were infused with medium containing [1-^{13}C] glucose and were monitored with ^{13}C NMR spectroscopy. For the ADR cells, strong labelling was observed in glutamate 4 and a lesser amount was observed at glutamate 3. For the WT cells, a very small amount of label was detected at glutamate-4. These findings indicate that the ADR cells have significantly more TCA cycle activity, which is consistent with the belief that drug resistance requires higher levels of energy production.

4.2.2. Effects of chemotherapeutics on phosphate metabolism of T47D cells

The effects of chemotherapeutic agents on T47D breast cancer cells have been examined with NMR [88]. Some agents have been shown to cause a marked increase in NTP (with ^{31}P NMR) levels shortly after administration, while others have had no effect at all [89]. Typical changes in NTP levels of T47D cells caused by adriamycin (doxorubicin) are shown in Figure 9. A significant increase in NTP, but not phosphocholine (PC) was observed with 10^{-5} M adriamycin. (Note that the ordinate scale is logarithmic). HPLC analysis showed that the increase was primarily in ATP and GTP, while UTP and CTP were unchanged. Similar results were observed with actinomycin-D and daunomycin. In contrast, cytotoxic doses of cytosine arabinofuranoside or cis-platin did not have any effect on NTP levels. Also, tamoxifen dosages, which caused growth arrest, did not change NTP levels.

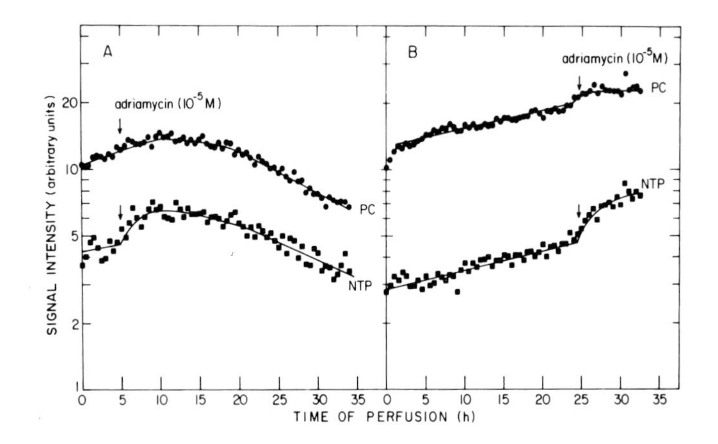

Figure 9. Effect of adriamycin on NTP and phosphocreatine levels in T47D cells during growth on non-porous polyacrolein-agarose microcarriers.

The effect of actinomycin-D and doxorubicin on T47D cells was further examined with ^{13}C NMR during infusion of [1-^{13}C] glucose. A slight (20%) reduction in the rate of [3-^{13}C] lactate formation was observed with actinomycin-D and no change was observed with doxorubicin. However, the rate of accumulation of [4-^{13}C] glutamate was attenuated by approximately 50% with both drugs.

The attenuation suggests that the TCA cycle rate was reduced, which should produce a reduction in the availability of cellular energy. The increased NTP observed with these drugs may reflect changes that occur within the cells as they attempt to repair their DNA, that are not directly related to the rate of their formation.

Increased ATP levels were also observed when human ovarian carcinoma or rat lymphoma cells embedded in agarose were exposed to cisplatin [90]. The increase was as high as 70% with wild-type cells and was significantly less with cis-platin resistant cells. These finding are in contrast to the findings of Neeman with cis-platin. They indicate that the high-energy phosphate response to chemotherapeutics is cell type dependent.

These results indicate that ^{31}P NMR spectroscopy may be a useful in understanding how cells are affected by therapeutic agents. They also may facilitate the design of NMR methods for clinical detection of response to therapy. With early spectroscopic indicators of response, treatment regiments can be tailored to best suit the needs of individual patients.

4.2.3. Effect of pH on phosphate metabolism of mammary tumour cells

Extracellular pH is an important parameter in tumour cell metabolism and can be measured directly with ^{31}P NMR spectroscopy [91, 92]. Modulation of extracellular pH has been used to enhance the effects of many chemotherapeutics [93, 94]. Studying the effects of pH *in vitro* could be important for understanding its effects *in vivo*. The effect of acidosis on phosphorous metabolism of mouse mammary tumour cells (EAT) have been examined during growth in MC1350 HFBRs (Microgon) with ^{31}P NMR spectroscopy [75]. Reducing pH from 7.2 to 6.7 resulted in a 3-fold increase in glyceryl-phosphocholine and a 40% reduction in phosphocholine level. Only a slight reduction was observed in NTP levels. These changes indicate that pH has a strong impact on phospholipid metabolism. Such changes could be used to monitor the response of cells during therapeutic modulation of pH.

4.3. ARTIFICIAL ORGANS

4.3.1. Pancreas

Both phosphorus and carbon spectroscopy have been used to examine primary metabolism of insulinomas, which could potentially be used in an artificial pancreas. Insulinomas are immortalised insulin secreting cells, derived from beta cells of the Islets of Langerhans. Normal pancreatic beta cells secrete insulin in response to an elevation in blood glucose levels. They also increase insulin secretion when other catabolites such as amino acids and lipids are elevated in blood, provided some glucose is present. The exact mechanism underlying the control of insulin secretion is not yet known for normal beta cells or insulinomas. The "fuels hypothesis" purports that the extracellular concentration of a metabolite is “sensed” when it is used to produce energy [95]. In this way, cells can simultaneously sense glucose, amino acid and fatty acid levels and produce an appropriate insulin response. However, the details of how increased catabolism of these substrates leads to increased insulin secretion is not well understood.

One hypothesis was that either ATP or ADP is the key link between energy metabolism and secretion of insulin. To examine this hypothesis, the effect of glucose on both βTC3 and βHC9 insulinoma metabolism was examined with ^{31}P NMR [95]. ATP levels were detected directly and ADP levels were detected indirectly by: adding creatine to the medium, determining the phosphocreatine level with NMR, and calculating the ADP concentration. In response to a glucose pulse, both cell lines produced a rapid increase in insulin secretion; an abrupt reduction in ADP level also occurred but no change in ATP was detected. How ADP levels change without a corresponding change in ATP was not determined. In a subsequent study, ATP levels in βTC3 levels were monitored during a change in glucose concentration from 16 mM to 0 mM and then back to 16 mM [96]. The results indicate that in the presence of normal levels of amino acids in the culture medium, ATP levels did not change significantly, but insulin secretion was markedly reduced. These results suggest that ATP levels are not directly involved in the control of insulin secretion, but ADP levels may be. The effect of oxygen on ^{31}P metabolites was also examined in both short-term and long-term studies of βTC3 cells entrapped in agarose/poly-L-lysine/agarose beads [97, 98]. The results indicated that for either time period, ATP levels were reduced by reduced oxygen concentration. For a sustained increase in oxygen concentration, a sustained increase was observed in ATP levels. These results suggest that cell growth within the beads was oxygen transfer limited.

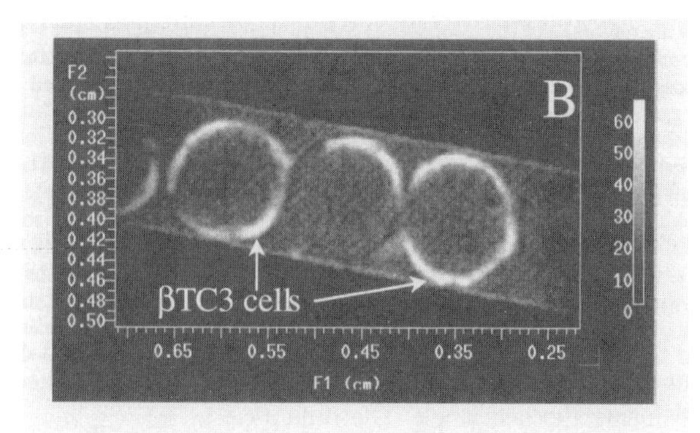

Figure 10. Diffusion-weighted images of βTC3 cells immobilised in calcium alginate/ poly-L-lysine. Hyperintense regions at the bead periphery contain the highest cell level [11].

Diffusion-weighted images of βTC3 cells in calcium alginate beads are shown in Figure 10 [11]. Diffusion-weighting is achieved with magnetic field gradients that selectively dephase the magnetisation of nuclei undergoing translational motion [33, 99]. Because the intracellular space is very structured, the mean free path of intracellular water is much shorter than it is in a hydrogel. Hence, the apparent diffusion coefficient of water (ADC) in regions that contain cells should be much lower. For the alginate bead images in Figure 10, the hyperintense regions at the bead perimeter are associated with low apparent diffusion rates. Histological examination of the beads indicated that these regions contain high concentrations of cells. These findings are consistent with the ^{31}P results described above and support the belief that growth in 1-mm alginate beads is mass transfer limited.

4.3.2. Cartilage

Chondrocytes cultured in HFBRs have been used as model for cartilage development that can be studied with MR imaging and spectroscopy [72]. The bioreactors used generally contain a very small number of hollow fibres, because *in vivo*, cartilage is in avascular structure that is passively nourished through the surrounding synovial fluid. Chondrocytes harvested from chick embryo sterna have been cultured in HFBRs with a 4-mm inner diameter and only six polypropylene fibres (630 micron O.D.) [72]. A number of different MRI microscopy methods were used to monitor growth and structural protein development over a 4-week period. These included T1-weighted, T2-weighted, magnetization transfer (MT)-weighted and diffusion-weighted imaging. Imaging was conducted with a 9.4 Tesla magnet and the in-plane resolution of 60 microns.

Histological evaluation revealed the following: With one week of growth, thin layers of tissue formed around the hollow fibres and along the inner wall of the reactor. Metachromasia were lacking, which was indicative of a lack of proteoglycan. By two weeks, proteoglycan deposition became significant and well-defined lacunae were formed. By three weeks, tissue volume increased and much of the extra capillary space was filled. Cartilage near the fibres was composed of cells and small lacunae, while in more distant regions chondrocytes occupied larger lacunae. During the fourth week, the quantity of metachromatic matrix was reduced, while collagen levels increased and acellular regions, which contained predominately collagen, were observed.

Water T1 values decreased during the first three weeks of growth as protein levels increased. Protein cause reduced T1 because water bound to its surface has a longer correlation time and hence a shorter T1. Since bound water is in rapid exchange with non-bound water, the overall T1 of the tissue drops. A more rapid reduction in T1 values occurred during the last week of growth when collagen levels increased while proteoglycan levels decreased. This result suggests that collagen may have a stronger impact on T1 than proteoglycan. T2 values also decreased during the first three weeks of growth but not during the last week. The reduction in T2 was due to the more rapid spin-spin relaxation for water molecules bound to proteins and as with T1, the overall T2 of the tissue dropped due to rapid exchange between bound and bulk water. The lack of change in T2 during the last week indicates that although T2 is sensitive to protein levels, it does not discriminate between proteoglycans and collagen.

MT contrast is produced by using narrow-banded pulses to selectively saturate or invert the magnetisation of water molecules bound to proteins, which have chemical shifts that are distinct from that of bulk water. Since these water molecules rapidly exchange with bulk water, the narrow-banded pulses will indirectly reduce the amplitude of the signal from the bulk water. During growth, MT values for tissue in the HFBR were found to increase with increasing collagen content. At the end of 4 weeks of growth, the *in vitro* MT values were comparable to values determined *in vivo*.

[56] Sonnewald, U.; Petersen, S. B.; Krane, J.; Westergaard, N. and Schousboe, A. (1992) Proton NMR study of cortex neurons and cerebellar granule cells on microcarriers and their PCA extracts: lactate production under hypoxia. Magn. Reson. Med. 23: 166-171.

[57] Merle, M.; Pianet, I.; Canioni, P. and Labouesse, J. (1992) Comparative ^{31}P and 1H NMR studies on rat astrocytes and C6 glioma cells in culture. J. Biochimie 74: 919-930.

[58] Pilatus, U.; Shim, H.; Artemov, D.; Davis, D.; van Zijl, P. C. and Glickson, J. D. (1997) Intracellular volume and apparent diffusion constants of perfused cancer cell cultures, as measured by NMR. Magn. Reson. Med. 37: 825-832.

[59] Culic, O.; Gruwel, M. L. and Schrader, J. (1997) Energy turnover of vascular endothelial cells. Amer. J. Physiol. 273: C205-C213.

[60] Neeman, M.; Rushkin, E.; Kadouri, A. and Degani, H. (1988) Adaptation of culture methods for NMR studies of anchorage-dependent cells. Magn. Reson. Med. 7: 236-242.

[61] Glickson, J. D. (1996) Cells and Cell Systems. In: Encyclopedia of nuclear magnetic resonance, John Wiley, New York.

[62] Mancuso, A.; Wehrli, S. Pickup. S.; Beardsley, N.J. and Glickson, J.D. (2002) Simultaneous determination of oxygen consumption and TCA cycle labeling kinetics detected with ^{13}C NMR spectroscopy. submitted.

[63] Abraha, A.; Shim, H.; Wehrle, J. P.; and Glickson, J. D. (1996) Inhibition of tumor cell proliferation by dexamethasone: ^{31}P NMR studies of RIF-1 fibrosarcoma cells perfused in vitro. NMR Biomed. 9: 173-178.

[64] Aiken, N. R.; McGovern, K. A.; Ng, C. E.; Wehrle, J. P. and Glickson, J. D. (1994) ^{31}P NMR spectroscopic studies of the effects of cyclophosphamide on perfused RIF-1 tumor cells. Magn. Reson. Med. 31: 241-247.

[65] Mazariegos, G. V.; Kramer, D. J.; Lopez, R. C.; Shakil, A. O.; Rosenbloom, A. J.; DeVera, M.; Giraldo, M.; Grogan, T. A.; Zhu, Y.; Fulmer, M. L.; Amiot, B. P. and Patzer, J. F. (2001) Safety observations in phase I clinical evaluation of the Excorp Medical Bioartificial Liver Support System after the first four patients. ASAIO J. 47: 471-475.

[66] Mancuso, A.; Sharfstein, S. T.; Tucker, S. N.; Clark, D. S. and Blanch, H. W. (1994) Examination of primary metabolic pathways in a murine hybridoma with carbon-13 nuclear magnetic resonance spectroscopy. Biotechnol. Bioeng. 44: 563-585.

[67] Sharfstein, S. T.; Tucker, S. N.; Mancuso, A.; Blanch, H. W. and Clark, D. S. (1994) Quantitative in vivo nuclear magnetic resonance studies of hybridoma metabolism. Biotechnol. Bioeng. 43: 1059-1074.

[68] Mancuso, A.; Sharfstein, S. T.; Fernandez, E. J.; Clark, D. S. and Blanch, H. W. (1998) Effect of extracellular glutamine concentration on primary and secondary metabolism of a murine hybridoma: an in vivo ^{13}C nuclear magnetic resonance study. Biotechnol. Bioeng. 57: 172-186.

[69] Gillies, R. J.; Scherer, P. G.; Raghunand, N.; Okerlund, L. S.; Martinez-Zaguilan, R.; Hesterberg, L. and Dale, B. E. (1991) Iteration of hybridoma growth and productivity in hollow fiber bioreactors using ^{31}P NMR. Magn. Reson. Med. 18: 181-192.

[70] Minichiello, M. M.; Albert, D. M.; Kolodny, N. H.; Lee, M. S. and Craft, J. L. (1989) A perfusion system developed for ^{31}P NMR study of melanoma cells at tissue-like density. Magn. Reson. Med. 10: 96-107.

[71] Gonzalez-Mendez, R.; Wemmer, D.; Hahn, G.; Wade-Jardetzky, N. and Jardetzky, O. (1982) Continuous-flow NMR culture system for mammalian cells. Biochim. Biophys. Acta, 720, 274-280.

[72] Potter, K.; Butler, J. J.; Adams, C.; Fishbein, K. W.; Mcfarland, E. W.; Horton, W. E. and Spencer, R. G. S. (1998) Cartilage formation in a hollow fiber bioreactor studied by proton magnetic resonance microscopy. Matrix Biol. 17: 513-523.

[73] Hrovat, M. I.; Wade, C. G. and Hawkes, S. P. (1985) A Space-Effecient Assembly for NMR Experiments on Anchorage-Dependent Cells. J. Magn. Res. 61: 409-417.

[74] Macdonald, J. M.; Grillo, M.; Schmidlin, O.; Tajiri, D. T. and James, T. L. (1998) NMR spectroscopy and MRI investigation of a potential bioartificial liver. NMR Biomed. 11: 55-66.

[75] Galons, J. P.; Job, C. and Gillies, R. J. (1995) Increase of GPC levels in cultured mammalian cells during acidosis. A 31P MR spectroscopy study using a continuous bioreactor system. Magn. Reson. Med. 33: 422-426.

[76] Drury, D. D.; Dale, B. E. and Gillies, R. J. (1988) Oxygen transfer properties of a bioreactor for use within a nuclear magnetic resonance spectrometer. Biotechnol. Bioeng. 32: 966-974.

[77] Piret, J. M. and Cooney, C. L. (1991) Model of oxygen transport limitations in hollow fiber bioreactors. Biotechnol. Bioeng. 37: 80-92.

PHYSIOLOGY OF IMMOBILISED MICROBIAL CELLS

RONNIE WILLAERT[1], VIKTOR NEDOVIĆ[2] AND GINO V. BARON[3]

[1]*Department of Ultrastructure, Flanders Interuniversity Institute for Biotechnology, Vrije Universiteit Brussel, Pleinlaan 2, B-1050 Brussel, Belgium – Fax 32-2-6291963 – Email: Ronnie.Willaert@vub.ac.be*
[2]*Departement of Food Technology and Biochemistry, University of Belgrade, Nemanjina 6, PO Box 127, 11081 Belgrade-Zemun, Yugoslavia – Fax: 381-11-193659 – Email: vnedovic@EUnet.yu*
[3]*Department of Chemical Engineering, Vrije Universiteit Brussel, Pleinlaan 2, B-1050 Brussel, Belgium – Fax: 32-2-6293248 – Email: gvbaron@vub.ac.be*

1. Introduction

The widespread development and application of gel immobilised cell technology has significantly increased the interest in the physiology of immobilised cells. A number of reports have appeared suggesting that the immobilisation has a profound effect on the metabolic behaviour of immobilised cells compared to free cells. On the other hand, results of other reports suggest that the observed changes are caused by a change in local concentrations due to mass transfer limitations.

In this review, the various techniques, which have been used to assess the physiology of immobilised cells, are discussed. Special attention is focussed on on-line and non-invasive techniques, which are very valuable to observe the behaviour of immobilised cells in their micro-environment without disturbing the cells. Next, some interesting reported physiological aspects of immobilised bacteria and fungi are discussed. Differences between immobilised and free cell systems are highlighted.

2. Physiology and experimental techniques

The study of immobilised cells requires specific experimental techniques. Detailed *in situ* information is needed to obtain accurate information of the physiology of the immobilised cells. Many biophysical techniques have been used to study immobilised cells *in situ*. Only a few of these techniques are non-invasive. Some involve the removal of the cells from the immobilisation matrix and/or destruction of the sample. Interest in non-invasive studies has increased rapidly as improved instrumentation with expanded

V. Nedović and R. Willaert (eds.), Fundamentals of Cell Immobilisation Biotechnology, 469-492.

utilisation by all cellular ATPase functions dropped from substantial control in suspension grown cells to negligible influence in alginate grown cells. The estimated change in PFK maximum velocity was also consistent with *in vitro* assays of PFK activity in extracts of suspension and alginate grown yeast cells. ^{31}P NMR studies with agarose immobilised *Candida tropicalis* have shown that *C. tropicalis* metabolising xylose was not capable of building up polyphosphate [80]. Moreover, this strain did not grow inside agarose beads when xylose was supplied. However, cell growth and polyphosphate metabolism resumed immediately when the perfusate was changed to a glucose containing medium. Although glucose metabolising *C. tropicalis* more closely resembled freely suspended cells in their metabolic response, they displayed a slight up field shift of the cytoplasmic inorganic phosphate (P_i) peak indicating an acidification of this compartment by 0.3 pH units. A pH drop close to one pH unit was observed for xylose metabolising *C. tropicalis*. In addition, ^{23}Na and ^{39}K NMR data showed that immobilised *C. tropicalis* cells had higher intracellular levels of Na^+ and K^+ than suspended cells. Moreover, measurements of the relaxation times of the intracellular cations suggested that the intracellular environment is more viscous in immobilised cells. Santos *et al.* [81] used NMR techniques to examine the phosphorus and carbon metabolism of free and κ-carrageenan immobilised *Propionibacterium acidi-propionici*. It was shown that polyphosphate was present both in the intracellular space and on the outside of the external membrane; and polyphosphate is used to phosphorylate glucose. Immobilised cells led to much higher yields of propionic acid compared to suspended cells. Taipa *et al.* [82] used ^{31}P NMR to probe the anaerobic metabolism of glucose by suspended and κ-carrageenan entrapped *S. bayanus* cells. It was shown that the intracellular pH was slightly higher in the case of immobilised cells (7.15 compared to 7.05) and is kept constant during the course of fermentation, in contrast with the suspended cells for which a steady state decrease in pH was observed. This difference was attributed to the differences in the extracellular pH that decreased much faster in the free cell fermentation medium compared to the gel matrix.

The metabolism of glucose and xylose as a function of oxygenation in free-cell and agarose-immobilised-cell cultures of *Pichia stipitis* and *S. cerevisiae* has been studied by ^{31}P and ^{13}C NMR [15]. *P. stipitis* was able to metabolise xylose or glucose for 24 to 60 h at rates and with theoretical yields of ethanol similar to those obtained with anoxic free-cell suspensions. In contrast to *P. stipitis*, immobilised *S. cerevisiae* showed a dramatic twofold increase in its ability to metabolise glucose relative to free-cell suspensions. This strain was also able to grow within beads, although the doubling time was by a factor 2 longer compared to log-phase batch cultures.

The intracellular metabolism of gel entrapped plant cells has been studied by ^{31}P NMR [80,83-85]. The spectra of freely suspended *Catharanthus roseus* cells, and agarose and alginate entrapped cells have been recorded under oxygenation conditions (incubation time of 48, 72 and 72 hours respectively). The same resonances have been observed in all three spectra and the chemical shifts of the cytoplasmic sugar phosphate and inorganic phosphate resonances were almost identical. Also, the ADP/ATP ratio was the same, within experimental error, indicating a similar energy status for the three cases. Although it is well known that changes in the intracellular pH exert regulatory effects on the metabolism of plant cells, it was clear from these NMR studies that the cytoplasmic pH as well as the vacuolar pH are not altered by the entrapment of the cells in agarose or alginate gel.

NMR spectroscopy has also been used as a probe of the physiological state of *Escherichia coli* contained by microporous hollow fibre membranes [11]. Two pH values were recorded during the anaerobic growth corresponding to the nutrient medium (pH 7.3) and the glucose deprived region of the cell layer (pH 6.5). The internal pH of non-starving *E. coli* could not be measured because only a small fraction of the reactor was populated by growing cells. The energy state of the cells was determined by measuring the concentration of nucleotide triphosphates.

Rehm and co-workers have used enzyme assays to measure enzyme activities of *S. cerevisiae* cells entrapped in Ca-alginate beads [6] and *Pichia farinosa* cells immobilised on sintered glass Raschig rings [7]. Cells grown in the form of micro-colonies in the alginate beads showed faster glucose uptake and ethanol productivity with simultaneously decreased product and cell yields. In these cells, increased specific hexokinase and phosphofructokinase activities could be determined. These alterations in physiology were not found in immobilised single cells. In the case of adsorbed *P. farinosa*, activities of some of the key enzymes involved in polyol and glycerol formation were determined. The course of the activities of glucose-6-phosphate dehydrogenase and glycerol-3-phosphate dehydrogenase showed differences between free and immobilised cells.

Concentrations of intracellular intermediary metabolites fructose 1,6-diphosphate, pyruvate, citrate and malate in free and Ca-alginate immobilised *S. cerevisiae* fermenting *D*-glucose anaerobically have been determined with NAD^+/NADH linked enzymatic assays [86,87] when the sugar uptake rate and ethanol production rate were constant [8]. No cell growth was observed and the fermentation yields and rates were the same in both types of cells. The concentrations of intermediary intracellular metabolites were also identical for immobilised and free cells.

Another non-invasive technique which can be used to probe cell metabolism in the immobilised state is radioisotope labelling/autoradiography. Radioisotope precursors for the synthesis of all major biological polymers are readily available. This technique has been applied to look at overall distributions of isotopes incorporated in cell mass in dense cell systems. By an appropriate choice of a pulse-chase labelling protocol, and in combination with mass balance measurements on activities in the inlet and outlet streams of a reactor, a variety of parameters of immobilised cell growth have been obtained [11,28-30]. These include: (i) the net rate of cell mass synthesis at a given time, (ii) the yield coefficients for cell mass synthesis, (iii) the degradation rate of cell mass in the immobilised cell system as a function of time after its synthesis, (iv) the size of the region in which cell growth takes place, (v) the specific rates of substrate consumption and cell growth within the growth region, (vi) the rate and pattern of cell movement within the cell layer. For the case of *E. coli* in membrane reactors and gel beads, these measurements have led to the conclusion that the intrinsic reactions are very similar to those for free cells.

Scanning microfluorimetry has been used for *in situ* investigations of spatial heterogeneities in biomass density, cell RNA content and DNA concentration [26,27,88,89]. The RNA content was related to the growth rate and used to infer the growth rate history of Ca-alginate immobilised *Zymomonas mobilis* cells [28]. A more accurate gauge of the cell growth rate is the rate of DNA synthesis. Immunofluorescent measurements of DNA synthesis rate of Sr-alginate entrapped *E. coli* cells were accomplished by bromodeoxyuridine (BrdU) pulse-labelling coupled with the use of anti-BrdU monoclonal antibodies [24,24]. By using this technique in combination with mathematical modelling, it was shown showed that gel immobilised *E. coli* cells behaved similar to free cells.

Al-Rubeai and Spier [90] used a cytometabolism diffusion test in conjunction with image analysis for assessing the viability and metabolism of mammalian hybridoma cells immobilised in agarose gel. The test is based on the capacity of mitochondrial enzymes of viable cells to transform the 3-(4,5-dimethylthiazol-2-yl)-2,5-diphenyl (MTT) tetrazolium salt (yellow) into MTT formazan (blue). It was observed that the activity of the cells decline with time in a manner consistent with oxygen diffusion limitation.

To investigate the effect of diffusional limitations and heterogeneous yeast cell distribution in a gel immobilised cell system, a Ca-alginate gel membrane reactor has been constructed [50,51,75]. The membrane reactor consists essentially of a gel membrane with immobilised *S. cerevisiae* cells, flanked by two well-mixed chambers. Substrate is pumped continuously through the first chamber, and this chamber can be considered as the equivalent of the space around a gel bead. The second closed measuring chamber contains a small quantity of liquid. Analysis of the liquid in this chamber gave direct information on substrate (glucose) concentrations at the gel surface, and is an indication of the situation in the centre of a gel bead. The results were used in a dynamic reaction-diffusion model to determine the kinetic parameters of the immobilised cells and revealed that they were comparable to those of free cells. Due to the diffusional limitations, yeast cells grew preferentially closer to the feed edge of the membrane. Hannoun and Stephanopoulos [74] have also used a membrane reactor to determine the intrinsic reaction rates of alginate-entrapped *S. cerevisiae* by using diffusion-reaction analysis. Under anaerobic conditions, the specific growth rate of the immobilised cells decreased by 20% compared to the growth rate for suspended cells. The specific glucose uptake rate and specific ethanol production rate increased by a factor 4 compared to those of suspended cells. The ethanol yield remained the same, and the biomass yield decreased to one-fifth of the yield for suspended cells. Further experiments were conducted under aerobic conditions to investigate the effects of dissolved oxygen: oxygen appeared to affect the immobilised cells in a way similar to suspended cells.

Dielectric spectroscopy (or capacitance measurement) is a tool for the on-line and real-time monitoring of immobilised cells. Noll and Biselli [33] studied the cultivation of hybridoma cells in porous glass carriers. The capacitance signal proved to be a lumped parameter influenced by cell concentration, cell size and culture conditions. An excellent correlation between the specific capacitance and the specific amount of nucleotidetriphosphates in the cells could be shown. The cell attachment and growth could be measured. In this system, only an approximate correlation with viable cell concentration appeared, whereas an exact correlation with the glutamine consumption rate, a measure of the metabolic activity of the cells could be shown.

Microelectrodes can be inserted into immobilised cell systems to measure concentration profiles. This technique is necessarily invasive, and limited by the fabrication of appropriate electrodes. Microelectrodes have been frequently used for biofilm research and especially oxygen microelectrodes are very popular (*e.g.*, [34-36,40,48,49,91-93]). An oxygen microsensor in combination with mathematical modelling has been used to determine the behaviour of *Thiosphaera pantotropha* immobilised in agarose [38] and of *E. coli* immobilised in Ca-alginate [37,94]. Schrezenmeir *et al.* [46] used a oxygen microelectrode to measure O_2 concentration profiles in Ba-alginate beads with entrapped Brockmann bodies (islet organs) of *Osphronemus gorami.* Müller and co-workers [45] investigated the oxygen monitoring technique with microelectrodes in Ca-alginate with and without entrapped *S. cerevisiae.* To obtain real Po_2-profiles it was important to be exactly informed about the physical, chemical and biological properties of the material to be investigated. It was recommended to apply a special stepwise puncture technique with distinct step-in/step-out movements of the electrode. Improper use of the electrode can result in serious errors (pseudo-Po_2-gradients), which could be explained by formation of artefacts and diffusion barriers in front of the electrode tip or oxygen "availability" at the tip and consumption of oxygen by the electrode itself.

A membrane mass spectrometric inlet reactor has been constructed which allowed the measurement of oxygen and carbon dioxide "in" an immobilisation matrix [50-52,95]. This reactor has a thin but very low permeability barrier membrane (PET) on which a layer of immobilised cells is deposited, so that the sample withdrawal is very low and the micro-environment of the immobilised cells is practically not disturbed. The reactor has been used to monitor the behaviour of gel and porous glass immobilised *S. cerevisiae* by measuring oxygen and carbon dioxide. The results in combination with mathematical modelling were used to assess the influence of the immobilisation on the kinetics and showed that the immobilised and free cell kinetics were comparable.

It has already been proven that light microscopy as well as electron microscopy (*e.g.*, [53-58,98]) can serve as a valuable tool in studying the morphology of immobilised cells, the surface coverage of carriers and the dynamic behaviour of immobilised cells (*cfr.* scanning microfluorimetry and autoradiography). κ-Carrageenan gel beads with immobilised *Nitrosomonas europaea* and *Nitrobacter agilis* cells, were sliced in thin cross sections after fixation and embedding; and the spatial biomass distribution was determined by a specific fluorescent-antibody labelling technique [99]. The disadvantage of most of these techniques is that they are unsuitable for on-line analysis. By constructing a continuous "microscope reactor", microscopy could be used to investigate on-line and *in situ* the growth kinetics of *S. cerevisiae* in 2% Ca-alginate [59]. The specific growth rate of single immobilised cells and free cells were measured (see Table 2). The growth of a microcolony in Ca-alginate was followed on-line and the specific growth rate of the yeast cells in the microcolony determined. Using a simple growth model for the microcolony, the cell volume fraction of the cells in a colony was estimated: the packing of yeast cells in a microcolony was comparable to the hexagonal close packing of spheres. Immobilised growth rates were found to be identical to those of free cells when internal and external mass transfer limitations are negligible. Willke and Vorlop [60] used the same technique for the long term observation of the growth of poly(carbamoylsulphonate)-entrapped *Paracoccus denitrificans* cells. Hüsken *et al.* [61]

used also this technique to study the growth and eruption of microcolonies of *Nitrosomonas europaea* that were immobilised in a 2.6% κ-carrageenan gel.

Table 2. Comparison of the specific growth rates for gel immobilised (μ_I) and free (μ_f) cells.

Micro-organism	Gel-system	μ_I (h^{-1})	μ_f (h^{-1})	Reference
Bacteria				
Escherichia coli	Carrageenan (2%)	2.04[a]	2.08	[56]
		1.69[b]	1.63	
E. coli B/pTG201	Carrageenan (2%)	0.24[c]	0.30	[37]
		0.18[d]		
E. coli B/pTG201	Carrageenan (2%)	0.18	0.36	[94]
Fungi				
Candida guilliermondii	Ba-alginate	0.021	0.029	[63]
Saccharomyces cerevisiae	Ca-alginate (2%)	0.30[e]	0.31	[59]
		0.27[f]		
S. cerevisiae	Gelatin (25-30%)	0.28	0.51	[1]
S. cerevisiae	Ca-alginate (1.5%)	0.115	0.126	[96]
	Carrageenan (2.5%)	0.100		
S. cerevisiae	Ca-alginate (2%)	0.25	0.41	[3]
S. cerevisiae	Ca-alginate (2%)	0.46	0.50	[97]
Thiosphaera pantotropha	Agarose (5%)	0.45[g]	0.45	[38]
		0.58[h]		

[a] supply of 21% oxygen; [b] supply of 100% oxygen
[c] growth in gel slabs; [d] growth in gel beads
[e] single immobilised cells; [f] cells in a microcolony
[g] growth in stirred tank reactor; [h] growth in Kluyver flask

Recently, the intracellular pH of *S. cerevisiae*, which was attached to ferric nitrate pretreated glass slides, was measured employing fluorescence ratio imaging microscopy [62]. The yeast cells were fluorescently labelled with a pH dependent probe – 5(and -6-)-carboxyfluorescein (cF) or 5(and -6-)-carboxyfluorescein succinimidyl ester (cFSE) – and subsequently attached on the glass slides. The measurement of the internal pH was performed during a continuous perfusion of the cells with buffer or medium. The continuous perfusion in combination with the cFSE labelling of the immobilised cells was successfully applied to determine the effect of low and high internal pH and addition of glucose on the internal pH of individual yeast cells over a long time period.

Information of additional experimental techniques, which have been used to study the behaviour of immobilised cells, can be found in Table 1.

3. Specific physiological aspects of bacteria and fungi

3.1 BACTERIA

3.1.1. Plasmid stability

Recombinant plasmid stability in host cells can be increased upon gel immobilisation [100]. Plasmid pTG201 (ampicillin resistant, tetracycline-sensitive derivative of pBR322 containing the *Pseudomonas putida Xyl E* gene that codes for catechol 2,3-dioxygenase) was unstable in free *E. coli* cells, but stable when immobilised in κ-carrageenan gel [56,101-104]. The maintenance of the plasmid of immobilised cells was always accompanied by the stabilisation of the plasmid copy number [105-107]. Enhanced plasmid stability has also been observed for other plasmids in various immobilisation matrices [108-113] and micro-organisms such as *Myxococcus xanthus* [114], yeast cells [115,116], *Lactococcus lactis* [117] and *Bacillus subtilis* [118]. The increased plasmid stability may have resulted from the mechanical properties of the gel bead system that allows only a limited number of cell divisions to occur in each microcolony before the cells escape from the gel bead [103,119,120].

3.1.2. Protective micro-environment

Immobilisation can confer protection to cells exposed to toxic or inhibitory substrates or environments. It has been demonstrated that better phenol degradation rates are obtained with *Pseudomonas putida* cells immobilised in Ca-alginate or polyacrylamide-hydrazide gel than free cells, and that the immobilised bacteria could be exposed to higher phenol concentrations without loss of cell viability [121]. The phenol tolerance for bacteria lacking the potential for phenol degradation (*E. coli* and *Staphylococcus aureus*) was also enhanced when they were grown in Ca-alginate [122]. It was found that the strength of the effect was correlated with the formation of colonies in the gel matrix [122,123]. The membranes of *E. coli* cells grown entrapped in Ca-alginate, showed low lipid-to-protein ratios, which were also encountered in the membranes of free-grown cells in the presence of phenol [124]. Immobilisation of *E. coli* cells also markedly changed the protein pattern of the outer membrane. Heipieper *et al.* [125] observed that cells immobilised and grown in alginate suffered a small loss of cations when exposed to phenols. The re-establishment of gradients was observed at a higher phenol concentration with immobilised cells compared to free cells and less membrane damage was observed.

Alginate-immobilised *Trichosporium* sp. LE3 cells were able to degrade up to 30 mM phenol [126]. Free cells did not completely degrade phenol at concentrations above 20 mM. The maximum phenol degradation rate was a strong function of initial phenol concentrations.

Bacillus sp. – capable of degrading dimethylphtalate (DMP) – was immobilised in calcium alginate and polyurethane foam [127]. Freely suspended cells degraded a maximum of 20 mM DMP. Whereas, alginate- and polyurethane foam-entrapped cells degrade a maximum of 40 mM DMP.

of the two antibiotics was detected on bacterial suspensions but not on biofilm-like structures. Effective diffusivity measurements showed that the diffusion of imipenem in the alginate layer was not hindered. A slight but insignificant enhancement of β-lactamase induction in immobilised cells as compared to their suspended counterparts was insufficient to explain the high resistance of sessile-like bacteria.

3.2.4. Diffusional effects

The expression of the *SUC2* gene encoding invertase was studied using free and gelatin-immobilised yeast cells to try to explain the high activity of this enzyme exhibited by immobilised cells when allowed to grow in a nutrient medium [183]. The results indicate that at least two factors are probably responsible for the accumulation of invertase in immobilised cells. First, the expression of the *SUC2* gene was maintained throughout growth in immobilised cells, whereas its expression was only transient in free cells. Second, invertase of immobilised cells was shown to be less susceptible to endogenous proteolytic attack than that of the corresponding free cells. These results have been interpreted, respectively, in terms of diffusional limitations and changes in the pattern of invertase glycosylation due to growth of yeast in an immobilised state.

3.2.5. Effect on enzyme stability

The zygomycete *Mortierella isabellina* has been entrapped in alginate beads to transform dehydroabietic acid into non toxic metabolites [184]. It was shown that immobilisation resulted in greater long term stability of enzyme activity compared with free mycelia and the breakdown products differed, which was attributed to a greater level of secondary metabolism.

References

[1] Doran, P. and Bailey, J.E. (1986) Effects of immobilization on growth, fermentation properties, and macromolecular composition of *Saccharomyces cerevisiae* attached to gelatin. Biotechnol. Bioeng. 28: 73-87.

[2] Bailey, J.M.; Axe, D.D.; Doran, P.M.; Galazzo, J.L.; Reardon, K.F.; Seressiotis, A. and Shanks, J. (1987) Redirection of cellular metabolism: analysis and synthesis. Ann. N.Y. Acad. Sci. 506: 1-23.

[3] Galazzo, J.L. and Bailey, J.E. (1990) Growing *Saccharomyces cerevisiae* in calcium-alginate beads induces cell alterations which accelerate glucose conversion to ethanol. Biotechnol. Bioeng. 36: 417-426.

[4] de Alteriis, E.; Porro, D.; Romano, V. and Parascandola, P. (2001) Relation between growth dynamics and diffusional limitations in *Saccharomyces cerevisiae* cells growing as entrapped in an insolubilised gelatin gel. FEMS Microbiol. Lett. 20: 245-251.

[5] Jones, A.; Razniewska, T.; Lesser, B.H.; Siqueira, R.; Berk, D.; Behie, L.A.; Gaucher, G.M. (1984) An assay for the measurement of protein content of cells immobilized in carrageenan. Can. J. Microbiol. 30: 475-481.

[6] Hilge-Rotmann, B. and Rehm, H.-J. (1990) Comparison of fermentation properties and specific enzyme activities of free and calcium-alginate-entrapped *Saccharomyces cerevisiae*. Appl. Microbiol. Biotechnol. 33: 54-58.

[7] Höötmann, U.; Bisping, B. and Rehm, H.-J. (1991) Physiology of olyol fermation by free and immobilized cells of the osmotolerant yeast *Pichia farinosa*. Appl. Microbiol. Biotechnol. 35: 258-263.

[8] Senac, T. and Hahn-Hägerdal, B. (1991) Concentrations of intermediary metabolites in free and calcium alginate-immobilized cells of D-glucose fermenting *Saccharomyces cerevisiae*. Biotechnol. Techn. 5: 63-68.

[9] Galazzo, J.L.; Shanks, J.V. and Bailey, J.E. (1987) Comparison of suspended and immobilized yeast metabolism using ^{31}P Nuclear Magnetic Resonance spectroscopy. Biotechnol. Techn. 1: 1-6.

[10] Galazzo, J.L. and Bailey, J.E. (1989) *In vivo* Nuclear Magnetic Resonance analysis of immobilization effects on glucose metabolism of yeast *Saccharomyces cerevisiae*. Biotechnol. Bioeng. 33: 1283-1289.

[11] Karel,S.F.; Briasco, C.A. and Robertson, C.A. (1987) The behavior of immobilized living cells. Ann. N.Y. Acad. Sci. 506: 84-105.

[12] Neeman, M., Rushkin, E., Kadouri, A., Degani, H. (1987) Modern approaches to animal cell technology, ESCAT Meeting, Butterworths, London.

[13] McGovern, K.A.; Schoeniger, J.S.; Wehrle, J.P.; Ng, C.E. and Glickson, J.D. (1993) Gel-entrapment of perfluorocarbons: a fluorine-19 NMR spectroscopic method for monitoring oxygen concentration in cell perfusion systems. Magn. Reson. Med. 29: 196-204.

[14] Fernandez, E.J. (1996) Nuclear magnetic resonance spectroscopy and imaging. In: Willaert R.G., Baron G.V., De Backer L. (Eds.) Immobilised living cell systems: modelling and experimental techniques. John Wiley & Sons, Chichester, UK; pp. 117-146.

[15] Lohmeier-Vogel, E.M.; McIntyre, D.D. and Vogel, H.J. (1996) Phosphorus-31 and carbon-13 nuclear magnetic resonance studies of glucose and xylose metabolism in cell suspensions and agarose-immobilized cultures of *Pichia stipitis* and *Saccharomyces cerevisiae*. Appl. Environ. Microbiol. 62: 2832-2838.

[16] Hesse, S.J.; Ruijter, G.J.; Dijkema, C. and Visser, J. (2000) Measurement of intracellular (compartmental) pH by ^{31}P NMR in *Aspergillus niger*. J. Biotechnol. 77: 5-15.

[17] Barbotin, J.N.; Portais, J.-C.; Alves, P.M. and Santos, H. (2001) NMR and immobilized cells. In: Wijffels, R.H. (Ed.) Immobilized cells. Springer-Verlag, Berlin Germany; pp. 123-138.

[18] Capuani, G.; Miccheli, A.; Tomassini, A.; Falasca, L.; Aureli, T. and Conti, F. (2000) Cellular volume determination of alginate-entrapped hepatocytes by MRI diffusion measurements. Artif. Cells Blood Substit. Immobil. Biotechnol. 28: 293-305.

[19] Doran, P.M. and Bailey, J.E. (1987) Effects of immobilization on the nature of glycolytic oscillation. Biotechnol. Bioeng. 29: 892-897.

[20] Müller, W.; Wehnert, G. and Scheper, T. (1988) Fluorescence monitoring of immobilized microorganisms in cultures. Anal. Chim Acta 213: 47-53.

[22] Scheper, T.; Anders, K.-D.; Busch, M.; Müller, W. and Reardon K.F. (1990) Culture fluorescence monitoring of immobilised cells. In: de Bont, J.A.M.; Visser, J.; Mattiasson, B. and Tramper, J. (Eds.) Physiology of immobilized cells. Elsevier, Amsterdam, The Netherlands; pp. 625-636.

[23] Monbouquette, H.G.; Sayles, G.D. and Ollis, D.F. (1990) Immobilized cell biocatalyst activation and pseudo-steady-state behavior: model and experiment. Biotechnol. Bioeng. 35: 609-629.

[24] Kuhn, R.H.; Peretti, S.W. and Ollis, D.F. (1991) Microfluorometric analysis of spatial and temporal patterns of immobilized cell growth. Biotechnol. Bioeng. 38: 340-352.

[25] Kuhn, R.H.; Peretti, S.W. and Ollis, D.F. (1993) Acid inhibition of immobilized cells: quantitative comparison of model and experiment. Appl. Biochem. Biotechnol. 39/40: 401-413.

[26] Ollis, D.F.; Monbouquette, H.; Kuhn, R. and Peretti, S.W. (1990) Scanning microphotometry and microfluorimetry of immobilized cells: an emerging quantitative tool. In: de Bont, J.A.M.; Visser, J.; Mattiasson, B. and Tramper, J. (Eds.) Physiology of immobilized cells. Elsevier, Amsterdam, The Netherlands; pp. 637-648.

[27] Ollis D.F. (1996) Scanning microfluorimetry. In: Willaert R.G.; Baron G.V. and De Backer L. (Eds.) Immobilised living cell systems: modelling and experimental techniques. John Wiley & Sons, Chichester, UK; pp. 99-116.

[28] Stewart, P.S. and Robertson, C.R. (1988) Product inhibition of immobilized *Escherichia coli* arising from mass transfer limitations. Appl. Environ. Microbiol. 54: 2464-2471.

[29] Karel, S.F. and Robertson, C.R. (1989) Autoradiographic determination of mass-transfer limitations in immobilized cell reactors. Biotechnol. Bioeng. 34: 320-336.

[30] Karel, S.F. and Robertson, C.R. (1989) Cell mass synthesis and degradation by immobilized *Escherichia coli*. Biotechnol. Bioeng. 34: 337.

[31] Harris, C.M.; Todd, R.W.; Bungard, S.J.; Lovitt, R.W.; Morris, J.G. and Kell, D.B. (1987) Dielectric properties of microbial suspensions at radio frequencies: a novel method for the real-time estimation of microbial biomass. Enzyme Microb. Technol. 9: 181-186.

[32] Kell, D.B.; Markx, G.H.; Davey, C.L. and Todd, R.W. (1990) Real-time monitoring of cellular biomass: methods and applications. Trends Anal. Chem. 9:190-194.

[33] Noll, T. and Biselli, M. (1998) Dielectric spectroscopy in the cultivation of suspended and immobilized hybridoma cells. J. Biotechnol. 63: 187-198.

[34] de Beer, D. and Sweerts, J.-P.R.A. (1989) Measurement of nitrate gradients with a ion-selective microelectrode. Anal. Chim. Acta 219: 351-356.

[81] Santos, H.; Pereira, H.; Crespo, J.P.S.G.; Moura, M.J.; Carrondo, M.J.T. and Xavier, A.V. (1990) *In vivo* NMR studies of propionic-acid fermentation by *Propionibacterium acidi – propionici*. In: de Bont, J.A.M.; Visser, J.; Mattiasson, B. and Tramper, J. (Eds.) Physiology of immobilized cells. Elsevier, Amsterdam, The Netherlands; pp. 685-687.
[82] Taipa, M.A.; Cabral, J.M.S. and Santos, H. (1990) Application of *in vivo* ^{31}P-NMR to compare the metabolism of free and gel entrapped yeast cells. In: de Bont, J.A.M.; Visser, J.; Mattiasson, B. and Tramper, J. (Eds.) Physiology of immobilized cells. Elsevier, Amsterdam, The Netherlands; pp. 689-691.
[83] Brodelius, P. (1984) Immobilized viable plant cells. Ann. N.Y. Acad. Sci. 434: 382-393.
[84] Brodelius, P. and Vogel, H.J. (1984) Noninvasive ^{31}P NMR studies of the metabolism of suspended and immobilized plant cells. Ann. N.Y. Acad. Sci. 434: 496-500.
[85] Brodelius, P. (1988) Immobilized plant cells as a source of biochemicals. In: Moo-Young, M. (Ed.) Bioreactor immobilized enzymes and cells: fundamentals and applications. Elsevier Applied Science, London, UK; pp. 167-196.
[86] Skoog, K. and Hahn-Hägerdal, B. (1989) Biotechnol. Techn. 3: 1-6.
[87] Senac, T. and Hahn-Hägerdal, B. (1990) Appl. Environ. Microbiol. 56: 120-126.
[88] Monbouquette, H.G. and Ollis, D.F. (1988) Scanning microfluorimetry of Ca-alginate immobilized *Zymomonas mobilis*. Bio/Technol. 6: 1076-1079.
[89] Monbouquette, H.G. and Ollis, D.F. (1988) Structured modelling of immobilized cell kinetics and RNA content. In: Moo-Young M. (Ed.) Bioreactor immobilized enzymes and cells: fundamentals and applications. Elsevier, London, UK; pp. 9-31.
[90] Al-Rubeai, M. and Spier, R. (1989) Quantitativecytochemical analysis of immobilised hybridoma cells. Appl. Microbiol. Biotechnol. 31: 430-433.
[91] Lee, Y.H. and Tsao, G.T. (1979) Dissolved oxygen electrodes. Adv. Biochem. Eng. 13: 35-86.
[92] Lewandowski, Z.; Walser, G. and Characklis, W.G. (1991) Reaction kinetics in biofilms. Biotechnol. Bioeng. 38: 877-882.
[93] Lens, P.N.L.; de Beer, D.; Cronenberg, C.C.H.; Houwen, F.P.; Ottengraf, S.P.P. and Verstraete, W.H. (1993) Appl. Environ. Microbiol. 59: 3803-3815.
[94] Huang, J.; Hooijmans, C.M.; Briasco, C.A.; Geraats, S.G.M.; Luyben, K.Ch.A.M.; Thomas, D. and Barbotin, J.-N. (1990) Effect of free-cell growth parameters on the oxygen concentration in gel-immobilized recombinant *Escherichia coli*. Appl. Microbiol. Biotechnol. 33: 619-632.
[95] Willaert, R. and Baron, G.V. (1994) The dynamic behaviour of yeast cells immobilised in porous glass studied by membrane mass spectrometry. Appl. Mirobiol. Biotechnol. 42: 664-670.
[96] Agrawal, D. and Jain, V.K. (1986) Kinetics of repeated batch production of ethanol by immobilized growing yeast cells. Biotechn. Lett. 8: 67-70.
[97] Vives, C.; Casas, C.; Godia, F. and Sola, C. (1993) Determination of the intrinsic fermentation kinetics of *Saccharomyces cerevisiae* cells immobilized in Ca-alginate beads and observations on their growth. Appl. Microbiol. Biotechnol. 38: 467-472.
[98] Larreta Garde, V.; Thomasset, B. and Barbotin, J.-N. (1981) Electron microscopic evidence of an immobilized living cell system Enzyme Microb. Technol. 3: 216-218.
[99] Hunik, J.H.; van den Hoogen, M. P.; de Boer, W.; Smit, M. and Tramper, J. (1993) Quantitative determination of the spatial distribution of *Nitrosomas europaea* and *Nitrobacter agilis* cells immobilized in κ-acrrageenan gel beads by specific fluorescent-antibody labelling technique. Appl. Environ. Microbiol. 59: 1951-1954.
[100] Barbotin, J.N. (2001) Plasmid stability in immobilized cells. In: Wijffels, R.H. (Ed.) Immobilized cells. Springer-Verlag, Berlin, Germany; pp. 235-246.
[101] De Taxis du Poët, P.; Dhulster, P.; Barbotin, J.-N. and Thomas, D. (1986) Plasmid inheritability and biomass production: comparison between free and immobilized cell cultures of *Escherichia coli* BZ18(pTG201). J. Bacteriol. 165: 871-877.
[102] Nasri, M.; Sayadi, S.; Barbotin, J.N. and Thomas, D. (1987) Influence of immobilization on the stability of pTG201 recombinant plasmid in some strains of *Escherichia coli*. Appl. Environ. Microbiol. 53: 740-744.
[103] Nasri, M.; Sayadi, S.; Barbotin, J.N. and Thomas, D. (1987) The use of the immobilization of whole living cells to increase the stability of recombinant plasmids in *Escherichia coli*. J. Biotechnol. 6: 147-157.
[104] Briasco, C.A.; Barbotin, J.-N. and Thomas, D. (1990) Spatial distribution of viable cell concentration and plasmid stability in gel-immobilized recombinant *E. coli*. In: de Bont, J.A.M.; Visser, J.; Mattiasson, B. and Tramper, J. (Eds.) Physiology of immobilized cells. Elsevier, Amsterdam, The Netherlands; pp. 393-398.

[105] Sayadi, S.; Berry, F.; Nasri, M.; Barbotin, J.N. and Thomas, D. (1988) Increased stability of pBR322-related plasmids in *Escherichia coli* W3101 grown in carrageenan gel beads. FEMS Microbiol. Lett. 56: 307-312.
[106] Sayadi, S.M.; Nasri, M.; Barbotin, J.N. and Thomas, D. (1989) Effect of environmental growth conditions on plasmid stability, plasmid copy number and catechol 2,3-dioxygenase activity in free and immobilized *Escherichia coli* cells. Biotechnol. Bioeng. 33: 801-808.
[107] Ollagnon, G.; Truffaut, N.; Thomas, D. and Barbotin, J.N. (1993) Effects of anaerobic conditions on biomass production and plasmid stability in an immobilized recombinant *E. coli*. Biofouling 6: 317-331.
[108] De Taxis du Poët, P.; Arcand, Y.; Barbotin, J.-N. and Thomas, D. (1987) Plasmid stability in immobilized and free recombinant *E. coli* JM105 (pKK223-200): Importance of oxygen diffusion, growth rate and plasmid copy number. Appl. Environ. Microbiol. 53: 1548-1555.
[109] Joshi, S. and Yamazaki, H. (1987) Film fermentor for ethanol production by yeast immobilized on cotton cloth. Biotechnol. Lett. 9: 825-830.
[110] Oriel, P. (1988) Amylase production by *Escherichia coli* immobilized in silicone foam. Biotechnol. Lett. 10: 113-116.
[111] Oriel, P. (1988) Immobilization of recombinant *Escherichia coli* in silicone polymer beads. Enzyme Microb. Technol. 10: 518-523.
[112] Walls, E.L., Gainer, J.L. (1989) Retention of plasmid bearing cells by immobilization. Biotechnol. Bioeng. 34: 717.
[113] Walls, E.L., Gainer, J.L. (1991) Increased protein productivity from immobilized recombinant yeast. Biotechnol. Bioeng. 37: 1029-1036.
[114] Jaoua, S.; Breton, A.M.; Younes, G. and Guespin-Michel, J.F. (1986) J. Biotechnol. 4: 313.
[115] Sode, K.; Brodelius, P.; Meussdoerffer, F.; Mosbach, K. and Ernst, J.F. (1988) Continuous production of somatomedin C with immobilized transformed yeast cells. Appl. Microbiol. Biotechnol. 28: 215-221.
[116] Sode, K.; Morita, T.; Peterhans, A.; Meussdoerffer, F.; Mosbach, K. and Karube, I. (1988) J. Biotechnol. 8: 113.
[117] D'Angio, C.; Béal, C.; Boquien, C.-Y. and Corrieu, G. (1994) Influence of dilution rate and cell immobilization on plasmid stability during continuous cultures of recombinant strains of *Lactococcus lactis* subsp. *lactis*. J. Biotechnol. 34: 87-95.
[118] Craynest, M.; Barbotin, J.N.; Truffaut, N. and Thomas, D. (1996) Stability of plasmid pHV1431 in free and immobilized cell cultures. Effect of temperature. Ann. N.Y. Acad. Sci. 782: 311-322.
[119] Barbotin, J.N.; Sayadi, S.; Nasri, M.; Berry, F. and Thomas, D. (1990) Improvement of plasmid stability by immobilization of recombinant micro-organisms. Ann. N.Y. Acad. Sci. 589: 41-53.
[120] Barbotin, J.N. (1994) Immobilization of recombinantbacteria. Ann. N.Y. Acad. Sci. 721: 303-309.
[121] Bettmann, H. and Rehm, H.J. (1984) Degradation of phenol by polymer entrapped microorganisms. Appl. Microbiol. Biotechnol. 20: 285-290.
[122] Keweloh, H.; Heipieper, H.-J. and Rehm, H.-J. (1989) Protection of bacteria against cytotoxicity of phenol by immobilization in calcium alginate. Appl. Microbiol. Biotechnol. 31: 383-389.
[123] Keweloh, H.; Heipieper, H.-J. and Rehm, H.-J. (1990) Phenol tolerance of immobilized bacteria. In: de Bont, J.A.M.; Visser, J.; Mattiasson, B. and Tramper, J. (Eds.) Physiology of immobilized cells. Elsevier, Amsterdam, The Netherlands; pp. 545-550.
[124] Keweloh, H.; Weyrauch, G. and Rehm, H.-J. (1990) Phenol-induced membrane changes in free and immobilized *Escherichia coli*. Appl. Microbiol. Biotechnol. 33: 66-71.
[125] Heipieper, H.-J.; Keweloh, H. and Rehm, H.-J. (1991) Influence of phenols on growth and membrane permeability of free and immobilized *Escherichia coli*. Appl. Environ. Microbiol. 57: 1213-1217.
[126] Santos, V.L.; Heilbuth, N.M. and Linardi, V.R. (2001) Degradation of phenol by *Trichosporon* sp. LE3 cells immobilized in alginate. J. Basic Microbiol. 41: 171-178.
[127] Niazi, J.H. and Karegoudar, T.B. (2001) Degradation of dimethylphtalate by cells of *Bacillus* sp. immobilized in calcium alginate and polyurethane foam. J. Environ. Sci. Health 36: 1135-1144.
[128] Steenson, L.R.; Klaenhammer, T.R. and Swaisgood, H.E. (1987) Calcium alginate-immobilized cultures of lactic streptococci are protected from bacteriophages. J. Dairy Sci. 70: 1121-1127.
[129] Perrot, F.; Hebraud, M.; Charlionet, R.; Junter, G.A. and Jouenne, T. (2001) Cell immobilization induces changes in the protein response of *Escherichia coli* K-12 to a cold shock. Electrophoresis 22: 2110-2119.
[130] Bonin, P.; Rontani, J.F. and Bordenave, L. (2001) Metabolic differences between attached and free-living marine bacteria: inadequacy of liquid cultures for describing in situ bacterial activity. FEMS Microbiol. Lett. 194: 111-119.

[131] Sun, W. and Griffiths, M.W. (2000) Survival of bifidobacteria in yogurt and simulated gastric juices following immobilization in gellan-xanthan beads. Int. J. Food 61: 17-25.
[132] Lee, K.Y. and Heo, T.R. (2000) Survival of *Bifidobacterium longum* immobilized in calcium alginate beads in simulated gastric juices and bile salt solution. Appl. Environ. Microbiol. 66: 869-873.
[133] Norton, S.; Lacroix, C. and Vuillemard, J.C. (1993) Effect of pH on the morphology of *Lactobacillus helveticus* in free-cell batch and immobilized-cell continuous fermentation. Food Biotechnol. 7: 235-251.
[134] Cachon, R. and Divies, C. (1993) Localization of *Lactococcus lactis* ssp. *lactis* bv. *diacetylactis* in alginate gel beads affects biomass density and synthesis of several enzymes involved in lactose and citrate metabolism. Biotechnol. Techn. 7: 453-456.
[135] Klinkenberg, G.; Lystad, K.Q.; Levine, D.W. and Dyrset, N. (2001) pH-controlled cell release and biomass distribution of alginate-immobilized *Lactococcus lactis* subsp. *lactis*. J. Appl. Microbiol. 91: 705-714.
[136] White, G.F. and Thomas, O.R.T. (1990) Immobilization of the surfactant-degrading bacterium *Pseudomonas* C12B in polyacrylamide gel beads: I. Effect of immobilization on the primary and ultimate biodegradation of SDS, and redistribution of bacteria within beads during use. Enzyme Microb. Technol. 12: 697-705.
[137] Smith, M.R.; de Haan, A. and de Bont, J.A.M. (1990) The physiology of epoxyalkane-producing immobilized bacterial cells. In: de Bont, J.A.M.; Visser, J.; Mattiasson, B. and Tramper, J. (Eds.) Physiology of immobilized cells. Elsevier, Amsterdam, The Netherlands; pp. 499-501.
[138] Pepeljnjak, S.; Filipovic-Grcic, J. and Jalsenjak, V. (1994) Alginate micropheres of microbial spores and viable cells of *Bacillus subtilis*. Pharmazie 49: 436-437.
[139] El-Sayed, A.M.M.; Mahmoud, W.M. and Coughlin, R.W. (1990) Production of dextransucrase by *Leuconostoc mesenteroides* immobilized in calcium-alginate beads: I. Batch and fed-batch fermentations. Biotechnol. Bioeng. 36: 338-345.
[140] El-Sayed A.M.M.; Mahmoud W.M. and Coughlin R.W. (1990) Production of dextransucrase by *Leuconostoc mesenteroides* immobilized in calcium-alginate beads: II. Semi-continuous fed-batch fermentations. Biotechnol. Bioeng. 36: 346-353.
[141] Shinmyo A., Kimura H., Okada H. (1982) Physiology of α-amylase production by immobilized *Bacillus amyloliquefaciens*. Eur. J. Appl. Microbiol. Biotechnol. 14: 7-12.
[142] Chevalier, P. and de la Noüe, J. (1987) Enhancement of a-amylase production by immobilized *B. subtilis* in an airlift fermenter. Enzyme Microb. Technol. 9: 53-56.
[143] Chevalier, P. and de la Noüe, J. (1988) Behavior of algae and bacteria co-immobilized in a fluidized bed. Enzyme Microb. Technol. 10: 19-23.
[144] Kokubu, T.; Karube, I. and Suzuki, S. (1978) α-Amylase production by immobilized whole cells of Bacillus subtilis. Eur. J. Appl. Microbiol. Biotechnol. 5: 233-240..
[145] Argirakos, G.; Thayanithy, K. and John Wase, D.A. (1992) Effect of immobilisation on the production of α-amylase by an industrial strain of *Bacillus*. J. Chem. Technol. Biotechnol. 53: 33-38.
[146] Vuillemard, J.C. and Amiot, J. (1988) In: Moo-Young M. (Ed.) Bioreactor immobilised enzymes and cells: fundamentals and applications. Elsevier Applied Science, London, UK; pp. 213.
[147] Aleksieva, P.; Petricheva, E.; Konstantinov, E.; Robeva, C. and Mutafov, S. (1991) Acid proteinases production by *Humicola lutea* cells immobilized in polyhydroxyethylmethacrylate gel. Acta Biotechnol. 11: 255.
[148] Kanasawud, P.; Teeyapan, S.; Lumyong, S.; Holst, O. and Mattiasson,B. (1992) *Thermus* 2S from Thai hot springs: isolation and immobilization. World J. Microbiol. Biotechnol. 8: 137.
[149] Fortin, C. and Vuillemard, J.C. (1990) Elucidation of the mechanism involved in the regulation of protease production by immobilized Myxococcus xanthus cells. Biotechnol. Lett. 12: 913.
[150] Zhang, X.; Bury, S.; Dibiasio, D. and Miller, J. (1989) Effects of immobilization on growth, substrate consumption, β-galactosidase induction, and byproduct formation in *Escherichia coli*. J. Ind. Microbiol. 4: 239.
[151] Brito, L.C., Vieira; A.M., Leitão,J.G.; Sá-Correia, I.; Novais, J.M. and Cabral, J.M.S. (1990) Effect of the aqueous soluble components of the immobilization matrix on ethanol and microbial exopolysaccharides production. In: de Bont, J.A.M.; Visser, J.; Mattiasson, B. and Tramper, J. (Eds.) Physiology of immobilized cells. Elsevier, Amsterdam, The Netherlands; pp. 399-404.
[152] Krieg, N.R. and Holt, J.G. (1984) Bergey's manual of systematic bacteriology, Williams and Wilkings, Baltimore, USA.
[153] Ulitzur, S. and Kessel, M. (1973) Giant flagellar bundles of *Vibrio alginolyticus* (NCMB 1803). Arch. Mikrobiol. 94: 331-339.

[154] Belas, R.; Simon, M. and Silverman, M. (1986) Regulation of lateral flagella gene transcription in *Vibrio parahaemolyticus.* J. Bacteriol. 167: 210-218.
[155] Shapiro, J.A. (1987) Organisation of developing *Escherichia coli* colonies viewed by scanning electron microscopy. J. Bacteriol. 169: 142-156.
[156] Shapiro, J.A. (1985) Scanning electron microscope study of *Pseudomonas putida* colonies. J. Bacteriol. 164: 1171-1181.
[157] Shim, H. and Yang, S.T. (1999) Biodegradation of benzene, toluene, ethylbenzene, and o-xylene by a coculture of *Pseudomonas putida* and *Pseudomonas fluorescens* immobilized in a fibrous-bed bioreactor. J. Biotechnol. 67: 99-112.
[158] Fang, B.S.; Fang, H.Y.; Wu, C.S. and Pan, C.T. (1983) Biotechnol. Bioeng. Symp. No. 13: 457.
[159] Vieira, A.M.; Sa-Correia, I.; Novais, J.M. and Cabral, J.M.S. (1989) Could the improvements in the alcoholic fermentation of high glucose concentrations by yeast immobilization be explained by media supplementation? Biotechnol. Lett. 11: 137.
[160] Parascandola, P.; de Alteriis, E.; Sentandreu, R. and Zueco, J. (1997) Immobilization and ethanol stress induce the same molecular response at the level of cell wall in growing yeast. FEMS Microbiol. Lett. 150: 121-126.
[161] Gilson, C. and Thomas, A. (1995) Ethanol production by alginate immobilised yeast in a fluidised bed bioreactor. J. Chem. Technol. Biotechnol. 62: 38-45.
[162] Mussenden, P.; Keshavarz, T.; Saunders, G. and Bucke, C. (1993) Physiological studies related to the immobilization of *Penicillium chrysogenum* and penicillin production. Enzyme Microb. Technol. 15: 2-7.
[163] Lohmeyer, M.; Dierkes, W. and Rehm, H.J. (1990) Alkaloid production by high-performance strains of *Claviceps purpurea*. In: de Bont, J.A.M.; Visser, J.; Mattiasson, B. and Tramper, J. (Eds.) Physiology of immobilized cells. Elsevier, Amsterdam, The Netherlands; pp. 503-512.
[164] Kren, V.; Ludvík, J.; Kofronová, O.; Kozová, J. and Rehácek, Z. (1987) Physiological activity of immobilized cells of *Claviceps fusiformis* during long-term semicontinuous cultivation. Appl. Microbiol. Biotechnol. 26: 219-226.
[165] Nava Saucedo, J.E.; Barbotin, J.-N. and Thomas, D. (1989) Physiological and morphological modifications in immobilized Gibberella fujikuroi mycelia. Appl. Environ. Microbiol. 55: 2377-2384.
[166] Nava Saucedo, J.E.; Barbotin,J.-N. and Thomas, D. (1990) *Fusarium moniliforme*: A paradigm for physiological studies of immobilized filamentous microorganisms. In: de Bont, J.A.M.; Visser, J.; Mattiasson, B. and Tramper, J. (Eds.) Physiology of immobilized cells. Elsevier, Amsterdam, The Netherlands; pp. 577-582.
[167] Svoboda, A. and Ourednicek, P. (1990) Yeast protoplasts immobilized in alginate: cell wall regeneration and reversion to cells. Curr. Microbiol. 20: 335-338.
[168] Necas, O. and Svoboda, A. (1985) In: Peberdy, J.F. and Ferenczy, L. (Eds.) Fungal protoplasts. Marcel Dekker Inc., New York, USA; pp. 115.
[169] Nava Saucedo, J.E.; Barbotin, J.-N.; Velut, M. and Thomas, D. (1989) Ultrastructure examination of *Gibberella fujikuroi* mycelia: effect of immobilization in calcium alginate beads. Can. J. Microbiol. 35: 1118-1131.
[170] Tanaka, H.; Ohta, T.; Harada, S.; Ogbonna, J.C. and Yajima, M. (1994) Development of a fermentation method using immobilized cells in unsterile conditions. 1. Protection of immobilized cells against anti-microbial substances. Appl. Microbiol. Biotechnol. 41: 544-550.
[171] Ohta, T.; Ogbonna, J.C.; Tanaka, H. and Yajima, M. (1994) Development of a fermentation method using immobilized cells under unsterile conditions. 2. Ethanol and L-lactic acid production without heat and filter sterilization. Appl. Microbiol. Biotechnol. 42: 246-250.
[172] Melzoch, K.; Rychtera, M. and Habova, V. (1994) Effect of immobilization upon the properties and behaviour of *Saccharomyces cerevisiae* cells. J. Biotechnol. 32: 59-65.
[173] Simon, J.-P.; Benoot, T.; Defroyennes, J.-P.; Deckers, B.; Dekegel, D. and Vandegans, J. (1990) Physiology and morphology of Ca-alginate entrapped *Saccharomyces cerevisiae*. In: de Bont, J.A.M.; Visser, J.; Mattiasson, B. and Tramper, J. (Eds.) Physiology of immobilized cells. Elsevier, Amsterdam, The Netherlands; pp. 583-590.
[174] Chamy, R.; Nunez, M.J. and Lema, J.M. (1994) Product inhibition of fermentation of xylose to ethanol by free and immobilized *Pichia stipitis*. Enzyme Microb. Technol. 16: 622-626.
[175] Norton, S.; D'Amore, T. and Watson, K. (1993) Ethanol tolerance of immobilized brewer's yeast. In: Proc. Bioencapsulation III, the reality of a new industrial tool, Brussel, Belgium, Bioencapsulation Research Group; pp. 17-22.
[176] Norton, S.; Watson, K. and D'Amore, T. (1995) Ethanol tolerance of immobilized brewers' yeast cells. Appl. Microbiol. Biotechnol. 43: 18-24.

[177] Dale, M.C.; Eagger, A. and Okos, M.R. (1994) Osmotic inhibition of free and immobilized *K. marxianus* anaerobic growth and ethanol productivity in whey permeate concentrate. Appl. Microbiol. Biotechnol. 29: 535-544.

[178] Hilge-Rotmann, B. and Rehm, H.-J. (1991) Relationship between fermentation capability and fatty acid composition of free and immobilized *Saccharomyces cerevisiae*. Appl. Microbiol. Biotechnol. 34: 502-508.

[179] Marwaha, S.S.; Kennedy, J.F.; Khanna, P.K.; Tewari, H.K. and Redhu, A. (1990) Comparative investigations on the physiological parameters of free and immobilized yeast cells for effective treatment of dairy effluents. In: de Bont, J.A.M.; Visser, J.; Mattiasson, B. and Tramper, J. (Eds.) Physiology of immobilized cells. Elsevier, Amsterdam, The Netherlands; pp. 265-273.

[180] van der Sluis, C.; Stoffelen, C.J.; Castelein, S.J.; Engbers, G.H.; ter Schure, *E.G.*; Tramper, J. and Wijffels, R.H. (2001) Immobilized salt-tolerant yeasts: application of a new polyethylene-oxide support in a continuous stirred-tank reactor for flavour production. J. Biotechnol. 88: 129-139.

[181] Honecker, S.; Bisping, B.; Yang, Z. and Rehm, H.-J. (1989) Influence of sucrose concentration and phosphate limitation on citric acid production by immobilized cells of *Aspergillus niger*. Appl. Microbiol. Biotechnol. 31: 17-24.

[182] Coquet, L.; Junter, G.A. and Jouenne, T. (1999) Resistance of artificial biofilms of Pseudomonas aeruginosa to imipenem and tobramycin. J. Antimicrob. Chemother. 42: 755-760.

[183] de Alteriis, E.; Alepuz, P.M.; Estruch, F. and Parascandola, P. (1999) Clues to the origin of high external invertase activity in immobilized growing yeast: prolonged *SUC2* transcription and less susceptibility of the enzyme to endogenous proteolysis. Can. J. Microbiol. 45: 413-417.

[184] Kutney, J.P.; Choi, L.S.L.; Hewitt, G.M.; Salisbury, P.J.; Singh, M. (1985) Biotransformation of dehydroabietic acid with resting cell suspensions and calcium alginate-immobilized cells of *Mortierella isabellina*. Appl. Environ. Microbiol. 49: 96-100.

MODELLING IMMOBILISED-CELL PROCESSES

Application to integrated nitrogen removal with co-immobilised microorganisms

VITOR A. P. MARTINS DOS SANTOS[1,2], JOHANNES TRAMPER[2] AND RENE H. WIJFFELS[2]

[1]Dept. Process Engineering, Wageningen University, PO.Box 8129, 6700 EV Wageningen, The Netherlands – Fax: +31 317 4 82237 – Email: rene.wijffels@algemeen.pk.wau.n l

[2]Dept. Microbiology, German National Centre for Biotechnology Research, Mascheroder Weg 1, D-38124 Braunschweig, Germany – Fax: +(0)531-6181-411 – E-mail: vds@gbf.de

1. Introduction

Mechanistic models are proven, valuable tools for both the understanding and for the design, control and scale-up of complex systems, and are as well essential for the prediction of the responses of a given system to changes in environmental and operating conditions. Also, they give quite much insight into the mechanisms underlying the processes under consideration. This insight is extraordinarily important in the sense that it allows us to understand into great detail the basic phenomena regulating the processes involved, and gives us a powerful tool to control, modify or extend them to other systems in which these or similar phenomena play a role. Finally, they provide us with a solid knowledge framework in which information regarding complex interactions can be organised and systematised.

These arguments regarding modelling of bioprocesses in general are particularly relevant when applied to immobilised-cell processes, which are intrinsically very complex due to the interplay of many different factors. Indeed, understanding and identifying the most relevant interactions among the different microbial types, their metabolisms and activities, as well as determining the influence of the many different physical, chemical and mechanical factors on their behaviour and on the system's performance as a whole, requires a great deal systematisation and fundamental knowledge.

The main feature of immobilised-cell systems is the high attainable concentration of biocatalyst in a solid support, which, combined with a high reactor load, can lead to smaller reactor volumes as compared to suspended-cell processes.

V. Nedović and R. Willaert (eds.), Fundamentals of Cell Immobilisation Biotechnology, 493-529.

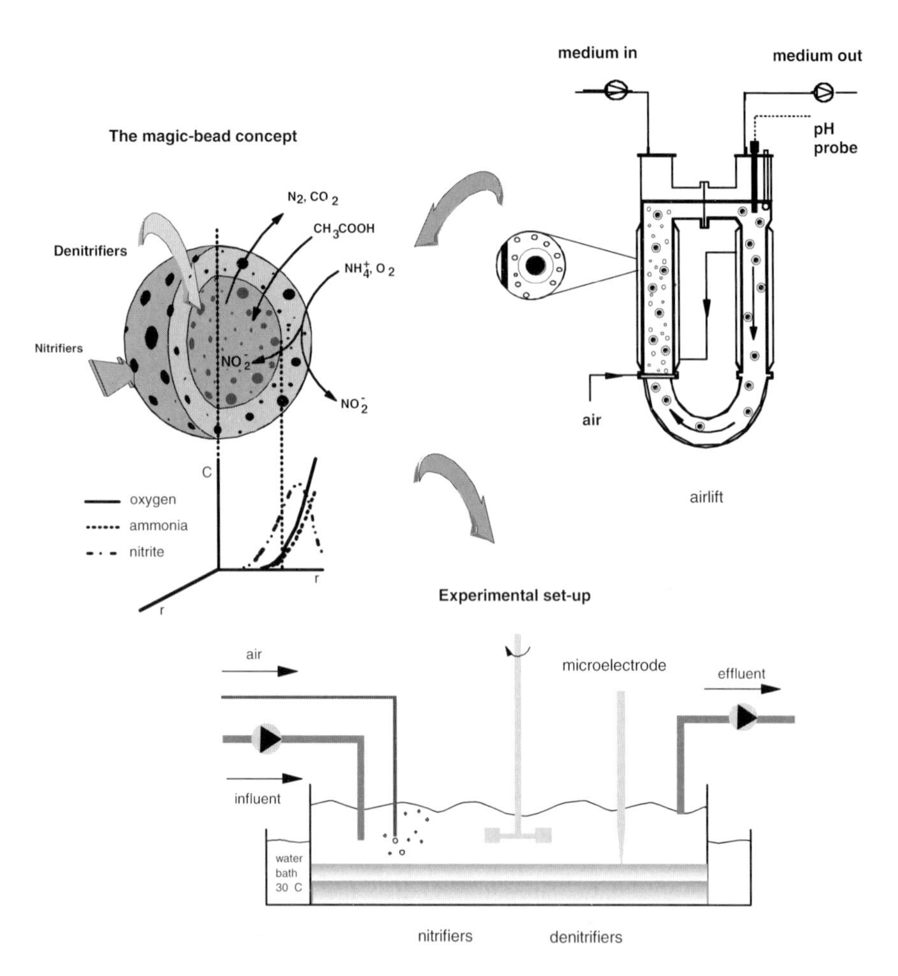

Figure 1. Schematic representation of the magic-bead concept and the experimental set up used for model validation.

Nowadays, many of the modern wastewater treatment plants rely on the use of immobilised microorganisms (either artificially entrapped into matrices or naturally attached onto carriers) for compact and effective removal of both inorganic and organic compounds. Unlike in well-mixed suspended-cell processes (such as activated-sludge basins), in immobilised-cell processes (biofilters, fluidized beds, gas-lift reactors, among others) cell growth and sequential elimination of pollutants may take place independently of the wastewater flow because the intervening microorganisms are retained in the treatment unit. These characteristics allow much more compactness and treatment effectiveness. Nevertheless, owing to additional mass-transfer resistance, pronounced gradients of substrates, products and, consequently, of biomass throughout the biocatalytic supports tend to develop in most immobilised-cell systems. In many aerobic treatment processes, as in nitrification, for instance, these gradients may ultimately result in that up to 95% of the biocatalytic support remains unused (*e.g.* entrapped systems) because the relevant microorganisms grow in a rather thin layer (typically 100 to 200 μm thick) just underneath (or just above) the carrier surface [1-4]. Aiming at achieving a greater compactness and treatment efficiency, we have previously described an integrated nitrification-denitrification with co-immobilised microorganisms [5]. A schematic representation of the system (the "magic-bead concept") is given in Figure 1. In the outer layer of the double-layered gel beads (placed in a continuous air-lift loop reactor), nitrifiers such as *Nitrosomas europaea* convert aerobically ammonium into nitrite, and this is subsequently (anaerobically) reduced to molecular nitrogen by denitrifiers such as *Pseudomonas sp.* or *Pseudomonas denitrificans* with simultaneous consumption of acetate. In attempt to describe the complex interdependencies underlying this system, a mechanistic mathematical model was developed accordingly [5]. This dynamic model accounted for diffusion of components, substrate utilization and growth, all occurring simultaneously in the beads. To be of practical use, any mathematical model, irrespectively of its conceptual importance or predictive capacity, needs necessarily to be validated experimentally. Only then can its outcome or predictions be seriously taken into account for the description of a system. A thorough validation of a model aiming at describing such processes as well as their dynamics requires measuring local (and time-dependent) concentrations of substrates, intermediates and products for a given range of relevant operating conditions. Hence, in this paper, the model is extended and validated by comparing the predicted substrate and product profiles throughout the biocatalysts with those measured with ion-selective and oxygen microelectrodes under different operating conditions. For technical reasons and to allow higher reproducibility, the experimental system was modified so that a double-flat plate was used instead of a double-layered bead (see Materials & Methods). This work aimed at gaining more insight and understanding into the complex mechanisms underlying these processes so that this knowledge can be used for the design, optimization and scale up of compact nitrogen removal systems. Furthermore, it constitutes a detailed case study for the broader scope of immobilised-cell processes.

2. Model development

2.1. STOICHIOMETRY

The following stoichiometric reactions are used in the modelling. These equations are based on elemental balances both for substrate conversion and synthesis of biomass. For nitrification by *Nitrosomonas europaea* the energy-yielding reaction (outer layer) comes as:

$$NH_4^+ + 1.5\ O_2 \Rightarrow NO_2^- + H_2O + 2\ H^+ \qquad \text{(a)}$$

Growth of *Nitrosomonas europaea*, using the empirical cell composition $C_5H_7NO_2$ (EPA, 1975) is described as:

by Wijffels *et al.* [10]. For *Pseudomonas* sp., a value of 170 kg m^{-3} was used according estimates of Gujer & Boller [45] for heterotrophic bacteria in a biofilm.

2.5.2. Transport parameters

Comparison of the characteristic times for the different mechanisms indicates that both external mass transfer and diffusion across the support have to be taken into account (Tables 1 and 2). A further sensitivity analysis of the model shows also that the model outcome is especially influenced by the accuracy of the effective diffusion coefficients of the solutes and the external mass-transfer coefficients between fluid and solid. Therefore, these were experimentally determined. The effective diffusion coefficients for NH_4^+ and NO_2^- in agar were measured experimentally as above described.

Table 3. Model parameters.

Parameter	Value	Dimensions
$\mu_{max,O2}^{Ps}$, $\mu_{max,NO2}^{Ps}$	$2.78\ 10^{-5}$, $3.33\ 10^{-5}$	s^{-1}
μ_{max}^{Ns}	$1.52\ 10^{-5}$	s^{-1}
m^{Ns}; m^{Ps}	$5.44\ 10^{-4}$; $3.84\ 10^{-4}$	mol kg^{-1} s^{-1}
m_{O2}^{Ns}	$9.4\ 10^{-4}$	mol O_2 kg^{-1} s^{-1}
Y_{NH4}^{Ns}; Y_{O2}^{Ns}	$2.16\ 10^{-3}$; $1.57\ 10^{-3}$	kg mol^{-1} NH_4^+, O_2
Y_{O2}^{Ps}; Y_{NO2}^{Ps}; Y_{HAC}^{Ps}	$9.86\ 10^{-3}$; $7.39\ 10^{-3}$; $1.37\ 10^{-3}$	kg mol^{-1} NO_2, Hac
K_{O2}^{Ns}; K_{NH4}^{Ns}	$5.05\ 10^{-3}$; 1.25	mol m^{-3}
K_{O2}^{Ps}; K_{NO2}^{Ps}; K_{HAC}^{Ps}	$2\ 10^{-3}$; 0.31; 0.73	mol m^{-3}
X^{Ns}_{max}; X^{Ps}_{max}	36.7; 90	kg m^{-3}
$K_{l,s}^{NH4}$; $K_{l,s}^{NO2}$	$2.21\ 10^{-5}$	m s^{-1}
$K_{l,s}^{O2}$; $K_{l,s}^{N2}$; $K_{l,s}^{HAC}$	$2.63\ 10^{-5}$; $2.07\ 10^{-5}$	m s^{-1}
D_w^{NH4}; D_w^{O2}; D_w^{HAC}	$2.20\ 10^{-9}$; $2.00\ 10^{-9}$	m^2 s^{-1}
D_w^{O2}; D_w^{O2}	$2.83\ 10^{-9}$	m^2 s^{-1}
D_{gel}^{NH4}; D_{gel}^{NO2}; D_{gel}^{HAC}	$1.78\ 10^{-9}$; $1.62\ 10^{-9}$	m^2 s^{-1}
D_{gel}^{N2}, D_{gel}^{O2}	$2.30\ 10^{-9}$	m^2 s^{-1}

The diffusion coefficient of oxygen in agar was that measured by Hulst *et al.* [46]. The same value is taken for the diffusion coefficient of molecular nitrogen in the gels. The mass-transfer coefficients between fluid and agar that were measured experimentally as above described. The dissociation constants and effective diffusion coefficients of the ions involved are those in Table 1. Strictly speaking, the diffusion rates of ions influence each other mutually [47]. However, since the exact values for each ionic species is mostly not available and since the concentration levels handled here are quite low (activity coefficients close to unit), we considered these mutual interferences to be negligible. Consequently, it is assumed that the diffusivities of both the dissociated and undissociated forms of a given electrolyte are the same.

3. Model validation: results and discussion

3.1. EXPERIMENTAL STRATEGY

As initial microelectrode studies with double-layered gel beads have shown to be very cumbersome and poorly reproducible, the experimental system was modified so that a double-flat plate was used instead of a double-layered beads (see experimental set up in Figure 1). This approximation is valid given the small size of the microelectrode (1-5 μm) as compared to the beads (2-3 mm, see [4,5]). The detailed methodologies and equipment used are described in the section Materials & Methods at the end.

Initially, the flow-through reactor with *Ns. europaea* and *Pseudomonas* sp. co-immobilised in the double-layered slab was subjected to predominantly denitrifying conditions. In practice this was done by sparging nitrogen instead of air at the same flow rate (10 ml min^{-1}) as in the measurement of mass-transfer coefficients to avoid any change in the hydrodynamic conditions (see above). NO_2^- and acetate were added to the medium at a final concentration of 0.2 and 0.5 mM, respectively. Ammonium at 0.4 mM was supplied as well to account for residual nitrification. In the model, these predominantly denitrifying conditions were accounted for by setting a bulk oxygen concentration at a value 4 times lower than that of its saturation in water. At day 6 after start-up, the acetate supply was interrupted and nitrogen was substituted by air. The system was thereafter run under purely nitrifying conditions. Six days later acetate was again continuously added to the influent medium to a final concentration of 0.5 mM so that coupled nitrification and denitrification could take place. These experiments were carried out both with medium with 5 mM (well-buffered) and 0.5 mM (poorly buffered) phosphate.

To study the dynamics of nitrification and coupled nitrification and denitrification a similar experiment was set and the corresponding ammonium, nitrite and oxygen profiles were measured along time.

3.2. MODEL PARAMETERS

3.2.1. Effective diffusion coefficients

Following the calculation procedure described by Beuling *et al.* [48], and Westrin *et al.* [49] for a pseudo-steady-state diffusion-cell (assuming Fickian diffusion), we arrived at an effective diffusion coefficient D_{eff} of 1.96 ± 0.06 10^{-9} m^2 s^{-1} for NO_2^- and 2.04 ± 0.05 10^{-9} m^2 s^{-1} for NH_4^+. This represents about 0.90 of the diffusion coefficient (D_0) of these ions in water at infinite dilution corrected for 303 K and viscosity of 0.795 mPa s through the Stokes-Einstein equation [59]. Such values agree well with those obtained using the equation $D_{eff} = D_0 (1 - \Phi_p)^3 / (1 + \Phi_p)^2$ of Makie and Meares [51], in which Φ_p represents the volume fraction of gel that is inaccessible to the diffusant due to the presence of polymer (here assumed to equal the volume fraction of gel, 2% (v/v)). This empirical relation implicitly assumes "a gel with a fine polymer structure with a chain thickness comparable to the molecular jump distance" in which only physical hindrance of the ions is considered [49]. This approach seems valid for this range of concentrations because the calculated values for the activity coefficients are close unity.

3.2.2. *Liquid-solid mass transfer coefficients*

Although in many studies the influence of the external diffusion limitation on the performance of immobilised-cell systems has been neglected (see Wijffels and Tramper, [52] for an overview), other detailed investigations have shown that this parameter often determines both their kinetic behaviour and the maximum attainable reaction levels [3,9,13,53a,53b]. Indeed, a sensitivity analysis of the model here described shows that the transfer of oxygen from the liquid to the solid phase is, under a wide range of operating conditions, the rate-limiting step in the overall conversion of ammonium to nitrogen gas [5].

3.2.3. *Measurement of the thickness of the Diffusive Boundary Layer (DBL)*

It is clear from Figure 2 that the stirring rate influenced the thickness of the boundary layer. However, this influence diminished strongly for higher rates, for which a difference lower than 8 μm in the thickness of the DBL was found. Hence, all subsequent measurements were carried out at a stirring rate of 150 rpm (at rates higher than 200 rpm, the turbulence affected negatively the measurements with ion-selective microelectrodes and caused vibration of the gel slab). The thickness of the boundary layers calculated at this rate was similar to those obtained by Wijffels *et al.* [3] for a similar set up, but were considerably lower than those found around gel beads in other microelectrode studies [27,53a,53b,54]. This is likely due to the different flow regimes applied and/or the different geometry of the biocatalytic support.

It is also noteworthy that estimations from an empirical relation (based on a rapidly rotating disk immersed in a fluid as defined by Sherwood, [55]) gave thicknesses consistently higher than those measured with the microelectrodes (points in Figure 3). The tendency is, however, quite similar.

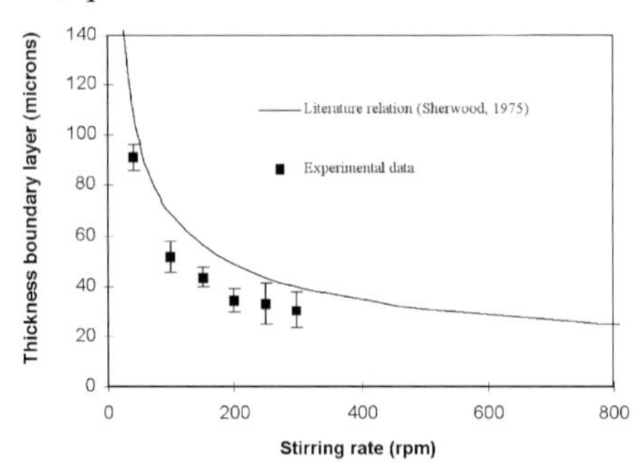

Figure 3. Experimental (points) and estimated (solid line) thickness of the boundary layer as a function of the stirring rate. Error bars for standard deviation refer to an average of four measurements.

This discrepancy between estimated and measured values is likely due to both the fact that the empirical relation is in fact a rough simplification of reality (gel slab as a rapidly rotating disk immersed in a fluid) and, perhaps more important, due to the turbulence introduced through aeration.

.3. SUBSTRATE AND PRODUCT PROFILES

.3.1. *Denitrification*

Under sufficiently buffered conditions one would not expect strong pH gradients to occur within the biocatalytic support. For weak buffering, however, pH profiles are likely to develop as a result of the denitrifying activity. Figure 4 shows the measured (symbols) and predicted (lines) concentration profiles of oxygen, ammonium, nitrite and pH across a double-layered gel slab during denitrification under well (A) and poorly (B) buffered conditions.

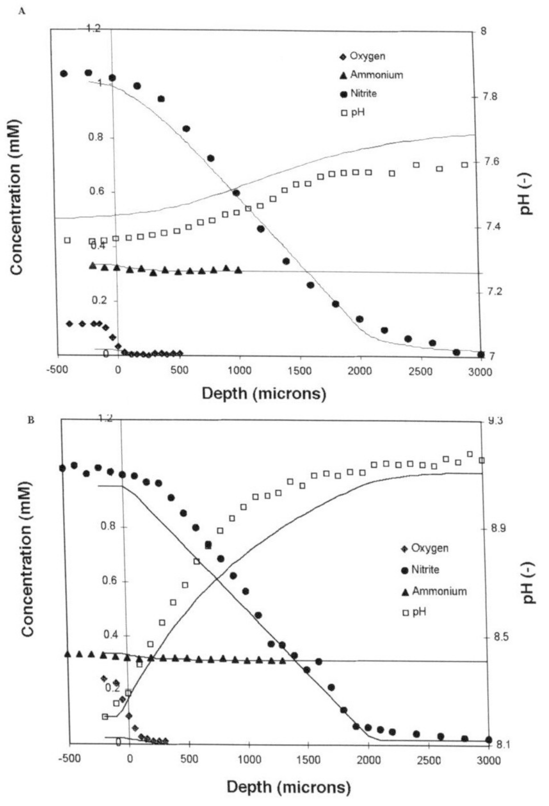

Figure 4. Predicted (lines) and measured (points) NH_4^+, NO_2^-, pH and O_2 gradients across the doubled-layer gel slab under denitrifying conditions in the presence of 5 mM (A) and 0.5 mM (B) phosphate buffer. Zero represents the slab surface. From left to right: bulk liquid: -500 to –52 μm; diffusive boundary layer: -52 to 0 μm; nitrifying layer: 0-2000 μm; denitrifying layer: beyond 2000 μm.

Under both buffering conditions, model predictions and experimental measurements agreed reasonably well. In both A and B a steep nitrite profile developed across the slab, whereas the ammonium concentration remained practically at the same level. Due to the low oxygen concentration, there was only a little nitrification within the very first outer layers. Nitrite, on the contrary, was consumed at a considerable rate within the first layers of the denitrifying zone. The shape of the nitrite profile reflects its diffusion across the nitrifying zone, being driven by the concentration gradient between the bulk and the denitrifying layer, which virtually works as a "sink". From the Figure it seems also clear that the *Pseudomonas* sp. within the first layers of the denitrifying zone would have accounted for most of the denitrification.

The actual oxygen profiles were sharper than predicted likely, as reported earlier [56], due to its uptake at the slab surface by undefined heterotrophic populations. The oxygen level in the bulk (about 0.1 mM) in both Fig. 4A and 4B was higher than predicted because it was difficult to maintain a constant (low) oxygen tension in such an open vessel. In any case, these discrepancies did not affect the denitrification process as a whole. Although the general trend in ammonium, oxygen and nitrite profiles was similar under both buffering conditions the shape of the respective pH profiles differed considerably. Under relatively well-buffered conditions (Fig. 4A), the pH increased only about 0.2 units across the whole slab, whereas for the lower buffering capacity (Fig. 4B) this increase amounted more than one pH unit.

This difference was due to the weaker capacity of the buffer in Fig. 4B, which was unable to compensate for the alkalinity generated by reduction of nitrite to nitrogen. From the profiles measured, however, it was unclear whether the pH increase did indeed result in a lower nitrite reduction rate, as predicted by Eq. (6). Nevertheless, and despite the simplifications done, the model was able to account for the interactions among all the charged species with a reasonable accuracy under the conditions tested. These results confirm those predicted long ago by Riemer & Harremoes [57] and measured indirectly by Arvin & Kristensen [58].

3.3.2. Nitrification

Figure 5 shows the predicted (lines) and measured (symbols) oxygen, ammonium, nitrite and pH profiles across the gel slab under purely nitrifying conditions (*i.e.* with no organic carbon supplied) for a well (A) and poorly (B) buffered influent. To increase accuracy and minimize errors we chose bulk N-limiting conditions so that the profiles could be measured throughout a large part of the biocatalytic support for relatively long periods of time.

In both Fig. 5A and 5B, ammonium was consumed at about the stoichiometric rate of nitrite accumulation. All ammonium that was consumed was thus transformed nitrite. The general shape of the curves (regardless of their absolute values) indicates that nitrifiers were active throughout the whole 2000-μm layer and that nitrite was not further converted into nitrogen through denitrification. Note also that, as shown earlier [56], the predictions for oxygen took into account heterotrophic oxygen uptake by the denitrifying bacteria in the inner slab. The grey line indicates the predicted oxygen gradient (which flattens at 0.05 mM oxygen) if there was no such uptake. In well buffered medium (Fig. 5A), the pH profile was relatively smooth (pH difference of about 0.2 units), whereas under poorly buffered (Fig. 5B) conditions a steep pH gradient of about 1 pH unit developed across the slab. The gradient was far more pronounced in the nitrifying slab, after which it seemed to increase slightly. Conversely, the oxygen, ammonium and nitrite profiles were much less steep in the poorly buffered medium than for that well buffered. In fact, the ammonium consumed and ammonium produced in Fig. 5B were less than half of that in Fig. 5A. Likely, the reduced nitrification in Fig. 5B resulted from the inhibitory effect of the pH decrease that arose from the nitrifying activity itself). Evidently, the buffer capacity of medium Fig. 5B was not sufficient to compensate for the acid production. Globally, this reflected as well on a lower bulk pH (7.1) than that (7.5) of the influent.

These effects were correctly described by the model, which predicted fairly well a reduced nitrifying activity as a result of a pH decrease. Also, not only the oxygen profile in Fig. 5B was less steep than in Fig. 5A, but also oxygen was not completely depleted even after 3000 microns. This likely means that the pH inhibited as well the heterotrophic activity of the denitrifiers in the denitrifying slab. The model accounted for this as well.

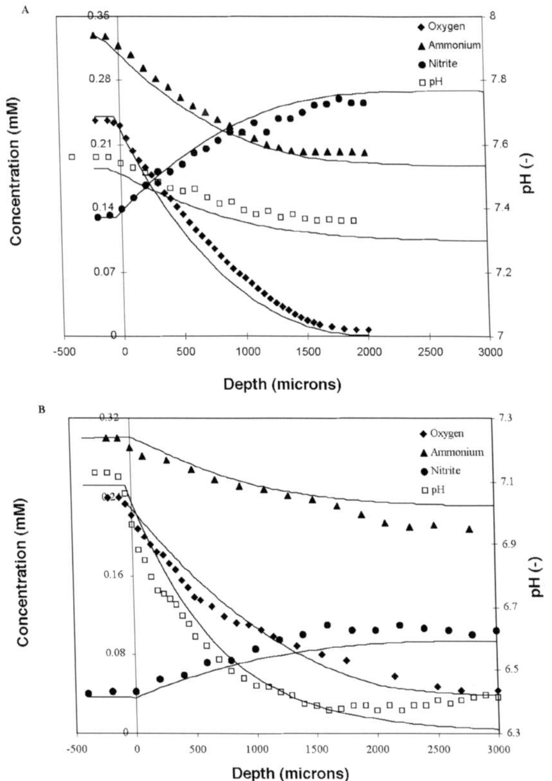

Figure 5. Predicted (lines) and measured (points) NH_4^+, NO_2^-, pH and O_2 gradients across the doubled-layer gel slab nitrifying conditions in the presence of 5 mM (A) and 0.5 mM (B) phosphate buffer.

The steepness of the pH profile across the slab in the poorly buffered medium also clearly illustrates the importance of external resistance to mass transfer on the development of substrate and product gradients and on the kinetics of nitrogen conversion. Indeed, although the system was well stirred and the thickness of the diffusive boundary layer was estimated to be only about 50 microns [56], within this layer the pH dropped between 0.2 and 0.3 pH units. External mass transfer resistance influences the in/efflux of acids and bases into/out-of a catalytic particle and therefore, in weakly buffered systems, internal pH values may be suboptimal despite of influent pH corrections. These results as a whole generally agree with those presented earlier by de Beer *et al.* [54] and Zhang & Bishop [59,60], who measured directly pH profiles in biofilms and with those of Szwerinski *et al.* [62] and Siegrist & Gujer [62], who predicted and estimated indirectly such gradients.

3.3.3. Simultaneous nitrification and denitrification

The shape and magnitude of the ammonium and oxygen profiles are similar to those in Fig. 5A but that of nitrite shows now a broad peak with a maximum at about 600 μm inside the slab. Also, its peak concentration is lower than that in Fig. 5A.

These features in the nitrite profile of Fig. 6B reflect the simultaneous production of nitrite by *Nitrosomonas* within the nitrifying layer, and its consumption by *Pseudomonas* in the denitrifying one (beyond 2000 μm, see Figure 6). This sequential process for nitrification and denitrification was hence proceeding simultaneously within the slab and at rates and magnitudes that were appropriately predicted by the model.

One of the advantages of coupling nitrification and denitrification within a single biocatalytic particle is, as claimed by several authors [1,3,63-65], the possibility of partially compensating the acid produced in nitrification (Eq. (c)) by the base formed in the denitrifying step (Eq. (b)). pH control would be thereby greatly facilitated, which is particularly relevant for wastewater treatment at large scales. Hence, if that is indeed so, one would expect that the pH gradient throughout a biocatalyst support such as that studied here would be a "combination" of the profiles of Figures 4 and 5 above. Figure 6 shows the oxygen, ammonium, nitrite and pH profiles across the double-layered slab under well (A) and poorly (B) buffered conditions for coupled nitrification and denitrification. As expected, at higher buffer capacity (Fig. 4A), the pH profile across the slab was fairly flat, with only a slight "bow" (lower than 0.1 pH unit) within the nitrifying zone, and was followed by a small increase thereafter. This profile (symbols) was actually flatter than that predicted by the model (line) for both the nitrifying and denitrifying zones. The lowest pH value corresponded to the point where (by taking into account mass transfer and the influence of all other ionic species) the acid and base resulting from nitrification and denitrification compensated each other. The corresponding experimental oxygen, ammonium and nitrite profiles agreed fairly well with the model predictions and reflected a relatively high nitrogen transformation rate (see inset in Fig. 6A).

For lower buffer capacity (Fig. 6B), there was a slight pH decrease within the first 500 microns and a progressive increase in the following 2000. However, these changes were less pronounced than those anticipated by the model, which had predicted a maximal pH drop of about 0.2 within the first 100 microns and an increase of about 0.4 units thereafter. Thus, in fact, the system seemed to have a buffer capacity higher than that predicted, even by taking into account that nitrogen removal activity was slightly lower than foreseen (as indicated by the measured ammonium and nitrite profiles). The slight deviation between the predicted and measured nitrogen removal activities is likely to be explained by the somewhat higher-than-predicted oxygen consumption at the bead surface by undefined heterotrophic populations. Indeed, we could observe a very thin layer of cells covering the slab surface after supplying continuously acetate to the reactor (to allow denitrification to occur) for a longer period of time. Likely, these cells were taking up some of the oxygen in the bulk for acetate oxidation. This has been observed earlier in both nitrifying [66] and nitrifying /denitrifying systems [5,56].

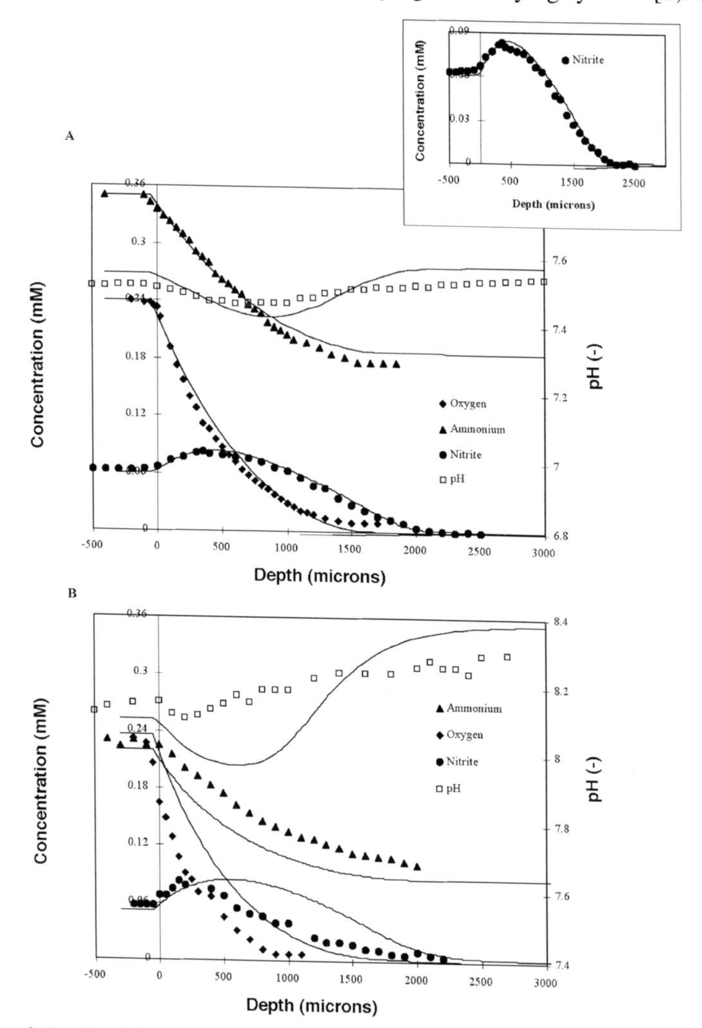

Figure 6. Predicted (lines) and measured (points) NH_4^+, NO_2^-, pH and O_2 gradients across the doubled-layer gel slab under coupled nitrifying and denitrifying conditions in the presence of 5 mM (A) and 0.5 mM (B) phosphate buffer.

A careful comparison between Figure 5B (poorly buffered nitrification) and 6B (poorly buffered nitrification and denitrification) shows that both the ammonium oxidation rate and its extent were higher in the co-immobilised system than in that with nitrifiers alone. Indeed, although the ammonium oxidation under poorly buffered conditions was about two thirds of that under good buffering, this rate was still almost twice as high as that for nitrification alone in weakly buffered medium (Figure 5B). Also, in Fig. 6B there is barely any oxygen beyond 2000 microns, which indicates a higher oxygen consumption activity both by the nitrifiers and denitrifiers. Both this enhanced ammonium oxidation and the higher oxygen uptake with regard to unbuffered nitrification are likely to be explained by the denitrifying activity in the inner slab, in which the denitrifiers partly counteract the acidity produced in the nitrifying step. These results confirm experimentally the qualitative predictions made earlier by Kokufuta *et al.* [63,64], Martins dos Santos *et al.* [5] and Tartakovski *et al.* [65].

Although advantageous with regard to pH control, it was not entirely clear why the system in Figure 6B seemed to have a buffer capacity higher than predicted. One plausible explanation is simply that the interactions among the various ionic species involved were not correctly handled by the model, which is actually very likely given their complexity (such as that regarding the role of the carbonate/bicarbonate pair, which is a particularly "difficult" pair) and the simplifications made here (see Model development above). Discrepancies in the pH evolution in methanogenic and nitrifying microbial aggregates observed by other authors were partly explained by the buffer capacity of the biomass itself (which can be quite significant, as shown by De Beer *et al.* [54], and by processes related to cell lysis or ammonification of proteins [26]. Neither of these effects, however, is likely to have been important here because the biomass concentrations involved were low and the experiments did not last long enough so as to expect significant cell lysis.

3.3.4. Rate-limiting substrates

The detailed analysis of Figure 5 (both A and B) showed that ammonium and not oxygen was actually limiting nitrification. The "Monod" term [$(S_i/(S_i + Ks_i)$ in which S_i is the local concentration measured for compound i and Ks_i the respective affinity constant by the cell, see Mankad & Nauman, [18]] for NH_4^+ is much lower than that for oxygen throughout the whole nitrifying layer. The "Monod" term gives an indication of what the cell "sees" and of what it is able to take up. For instance at 1500 μm, a depth at which flattening is almost complete, the oxygen concentration of 0.035 mM is still about 7 times higher than 0.005 mM, which is the value for the affinity constant of *Nitrosomonas* for O_2. On the contrary, at this depth the local ammonium concentration (0.24 mM) is only one fifth of *Nitrosomonas*'s calculated affinity constant (1.25 mM).

Hence, regardless whether an interactive (growth and consumption are dependent on the product of the Monod terms for oxygen and ammonium) or a non-interactive (growth and consumption dependent on the substrate for which the Monod term is the lowest) approach is used for analyzing such results (a controversial dichotomy widely discussed in the relevant literature), ammonium remains the substrate that determines the most, under these conditions, the performance and effectiveness of nitrification. In fact, the availability of ammonium determined not only the nitrification level but also the overall nitrogen removal rate as a whole since the rate of denitrification depended directly on the nitrite produced within the nitrifying layer. This became clear as well from the analysis of nitrogen, carbon and oxygen fluxes for coupled nitrification and denitrification in Figure 5B (not shown). Here too, the Monod term for ammonium was

[9] de Gooijer, C.D.; Wijffels, R.H. and Tramper, J. (1991) Growth and substrate consumption of *Nitrobacter agilis* cells immobilised in carrageenan: part 1. Dynamic modeling. Biotechnol. Bioeng. 38: 224-231.

[10] Wijffels, R.H.; Schepers, A.W.; Smit, M.; de Gooijer, C.D.; Tramper, J. (1994) Effect of initial biomass concentration on the growth of immobilised *Nitrosomonas Europaea*. Appl. Microbiol. Biotechnol. 42: 153-157.

[11] Beefink, H.H.; van der Heijden, R.T.J.M. and Heijnen, J.J. (1990) Maintenance requirements: energy supply from simultaneous respiration and substrate consumption. FEMS Microbiol. Ecol. 73: 203-210.

[12] Hunik, J.H.; Bos, C.G.; Van Den Hoogen, P.; De Gooijer, C.D. and Tramper, J. (1994) Co-immobilised *Nitrosomonas euroapaea* and *Nitrobacter agilis* cells: validation of a dynamic model for simultaneous substrate conversion and growth in κ-carrageenan gel beads. Biotechnol. Bioeng. 43: 1153-1163.

[13] Wijffels, R.H.; De Gooijer, C.D.; Kortekaas, S. and Tramper, J. (1991) Growth and substrate consumption of *Nitrobacter agilis* cells immobilised in k-carrageenan. Part 2: Model evaluation. Biotechnol. Bioeng. 38: 232-240.

[14] Wijffels, R.H.; de Gooijer, C.D.; Schepers, A.W.; Beuling, E.E.; Mallée, L.F. and Tramper, J. (1996) Dynamic modeling of immobilised *Nitrosomonas europaea*: implementation of diffusion limitation over expanding micro-colonies. Enzyme Microb. Technol. 17: 462-471.

[15] Willaert, R.G.; Baron, G.V. and De Backer, L. (1996) Immobilised living cell systems. John Wiley & Sons, New York, USA.

[16] Ryder, D.N. and Sinclair, C.G. (1972) Model for the growth of aerobic micro-organisms under oxygen limiting conditions. Biotechnol. Bioeng. 14: 787-798.

[17] Bader, F.G. (1978) Analysis of double-substrate limited growth. Biotechnol. Bioeng. 20:183-202.

[18] Mankad, T and Nauman, E.B. (1992) Modeling of microbial growth under dual limitations. Chem. Eng. J. 48: B9-B11.

[19] Robertson, L.A. and Kuenen, J.G. (1990) Combined heterotrophic nitrification and aerobic denitrification in *Thiosphera pantotropha* and other bacteria. Antonie van Leeuwenhoek 57: 139-152.

[20] Robertson, L.A.; Corneline, R.; De Vos, P.; Hadioetomo, R. and Kuenen, J.G. (1989) Aerobic denitrification in various heterotrophic nitrifiers. Antonie van Leeuwenhoek 56: 289-299.

[21] Hunik, J.H.; Tramper, J. and Meijer, H.J.G. (1992) Kinetics of *Nitrosomonas europaea* at extreme substrate, product and salt concentrations. Appl. Microbiol. Biotechnol. 37: 802-807.

[22] Wang, J.H.; Baltzis, B.C. and Lewandowski, G.A. (1995) Fundamental denitrification studies with *Pseudomonas denitrificans*. Biotechnol. Bioeng. 47: 26-41.

[23] Almeida, J.S.; Reis, M.A.M. and Carrondo, M.J.T. (1995) Competition between nitrate and nitrite reduction in denitrification by *Pseudomonas fluorescens*. Biotechnol. Bioeng. 46: 476-484.

[24] Press, W.H. (1989) Numerical recipes in Pascal: the art of scientific computing. Cambridge University Press, Cambridge, UK.

[25] Flora, J.RV.; Suidan, M.T.; Biswas, P. and Sayles, G.D. (1993) Modeling substrate transport into biofilms: role of multiple ions and pH effects. J. Environm. Eng. 119: 908-930.

[26] Lens, P.; de Bee,r D.; Cronenberg, C.; Ottengraf, S. and Verstaete, W. (1995) The use of microsensors to determine population distributions in UASB aggregates. Wat. Sci. Technol. 31: 273-280.

[27] Cronenberg, C.C.H. and Van den Heuvel, J.C. (1991) Determination of glucose diffusion coefficients in biofilms with microelectrodes. Biosens. Bioelectron. 6: 255-262.

[28] Flora, J.RV.; Suidan, M.T.; Biswas, P. and Sayles, G.D. (1995) A modeling study of ananerobic biofilm systems: I. Detailed biofilm modeling. Biotechnol Bioeng. 46: 43-53.

[29] Bailey, J.E. and Ollis, D.F. (1992) Biochemical Engineering Fundamentals. 2nd ed., McGraw-Hill, New York.

[30] Quinlan, A.V. (1984) Prediction of the optimum pH for ammonia-N oxidation by *Nitrosomonas europaea* in well-aerated natural and domestic-waste waters. Wat. Res. 18: 561–566.

[31] Antoniou, P.; Hamilton, J.; Koopman, B.; Jain, R.; Holloway, B.; Lyberatos and Svoronos, S.A. (1990) Effect of temperature and pH on the effective maximum specific growth rate of nitrifying bacteria. Wat. Res. 24: 97-101.

[32] Laudelout, H.; Lambert, R. and Pham, M.L. (1976) Influence du pH et la pression partielle d'oxygene sur la nitrification. Ann. Microbiol. (Inst. Pasteur) 127A: 367-382.

[33] Suzuki, I. ; Dular, U. and Kwok, S.C. (1974) Ammonia or ammonium as substrate for oxidation by *Nitrosomonas europaea* cells and extracts. J. Bacteriol. 120:556-558

[34] Thomsen, J.K.; Geest, T. and Cox, R.P. (1994) Mass spectrometric sudies of the effect of pH on the accumulation of intermediates in denitrification by *Paracoccus denitrificans*. Appl. Environm. Microbiol. 60: 536-541.

[35] Wiesmann, U. (1994) Biological nitrogen removal from wastewaters. In: Fletcher, A. (Ed.), Advances in Biochemical Engineering and Biotechnology. Springer-Verlag, 51: 113-154.
[36] Sheintuch, M.; Tartakovski, B.; Narkis, N. and Rebhun, M (1995) Substrate inhibition in a continuous nitrification process. Wat. Res. 29: 953-963.
[37] Dombrowski, T. (1991) Kinetik der Nitrifikation und Reaktionstechnik der Stickstoffeliminierung aus hochbelasteten Abwässern. VDI-Fortschrittsberichte, Reihe 15: Umwelttechnik no. 87.
[38] Anthonissen, A.C.; Loehr, R.C.; Prakasmam, T.B.S. and Srinath, E.G. (1976) Inhibition of nitrification by ammonia and nitrous acid. J. WPCF 48: 835-852.
[39] Prosser, J.I. (1989) Autotrophic nitrification in bacteria. Adv. Microbiol. Physiol. 30: 125-181.
[40] Van Niel, E.W.J.; Braber, K.J.; Robertson, L.A. and Kuenen, J.G. (1992) Heterotrophic nitrification and denitrification in *Alcaligenes faecalis* strain TUD. Antonie van Leeuwenhoek. 62: 231-237.
[41] Robertson, L.A. and Kuenen, J.G. (1990) Combined heterotrophic nitrification and aerobic denitrification in *Thiosphera pantotropha* and other bacteria. Antonie van Leeuwenhoek 57: 139-152.
[42] van Niel, E.W.J.; Arts, P.A.M.; Wesselink, B.J.; Robertson, L.A. and Kuenen, J.G. (1993) Competition between heterotrophic and autotrophic nitrifiers for ammonia in chemostat cultures. FEMS Microbiol. Ecol. 102: 109-118.
[43] Koike, I. and Hattori,A. (1975) Growth yield of a denitrifying bacterium, *Pseudomonas denitrificans*, under aerobic and denitrifying conditions. J. Gen. Microbiol. 88: 1-10.
[44] Zumft, W.G. (1991) The denitrifying prokaryotes. In: Balows, A.; Truper, H.; Dworkin, M; Harder, W. and Schleifer, K. (Eds), The prokaryotes. Springer-Verlag, New York, USA, pp. 555-556.
[45] Gujer, W. and Boller, M. (1989) A Mathematical model for rotating biological contactors. In: Proc. EWPCA Conference on Technical advances in biofilm reactors, Nice, France; pp. 69-89.
[46] Hulst, A.C.; Hens, H.J.H.; Buitelaar, R.M. and Tramper, J. (1989) Determination of the effective diffusion coefficient of oxygen in gel materials in relation to concentration. Biotechnol. Techn. 3: 199-204.
[47] Wesselingh, J.A. and Krishna, R. (1990) Mass Transfer. Ellis Harwood, Chichester, UK.
[48] Beuling, E.E.; Van den Heuvel, J.C. and Ottengraf, S.P.P. (1996) Determination of biofilm diffusion coefficients using microelectrodes. In: Wijffels, R.H.; Buitelaar, R.; Wessels, H.; Tramper, J. and Bucke, C. (Eds.), Immobilised Cells: Basics and Applications. Elsevier Science, Amsterdam, The Netherlands, pp. 31- 38.
[49] Westrin, BA. (1991) Diffusion measurements in gels: a methodological study. PhD dissertation, Lund University, Lund, Sweden.
[50] Lide, D.R. (Ed.) (1992) Handbook of Chemistry and Physics. 73rd edition. CRC Press, Boca Raton, U.S.A.
[51] Mackie, J.S and Meares, P. (1955) The diffusion of electrolytes in a cation-exchange resin membrane. Theoretical. Proc. Roy. Soc. London. A232: 498-509.
[52] Wijffels, R.H. and Tramper, J. (1992) Nitrification by immobilised cells. Enzyme Microb. Technol. 17: 482-492.
[53a] Hooijmans, C.M.; Gerats, S.G.M., van Niel, E.W.J.; Robertson, R.A.; Heijnen, J.J. and Luyben, K. Ch.A.M. (1990) Determination of growth and coupled nitrifcation/dewnitrification by immobilised *Thiophaera pantotropha* using measurements and modeling oxygen profiles. Biotechnol. Bioeng. 36: 931-939.
[53b] Hooijmans, C.M.; Gerats, S.G.M.; Potter, J.J.M. and Luyben, K.Ch.A.M. (1990) Experimental determination of mass transfer boundary layer around a spherical biocatalyst particle. Biochem. Eng. J. 44: B41.
[54] de Beer, D.; van den Heuvel, J.C. and Ottengraaf, S.P.P. (1993) Microelectrode measurements of activity distribution in nitrifying bacterial aggregates. Appl. Environ. Microbiol. 59: 573-579.
[55] Sherwood, T.K.A.; Pigford, R.L. and Wilke, C.R. (1975) Mass transfer. McGraw-Hill Book Company, London, UK.
[56] Martins dos Santos, V.A.P.; Verschuren, P.; van den Heuvel, H.; Tramper, J. and Wijffels, R.H. Substrate and product profiles across double-layer gel beads: modelling and experimental evaluation. Submitted.
[57] Riemer, M. and Harremoes, P. (1978) Multi-component diffusion in denitrifying biofilms. Prog. Wat. Technol. (presently Wat. Sci. Technol.) 10: 149-163.
[58] Arvin, E. and Kristensen, G.H. (1982) Effect of denitrification on the pH in biofilms. Wat. Sci. Technol. 14: 833-848.
[59] Zhang, T.C. and Bishop, P. (1996) Evaluation of substrate and pH effects in a nitrifying biofilm. Wat. Environm. Res. 68: 1107-1115.

[60] Zhang, TC; Fu, Y. and Bishop, P. (1995) Competition for substrate and space in biofilms. Wat. Environm. Res. 67: 992-1003.
[61] Szwerinski, H.; Arvin, E. and Harremoes, P. (1986) pH decrease in nitrifying biofilms. Wat. Res. 20: 971-976.
[62] Siegrist, H. and Gujer, W. (1987) Demonstration of mass transfer and pH effects in a nitrifying biofilm. Wat. Res. 21: 1481-1487.
[63] Kokufuta E; Yukishige M and Nakamura I. (1987) Coimmobilisation of *Nitrosomonas* and *Paracoccus denitrificans* cells polyelectrolyte complex-stabilized calcium alginate gel. J. Ferment. Bioeng. 6: 659-664.
[64] Kokufuta, E.; Shimohashi, M. and Nakamura, I. (1988) Simultaneously occurring nitrification and denitrification under oxygen gradient by polyelectrolyte complex-coimmobilised *Nitrosomonas* and *Paracoccus denitrificans* cells. Biotechnol. Bioeng. 31: 382-384.
[65] Tartakovsky, B.; Kotlar, E.; Sheintuch, M. (1996) Coupled nitrification-denitrification processes in a mixed culture of co-immobilised cells:Analysis and experiments. Chem. Eng. Sci. 51: 2327-2336.
[66] van Benthum, W.A.J.; van Loosdrecht, M.C.M. and Heijnen, J.J. (1997) Control of heterotrophic layer formation on nitrifying biofilms in a biofilm airlift suspension reactor. Biotechnol. Bioeng. 53: 397-405.
[67] Meyerhoff, J.; John, G.; Bellgardt, K.H. and Schugerl, K. (1997) Characterization and modelling of coimmobilised aerobic/anaerobic mixed cultures Chem. Eng Sci.52 (14): 2313-2329.
[68] Tartakovsky, B.; Guiot, S.R.; Sheintuch, M. (1998) Modeling and analysis of co-immobilised aerobic/anaerobic mixed cultures. Biotechnol. Progr. 14: 672-679.
[69] Peng, C.A. and Bly, M.J. (1998) Analysis of xenobiotic bioremediation in a co-immobilised mixed culture system. Biochem. Eng. J. 1: 63-75.
[70] Hellendoorn, L.; Ottengraaf, S.P.; Pennings, J.A.M.M.; van den Heuvel, J.C.; Martins dos Santos, V.A.P. and Wijffels, R.H. (1999) Kinetic behaviour of and performance of a co-immobilised system of amyloglucosidase and *Zymomonas mobilis*. Biotechnol. Bioeng. 63: 694-704.
[71[Martins dos Santos, V.A.P.; Jacobs, M; Tramper, J.; Jetten, M.; Kuenen, G. and Wijffels, R.H. simultaneous autotrophic nitrification and anaerobic ammonium oxidation with co-immobilised micro-organisms. Appl. Environm. Microbiol., submitted.
[72] Martins dos Santos, V.A.P.; Verschuren, P.; van den Heuvel, H.; Tramper, J., and Wijffels, R.H. pH effects on coupled nitrification and denitrification measured by specific microelectrodes. Biotechnol. Bioeng., submitted.
[73] Goodall, J.L. and Peretti, S.W. (1998) Dynamic modeling of meta- and para-nitrobenzoate metabolism by a mixed co-immobilised culture of *Comamonas* spp. JS46 and JS47. Biotechnol Bioeng. 59: 507-16.
[74] Goodall, J.L.; Thomas, S.M.; Spain, J.C. and Peretti, S.W. (1998) Operation of mixed-culture immobilised cell reactors for the metabolism of meta- and para-nitrobenzoate by *Comamonas* sp. JS46 and *Comamonas* sp. JS47. Biotechnol. Bioeng. 59: 21-27.
[75] Tsien, R.Y. and Rink, T.J. (1980) Neutral carrier ion-selective microelectrodes for the measurement of intacellular free calcium. Biochim. Biophys. Acta 599: 623-628.
[76] Schaller, U.; Bakker, E.; Spichiger, U.E. and Pretsch, E. (1994) Nitrite-selective microelectrodes, Talanta, 41: 1001-1005.
[77] de Beer, D.; Schramm, A.; Santegoeds, C.M. and Khul, M. (1997) A nitrite microsensor for profiling environmental biofilms. Appl. Environm. Microbiol. 63: 973-977.
[78] Revsbech, N.P. and Ward, D.M. (1980) Oxygen microelectrode that is insensitive to medium chemical composition: use in an acid microbial mat dominated by *Cyanidium caldarium.* Appl. Environ. Microbiol. 45: 755-759.
[79] Axelsson, A.; Westrin, B.; and Loyd D. (1991) Application of the diffusion cell for the measurement of diffusion in cells. Chem. Eng. Sci. 46: 913-915
[80] Bird, R.B.; Stewart, W.E. and Lightfoot, E.N. (1960) Transport phenomena. Wiley International Edition, New York, USA

BIOFILM MODELLING

BORIS TARTAKOVSKY AND SERGE R. GUIOT
Biotechnology Research Institute, National Research Council of Canada, 6100 Royalmount Ave, Montreal, Quebec, Canada H4P 2R2 – Fax: (514)496-6265 – Email: Boris.Tartakovsky@nrc.ca; Serge.Guiot@nrc.ca

1. Introduction

Attachment of microorganisms to solid surfaces followed by biofilm formation is a well known phenomenon, which must be accounted for both in the design and operation of biotechnological processes. Sometimes biofilm formation is undesirable because it may either decrease process performance or cause damage to equipment [1]. Examples include biologically-assisted corrosion of metals and biofilm growth in water distribution and heat transfer systems. In health care, biofilm formation is linked to tooth decay and contamination of medical implants and catheters. In contrast, a large number of industrial-scale microbiological processes are dependent on biofilms. Examples of biofilm applications include such diverse areas as wastewater treatment and the food and pharmaceutical industries. Most wastewater treatment processes use biofilms (trickling filters, rotating biological contactors, fixed film reactors, anaerobic granular bed reactors) to improve biomass retention and volumetric removal rates. In bioprocessing, biofilm reactors provide high volumetric biomass density and improved operational stability [2]. Overall, the importance of biofilms in industrial processes and health care has prompted extensive experimental and theoretical studies of biofilm systems.

Initial attempts at biofilm modelling used material balances based on suspended cell systems. These models neglected the differences in phenotypic characteristics of suspended and immobilised microorganisms as well as the existence of substrate gradients within the biofilm. It was quickly recognised that diffusion limitations in the transport of dissolved components and physiological changes of the microorganisms in the biofilm significantly influenced the process. As a result, the first realistic biofilm models were developed.

Early models used a single microorganism – single substrate system and were focused on the diffusion limitation of substrate transformations in the biofilm [3]. Experimental evidence of the existence of mixed microbial consortia in biofilms and the sequential nature of substrate transformations by microbial consortia prompted the development of multispecies - multisubstrate biofilm models. As a consequence, the predictive power of biofilm models was improved significantly and a number of

V. Nedović and R. Willaert (eds.), Fundamentals of Cell Immobilisation Biotechnology, 531-545.

comprehensive biofilm models were developed. This progress can be attributed to several factors. Recent advances in molecular techniques have allowed for detailed studies of the distribution of microorganisms within biofilms, providing experimental data for model verification. As well, newly developed microsensors have been used for *in-situ* studies of substrate distribution profiles within biofilms. This progress in experimental studies has been accompanied by rapid advances in computer technologies allowing for the development of more complex numerical models that can be used to model these experimental results.

2. Observed biofilm dynamics

Experimental studies have shown that biofilm formation starts with the attachment of free-floating cells to a solid substratum. Upon attachment, microorganisms undergo changes in physiology and morphology [4] and begin to produce significant amounts of extracellular polymeric substances (EPS), which assist in biofilm formation [5-7]. Other mechanisms involved in biofilm development at initial stages include redistribution of attached cells by surface motility, recruitment of single cells or cell flocs from the bulk liquid, and binary division of attached cells [8].

With time, the growth of microorganisms results in the formation of a biofilm. Studies of cross-sections of well established biofilms have shown the existence of complex three-dimensional structures with numerous pores and channels. The use of biomolecular tools has demonstrated that in thick biofilms most biologically active cells are often adjacent to the biofilm-liquid interface due to diffusion limitations of substrate transport [9]. Even more interestingly, a combination of biomolecular probes and of confocal and epifluorescent microscopy has shown that the distribution of microorganisms in a mixed culture biofilm is far from homogeneous [10-12]. The distribution depends on environmental factors, such as concentrations of substrates, temperature, populations of microorganisms in the bulk liquid, etc. Depending on these factors, microorganisms can be homogeneously distributed in the biofilm or form biofilms with distinct layers of specific groups of microorganisms. The process of biofilm growth is accompanied by a constant biofilm detachment until eventually a steady-state thickness of biofilm is reached. Several categories of detachment processes are distinguished: sloughing (detachment of large biofilm particles), erosion (removal of individual cells), abrasion, predator grazing, and human intervention [13]. Overall, the complexity of physical and biological interactions in biofilms is reflected in the complexity of biofilm models. In order to provide an adequate mathematical description of the phenomena that are observed in experiments, biofilm models should account for the multiplicity of dissolved components and microorganisms and the complex dynamics of biotransformations.

3. Biofilm models

Early biofilm models were developed for the practical purposes of bioreactor design. Consequently, biofilm reactors were modelled using conventional reactor mass balances, which neglected biofilm properties and substrate mass transfer limitations.

Most often these models used Monod-like kinetics to describe both substrate transformations and the dynamics of biomass growth [14-16]. Because of diffusion limitation of substrate transport in the biofilm, the use of Monod-like kinetics in such models resulted in increased values of the Monod half-saturation constants [17]. This approach only provided a reasonable description of the experimental results in a narrow interval of operational parameters used for model identification. In addition, suspended biomass models did not give insight into the dynamics of biofilm formation and structure. Nevertheless, this approach remains acceptable for certain engineering applications, in which substrate gradients are negligible, as in the case of either thin biofilms or high substrate concentration in the bulk liquid.

An increasing amount of experimental evidence of complex biofilm structure and significant disagreement between experimental results and predictions from suspended biomass models have prompted the development of models that account for transport of dissolved components in the biofilm. In these models biofilms are described as a two-phase solid-liquid system with a stagnant liquid layer separating the two phases. Concentrations of chemical and biological species in the liquid phase determine boundary conditions for the biofilm, which is represented as a solid phase. Biofilm thickness and composition are determined by co-existing processes of biomass growth and detachment.

Two main approaches to biofilm modelling can be identified. In the first approach, mass conservation balances are used to develop a system of partial differential reaction-diffusion equations to describe the distribution of dissolved chemical species and microorganisms as a function of biofilm depth, *i.e.* biofilms are modelled as a continuum. In the second approach, which uses cellular automata models, a biofilm is represented by a two- or three-dimensional array of compartments, each of which can contain either cells or substrates or both, *i.e.* the biofilm is described by a spatially and temporally discrete system. Both approached are discussed in greater details in the following sections.

4. Reaction-diffusion biofilm models

As previously stated, reaction-diffusion models use a set of partial differential equations to describe a biofilm. Assuming the absence of axial dispersion in the biofilm, a typical mass balance of a dissolved component s (*e.g.* substrate) can be written as follows:

$$\frac{\partial s}{\partial t}+\frac{\partial j}{\partial z}=r_s(K,S,X) \qquad (1)$$

with boundary conditions $\partial s/\partial z=0$ at the centre of the biofilm and $\partial s/\partial z=\phi$ at the surface of the biofilm, where j is the flux of dissolved component s, z is the radial position, $r_s(K,S,X)$ is the transformation rate of the component s, and ϕ is the flow of s from the bulk to the biofilm through the stagnant layer. Notably, the transformation rate $r_s(K,S,X)$ depends on the vectors of dissolved component concentrations S, biomass densities X, and model parameters K.

The flux j is defined as $j = D_e \partial s / \partial z$, where D_e is the effective diffusivity, and the flow ϕ can be calculated as

$$\phi = k_l a (s^b - s_L) \tag{2}$$

where $k_l a$ is the specific mass transfer coefficient, s^b is the bulk concentration of s, and s_L is the concentration of dissolved component s at the biofilm surface.

Mass transfer limitations due to the presence of the stagnant layer are important at low flow rates of the bulk liquid, however in systems with sufficient mixing the existence of the stagnant layer can be neglected, *i.e.* $s_L = s^b$.

Although biomass density affects the transformation rate r_s, reaction - diffusion models dedicated to modelling the substrate distribution often use the assumptions of constant biofilm density and composition. The characteristic time of the substrate transformations is significantly shorter than the typical doubling times of the microorganisms, which justifies these assumptions. Moreover, with respect to relatively fast substrate transformation rates, a steady-state form of the reaction-diffusion model is often used in model analysis, *i.e.* Eq (1) takes the form

$$\frac{dj}{dz} = r_s(K, S, X) \tag{3}$$

For purposes of model analysis it can be convenient to use a dimensionless form of Eq. (3), which can be obtained by using dimensionless variables, *e.g.* concentrations ($S = s/s^b$), radial position ($Z = z/L$), where L is the biofilm thickness, and dimensionless Thiele modulus ($\phi^2 = z^2 (\mu_{max} x/(s^b Y D_e))$, which characterises the reaction to diffusion ratio. Here Y is the yield of microorganism x on substrate s.

A similar approach can be used to model the growth of the microorganisms. As in the case of dissolved component modelling, the biomass is represented as a continuum and homogeneous distribution of the biomass in the axial direction is assumed [18-21]. Hence, a one dimensional biomass distribution model takes the form

$$\frac{\partial x}{\partial t} + \frac{\partial j}{\partial z} = r_x(K, S, X) \tag{4}$$

with conventional boundary conditions given above. The biomass flux j is defined as $j = D_{xe} \partial x / \partial z$ and r_x is the biomass growth rate.

Most often biofilm models assume zero flux of biomass (D_{xe}=0), and cell transport limitations through the stagnant layer are neglected [21]. To simulate the processes of biofilm growth and decay, which change the overall thickness of the biofilm, a moving boundary method as suggested by Wanner and Gujer [18], can be used.

As in the modelling of substrate distribution, early biomass growth models considered only one microbial population. The need for modelling natural biofilms, which consist of a multiplicity of bacterial species, led to the development of multi-species biofilm models. As in the single-species model, conservation laws in the form of equations (1) and (4) are used for each microorganism and for each of the dissolved components (substrates, products, and intermediates).

The use of conventional growth dependencies (*e.g.* Monod and Andrews) in Eq. (4) results in the prediction of an unrealistically high biofilm density. In fact, biofilm density is limited by maximal attainable cell density and this limitation often is accounted for by using a threshold biofilm density, after which the rate of biomass growth is zeroed [22]. This approach results in the discontinuity in Eq. (4), which negatively affects its numerical solution. To avoid the discontinuity problem, the biomass density can be limited by assuming growth rates to be inversely dependent on the biofilm density, in particular using a non-linear dependence in the form $k_x/(k_x+x)$, where k_x is the space limitation constant.

Since the physical properties of microorganisms, EPS and other organic materials, which make up the biofilm, differ from each other, the processes of biomass growth and decay will change the effective diffusion permeabilities in the biofilms. Experiments demonstrated a sharp decrease in effective diffusive coefficients as the biomass volume fraction increased [23]. A number of empirical dependencies have been suggested to describe the dependence of effective diffusion coefficient on cell density in biofilms and gels of immobilised microorganisms [24,25].

Experiments on the colonisation of inner biofilm areas by microorganisms that have been introduced into the bulk liquid [26] have prompted the development of biofilm models that account for cell transport. There are several explanations for the observed phenomenon of cell transport in biofilms. Biofilms are porous structures with numerous channels and wells that can easily serve as routes of entry and transport for cells. Cell transport can be mediated by bacterial motility, convection and/or diffusion [8]. By using a positive value for the biomass effective diffusivity coefficient (D_{xe}) in Eq. (4), a phenomenological model of cell transport can be developed.

Experimental observations have emphasised the importance of detachment processes in overall dynamics of biofilms [27]. Consequently, a number of detachment models have been developed [28-30]. Biofilm detachment models often use empirical dependencies, which correlate the rate of detachment with biofilm thickness, biofilm mass, shear stress, and biofilm growth rate. These empirical models of biofilm detachment have been summarised by Stewart [30]. This paper also proposes a general framework for modelling biofilm detachment.

The most common numerical algorithms used in solving biofilm models include a number of finite-difference spatial discretization approaches such as Galerkin methods [31,32], central difference methods [20,33,34], and shooting algorithms [35,36]. In multi-species biofilm models the overall growth or decay of the biofilm is simulated as a moving boundary [18] or defined as the sum of all gains and loses in biomass for each segment defined by the discretization algorithm [20].

5. Cellular automata models

While reaction-diffusion models depict biomass as a continuum, cellular automata models describe it as a discrete lattice of sites [37-39]. At the beginning of the simulation, random and discrete values of biomass and substrate concentrations are assigned to each site. At each discrete time step logical rules or conventional kinetic equations are used to describe cell growth and to update substrate concentrations while logical rules are used to describe cell division and to chose a proximal site for the new cell. Often the cellular automaton biomass growth model is used in conjunction with a system of partial differential equations to calculate the distribution of soluble materials, such as substrates and products [40,41].

The simplicity of the calculations allows for the simulation of complex biofilm structures in three dimensions (*i.e.* channels, pores, pits, etc). However, due to the random initial distribution of cells, this method is difficult to use for quantitative calculations, which are required for practical purposes. The cellular automaton approach is more suited for use as a tool in understanding fundamental biofilm properties [42].

6. Comprehensive *versus* simplified models

Another classification of biofilm models is based on model complexity. Two model types can be considered: comprehensive and simplified. Comprehensive biofilm models account for the multiplicity of dissolved components and microorganisms in the biofilm and non-linear kinetics of growth and biotransformations. However, this complexity comes at a cost. The use of non-linear kinetic dependencies in the reaction-diffusion models requires a numerical solution to partial differential equations. Although numerical methods for solving of large-dimensional sets of non-linear partial differential equations are well developed, the computational time required is still considerable, being in the range of hours. Cellular automata models may require simpler computational algorithms, but these models use large two or three dimensional lattices and involve repetitive computations thus also requiring significant computational times.

Comprehensive reaction-diffusion biofilm models can accurately describe a number of experimentally observed phenomena, however these models contain a large number of parameters that are difficult to estimate. Consequently, these models provide qualitative rather than quantitative predictions. Moreover, model predictions depend largely on the choice of kinetic equations and parameters, which in turn depend on the experience of the researcher. A comprehensive model often reflects a researcher's understanding of the biofilm system and interpretation of the experimental data, which can lead to biased model predictions.

Cellular automata models appear to be more convenient for modelling complex two and three dimensional biofilm structures. The computational algorithms of cellular automata models allows for parallel computations, which may significantly reduce computational time if a multiprocessor computer capable of parallel computations is used. Nevertheless, a set of logical rules, which defines these models is also set by the researcher, resulting in biased simulations. In addition, the initial distribution of species is stochastic and the use of random values in simulating biofilm dynamics makes the results of simulations rather qualitative.

In general, comprehensive models can be regarded as cognitive tools for overall biofilm analysis. These models are not suitable for engineering applications, which require an ease of parameter estimation, straightforward analysis, and reproducibility of simulation results.

Contrary to comprehensive models, simplified biofilm models only account for the limiting steps in substrate and/or microorganisms dynamics, which results in a significantly smaller number of equations and parameters. These models are based on the assumption of a biomass continuum and use simplified kinetic dependencies to describe biomass growth and substrate transformations, *i.e.* zero or first order kinetic dependencies, which can be solved analytically [41,43].

Although simplified models are convenient for engineering analysis, they are often oversimplified and valid only in a narrow range of operating parameters. In this regard, an interesting compromise between the model predictive power and complexity was suggested in a simplified mixed-culture model developed by Rauch et al [44]. This model is based on segregating the processes of substrate diffusion and biotransformations. The depth of substrate penetration is calculated using a zero order reaction-diffusion model and the thickness of active biomass layer is assumed to be equal to the substrate penetration depth. After the fraction of each biomass species in the bioreactor is estimated, all biotransformation processes are calculated using material balances of an ideally mixed reactor.

7. Analysis of biofilm models

7.1. SUBSTRATE DISTRIBUTION

Most of the published substrate distribution models use a steady state assumption for analysis [21,33,35,43,45]. Examination of substrate concentration profiles showed that for thin biofilms or high bulk substrate concentrations, *i.e.* for small values of Thiele moduli, the entire biofilm is active [33,46]. In this case, the biotransformation rates are kinetic-limited, which implies that for practical applications the influence of the biofilm on the process dynamics can be neglected. For larger values of Thiele moduli, there exists a penetration depth at which the substrate concentration approaches a value near zero and no significant biotransformations take place beyond this depth. For these biofilms, the substrate transformation process is diffusion limited and only a part of the biofilm participates in the biotransformation resulting in a substrate limited core. Notably, the presence of a significant substrate limited core is undesirable for aerobic biofilm-based processes as it leads to decreasing volumetric efficiency.

While substrate gradients limit process efficiency for non-toxic substrates, diffusion limited transport can be beneficial if substrate transformation kinetics exhibits self-inhibition or if a toxic yet biodegradable compound is present in the bulk liquid. In this case, a decrease in the concentration of the dissolved compound of interest reduces the toxicity [34,45]. Anaerobic biofilm-based processes provide a number of examples because of a higher average biofilm thickness and relatively high bulk substrate concentrations. In particular, mathematical simulations of pentachlorophenol (PCP) degradation under anaerobic conditions by a biofilm system predicted the existence of a PCP gradient in the biofilm [34]. PCP degradation in the outer part of the biofilm protected the biofilm interior from the PCP-related toxicity allowing the intermediates of PCP degradation inside the biofilm to be mineralised.

The wealth of biotransformations in natural anaerobic biofilms offers two more examples illustrating the importance of substrate distribution modelling. Analysis of the detailed model of an acetate-utilising methanogenic biofilm demonstrated an increase in pH with increasing biofilm depth. Consequently, the optimal pH for acetate-utilising biofilms was found to be lower than the optimal pH found for suspended growth anaerobic reactors [47]. This finding is important in designing pH control strategies of anaerobic digesters.

Modelling of anaerobic biofilms also helped to explain the experimentally observed phenomenon of propionate degradation under thermodynamically unfavourable bulk hydrogen concentrations [48]. Analysis of glucose, hydrogen, and propionate distribution profiles in the biofilm suggested low hydrogen concentrations occurred in the biofilm core due to its intensive consumption by hydrogenotrophic microorganisms in the outer layers of the biofilm. Consequently, propionate transformation occurred in the biofilm core in spite of high concentrations of dissolved hydrogen in the bulk liquid [49,50].

The diffusion-limited transport of soluble components in combination with either substrate self-inhibition kinetics or the presence of toxic intermediates, has been shown to result in the existence of multiple steady states in biofilms [35,45]. Computer simulations showed the existence of several steady states for a range of physically meaningful parameters. Although steady state multiplicity in biofilm reactors is difficult to observe due to the large number of factors affecting overall reactor performance, instability of anaerobic processes is well known and can be related to steady state multiplicity.

As mentioned above, simplified biofilm models can be conveniently used to estimate the depth of substrate penetration in biofilm. In particular, a simple analytical solution of substrate distribution can be obtained for a zero-order reaction-diffusion model at steady state:

$$D_e \frac{d^2 s}{dz^2} = \frac{\mu(s)x}{Y} \tag{5}$$

with conventional boundary conditions $s(0) = s^b$, $\left. {}^{ds}\!/_{dr} \right|_{z^*} = 0$ and a zero-order rate of substrate consumption $\mu(s)$ defined as

$$\mu(s) = \begin{vmatrix} \mu_{\max}, & \text{if } s > 0 \\ 0, & \text{if } s \leq 0 \end{vmatrix} \tag{6}$$

where $\mu_{\max}$ is the biomass maximal growth rate, x is the constant biomass density, Y is the yield coefficient of microorganism x on substrate s, and z^* is the substrate penetration depth.

The analytical solution of the model defined by Eqs. (5-6) is given by a simple dependence $s = s^b(1 - z / z^*)$ [51], where the substrate penetration depth is defined as $z^* = (2sYD_e /(\mu_{\max} x))^{1/2}$.

This simple analytical solution can be conveniently used in engineering applications for estimations of active biofilm thickness. Solutions for first order kinetics of substrate consumption can be found elsewhere [52].

7.2. DISTRIBUTION OF MICROORGANISMS

The distribution of microorganisms in biofilms is strongly affected by substrate gradients. Oxygen-limited growth of aerobic biofilms is an obvious example. Due to the low solubility of oxygen in water a steep oxygen gradient in a biofilm was measured experimentally [53] and predicted in model simulations [22,41]. Consequently, an oxygen deficient core exists in aerobic biofilms with thicknesses exceeding the oxygen penetration depth. Biomass decay in the oxygen-deficient core results in biofilm sloughing, which can be considered as one of the mechanisms for maintaining a quasi-constant biofilm thickness.

Experimental observations of different microorganisms coexisting in natural biofilms have prompted the development of multi-species models. Computer simulations of biofilm growth have shown that the choice of growth kinetics strongly affects the steady state distribution of the microorganisms. The differences in growth and decay rates of different microorganisms due to oxygen limitation result in the development of mixed culture biofilms [18]. If species with similar growth kinetics but differing maximal specific growth rates are competing for a common substrate, the faster growing species will suppress the others and form a single-species biofilm. Thus, the coexistence of different species was predicted in sufficiently deep biofilms and for species with different affinities to a growth-limiting substrate [18,36]. In addition, the biomass decay rate was demonstrated to play an important role in determining the steady state distribution of microorganisms in mixed culture biofilms. Simulation results of Wanner and Gujer [18] showed that the spatial distribution of microbial species can be modelled by segments, each being either homogeneous or mixed. This conclusion provided a basis for a number of biofilm models, which represent the biofilm as a set of discrete intervals.

In anaerobic biofilms, bulk substrate concentrations are on average higher thus allowing for the existence of thick biofilms. It is not unusual to observe 3-5 mm biofilm granules in upflow anaerobic sludge bed reactors [54]. This significant thickness of biofilm combined with the complexity of biotransformations involved in the anaerobic degradation of carbohydrates provides a rich population dynamics, which has been the subject of several modelling efforts.

To explain observations of the multi-layered anaerobic biofilms, a multi-species dynamic biofilm model was developed [48]. Numerical solutions to this model agreed well with the observed biofilm structure. The simulations showed the development of layers of different trophic groups of microorganisms, which mirrored profiles of corresponding rate-limiting substrates, *e.g.* glucose for the fermentative bacteria and acetate for the methanogenic bacteria. A steady state model of a mixed anaerobic biofilm was used to compare the efficiencies of layered and homogeneous anaerobic biofilms [21]. The analysis suggested that the layered biofilm structure was superior in achieving complete degradation of the organic compounds.

Development of a layered biofilm structure was also predicted by a dynamic model of pentachlorophenol degradation under anaerobic conditions [34]. In this work, experimental observations of biofilm colonisation by a laboratory strain of PCP-degraders was accompanied by dynamic simulations of biofilm formation using a biofilm growth model which accounted for growth and transport of microorganisms in the biofilm. Sensitivity analysis of the model parameters suggested that an inward propagation of microorganisms from the biofilm surface can be expected for slow growing species in porous biofilms or supports.

The existence of an oxygen-depleted core in sufficiently thick biofilms was exploited in the development of a combined aerobic-anaerobic biofilm system. A combination of oxidative and reductive metabolic pathways can be required for conducting two or more consecutive biotransformations. Examples of bioprocesses that benefit from aerobic-anaerobic coupling include ethanol production from starch [55], nitrogen removal [56], and mineralization of chlorinated aliphatics [57].

A general model of a coupled aerobic-anaerobic system was developed in [51]. Model analysis demonstrated the advantages of coupling over sequential process organisation, such as complete substrate transformation and maximisation of biotransformation rates due to the shortening of diffusion pathways and the rapid elimination of toxic intermediates. Optimally designed coupled biofilm systems featured higher biotransformation rates without the release of intermediates. An optimal biofilm structure was shown to consist of two segregated layers of aerobic and anaerobic bacteria. Although the two-layer structure can be manufactured if cell immobilisation techniques are used, the existence of an oxygen gradient in the biofilm can be exploited to form a natural two-layer biofilm. However, in the case of competition between aerobic bacteria this natural biofilm structure may not provide stable process operation due to species competition for oxygen.

Biofilm simulations using cellular automata models were focused mostly on the modelling of biofilm structure. The simulations showed biofilm patterns qualitatively similar to those observed in the experiments. In particular, dense biofilms were observed in the absence of substrate-related limitations of biomass growth, while substrate limited growth conditions resulted in the development of porous biofilms [37,39,58]. At the lowest substrate concentrations dendritic or rhizoid structures were predicted in agreement with observations.

A cellular automaton model was also used in simulating the formation of microcolonies in a gel support [41]. In this case, the initial cell density was found to play an important role. An initially high density of cells resulted in dense surface colonisation of the solid support with a steep substrate gradient and low interior cell density. At the same time, an initially low number of cells led to the formation of distinctly spaced microcolonies distributed almost evenly across the biofilm, due to improved substrate diffusivity. Consequently, a higher steady state cell density was achieved when starting with an initial low number of cells. Based on the results of this work, it can be concluded that in an attempt to introduce laboratory strains to natural biofilms, the use of a high initial cell density provided no advantages to the steady state performance of the biofilm.

8. Applications of biofilm models in process design and control

As mentioned above, reaction-diffusion and cellular automata biofilm models use a large number of tuneable parameters. Taking into account the large variability of biofilm processes and the limited number of measurable process variables, comprehensive biofilm models describe an average process. Consequently, practical applications of comprehensive biofilm models are limited to model analysis for process design and troubleshooting.

A number of biofilm models have been developed for the simulation of water treatment processes. The model of [59] allows for the calculation of COD and ammonium profiles along the bioreactor height and in the biofilm. It is capable of simulating filter clogging due to biomass growth. This model can be used to understand the influence of operating conditions on process efficiency. Similarly, a trickling filter model was developed to study the effects of different operating conditions such as VOC loads and air flow rates on process performance [60].

A comprehensive dynamic multi-species biofilm model of the biofiltration process developed by [20] included four biomass types and seven chemical species. Also, the model simulated biomass decay and constant or periodic detachment of biofilm. The simulations demonstrated transient development of multiple-species biofilms and the role of soluble microbial products and detachment in controlling the distribution of microorganisms and process performance.

Biofilm models may also be useful for the design of bioprocesses aimed at the removal of toxic compounds. Examples include analysis of process limitations of an aerobic fluidized-bed biofilm reactor (FBBR) used for degrading dinitrotoluene, which showed that process limitations were related to growth potential [61] and biofilm augmentation with a pentachlorophenol degrading strain [34]. In the latter study, mathematical simulations showed dynamics of biofilm colonisation and suggested conditions for optimal engineering of stable multispecies consortia.

In health care, modelling efforts have been focussed on studying biofilm formation mechanisms [58,62] and the effects of antimicrobial agents on the biofilm development [63,64]. The reduced susceptibility of biofilms to antimicrobial agents was simulated using two different mechanisms, depletion of the antibiotic by reaction with biomass and physiological resistance due to reduced bacterial growth rates [63]. Computer simulations predicted differences between the two cases, thus allowing these resistance mechanisms to be experimentally distinguished.

Another possible application of biofilm models is in the design of complex microbial consortia capable of multi-step biotransformations. Multi-species and in particular multi-layered biofilms can be used in a variety of bioconversions with a sequential transformation pathway. The general biofilm model of a combined aerobic-anaerobic biofilm system described above can be used in the design of novel bioprocesses that require sequential biotransformations.

Large variations in kinetic parameters of biofilm models, which can be attributed to natural variability of biological processes and a large number of uncontrollable factors, limits the application of biofilm models for process control. At the same time, simplified biofilm models are only valid for a narrow range of operational parameters and may not be suitable for process control. One possible solution to this dilemma can be found in using variable structure models (VSM) for process control. The VSM approach is based on the use of a set of simple non-linear models [65,66]. Each model adequately describes the dynamics of a certain process state, although the models cannot describe phase to phase transitions. A transition between submodels is handled by a set of rules. The VSM approach combines the flexibility of a knowledge-based system with the precision of model-based control. A set of simple biofilm models, each of which describes biofilm dynamics in a narrow range of operational conditions, can be developed (*e.g.* diffusion limited model, substrate – limited model, biomass growth model, etc). A knowledge based system can be used to analyse measurable process outputs and select an active submodel. Existing biofilm models can be used to develop VSM submodels and the knowledge of biofilm processes can be converted into a knowledge-based system.

9. Conclusion

Overall, the modelling of biofilms has advanced significantly in recent years. Although biofilm models are not currently a routine tool in process design and their use in practical engineering is limited, growing interest in complex multi-step bioconversion processes makes biofilm modelling an essential component of successful process design.

Several potential avenues in the development of biofilm models can be outlined. Because of the complexity of biofilm processes, comprehensive models will receive more attention in the future and it is likely to expect an emergence of structural comprehensive biofilm models, which combine detailed modelling of metabolic pathways, cell growth, dissolved component transport, and reactor mass balances. In addition, a greater number of factors influencing biofilm dynamics will be included in the model formulation. Considering the ease of computations and flexibility of cellular automata models in modelling biofilm development, this approach will also receive further attention. New structural biofilm models can be based on a convergence of reaction-diffusion and cellular automaton models. In this case, spatial discretization methods for solving material balances of reaction-diffusion substrate distribution models provide a discrete lattice for modelling of cell growth using a cellular automaton.

Along with the development of new comprehensive models, simplified models will be still required for practical purposes of process design and control. These simplified models perhaps should be based on comprehensive models, while incorporating a number of simplifying assumptions that permit parameter estimation in a target range of operational conditions. More attention can be anticipated for the analysis of the sensitivity of model parameters and the development of general approaches in the reduction of model dimension [42]. The simplified models can be used alone, or as a part of newer modelling approaches, such as the variable structure model approach, to develop models that are exploitable in process control applications.

References

[1] Costerton, J.W.; Marrie, T.J. and Cheng, K.-J. (1985) Phenomena of bacterial adhesion. In: Savage, D. C. and Fletcher, M. (Eds.) Bacterial adhesion, Plenum Publishing Corporation, New York, NY, USA.

[2] Iwai, S. and Kitao, T. (1994) Wastewater treatment with microbial films. Technomic Publishing Co. Inc. Lancaster, PA, USA.

[3] Characklis, W.G. (1983) Process analysis in microbial systems: biofilms as a case study. In: Bazin, M. (Eds.) Mathematics in microbiology, Academic Press. London.

[4] Rehm, H.-J. and Omar, S.H. (1993) Special morphological and metabolic behaviour of immobilized microorganisms. In: Sahm, H. (Ed.) Biotechnology, VCH Verlagsgesellschaft mbH. Weinheim, Germany.

[5] Fletcher, M. and Floodgate, G.D. (1976) The adhesion of bacteria to solid surfaces. In: Fuller, R. and Loverfock, D. W., (Eds.), Microbial ultrastructure, Academic Press. London.
[6] Marshall, K.C. (1992) Biofilms: an overview of bacterial adhesion, activity and control at surfaces. ASM News, 58, 202-207.
[7] van Loosdrecht, M.C.K. and Heijnen, J.J. (1996) Biofilm processes. In: Willaert, R. G.; Baron, G. V. and De Backer, L. (Eds.) Immobilised living cell systems, J. Wiley & Sons. Chichester, UK.
[8] Hall-Stoodley, L. and Stoodley, P. (2002) Developmental regulation of microbial biofilms. Curr. Opin. Biotech. 13: 228-233.
[9] Schonduve, P.; Sara, M. and Friedl, A. (1996) Influence of physiologically relevant parameters on biomass formation in a tricle-bed bioreactor used for waste gas cleaning. Appl. Microbiol. Biot. 45: 286-292.
[10] Hibiya, K.; Terada, A.; Tsuneda, S. and Hirata, A. (2003) Simultaneous nitrification and denitrification by controlling vertical and horizontal microenvironment in a membrane-aerated biofilm reactor. J. Biotechnol. 100: 23-32.
[11] Okabe, S.; Santegoeds, C.M.; Watanabe, Y. and D., B. (2002) Successional development of sulfate-reducing bacterial populations and their activities in an activated sludge immobilized agar gel film. Biotechnol. Bioeng. 78: 119-130.
[12] Rocheleau, S.; Greer, C.W.; Lawrence, J.R.; Cantin, C.; Laramee, L. and Guiot, S.R. (1999) Differentiation of *Methanosaeta concilii* and *Methanosarcina barkeri* in anaerobic mesophilic granular sludge by fluorescent in situ hybridization and confocal scanning laser microscopy. Appl. Environ. Microbiol. 65: 2222-2229.
[13] Bryers, J.D. (1988) Modeling biofilm accumulation. In: Bazin, M. J. and Prosser, J. I. (Eds.) Physiological models in microbiology, CRC. Boca Raton, FL.
[14] Kappeler, J. and Gujer, W. (1994) Development of a mathematical model for aerobic bulking. Water Res. 28: 303-310.
[15] Sanz, J.P.; Freund, M. and Hother, S. (1996) Nitrification and denitrification in continuous upflow filters - process modelling and optimization. Water Sci. Technol. 34: 441-448.
[16] Masse, D.I. and Droste, R.L. (2000) Comprehensive model of anaerobic digestion of swine manure slurry in a sequencing batch reactor. Wat. Res. 34: 3087-3106.
[17] Gonzalez-Gil, G.; Seghezzo, L.; Lettinga, G. and Kleerebezem, R. (2001) Kinetics and mass-transfer phenomena in anaerobic granular sludge. Biotechnol. Bioeng. 73: 125-134.
[18] Wanner, O. and Gujer, W. (1986) A multispecies biofilm model. Biotechnol. Bioeng. 28: 314-328.
[19] Wanner, O. (1995) New experimental findings and biofilm modelling concepts. Water Sci. Technol. 32: 133-140.
[20] Rittmann, B.E.; Stilwell, D. and Ohashi, A. (2002) The transient-state, multiple-species biofilm model for biofiltration processes. Water Res. 36: 2342-2356.
[21] Tartakovsky, B. and Guiot, S.R. (1997) Modeling and analysis of layered stationary anaerobic granular biofilms. Biotechnol. Bioeng. 54: 122-130.
[22] Hunik, J.H.; Bos, C.G.; Hoogen, M.P.; DeGooijer, C.D. and Tramper, J. (1994) Co-immobilized *Nitrosomonas europea* and *Nitrobacter agilis* cells: validation of a dynamic model for simultaneous substrate conversion and growth in k-carrageenan gel beads. Biotechnol. Bioeng. 43: 1153-1163.
[23] Stewart, P.S. (1998) A review of experimental measurements of effective diffusive permeabilities and effective diffusion coefficients in biofilms. Biotechnol. Bioeng. 59: 261-272.
[24] Hinson, R.K. and Kocher, W.M. (1996) Model for Effective diffusivities in aerobic biofilms. J. Environ. Eng. 122: 1023-1030.
[25] Westrin, B.A. and Axelsson, A. (1991) Diffusion in gels containing immobilized cells: A critical review. Biotechnol. Bioeng. 38: 439-446.
[26] Horber, C.; Christiansen, N.; Arvin, E. and Ahring, B. (1998) Improved dechlorinating performance of upflow anaerobic sludge blanket reactors by incorporation of *Dehalospirillum multivorans* into granular sludge. Appl. Environ. Microb. 64: 1860-1863.
[27] Trulear, M.G. and Characklis, W.G. (1982) Dynamics of biofilm processes. J. WPCF 54: 1288-1301
[28] Morgenroth, E. and Wilderer, P.A. (2000) Influence of detachment mechanisms on competition in biofilms. Water Res. 34: 417-426.
[29] Nicolella, C.; Di Felice, R. and Rovatti, M. (1996) An experimental model of biofilm detachment in liquid fluidized bed biological reactors. Biotechnol. Bioeng. 51: 713-719.
[30] Stewart, P.S. (1993) A model of biofilm detachment. Biotechnol. Bioeng. 41: 111-117.
[31] Horn, H.; Neu, T.R. and Wulkow, M. (2001) Modelling the structure and function of extracellular polymeric substances in biofilms with new numerical techniques. Water Sci. Technol. 43: 121-127.

[32] Soyupak, S.; Nakiboglu, H. and Surucu, G. (1990) A finite element approach for biological fluidized bed modelling. Appl. Math. Modelling 14: 258-263.
[33] Buffiere, P.; Steyer, J.-P.; Fonade, C. and Moletta, R. (1995) Comprehensive modeling of methanogenic biofilms in fluidized bed systems: mass transfer limitations and multisubstrate aspects. Biotechnol. Bioeng. 48: 725-736.
[34] Lanthier, M.; Tartakovsky, B.; Villemur, R.; DeLuca, G. and Guiot, S.R. (2002) Microstructure of anaerobic granules bioaugmented with *Desulfitobacterium frappieri PCP-1*. Appl. Environ. Microb. 68: 4035-4043.
[35] Flora, J.R.V.; Suidan, M.T.; Biswas, P. and Sayles, G.D. (1995) A modeling study of anaerobic biofilm systems: I. Detailed biofilm modeling. Biotechnol Bioeng, 46, 43-53.
[36] Wanner, O. and Reichert, P. (1996) Mathematical modeling of mixed-culture biofilm. Biotechnol Bioeng, 49, 172-184.
[37] Hermanowicz, S.W. (2001) A simple 2D biofilm model yields a variety of morphological features. Math. Biosci. 169: 1-14.
[38] Kreft, J.U.; Picioreanu, C.; Wimpenny, J.W. and van Loosdrecht, M.C. (2001) Individual-based modelling of biofilms. Microbiol. (Reading, England) 147: 2897-2912.
[39] Picioreanu, C.; van Loosdrecht, M.C. and Heijnen, J.J. (1998) Mathematical modeling of biofilm structure with a hybrid differential-discrete cellular automaton approach. Biotechnol. Bioeng. 58: 101-116.
[40] Picioreanu, C.; van Loosdrecht, M.C. and Heijnen, J.J. (1998) A new combined differential-discrete cellular automaton approach for biofilm modeling: application for growth in gel beads. Biotechnol. Bioeng. 57: 718-731.
[41] Greenberg, N.; Tartakovsky, B.; Yirme, G.; Ulitzur, S. and Sheintuch, M. (1996) Observations and modeling of growth of immobilized microcolonies of luminous *E.coli*. Chem. Eng. Sci. 51: 743-756.
[42] Noguera, D.R.; Okabe, S. and Picioreanu, C. (1999) Biofilm modeling: present status and future directions. Water Sci. Technol. 39: 273-278.
[43] Pritchett, L.A. and Dockery, J.D. (2001) Steady State Solutions of a One-Dimensional Biofilm Model. Math. Computer Modelling 33: 255-263.
[44] Rauch, W.; Vanhooren, H. and Vanrolleghem, P.A. (1999) A simplified mixed-culture biofilm model. Water Res. 33: 2148-2162.
[45] Gupta, N.; Gupta, S.K. and Ramachandran, K.B. (1997) Modelling and simulation of anaerobic stratified biofilm for methane production and prediction of multiple steady states. Chem. Eng. J. 65: 37-46.
[46] Droste, R.L. and Kennedy, K.J. (1986) Sequential substrate utilization and effectiveness factor in fixed bofilms. Biotechnol. Bioeng. 28: 1713-1720.
[47] Flora, J.R.V.; Suidan, M.T.; Biswas, P. and Sayles, G.D. (1995) A modeling study of anaerobic biofilm systems: II. Reactor modeling. Biotechnol. Bioeng. 46: 54-61.
[48] Arcand, Y.; Chavarie, C. and Guiot, S. (1994) Dynamic modelling of the population distribution in the anaerobic granular biofilm. Wat Sci. Technol. 30: 63-73.
[49] Pauss, A.; Samson, R. and Guiot, S. (1990) Thermodynamic evidence of trophic microniches in methanogenic granular sludge-bed reactors. Appl. Microb. Biotechnol. 33: 88-92.
[50] Guiot, S.R.; Pauss, A. and Costerton, J.W. (1992) A structured model of the anaerobic granule consortium. Wat Sci. Technol. 25: 1-10.
[51] Tartakovsky, B.; Guiot, S. and Sheintuch, M. (1998) Modeling and analysis of co-immobilized aerobic/anaerobic mixed cultures. Biotechnol. Prog. 14: 672-679.
[52] Beg, S.A. and Hassan, M.M. (1985) A biofilm model for packed bed reactors considering diffusional resistances and effects of backmixing. Chem. Eng. J. 30: B1-B8.
[53] Lens, P.; de Beer, D.; Cronenberg, C.; Ottengraf, S. and Verstraete, W. (1995) The use of microsensors to determine population distributions in UASB aggregates. Wat Sci. Technol. 31: 273-280.
[54] Kosaric, N. and Blaszczyk, R. (1990) The morphology and electron microscopy of microbial aggregates. In: Tyagi, R. D. and Vembu, K. (Eds.), Wastewater treatment by immobilized cells, CRC Press. Boca Raton.
[55] Tanaka, H.; Kurosawa, H. and Murukami, H. (1986) Ethanol production from starch by coimmobilized mixed culture system of *Aspergillus awamori* and *Zymomonas mobilis*. Biotechnol. Bioeng. 28: 1761-1768.
[56] Kotlar, E.; Tartakosky, B.; Argaman, Y. and Sheintuch, M. (1996) The nature of interaction between immobilized nitrification and denitrification bacteria. J. Biotechnol. 51: 251-258.
[57] Beunink, J. and Rehm, H.-J. (1988) Synchronous anaerobic and aerobic degradation of DDT by an immobilized mixed culture system. Appl. Microbiol. Biotechnol. 29: 72-80.
[58] Wimpenny, J.W.T. and Colasanti, R. (1997) A unifying hypothesis for the structure of microbial biofilms based on cellular automaton model. FEMS Microbiol. Ecol. 22: 1-16.

[59] Viotti, P.; Eramo, B.; Boni, M.R.; Carruci, A.; Leccese, M. and Sbaffoni, S. (2002) Development and calibration of a mathematical model for the simulation of the biofiltration process. Adv. Environ. Res. 7: 11-33.

[60] Vanhooren, H.; Verbrugge, T.; Boeije, G.; Demey, D. and Vanrolleghem, P.A. (2001) Adequate model complexity for scenario analysis of VOC stripping in a trickling filter. Water Sci. Technol. 43: 29-38.

[61] Smets, B.F.; Riefler, R.G.; Lendenmann, U. and Spain, J.C. (1999) Kinetic analysis of simultaneous 2,4-dinitrotoluene (DNT) and 2, 6-DNT biodegradation in an aerobic fluidized-bed biofilm reactor. Biotechnol. Bioeng. 63: 642-653.

[62] Van Loosdrecht, M.C.M.; Picioreanu, C. and Heijnen, J.J. (1997) A more unifying hypothesis for biofilm structures. FEMS Microbiol. Ecol. 24: 181-183.

[63] Stewart, P.S. (1994) Biofilm accumulation model that predicts antibiotic resistance of *Pseudomonas aeruginosa* biofilms. Antimicrob. Agents Ch. 38: 1052-1058.

[64] Dodds, M.G.; Grobe, K.J. and Stewart, P.S. (2000) Modeling biofilm antimicrobial resistance. Biotechnol. Bioeng. 68: 456-465.

[65] Dainson, B.E.; Tartakovsky, B.; Scheintuch, M. and Lewin, D.R. (1995) Variable structure models in process observation and control. Ind. Eng. Chem. Res. 34: 3008-3013.

[66] Tartakovsky, B.; Morel, E.; Steyer, J.-P. and Guiot, S.R. (2002) Application of a variable structure model in observation and control of an anaerobic digestor. Biotechnol. Prog. 18: 898-903.

INDEX